Genetic Analysis of Animal Development

Second Edition

Genetic Analysis of Animal Development

Second Edition

Adam S. Wilkins

A JOHN WILEY & SONS, INC., PUBLICATION

New York • Chichester • Brisbane • Toronto • Singapore

Address All Inquiries to the Publisher
Wiley-Liss, Inc., 605 Third Avenue, New York, NY 10158-0012

Library of Congress Cataloging-in-Publication Data

Wilkins, A.S. (Adam S.), 1945-
    Genetic analysis of animal development / by Adam S. Wilkins. --
2nd ed.
        p.   cm.
    Includes bibliographical references and index.
    ISBN 0-471-50271-5 (cloth). -- ISBN 0-471-50270-7 (paper)
    1. Developmental genetics. 2. Developmental biology.
3. Embryology. I. Title.
QH453.W55   1992
591.3--dc20   92-5570
                                                                  CIP

The text of this book is printed on acid-free paper.

To Salome Gluecksohn-Waelsch,
one of the first to see the possibilities

CONTENTS

PREFACE

The first edition of this book grew out of an undergraduate developmental genetics course that I taught in the late 1970s. When the book was first planned in detail in 1980, the subject of developmental genetics was only available to students through the primary literature and in rather abbreviated and anecdotal discussions in introductory genetics texts. To undergraduates unfamiliar with the assumptions and methods of genetics, the original research papers often seemed esoteric and obscure, while the accounts in first year genetics textbooks were usually too brief and general to convey much insight into developmental genetics or its possibilities. In marked contrast to this neglect in the secondary literature of genetic approaches to development, there were a few excellent treatments of eukaryotic developmental biology as viewed from a molecular perspective, chief among these being Eric Davidson's *Gene Activity in Early Development*.

My initial aim, therefore, was to provide a book that would complement the molecular biological texts by explaining the genetic strategies used for analyzing animal development and the kinds of information that genetic approaches specifically can reveal. To do so, and to give the material a strong focus, the discussion was centered on the developmental genetics of the three animals whose genetics were the best characterized, namely the fruit fly (*Drosophila melanogaster*), the nematode (*Caenorhabditis elegans*), and the mouse (*Mus musculus* and its near relatives). Although in 1980–1981, it was obvious that there was not yet in-depth knowledge about the genetic basis of development in any of these animals (despite a large body of interesting and suggestive work), it was also clear that genetic approaches were going to make substantial contributions in the not-too-distant future. As a foundation for what was to come, it seemed worthwhile to construct a book that would describe these three animal systems and their advantages, and, at the same time, illustrate the methodologies of developmental genetics and their potential outcomes.

The impetus for writing the second edition, which was begun in 1989, was different. By the late 1980s, it was clear that a great deal could now be said about the genetic basis of development. (Just how much more only became fully apparent as the writing progressed.) The very abundance of material, and the fact that some major advances had been made in specific areas, demanded a treatment that would, at least, attempt a more integrated view of the underlying biology, both within and between the various model systems, than the first edition had achieved. Moreover, the field of developmental genetics, and perceptions about it, had not only grown and matured as a body of research, but its broader context had changed as well. In particular, among both students and researchers outside the recognized network of developmental geneticists, there was a generally wider understanding of genetic methods, and their usefulness; furthermore, most students were now familiar, to greater or lesser degree, with the basics of gene cloning. Not least, the whole scientific framework of the field had been altered. The previously sharp border line between developmental genetics and molecular biology had all but vanished. So too had many of the other demarcations in biology, such as that between studies on growth and/or cancer, on the one hand, and development, on the other. If a general text on developmental genetics, begun in the late 1980s, were to be useful, it should reflect and accommodate these changes.

This edition has been written with such considerations in mind. The core material and its organization remain the same, namely the developmental genetics of the nematode, the fruit fly, and the mouse. Furthermore, as in the first edition, separate chapters have been devoted to the early and late developmental stages, respectively, of each. Although somewhat less stress has been placed on genetic techniques and approaches, I have, as in the first edition, illustrated these methods with particular, appropriate examples in the course of the narrative. The essential premise behind this approach is that in order to understand and evaluate conclusions, it is essential to comprehend the experimental procedures that led to those conclusions. There is, however, relatively more emphasis in this edition on the recent discoveries and ideas about the genetic basis of development than on methodology. Where possible, I have also tried to trace the growing number of connections, at the genetic and molecular levels, between developmental processes in different organisms. Finally, while in the first edition discussions of pattern formation and molecular approaches were segregated into separate chapters, this material is now diffused throughout the text.

Other changes include the addition of two new chapters: Chapter 2, which is intended to serve as an introduction to the key issues and questions in developmental biology as seen by geneticists, and chapter 9, the penultimate chapter, which deals with pattern formation phenomena on the external surface of developing animals and which attempts to encapsulate the basic issues in pattern formation. The final chapter, chapter 10, is a reprise of some of the fundamental questions about development raised in chapter 2, in light of the material covered in the intervening chapters, and attempts to provide a perspective on some of the fundamental issues in developmental biology today. As in the earlier edition, the book is completed by a glossary and an appendix which deals with several technical matters. (The previous appendix contained partial genetic maps of the three organisms, but these have been omitted here because of the growth of mapping information; references to sources about current maps have been included in the text, instead.)

A word about the scope of the coverage is required. As before, a certain background knowledge on the part of the reader is assumed, of the kind that is acquired in a first year university genetics or cell biology course; the basic fact of and the fundamentals of gene cloning are taken as part of that background. Although the text itself is about 15% longer than that of the first edition, and the focus has been kept to three model systems and their major areas of research, this book remains, even within its self-imposed limits, an incomplete and selective account of the developmental genetics of these three animals. Choices of topic and emphasis have been made and a great deal has had to be left out. For example, although there has been some attempt to name the individuals who made seminal discoveries, much of the investigative history of many topics has been slighted with most attention paid, inevitably, to more recent work. As a partial remedy for these gaps, I have tried to indicate, where appropriate, review articles that do justice to the history of the work and to the relevant background cellular and molecular biology of particular topics.

My hope is that the book will convey not only the major lines of research but both the excitement of development genetics today and the sense that, despite much recent progress, the major problems in developmental biology still await solution. The animal embryo may be small but it presents a miniature universe of ordered, unfolding change whose processes and mechanisms we are only just beginning to comprehend. If the reader comes away with the feeling that some of the most exciting days in developmental genetics still lie ahead, the book, whatever its deficiencies, will have achieved one of its primary aims.

Adam S. Wilkins
Cambridge, England

ACKNOWLEDGMENTS

The preparation of this edition has been greatly aided by individuals who have generously donated their time and expertise. Most importantly, I would like to acknowledge my debt to those who read drafts of the various chapters and sent me their comments and criticisms: Drs. Michael Ashburner, Jonathan B.L. Bard, Martin Chalfie, William F. Dove, Salome Gluecksohn-Waelsch, David Gubb, Lewis I. Held, Jr., Jonathan A. Hodgkin, Martin H. Johnson, Paul Lasko, Mary F. Lyon, Rolf Nothiger, Einhard Schierenberg, John E. Sulston, Azim Surani, Diethard Tautz, and Andrew Tomlinson. (In particular, I would like to thank Jonathan Bard for his very helpful comments on chapters 1 and 2 and his oft-repeated, but always needed, reminders that short sentences are usually preferable to long ones.) Furthermore, there were numerous friends and colleagues who contributed illustrations and reprints. Less tangibly, I have been the beneficiary of numerous informative conversations about developmental biology with many individuals over the years.

On the technical side, essential help with illustrations was provided by Mr. John Rodford, who did the new drawings for this book, and two colleagues at the Company of Biologists, Ltd., Mr. Thomas Galliers and Mr. Christopher Love, who supplied a number of needed photographic reproductions. Finally, I would like to express appreciation to all those who have put up with me during the past six months, usually with humor and forebearance, especially during those moments when I explained that I could not at present do whatever was being requested because of the demands of "the book." These tolerant people have included members of my family, various friends, my students at Queens College, and, not least, Richard Skaer and my other colleagues at the Company of Biologists. Trying to keep up with developmental genetics during the last three years, and to do it justice in words, has at times seemed like a long-distance race without a finish line, but it would have been an even more difficult experience without such a supportive group of friends.

A.S.W.

ONE

PAST AND PRESENT

...it follows that semen will be either blood or the analogous substance, or something formed out of these....And this, too, is why we should expect children to resemble their parents: because there is a resemblance between that which is distributed to the various parts of the body and that which is left over [for reproduction]. Thus, the semen of the hand or of a whole animal *is* hand or face or a whole animal though in an undifferentiated way; in other words, what each of these is *in actuality,* such the semen is *potentially....*

Aristotle, *Generation of Animals,* p. 91

INTRODUCTION

Questions about embryonic development have been intertwined with questions about heredity since the first written speculations on inheritance. To ask why children resemble their parents involves more than an inquiry about the passage of hereditary factors, the genes, from one generation to the next. It is equally a set of implicit questions about the ways in which genes produce their effects in development. These questions have persisted for millenia, but it is only in our era that answers of any precision have begun to emerge.

Although the core subject of 20th-century genetics has been the mechanisms and patterns of gene transmission from one generation to the next, a particular branch, developmental genetics, is devoted to the specific question of how genes produce their effects in development. This subdiscipline of genetics possesses its own distinctive methodologies and perspectives, yet during the past decade it has changed more than in the preceding eight decades combined. These changes have reflected both its growth, in terms of the number of investigators, and, even more profoundly, its fusion with molecular biology, a process that has accelerated enormously in recent years. Although any contemporary account of developmental genetics should emphasize recent findings and approaches, it would distort and impoverish the view to eliminate the historical perspective entirely. A glance backward at the history of developmental genetics can provide a useful context for appreciating the present state of the field and, perhaps, a feeling for how it is likely to evolve.

Any attempt to divide a continuous sequence into discrete periods is necessarily somewhat artificial, but the history of developmental genetics can be seen as a sequence of three fairly distinct phases, each characterized by its prevailing ideas and approaches. The first was the "classical" period, stretching from the first decade of this century to about 1960. This phase was dominated by the study of mutant phenotypes as a way of deducing the roles of wild-type genes in development. The second period, beginning in 1961 and lasting approximately two decades, witnessed both the growing influence of molecular concepts, though not yet its techniques, and the revival and increased application of the various techniques of "clonal analysis" from the previous period.

The third phase, the present one, lacks a distinct beginning but can be dated from approximately 1980. It has been characterized by the reintroduction of the technique of saturation mutagenesis (Nüsslein-Volhard and Wieschaus, 1980), a strategy for determining all the genes affecting a particular developmental process that was pioneered by Waddington (1940), and the increasing application of molecular techniques to developmental studies. The convergence of approaches has, by now, effectively erased the previously clear boundary line between developmental genetics and molecular biology. Without subscribing to the Whiggish view of history as a steadily ascending improvement over past conditions, one cannot but view the present period as the most fruitful in the history of developmental studies. To say that much, however, is not to deny the existence of blind spots and lacunae that must exist in our current perceptions, gaps that will, undoubtedly, become increasingly apparent in future years.

This chapter attempts to provide a capsule history of the field of developmental genetics, from its hazy origins in the first decades of this century to the present. In the following chapter we will look at the questions about development that are addressed by contemporary geneticists.

THE GORDIAN KNOT OF GENETICS AND DEVELOPMENT AND HOW MENDEL CUT IT

Today we realize that the fundamental connection between genes and development is that genes specify the proteins that determine cellular character; the progression of cell types and cell behaviors that comprise development reflects the changes in protein composition that follow from changing gene activities. Ultimately, the kind of organism that a particular fertilized egg gives rise to is determined by its genes, which will be those typical of a given species. The precise details of the individual's development are, however, influenced by the particular variants of the genes, or alleles, that comprise its own genetic constitution. Nevertheless, the relationships between the details of observable form, or phenotype, of the individual and its underlying genetic constitution, or genotype, are only rarely simple and obvious.

Two key facts that Gregor Mendel discovered, in the research that laid the foundations for modern genetics, are that each gene is present in two copies in the mature organism (the state of diploidy) and that the phenotype does not always reveal this dual genetic character. In particular, certain alleles of each gene can be dominant to, or mask, the expression of other alleles. In other cases, dominance may be partial or the combination of alleles can produce a new trait. In still more complicated genetic situations, the action of one gene can wholly or partially obscure the effect of a different gene, the phenomenon known as epistasis. Furthermore, many interesting traits are determined not by individual alleles at one genetic locus, but rather by the cumulative or synergistic action of many genes. This is known as polygenic or quantitative inheritance.

Finally, the relationships between genes and development are influenced, indeed conditioned, by the environment in which the developing organism is placed. Even very small alterations of environmental conditions can affect the degree of expression of a particular gene, while strong environmental shocks may produce wide-scale alterations of gene expression, which can cause embryos to develop in highly abnormal fashions (though generally the developmental syndrome will be characteristic of the kind and timing of the trauma administered). Furthermore, beyond all such deterministic effects there is the phenomenon of "developmental noise": small, random biochemical perturbations characteristic of any system.

It is this very complex set of interconnections between genes and ultimate developmental outcomes that so delayed the birth of genetics, a science that is based on procedures for deducing genotypes of parents from the phenotypes of their progeny. Given the enormous number and complexity of the processes that intervene between the creation of the offspring's genotype at fertilization and the ultimate production of the progeny's phenotype, the process of inferring genotypes from descendants' phenotypes is necessarily indirect and one dependent upon many contingent phenomena. In large part, the failure of Mendel's predecessors to understand the process of inheritance was because they chose to study traits with a complex genetic and developmental basis; they became lost in the wilderness of developmental complexity without realizing it (Olby, 1966; Sandler and Sandler, 1985).

Even more fundamentally, these earlier investigators, from Aristotle to Darwin (whose hypothesis of pangenesis was published in the same year as Mendel's paper), did not clearly distinguish between the phenomena of inheritance and of development; their hypotheses implicitly or explicitly attempted to account for both at once (Sandler and Sandler, 1985). Mendel not only sensed the potential pitfalls in the genetic analysis of complex traits, choosing to study hereditary factors whose developmental effects were as simple and direct as could be found, but he deliberately focused on how these factors were transmitted, bypassing speculations on how they achieved their effects. Mendel's awareness that factor transmission could be tackled as a distinct and separate problem, in conjunction with his novel statistical approach to inheritance, enabled him to deduce the basic rules by which genes are transmitted from one generation to the next. Yet it is also probable that as a result of this conceptual splitting, his solution may have seemed a highly incomplete one to those interested in heredity, a perception that may have contributed to the eclipse of his work (Sandler and Sandler, 1985).

Although for many biologists the question of how genes work continued to be seen as inextricably bound up with the manner of their transmission, even into the 1920s (Allen, 1986), during the 1890s a number of key investigators had begun to separate the two issues *in practice*, as had Mendel before them and by similar methods. The early 20th century would prove to be a considerably more receptive period for Mendelian genetics than Mendel's own era had been.

1900–1960: CLASSIC DEVELOPMENTAL GENETICS

A mutation that causes a certain malformation as the result of a developmental disturbance carries out an "experiment" in the embryo by interfering with the normal development at a certain point. By studying the details of the disturbed development it may be possible to learn something about the results of the "experiment" carried out by the gene. However, to discover anything about the nature of the action of the gene is a much harder task.

Salome Gluecksohn-Schoenheimer (1938)

It might have been anticipated that the rediscovery in 1900 of Mendelian inheritance would clear the way for a direct use of genetics to probe the role of genes in development. Early 20th-century geneticists were fully aware of the importance of hereditary factors in biological development, and indeed, many had a background in embryology. Foremost among them were William Bateson, the chief proponent of Mendelian genetics, and Thomas Hunt Morgan. It was Morgan who would, in the second decade of the century, synthesize Mendelian genetics and the chromosome theory of inheritance. Given the intense interest and involvement of these men and their colleagues with questions of development, one might have thought that they would combine their twin interests, genetics and embryology, to explore the ways that individual genes affect embryological development. Such a synthesis of interests and approaches, however, conspicuously failed to materialize.

There were several reasons for this hesitancy to address the genetic basis of development. The first was the controversy that arose between those around Bateson and the group centered around Karl Pearson, the biometricians, over the sufficiency of Mendelian genetics to explain the facts of heredity (Provine, 1971). At issue were the large number of traits that we now understand to have a polygenic basis of inheritance; these traits were, at the time, regarded by the biometricians as having a qualitatively different basis of inheritance than that of Mendelian traits. This issue was not resolved, in favor of the Mendelians, for more than 15 years following the rediscovery of Mendel's work (Provine, 1971). For nearly two decades, however, and for many of the participants, the debate effectively diverted attention from applying genetics to the study of development.

Furthermore, even for the convinced Mendelians, the conceptual gap between genes and phenotypes was still forbiddingly large. Although the connections between genes (which were not even named as such until 1909) and cellular metabolism were sensed by a few, including Bateson, and though exploration of these connections would begin long before there was a theoretical framework for understanding these relationships, the concepts were too primitive and the analytical methods inadequate for approaching the subject. The lever of Mendelian genetics was simply the wrong tool for prying apart the facts of embryonic development, and would have been so even had the conceptual framework been stronger (Hull, 1974). The second reason for the neglect of development had to do with the interests and perspectives of the early 20th-century geneticists. It is true that several had a background in embryology, but some of the more influential, including Bateson, were interested in embryos primarily as mirrors of the evolutionary process, a major preoccupation of post-Darwinian biology. Morgan was different in this respect; he was greatly interested in the developing embryo as a subject in its own right, but he evidently thought that a direct attack on the problem of how genes achieve their effects was an impossibility at the time, and even he tended to conflate transmission phenomena and developmental questions into the 1920s (Allen, 1986). Though he never lost his interest in embryology, and even explicitly addressed its issues again toward the end of his career

(Morgan, 1934), both he and most members of his school remained oddly remote from the question of how genes exert their effects in development.

Yet while the gaze of the early Mendelians was focused on the problems of gene transmission and evolution, they were soon tripping over the facts of development. It could hardly have been otherwise, given the genetic basis of development. The ever-wider search for genetic factors that showed typical Mendelian inheritance patterns made it inevitable that developmental phenomena would soon be intruding into studies whose goal was the elucidation of gene inheritance.

One of the earliest findings of this kind involved a test for Mendelian inheritance in a mammal, that of the *yellow* gene in the laboratory mouse. The initial studies were carried out by Lucien Cuénot, who discovered that the *yellow* allele (symbolized A^Y) was dominant to the standard agouti (A) allele, which produces a grayish color. Cuénot observed, to his surprise, that the yellow trait never bred true: intercrossed yellow mice always produced some wild-type progeny. Furthermore, when any yellow mouse was mated to a wild-type, roughly equal numbers of yellow and wild-type progeny were obtained. This finding was explicable if the yellow mice were always heterozygous for the *yellow* and wild-type alleles. However, yellow × yellow crosses never gave the expected 3:1 Mendelian ratio for a monohybrid (heterozygote × heterozygote) cross. Cuénot consistently observed both a deficit of yellow progeny relative to the expected 75% proportion and reduced litter sizes (Cuénot, 1908). To explain his result, he proposed a developmental hypothesis: that *yellow*-allele-carrying eggs could never be fertilized by *yellow*-bearing sperm but only by wild-type–bearing sperm. The explanation accounted for the permanently heterozygous condition of *yellow* and the relative deficit of yellow progeny in the monohybrid cross, but not for the reduction in total progeny numbers since there should always be sufficient wild-type sperm to affect fertilization.

In a later and more extensive study, W.E. Castle and C.C. Little (1910) showed that the true ratio in the monohybrid cross was 2 yellows: 1 agouti. They pointed out that this ratio was exactly the one expected if one-quarter of the progeny were homozygous yellow and if all of these were lost. Their hypothesis was that *yellow*, which acts as a simple visible dominant mutation when present in one dose ($A^Y/+$), acts as a recessive lethal when homozygous (A^Y/A^Y). This hypothesis of recessive lethality of homozygotes was soon confirmed by direct observation of the embryos by other workers.

As the testing of factors for Mendelian inheritance patterns progressed, comparable cases in other organisms soon came to light, in both plants and animals. In 1912, for instance, the Morgan *Drosophila* group reported the first recessive lethal in the fruit fly. As comparable observations accumulated, the nature of the interest that they evoked began to change. At first regarded as noteworthy because of the distortion they caused of Mendelian inheritance ratios, the mutants soon came to be seen as interesting in their own right, because of the developmental effects they produced. When examined closely, these effects were found to be specific for the mutant genes under study, each lethal mutant producing its own characteristic pattern of abnormalities and death. As the specificity of individual lethal action became increasingly apparent, it seemed ever more likely that each lethal was revealing the specific developmental effects of altering particular genes.

It was through such studies of mutant effects that the field of developmental genetics came into being. The growth was slow, piecemeal, and proceeded without a single dramatic discovery or landmark publication of the kind that has so often signaled new departures in genetics. Despite these hazy and undramatic origins, developmental genetics had acquired a recognizable unity by the late 1930s. Most of the early contributions came from European-born scientists, in particular, Curt Stern, Salome G. Waelsch, and Richard Goldschmidt, but the earlier work of Sewall Wright in the United States on the genetic and enzymatic basis of coat color in guinea pigs had also been of seminal importance.

The goal in each of the early genetic studies was to reconstruct the developmental function of a particular wild-type gene from the pattern of tissue or organ changes occurring in the mutant. The assumption was that the entire body of work would illuminate the basic biological rules of the genetic control of development (for reviews see Gruneberg, 1952; Hadorn, 1961). Since lethal mutants possess clear, readily observable effects, the primary emphasis in much of this work was on genetic changes that produce death. These included both single gene changes ("point mutants") and those chromosomal rearrangements, such as inversions and translocations, that were found to be associated with a lethal effect. The basic method of analysis consisted of three steps. The first was the careful delineation of the time of death, the lethal phase, of the mutant. The second step was the cataloging of all the terminal differences between the mutant and wild-type at the time of death, as revealed by the organs and tissues of the dead embryos. The third phase involved a careful tracing backward in development, using progressively earlier embryos, of these mutant characteristics or "phenes" to the earliest point at which abnormalities could be detected. This initial point of aberrancy was termed the "phenocritical phase." From the initial morpho-logical aberrancy, one would then attempt to reconstruct the function of the corre-sponding wild-type gene during the phenocritical phase. And, from the aggregate of studies, it was hoped that general principles of gene action in development would appear.

The analytical method is a retrospective one—working backward from the known terminal phenotype to the first effects of the mutant condition—and was the core technique of developmental genetics for more than 40 years, from the 1920s through the 1960s. Nevertheless, despite the wealth of observational data it produced, at the time the method was unsuccessful in either delineating general underlying mecha-nisms of development or even unambiguously reconstructing the roles of any wild-type genes (although a number of good approximate characterizations were achieved). Not surprisingly, therefore, the literature of developmental genetics from this period has about it an air of disappointment (as can be seen, for instance, in the concluding discussion of Hans Gruneberg's classic work, *The Genetics of the Mouse*, 1952). Instead of the emergence of general principles, the field produced a plethora of unrelated observations, each pertaining to the biology of a particular mutant and none definitively identifying the precise role of a wild-type gene in development.

If most experimental embryologists largely ignored (and even scorned) the contri-butions of geneticists to embryology during this period, the failure of developmental genetics to achieve its own programmatic goals must be held responsible in part. In retrospect, the program can be seen not to have been misguided but, rather, prema-ture; there simply wasn't the wealth of background biochemical and molecular knowledge to permit the kinds of characterization that were needed. However, in two respects, and belying the appearances, some critical foundations had been laid.

The first of these was the necessary delineation of the potential snares in the interpretation of mutant effects, the understanding of these being essential for all later work. One such snare is the phenomenon of pleiotropy. As observations accumulated, it became apparent that most mutants do not suffer single developmental changes but a whole range of abnormalities; "pleiotropy" means simply "many ways." Two different classes soon came to be distinguished, traditionally designated as "mosaic" (or "multiple") and "relational" (or "unitary") (Stern, 1955; Hadorn, 1961). In the former, the defects arise in two or more sites because the wild-type gene action is required for expression independently at those sites; in the latter, the multiple defects arise as a cascade of consequences from the mutant defect occurring at one site. In reality, and as subsequently appreciated, most instances of pleiotropy have both "mosaic" and "relational" components.

A second class of complications in the interpretation of mutant effects can occur if the characteristics of the mutation itself are not properly categorized (Muller, 1932). In general, most deleterious gene mutations involve a deficiency of the relevant gene activity—a deficiency in the amount of product or of its intrinsic activity. Such defects are usually recessive because one dose of wild-type gene activity is often sufficient for normal development. However, the precise degree of deficiency can make a dramatic difference in the degree or even the kind of developmental syndrome observed. If no phenotypically detectable wild-type gene activity remains, the mutant is termed an "amorph." In general, amorphs for a vital gene cause sharp, defined defects in development and often early death of the affected embryos. In contrast, mutant alleles which leave some residual wild-type gene activity, "hypomorphs," will not infrequently show a more normal course of development and variable points of developmental arrest in the mutant homozygotes.

As for dominant mutants, these too are not of a single type but fall into different categories that must be distinguished. A mutant may be dominant to wild-type simply because it reduces gene activity and because the development of the wild-type phenotype requires two full wild-type gene doses; this form of dominance is said to reflect "haplo-insufficiency." More commonly, dominants produce a genuinely novel gene activity. Most such dominants are termed "neomorphs"; if, however, the activity antagonizes and competes with the wild-type gene activity, it is said to be an "anti-morph." Just as with recessives, it is critically important to know the kind of dominant one is dealing with in order to interpret the mutant effect correctly. Muller's (1932) tests for categorizing mutations are described in chapter 4.

Despite the difficulties of interpretation inherent in the retrospective mode of mutant defect analysis, by the late 1940s two general conclusions could be formulated. These conclusions constituted the principal conceptual advances contributed by developmental genetics. The first was that of *site specificity of action* of genes: each gene appeared to be expressed and required in one or a few tissues but generally not in all of them. The second generalization was *phase specificity of gene action*, the idea that a particular gene is active only at particular times in development. One can condense these two conclusions into one: that biological development is accompanied, and probably driven by, an *ordered sequence of differential gene expression*. The idea of spatially and temporally patterned sequences of gene expression in development had been sensed by a few, including Morgan, in the 1930s but could only be expressed with confidence by the century's midpoint (Hadorn, 1948; Spiegelman, 1949).

Crucial as this idea of spatial and temporal sequences of gene expression is today, it failed to evoke much response among experimental embryologists. The fields of developmental genetics and embryology continued to go their separate ways. The reason is not hard to find. Developmental geneticists talked at length about genes and gene actions, yet they could not visualize or explain what the gene was or how it worked. There had been repeated speculations for decades that particular genes somehow determine particular enzymatic activities, a link that was definitively established by the work of George Beadle and Edward Tatum on *Neurospora* auxotrophs (Beadle and Tatum, 1941), but the exact nature of the gene–enzyme connection was very obscure. The very idea of changing and changeable gene activities during development, when the gene itself was so remote, sounded more like a figure of speech than a statement of biochemical fact, perhaps even to those who invoked the concept.

The ideas and methods of developmental genetics at midcentury were summarized by Ernst Hadorn in his book *Developmental Genetics and Lethal Factors*. In the 1930s, Hadorn had left the experimentally tractable amphibian embryo for the much more difficult terrain of *Drosophila*, with the explicit goal of applying genetic techniques to developmental questions. By the 1950s, he was one of the major figures in the field. His book was first written in German and published in 1955. Later it was translated, appearing in 1961 in its English-language edition. As chance would have it, the second edition appeared just as the whole perspective of the field described in the book was being superseded by a new outlook and a new set of approaches. The dates that span the two editions, 1955 to 1961, virtually bracket the period in which the field of genetics experienced its most fundamental change since its birth. The change, of course, was that brought about by the advent of molecular biology, initiated by the publication of the model of DNA structure by James Watson and Francis Crick in 1953 and completed by the Jacob-Monod model of gene regulation in 1961 and the elucidation of the genetic code in the early 1960s.

The new discipline was concerned with twin problems: What is the molecular nature of the gene and how does the gene work? The Watson-Crick model provided an answer to the first: Each gene is a stretch of chemical information, a sequence of nucleotides that "encodes" and specifies the unique amino acid sequence of a polypeptide. Until the early 1950s, genes had been thought of as entities possessing direct "activities" themselves, much as enzymes do. The concept that a gene works solely by specifying the structure of a polypeptide, through encoding its sequence of amino acids, was a major departure.

For those who worked in molecular biology, the post–Watson-Crick era of the 1950s was devoted to figuring out the "logic" of the genetic code—the number and sequence of nucleotides that specify particular amino acids—and to unraveling the biochemical mechanism of protein synthesis, from gene to finished polypeptide. The capstone of this edifice was the realization that there must be an intermediary molecule that exists between the gene and the synthesized protein product. This intermediary carries the information of the gene to the cellular sites of protein synthesis in the cytoplasm. From various considerations, the intermediary molecule was deduced to be a species of RNA, a class of molecules broadly similar to DNA but single-stranded instead of double-stranded. Because of its information-conveying role, this class of RNA was designated "messenger RNA" or simply mRNA.

How did the revolution of molecular biology affect developmental genetics? The first stages left the field untouched. In Hadorn's book, genes are discussed on every page but DNA is not mentioned once, although the identity of the genetic material with the molecular substance of DNA was well established. The response of developmental biologists to the final step, however the solution of the problem of protein synthesis, was electric. This suddenly heightened interest occurred because of a historical accident: The discovery of mRNA became bound up with the central problem of developmental genetics—the fact that gene activities are regulated.

1961–1980: THE IMPACT OF MOLECULAR IDEAS AND THE REVIVAL OF CLONAL ANALYTIC TECHNIQUES

It may be in the interpretation and analysis of differentiation that the new concepts derived from the study of microorganisms will prove to be of the greatest value. One point at least already seems to be quite clear: namely that biochemical differentiation (reversible or not) of cells carrying an identical genome does not constitute a "paradox," as it appeared to do for many years, to both embryologists and geneticists.

J. Monod and F. Jacob (1962)

Throughout the 1950s, when American and British molecular biologists were grappling with the general problem of protein synthesis, two French scientists at the Institut Pasteur, Jacques Monod and François Jacob, were wrestling with an equally mysterious but seemingly smaller problem. The question was this: How does the bacterium *Escherichia coli* turn on the synthesis of three proteins required for lactose metabolism in response to the addition of lactose "inducer" to the medium? Monod and Jacob and their colleagues had found that the genes encoding these enzymes, which they termed "structural genes," were clustered together on the bacterial chromosome and that their synthesis was controlled coordinately in response to inducer molecules. In recognition of this coordinate control, the three genes were said to comprise an "operon." A second discovery was that mutations in a nearby gene could permit massive synthesis of the three *lac* operon enzymes, even in the absence of inducer. The sole function of this "regulator" gene was to control the switching on and off of the three structural genes. The regulator gene was dubbed the *i* gene, for the *i*nducer-independent $(i-)$ expression of the enzyme-encoding genes of the operon. Several observations indicated that this gene specified a "repressor," a molecule that physically shut off the expression of the three structural genes, and that the inducer molecules antagonized the function of the repressor, thereby releasing synthesis of the three proteins.

However, several puzzles remained. One of these was that enzyme synthesis apparently proceeded through the synthesis of a short-lived, intermediary template molecular species. When inducer was removed, synthesis of this species was shut off abruptly but enzyme synthesis continued on the existing molecules, whose number then declined exponentially through a degradative process. From the idea of an intermediary template molecule in lactose enzyme synthesis to the idea of a general intermediary molecule in gene-to-protein synthesis is not a large leap—at least in retrospect. The concept of mRNA as a distinct molecular species was born in a conversation one afternoon in April 1960 between Jacob and Monod, on the one hand, and Francis Crick and Sydney Brenner, the chief theoreticians of protein synthesis, on the other (Judson, 1979).

The *lac* operon model, completed by the idea of mRNA, provided biology with a conception of how genes may be selectively turned on and off (Jacob and Monod, 1961). Generalized and stripped to its essentials, the idea was that for structural genes, those encoding proteins that play a direct part in the economy of the cell, there is a special regulatory apparatus consisting of a separately encoded regulator molecule and one (or more) special sites (*cis*-acting sites) to which it binds, adjacent to the structural genes. The binding or nonbinding of the regulator molecule is determined by the presence or absence of one or more environmental "signals," termed coeffectors, and the net result is the control of transcription of the structural genes.

In the case of the *lac* operon, and also in that of phage lambda, whose control had been investigated simultaneously by the Pasteur group, the regulator molecule acts solely to switch off or "repress" transcription and the system is said to be one of "negative control"; it is diagramed in Figure 1.1a. The discovery of negative regulation in a few systems led to the short-lived dogma, accompanied by some bitter debates, that all gene regulation involves control by repressors. However, within a few years after the publication of the Jacob-Monod model, evidence was obtained that at least one bacterial operon, the *ara* operon which encodes the enzymes that degrade arabinose, is regulated by "positive control," a system of selective direct *stimulation* of transcription by bound regulator molecule (Englesberg et al., 1965). Indeed, we now know that the economy of the bacterial cell is as dependent on positive control systems as on negative control systems (Englesberg and Wilcox, 1974); such a system is illustrated schematically in Figure 1.1b.

The diagrams shown in Figure 1.1 are highly idealized descriptions of general transcriptional control mechanisms; today it is appreciated that the sharp dichotomy between negative and positive control systems, posited in the 1960s, is to a large extent unreal. Many regulator molecules, both in prokaryotes and eukaryotes, possess both negative and positive activities with respect to transcription, depending upon the precise molecular context. These factors include the presence or absence of particular metabolite coeffectors in the cell, the particular sites of binding employed by a transcription control molecule (a given regulator may act as a repressor at one site, an activator at another), and the presence or absence of other regulator molecules at nearby binding sites (Englesberg and Wilcox, 1974; Straney and Crothers, 1987).

In the early 1960s, however, such considerations were still well in the future. The crucial fact was that a conceptual breakthrough had been made; it was now possible to think of sequences of developmental change as the products of an underlying but readily comprehensible molecular mechanism. Although the general idea of a distinctive genetic regulatory apparatus had been proposed before, most particularly in the pioneering work of Barbara McClintock (McClintock, 1951, 1956), the operon model provided a concrete mechanism for genetic regulation via cytoplasmic signals. In particular, it could explain the nature of the interplay between nucleus and cytoplasm, long acknowledged by biologists to be intrinsic to development (Wilson, 1925) but not understood in its details.

Furthermore, the model could address two central properties of biological development: ordered, progressive change and the existence of stable cell states. In several papers, Jacob and Monod detailed the principle of "regulatory circuitry," derived from the *lac* operon work, as the basis of developmental change. The idea is that an initial regulatory step can initiate a second regulated change in gene expression for one or

a) Negative Control

no inducer :

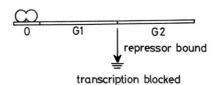

0   G1   G2

repressor bound

transcription blocked

addition of inducer ✱ :

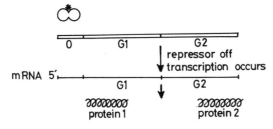

0   G1   G2

repressor off
transcription occurs

mRNA 5'

G1   G2

protein 1   protein 2

b) Positive Control

no inducer :

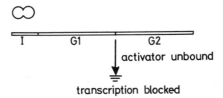

I   G1   G2

activator unbound

transcription blocked

addition of inducer ✱ :

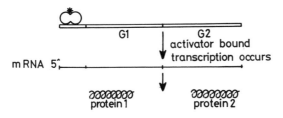

G1   G2

activator bound
transcription occurs

mRNA 5'

protein 1   protein 2

Fig. 1.1. Mechanisms for the control of gene expression in bacteria. **a**: Negative control of transcription: The inducer binds to the repressor, removing it from the operator and allowing transcription. **b**: Positive control of transcription: The inducer binds to the activator, which binds to the receptor region, facilitating transcription.

many genes which can, in turn, automatically produce a third set of changes and so on, the net effect being a highly ordered sequence of gene expression changes ensuing from an initial regulatory event (Monod and Jacob, 1962; Jacob and Monod, 1963).

The concept could also address the long-standing problem of the stability of differentiated states. In principle, regulatory circuitry can produce highly stable gene expression patterns, and correspondingly stable differentiated states, through the appropriate feedback loops. One can thus envisage a mechanism for stability that does not invoke genetic change.

How Important Is the Operon Model Today?

More than three decades have passed since the publication of the operon model, and it is pertinent to ask precisely what contemporary relevance this concept, and the subsequent findings on bacterial transcription, have for current studies on eukaryotic developmental biology. The question of "relevance," in fact, is most usefully broken down into two distinct questions (which are not always so distinguished). The first is of *applicability*, the second of *sufficiency*.

The question of applicability can be put as follows: Is the general mechanism embodied in the *lac* operon model (extended to include positive control switches) extensively employed by eukaryotic organisms during development to regulate transcription? Although genes in eukaryotes are considerably larger, are often split by introns, and are contained within large, complex chromosomal structures, which in turn are housed within true nuclei, the essential features of prokaryotic transcriptional control still apply to eukaryotic cells. Despite the greater complexity of these cells, the fundamental answer to the question of applicability of the Jacob-Monod model is "yes."

However, the question about sufficiency is a different matter. In essence, it asks how much of eukaryotic development can be explained by the prokaryotic transcriptional models. It was not uncommon in the 1960s and 1970s to hear molecular biologists state that understanding development *consisted of* understanding the transcriptional control events that underlie it. Today that position is clearly untenable (Brenner et al., 1990). In bacteria, the number of molecular events, what one may informally term the "distance" between the transcriptional control event and the phenotypic outcome, is small, with the latter often embodied directly in the newly synthesized proteins themselves (for example, β-galactosidase activity, phage coat components, etc.). In eukaryotes, the "distance" between transcriptional event and ultimate phenotype is often vast. Furthermore, the number of intermediary events involved in initiating the transcriptional event, a chain which often begins at the cell surface and whose signal is transmitted and transduced through many components, can be equally large. The greater the number of factors involved in a particular direct chain, and the greater the number of factors from other regulatory networks that can affect members of the chain, the greater the number of possible checkpoints and the less automatic any potential transcriptional cascade becomes. In the words of Brenner et al. (1990): "The operon paradigm survives, but with numerous twists."

Yet the work on bacterial regulation in the 1960s had two lasting effects on developmental genetics, moving the field, as it did, from a preoccupation with phenomenology to a much more precise focus on mechanism. The principal consequence for practitioners was to shift attention from the observable developmental effects of mutations to the placement of the corresponding wild-type genes within regulatory schemes. Under the terms of the prokaryotic models, genes could be classified into either structural genes or regulatory ones: The immediate challenge was to devise ways of making these assignments for genes affecting development that had been identified by their mutations. A secondary, subtler change concerned the way that biologists began to view the "logic" of developmental control. Development began to be seen as a "program" encoded in the genome. The result was a new emphasis on the genome as a source of information not just for the organism but for the investigator as well. The metaphor of the "developmental program" continues to be invoked in much of the literature of this field, and we shall look at its implications in the final chapter.

An Internal Development: Clonal Analysis

The story of developmental genetics in the 1960s and 1970s, however, was more than one of the fertilizing effect of ideas from molecular biology. The field had simultaneously begun to experience its own internal revolution, one that owed little to molecular biology. One of the most important of these was the use of "saturation" mutant hunting to obtain mutants in all of the genes that affect a particular developmental process. The technique was invented by C.H. Waddington, in an attempt to analyze wing development (Waddington, 1940), but was later adopted by microbial geneticists analyzing metabolic pathways and the construction of phage coats. When one carries the process to true "saturation," one can not only identify most or all of the genes in the pathway but also obtain several mutations for the same gene and delineate the "amorphic" mutant phenotype for that gene. In modern developmental genetics, this approach was first used in the isolation of early zygotic mutants affecting the segmentation process in *Drosophila* (Nüsslein-Volhard and Wieschaus, 1980).

In addition, a second development involved the revival and application, on a large scale, of techniques that were first developed during the earlier "classic" period, techniques that were designed to elucidate the growth and developmental potentials of individual cells and their descendants during development. The genetic methods that comprise the heart of this program of investigation have been collectively termed "clonal analysis" (Nöthiger, 1976). Clonal analysis began with the work of Sturtevant (1929) on fruit fly gyandromorphs—individual animals that are part male, part female—and was extended by the discovery of mitotic recombination in *Drosophila* by Curt Stern (1935). However, the methods of clonal analysis only began to be extensively applied in the late 1960s and early 1970s.

In essence, clonal analysis involves tracking, during development, the mitotic descendants of a cell—its *cell lineage*—made distinguishable from the surrounding cells by the presence of a distinctive genetic "marker" in those cells. An essential prerequisite is that the phenotypic character be expressed independently or "autonomously" solely within the genetically distinctive cells. Broadly, there are two different general methods of clonal analysis. The first involves physically introducing genetically marked cells from one individual embryo into a second embryo. Such genetically composite animals are termed "chimeras," after the legendary Greek monster which had the head of a lion, the body of a goat, and the tail of a dragon. The second method involves inducing a genetic difference in one or more cells of a single individual (usually by provoking chromosome loss or recombination). Such individuals, in which all cells derive from the same zygote, are said to be "genetic mosaics."

Both mosaics and chimeras can be employed to obtain either of two distinct sorts of information. The first is about the behavior of normal cells, and their descendants, during development. If the marker itself does not significantly affect the developmental process under study, the shape and location of the marked clone will provide information about the initial cell's developmental role or "fate" in terms of the number of its descendants and their spatial array. Furthermore, if the original marked cell has been placed by some means in a new location within the embryo or within a different embryo, the development of the clone can then provide a hint as to that cell's repertoire of developmental capacities, its "potencies."

The second general application of clonal analysis is to describe the behavior of particular mutations in terms of their localized cellular effects. If one includes in the marked cells a second mutation that does influence the developmental process under study, one can determine whether the effects of that mutation are expressed in all cells of that genotype ("autonomous" behavior), whether the mutant phenotype is erased or "rescued" by the surrounding cells ("nonautonomous" behavior), or whether the mutant cells can induce defects that spread to the surrounding cells ("domineering nonautonomy"). Such information is often invaluable in assessing the likely properties of the corresponding wild-type gene product.

Clonal analysis, in its various forms, emphasizes the cell as the unit of development and thus bridges the gap between the molecular analysis of developmental change, with its primary emphasis on subcellular events at the level of the genome, and that of traditional experimental embryology, with its emphasis on the aggregate behavior of cell groups, "embryonic fields" or "blastemas." Indeed, clonal analysis can remove much of the ambiguity associated with the older methods of experimental embryology, involving grafts of tissues, where it is often difficult, if not impossible, to separate the interactive effects from cell autonomous effects (see the discussion by Conway et al., 1980).

THE PRESENT

The reintroduction of techniques of clonal analysis, more than three decades after their development, was a major element in the rejuvenation and expansion of developmental genetics in the 1970s. The second major stimulus, also dating from the 1970s, but whose effects in developmental genetics were not felt until the 1980s, was the development of the techniques of recombinant DNA technology. The ability to clone genes of interest has immensely extended and, indeed, transformed developmental genetics.

At the most fundamental level, the ability to purify and analyze genes—and, today, virtually *any* can be cloned, with the requisite effort—has meant that the sequence and properties of the gene can be directly ascertained and, often, the likely biochemical properties of the gene's product can be inferred. Indeed, as more and more genes are cloned and sequenced, it becomes increasingly easy to assign probable biochemical functions, and developmental roles, to their products. Furthermore, with a cloned gene, one can study its transcriptional pattern, either in bulk within whole embryos during development or by means of *in situ* hybridizations.

Using cloned genes, one can not only analyze their normal gene expression patterns but can also make partial gene constructs *in vitro*, reinsert these constructs into the organism, and then determine which sequences are responsible for which aspects of expression. In addition, molecular markers are increasingly employed in conjunction with clonal analysis, considerably extending the range of what can be done. Finally, with a wide variety of DNA sequence constructs, one can either search for genes that give novel or desired expression patterns or make new mutations with constructs that have inserted into the genome and then proceed to isolate and identify the genes carrying these inserts. The number of useful recombinant DNA techniques employed in studying development is not only legion but, astonishingly, continues to grow with each passing year.

Nevertheless, despite the increasingly blurred distinction between genetic and molecular approaches to development, the methods and perspectives of developmental genetics have not been supplanted by molecular biological techniques but rather enriched and extended by them. Though developmental genetics changed more as a discipline in the 1980s than in the previous eight decades of this century, all told, genetic reasoning and approaches remain central to the analysis of developmental biology. The unique capacity of genetics is the ability to ascertain gene function in developmental processes; techniques for doing so will remain essential in developmental biology for the foreseeable future. Indeed, in the work taking place today, one of the main resources is proving to be the large mutant collections of mice and fruit flies assembled during the period of classical developmental genetics. Enriched by the precision of molecular techniques, and accumulated knowledge, this legacy from the pioneers in the field is about to be redeemed.

In the next chapter, some of the general phenomena that developing animal systems exhibit will first be described and then examined from the point of view of genetics and the specific kinds of questions that geneticists put to these developmental processes. In chapters 3–5, we will look at the analysis of early development, focusing on the three animal systems most susceptible to genetic approaches (the nematode *Caenorhabditis elegans*, the fruit fly *Drosophila melanogaster*, and the mouse *Mus musculus*). In chapters 6–8, the genetic analysis of later developmental stages in these same animal systems will be described. Chapter 9 provides a discussion of pattern formation phenomena, as seen in the surface features of these systems. Chapter 10, which concludes the book, will attempt to summarize the state of developmental genetics today and to sketch some of the areas that are likely to be emphasized in the future.

TWO

ANIMAL DEVELOPMENT
The Genetic Perspective

In one way, you could say, all the genetic and molecular biological work of the last sixty years could be considered as a long interlude—sixty years of following out Morgan's Deviation into the tractable genetic problems. And now that that program has been completed, we have come full circle—back to the problems that they left behind unsolved. How does a wounded organism regenerate to exactly the same structure it had before? How does the egg form the organism?

S. Brenner, quoted in *The Eighth Day of Creation,*
by H. Judson, p. 209

INTRODUCTION

Although the area of biology known as "developmental biology" did not emerge as a distinctive field until the 1950s, its precursor discipline, embryology, has had a comparatively long history. Embryology began in the 1820s with the seminal work of Karl von Baer and Christian Pander—it is to them that we owe the first careful descriptions of embryonic change and the concept of the "germ layers" (Churchill, 1991)—and evolved into an experimental science in the second half of the 19th century, in particular with the first surgical experiments of Wilhelm Roux in the late 1880s and his carefully delineated program of *Entwicklungsmechanik* (Sander, 1991).

Both kinds of work, the descriptive and the experimental, have immeasurably increased our knowledge about the details of embryonic development, but, in many respects, our grasp of the processes remains incomplete. The added wealth of molecular description in the last two decades has certainly deepened our understanding, but the underlying mechanisms still remain largely hidden. Even the constellation of ideas about progressive changes in gene expression, which is the closest approximation we have to a general explanation of the hidden dynamics of development, lacks for the most part *predictive* power; it has cogency as a partial general explanation and as a framework to think about development, but it is far from a theory of biological development. The fundamental problems of understanding developmental change are, therefore, to a large extent still very much with us, though the precision with which we can frame questions and attempt to answer them has increased considerably.

Genetic perspectives and methodologies provide a powerful approach to these problems. As discussed in the previous chapter, genetic techniques can provide additional descriptive data on development, particularly through clonal analysis, and,

16

more significantly, allow us to analyze causal mechanisms through ascertainment of the roles of particular gene products in specific processes. In this chapter we review some of the fundamental phenomena in animal development, in the context of the specific kinds of questions that geneticists ask about these phenomena. We then look at some of the properties of the genetic regulatory machinery that are relevant to developmental systems.

VISIBLE AND INVISIBLE EVENTS IN EMBRYOGENESIS: USING GENETICS TO PROBE MORPHOGENESIS, SPATIAL DIVERSIFICATION, AND STABLE STATES

The early events of embryogenesis possess a twofold significance for developmental biology. First, early embryogenesis involves special issues in development, issues of distinct interest in their own right. These aspects arise because of the particular properties of the precursor cells that form the embryo (the oocyte and sperm) and the way in which these cells, especially the oocyte, establish the initial properties and propensities of the embryo (reviewed in Wilson, 1925; Davidson, 1986). These first diversifications and states provide the foundations for the embryo, upon which all else is built. Second, early embryogenesis provides a microcosm of development and its basic phenomena as a whole. These phenomena include the capacities of somatic cells to pass on certain developmental capabilities to their descendants, cell–cell interactions that evoke new cellular properties, and the various processes that modify both stable and transient states. In considering these cellular and developmental changes, one is inevitably drawn into an analysis of both the details of cell structure and the genetic and regulatory machinery that mediate and facilitate such changes.

The Early Stages of Development

The inherent fascination of embryonic development is apparent to anyone who has watched a line embryo directly through a microscope or seen a time-lapse film of the events. It lies in the choreography of the embryo's cells, as they divide, move, and differentiate in a characteristic sequence to produce an increasingly more complex entity. Beginning with the entry of the sperm and the union of the male and female pronuclei to produce the diploid nucleus, the zygote commences a rapid series of mitotic divisions, the so-called cleavage divisions. These mitotic nuclear divisions are accompanied, for most animal embryo types, by complete cellular divisions, dividing the original egg cell into a cluster or ball of cells, the blastula (for instance, see Fig. 3.8). In insect embryos, such as *Drosophila*, however, an interesting variation takes place, in which cleavage consists of rapid nuclear divisions without cell divisions. Such nuclear divisions are accompanied by outward migration of the daughter nuclei within a single cytoplasmic mass to form a surface monolayer syncytium, which becomes a cellular monolayer as individual nuclei become enclosed by cell membranes (see Fig. 4.9).

Depending upon the type of embryo, the cleavage divisions are either succeeded or accompanied by a series of early cellular invaginations, the movements of gastrulation. The forms of gastrulation are nearly as diverse as the number of major phylogenetic divisions within the animal kingdom. The differences in modes of gastrulation include the times and total cell numbers in the blastula at which cell movement begins, the number of cells that invaginate, and the fates of these cells. For one of the embryo types that we will be looking at in the following chapters, *Drosophila*, the first stages of gastrulation are illustrated in Figure 2.1. In the *Drosophila* embryo, this first invagination takes place immediately after cellularization, following the syncytial nuclear cleavage stage. The figure illustrates a fact that is often skirted over lightly in introductory discussions of embryogenesis: that the inwardly moving cells in the gastrulating embryo are differentiated in composition and structure from those that do not move. Here the cells that invaginate are seen to be characterized by expression of the protein specified by the *twist* (*twi*) gene, whose

Fig. 2.1. Gastrulation in *Drosophila*. Stages in the infolding of the central cleavage furrow, with immunostaining for the protein encoded by the gene *twist*. (Reproduced from Leptin and Grunewald, 1990, with permission of the Company of Biologists, Ltd.)

gene product is one of several required for the subsequent development of the invaginated cells into mesodermal tissue (Thisse et al., 1987).

From the geneticist's perspective, the key to understanding the visible events of early embryogenesis is to identify the genes whose products are necessary for these processes to occur and to delineate the precise roles of the products of these genes. The essential technique is to isolate mutants defective in the process(es) under study and then to reconstruct the role of the wild-type gene, using the full panoply of current techniques. In the case of *Drosophila* gastrulation, for instance, one finding has been that *twi* is needed not only for mesoderm formation but for gastrulation as well; the latter defect provided part of the phenotype that led to the isolation of the first mutant (Nüsslein-Volhard et al., 1984). When two such correlated defects are found in a mutant, part of the ensuing investigation becomes directed toward understanding the nature of the connection and the role of the wild-type gene product in both of the affected processes. If other mutant conditions are subsequently found that uncouple the occurrence of the two defects, then one has established that there is no obligate link between the corresponding aspects of wild-type development; such situations illustrate the power of mutations as tools for dissecting normal development, beyond their informativeness about the roles of particular genes.

Furthermore, with one mutant in hand, it is often possible to isolate mutants in other genes that are involved in the same process; instances will be given in chapters 6 and 7. Indeed, whatever the intrinsic interest of a particular gene, if one wants to understand the entire process in which a gene product participates, it is of crucial importance to identify all of the genes required for that process (or as many as possible). This involves the technique of saturation mutagenesis, briefly described in chapter 1, and it has been extensively applied in both *Caenorhabditis* and *Drosophila* work (chapters 3, 4, 6, and 7). In addition, once a set of such genes has been identified, one can work out the "genetic pathway," the sequence in which they create their effects, for those whose mutants differ in some aspect of phenotype; the method will be described in chapter 4 and illustrated with a number of examples in both the nematode and the fruit fly.

Maternal Effect Mutants

When the focus is on early development, the genes of interest are often those of the mother, rather than of the embryo itself. The reason is that much of early development stems from the properties of the egg cell, which supplies both the bulk of the constituents and much of the initial cellular organization of the one-cell embryo. Because the egg is the product of the mother, its structure, and hence many of the properties of the early embryo, including its initial pattern, is determined by the genotype of the mother. (These properties are a function of the diploid maternal genotype, and not that of the haploid maternal gamete, the egg, because both the accessory cells that contribute to oocyte formation and the oocyte itself are diploid, the latter only completing meiosis just before or after fertilization, for most oocyte types.)

In consequence, a large part of the genetic analysis of early animal development has therefore concentrated on the maternal genes that are particularly important in egg and embryo development. If the mother is homozygous for a mutation that does not substantially reduce her own viability, but which affects egg development and subsequent development of the embryos derived from those eggs, she is said to carry a "maternal effect" mutation. For systems in which it is relatively easy to isolate new mutants, the identification and characterization of maternal effect mutants are thus at the heart of understanding much of the early development of those species (chapters 3 and 4).

One of the first maternal effect mutants to be discovered illustrates the basic genetic properties of this class. This mutation affects the direction of shell coiling in snails. When one views snail shells from the top, one can see that the shell spirals in either a clockwise pattern, *dextral coiling*, or a counterclockwise pattern, *sinistral coiling*. The internal organs follow these respective patterns of coiling. The two forms, when found in the same species, are thus mirror images. In the species *Limnaea peregra*, the most common form is dextral, while the rarer sinistral form is found to act as a simple Mendelian recessive. However, its inheritance pattern reveals that the direction of shell coiling is not a function of the individual snail's own genotype (the so-called zygotic genotype) but that of its maternal parent (Boycott and Diver, 1923; Sturtevant, 1923).

The two reciprocal crosses that reveal this maternal-dominated inheritance pattern are shown in Figure 2.2. When the eggs of dextral ($+/+$) females are fertilized by sperm bearing the sinistral (s) allele, all the progeny are dextral. When the reciprocal cross takes place, the fertilization of eggs from s/s females by $+$-bearing sperm, all the progeny are sinistral in phenotype, though dextral in genotype ($+/s$) as in the previous cross. When the dextral genotype F_1 progeny from both crosses self-fertilize (snails being hermaphroditic and capable of self- or cross-fertilization), both F_2 broods are dextral in phenotype (showing that the $+$ allele is dominant to the s allele in the maternal genotype). However, self-fertilization of these dextral phenotype F_2 individuals reveals a classic 3:1 segregation of dextral: sinistral broods in the F_3, each brood revealing the maternal genotype of the parent from which it derived.

The developmental basis of this maternal effect on shell coiling stems from a maternally based difference in the organization of cleavage, its system of "topogenesis" (Nigon, 1965). In gastropods, the topogenetic organization is spiral: The first two divisions produce a quartet of nearly equal-sized blastomeres within one plane, but the third cleavage divides each blastomere into a large cell, a "macromere," and a small cell, a "micromere," with the micromeres situated obliquely and in a plane above the macromeres. The micromeres are therefore offset with respect to their larger sister cells, thus giving a spiral pattern so that the dextral and sinistral alleles create mirror-image cleavage spindle placements with respect to one another. In third-cleavage embryos from dextral genotype mothers, the result is that the micromeres are placed clockwise with respect to the macromeres, while in embryos from sinistral mothers, the micromeres are placed counterclockwise. Since the internal organs are derived from the large, so-called "D" macromere, the mirror-image symmetrical placement of this macromere with respect to the other blastomeres determines the direction of coiling of these organs and, ultimately, the mirror-image shell coiling seen in the two forms. In *Limnaea*, the dextral form has been found to specify a component that can be obtained from egg extracts of dextral mothers, which can reverse, or "rescue," the sinistral character when injected into eggs of sinistral mothers (Freeman and Lundelius, 1982).

The topogenesis gene in *Limnaea* is an example of what may be termed a "strict" (Wood et al., 1980) or *exclusive* maternal effect gene. The mutant has no other effect on

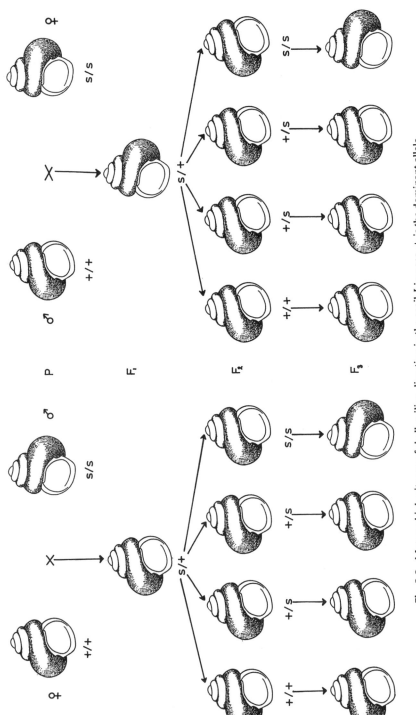

Fig. 2.2. Maternal inheritance of shell-coiling direction in the snail *Limnaea*. + is the dominant allele and gives dextral coiling; *s* is the allele for recessive sinistral shell coiling. (Adapted from Suzuki and Griffiths, 1981.)

the biology of the animal and, as far as can be determined, is not expressed at any stage other than oogenesis. It is also one of the few examples of a maternal effect mutation that, while altering embryogenesis in a distinctive fashion, allows full development.

There is, however, another, and much larger, category of oogenesis gene functions: those that are expressed both during oogenesis and in later stages of development. Completely deficient, or amorphic, mutants of such genes would be standard lethal mutants, but if the gene product for any such gene is required in much larger amounts in early development rather than later, a hypomorphic mutant of such a gene can produce a maternal effect phenotype (Bischoff and Lucchesi, 1971). Such "leaky" mutations permit the survival and development of homozygous female progeny, but these females subsequently produce defective progeny because of their inability to supply sufficient gene activity during oogenesis. In *Drosophila* and *Caenorhabditis*, at least, perhaps, the majority of loci can mutate to give rare hypomorphic maternal effect mutations (Perrimon et al., 1986; Kemphues et al., 1988b). A gene that can mutate to give a maternal effect but which is expressed at other stages of development, besides oogenesis, is denoted as a *partial* maternal effect gene (Wood et al., 1980).

Given the importance of maternal effect mutants in the analysis of early embryogenesis, one should note their genetic hallmarks. There are three crosses that serve to identify maternal effect mutants; these are as follows (where m stands for the mutation and + for the wild-type allele):

1. $+/+$ females $\times$ m/m males $\rightarrow$ $+/m$ progeny, with normal development

2. $m/+$ females $\times$ $+/+$ males $\rightarrow$ $+/m$ and $+/+$ progeny, with normal development

3. m/m females $\times$ $+/+$ males $\rightarrow$ $+/m$ progeny, with defective or altered development

While crosses 1 and 3 both yield embryos of identical heterozygous genotype, these heterozygotes have dramatically different phenotypes. Development is unaffected when the fathers are homozygous for the mutation but is defective when the mothers are. For heterozygous mothers, in contrast, the wild-type allele can usually supply enough of the normal component to permit normal development.

Employing Maternal Effect Mutations to Study Early Cellular States

The principal phenomena that have been addressed by most of the laboratories using maternal effect mutations to study early development have been the processes by which different fates and potencies are allocated to blastomeres, properties that in many animal cell types ultimately derive from the properties of the egg. Indeed, states of developmental capacity in the early embryo have been the central focus of much of experimental embryology during the past century, and their nature has been one of the central mysteries in developmental biology.

Some of these states are relatively labile and are created by interactions with other cells (as shown by their loss upon transplantation of the cells to new locations, either in culture or to other regions of the embryo), while others are stable and perpetuated as cell autonomous (intrinsic) states, usually because they inherit particular regions of the egg cytoplasm. The nature and extent of these conditional and autonomous cell commitments is characteristic of each egg type. Furthermore, the structure of the egg will often predetermine one or more axes of the embryo that develops from it; such axial specification can be found both in embryos with rigidly fixed cleavage patterns, such as *Caenorhabditis*, and in those with a much looser specification of cleavage mode, such as *Xenopus* (reviewed in Davidson, 1990).

The traditional descriptive term that denotes the allocation of particular fates and capacities to particular blastomeres or groups of blastomeres is "determination," and a critical question about this phenomenon concerns the nature of the "morphogenetic system" of the egg. Three different sorts of possibilities are sketched in Figure 2.3: A "mosaic" system of "morphogenetic determinants" (a), a concentration gradient of a "morphogen" (b), and a system that has only some degree of initial specification,

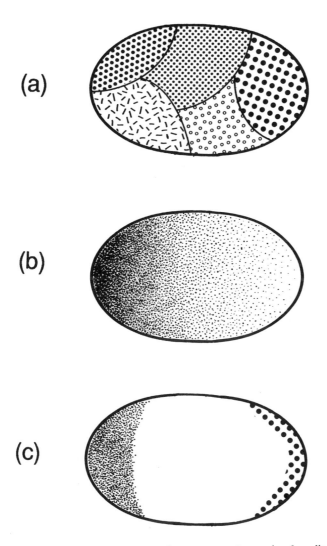

Fig. 2.3. Models of egg fate specification systems in oocytes. **a**: A mosaic of qualitatively different determinants. **b**: A monotonic concentration gradient. **c**: A partially specified system, in which the unspecified sectors become "instructed" as the result of reactions and/or interactions beginning in the specified portion.

with the rest being created by subsequent interactions (c). By isolating and analyzing maternal effect mutations that alter the pattern of blastomere fates, one can explore the nature of the morphogenetic system. The simple prediction for mosaic systems is that maternal effect mutations affecting just one part of the embryo should be found, while the other models predict that all mutants affecting the maternal specification system will have more global effects; testing these predictions is a core part of the material of chapters 3 and 4.

Stability and Reversibility of Developmental States

"Determination," as thought about by experimental embryologists, involves two distinct aspects, a differential "instruction" of blastomere fate and a set of stabilizing processes that fix or "commit" these blastomeres to these fates (thereby restricting their developmental potencies) (Sander, 1976). While much of the genetic analysis of determination has focused on the first part, differential instruction, and the answers for *Drosophila*, in particular, have been illuminating (chapter 4), the second aspect, the processes that "lock in" cells to follow specific developmental destinies, has received less attention from geneticists. These maintenance processes, however, are at the heart of the question of the extent to which cell inheritance, or "cell lineage," dictates particular developmental paths, a central question in both *Drosophila* (chapter 7) and *Caenorhabditis* development (chapters 3 and 6).

To analyze the commitment processes genetically, one would need to search for mutants that perform the initial fate settings accurately but which cannot then maintain them. Although there have been few systematic searches of this sort, a group of genes involved in the maintenance of developmental states, the so-called *Polycomb (Pc)* group, has been identified in *Drosophila* (reviewed in Paro, 1990) and some candidate maintenance genes in *C. elegans* have been found (Way and Chalfie, 1989; Au and Chalfie, 1989). These genes and the properties of some of their known products will be discussed in chapter 7.

The obverse developmental phenomenon to that of cell inheritance and stability is reversibility or flexibility of developmental "decisions"; such flexibility invariably involves cellular interactions. This topic is also a central one, pertaining to all animal systems, but is of special relevance to vertebrate development, in the extent to which cell lineage plays a comparatively small role and cellular interactions large ones; this will be a major theme in the two chapters on mouse development (chapters 5 and 8). However, some genetically based, and analyzed, instances of cell interactions in *Caenorhabditis* and *Drosophila* will also be discussed (chapters 3, 4, 6, 7, and 9).

Terminology

Before leaving the subject of hidden developmental states, for the moment, a few words are needed about the problem of the traditional terminology of cell states. Many of these terms, such as "determination," "competence," and "commitment," were deliberately borrowed from psychology in the first part of this century (Oppenheimer, 1967). Although they have become irreplaceable through frequent usage, there is often some ambiguity in the ways that these terms are applied to descriptions of particular situations. Perhaps the tradition of adopting terms from psychology was fated to cause trouble; to define one set of mysterious processes by analogy to an even more obscure set of phenomena would appear to be a guarantee of further confusion. The fundamental difficulty, however, is that each term represents a shorthand operational definition of a complex cellular and developmental property or process displayed under a particular set of experimental conditions. In general, when characterizing developmental states, the temptation is to equate, unconsciously, the label with some kind of fixed set of molecular properties, and to forget the contingent, operational nature of the classification.

The skepticism voiced here about the utility of these general classificatory terms for developmental states is not universally shared. A clear exposition of the distinctions implied by several of these terms can be found in Slack (1991). For the most part, however, in this book, we will simply refer to processes that "assign" particular developmental states and to those that "maintain" these states, rather than placing heavy reliance on the traditional distinctions. Where it is necessary or desirable to use some of the customary classifications, such as "determination," the operational basis for the use of the term will be made explicit.

PATTERN FORMATION: A SPECIAL PROBLEM

In addition to the "classic" problems of developmental biology (differentiation, morphogenesis, determination), there is one more phenomenon that is central to developmental biology. This is the problem of pattern formation. Interestingly, pattern formation was hardly even recognized by most biologists as a distinct problem until the 1960s. The term itself is often used loosely as a synonym for "development" as a whole, but it will be employed here to denote those mechanisms that produce the characteristic spatial arrays of cells and structures that arise during development. The problem of pattern formation is that of how typical spatial organization is achieved with such stunning regularity in the developing organism, and is both distinct from and, in a sense, prior to the question of how the organism forms particular cells of highly specific types (such as muscle, skin, or gut cells), the problem of differentiation.

The fact that recognizable animal morphologies arise during development does not in itself prove that general mechanisms of pattern formation exist. In principle, the final, complex end result might be produced by a summed series of independent, piecemeal events that happen to occur in an invariant order without direct coordinating links. The evidence for the existence of general mechanisms of pattern generation comes almost entirely from observations on the disruption of development. Such disruptions are often followed (depending upon the organism) by growth that regenerates either the original pattern or large sections of it. It is the phenomena of pattern *re*-formation that provide grounds for believing in the existence of coordinating mechanisms of pattern formation in normal development.

Two Distinct but Related Problems

We have already encountered one question of pattern formation, namely how the egg develops into an embryo whose cells have different developmental potentialities and capabilities. In the terms of experimental embryology, the egg is a "primary field" of pattern formation, subject to perturbations that affect the whole unit. In normal development, however, the embryo subsequently becomes divided into so-called "secondary fields," whose developmental processes are quasi-autonomous with respect to one another. The standard example of such secondary fields is the set of appendages, each developing from a group of particular precursor cells, each little affected by injury to the other appendage rudiments.

Appendage development raises the two principal questions about pattern formation, each with its genetic correlates. The first is, How do different regions, such as appendages, become different from one another? In genetic terms, What gene activities are responsible for these differences? The question is given point by the long-standing set of observations (see Bateson, 1894) that different appendages can be "transformed" into one another by mutations (the term "transformation" signifying that a particular rudiment which normally gives rise to an appendage of phenotype A can be altered by mutation to give rise to an appendage of phenotype B); Figure 2.4 illustrates such a change. Wild-type adult *Drosophila* possess two wings, developing from the second thoracic segment (T2 or mesothorax) and two small balancer organs termed "halteres," which develop from the third thoracic segment (T3 or metathorax). The figure shows a fly in which two additional wings have replaced the halteres, creating a four-winged fly, because of the cumulative effects of three mutations in genes of the bithorax complex.

Such mutations are termed "homeotic" mutations and much of the progress in recent years in developmental genetics has been in explorations of the nature of homeotic genes. These genes play major roles in the development of both

Fig. 2.4. A four-winged fly, produced by mutations in the bithorax complex (genotype: *abx pbx bx*). (Reproduced with permission from *Science*, **221**, cover, July 1, 1993. © AAAS.)

Caenorhabditis and *Drosophila* studies (chapters 6 and 7) and may prove to be equally important in mammalian development. Despite a relative paucity of homeotic mutants in the mouse, the homologs of *Drosophila* homeotic genes have been found in mouse and man and there is, increasingly, reason to believe that they play somewhat comparable roles in pattern formation in mammals (chapter 8).

Just as homeotic genes illustrate, and participate in, the establishment of differences in secondary fields, they can illustrate the second important question: What elements in pattern formation mechanisms are shared among or are common to different regions, such as appendages? A hint of shared properties is given by the homeotic mutation *Antennapedia (Antp)*, which causes the adult fly to have one or both antennae partially transformed into legs. (As shall be explained in more detail in chapter 4, the development of the surface of the adult *Drosophila* comes about during the final (pupal) stages through the differentiation and morphogenesis of large packets of cells called "imaginal discs"; a structure such as the antenna develops during the pupal metamorphosis from a part of the eye-antennal imaginal disc of the head.)

Despite the variability of the transformation produced by *Antp* (a common feature of many homeotic mutations), a crucial fact about the transformed antennae is that there is always a specific spatial pattern of corresponding replacements with respect to the proximodistal (p-d) axis (Fig. 2.5) (Postlethwait and Schneiderman, 1971). Thus the region that normally gives rise to the most distal segment of the antenna, the arista, is always converted to the most distal leg structures, namely the tarsal segments; the next most distal antennal segment, AIII, is replaced by tibia or femur; and the most proximal antennal segment, AI, is replaced by the most proximal leg segment, the coxa. In some manner, cells seem to "perceive" their relative position within the developing structure and respond appropriately with respect to this position. Such correspondences suggest that beneath the overt differences between antenna and leg there may be some significant commonalities, in the form of shared positional similarities, in turn underlain by a shared set of genetic activities.

Two General Sorts of Solution

Although many hypotheses about pattern formation have been launched, they divide into two general categories (see review by Martinez-Arias, 1989). The first group posits some form of global "instructions" throughout the secondary field, while the second group emphasizes the importance of local and sequential cell interactions that generate the "instructions."

Three global instruction hypotheses are schematized in Figure 2.6. The first, and perhaps the first general model for pattern formation, was the "prepattern hypothesis" of Curt Stern (1954). Stern's formulation was designed to explain the origin of certain two-dimensional patterns, those of the bristle arrays on the surface of the fruit fly. A striking feature of these bristle patterns is that each external region has its own characteristic arrangement. One example is the surface of the mesothorax (Fig. 2.7), which possesses both relatively short bristles (microchaetes) and large bristles (macrochaetes) in characteristic positions with respect to one another. Stern's hypothesis was that the distribution of a "morphogen" (literally a form-giving substance) in the epidermal sheet from which the bristles arise determines where bristles will appear. The prepattern consists of the spatial distribution of a morphogen, its peaks

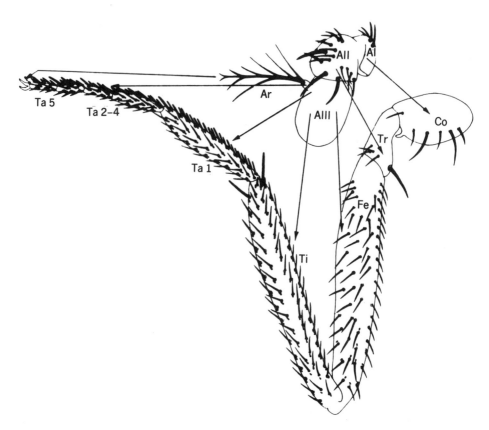

Fig. 2.5. Proximodistal positional correspondence between leg and antennal segments in the *Antennapedia* homeotic antenna-to-leg transformation. (Reproduced from Postlethwait and Schneiderman, 1971, with permission of the publisher.)

and valleys, with the visible pattern being isomorphic in large degree to the prepattern (Fig. 2.6a). In this view, both the differences and shared elements of pattern reduce to those respective elements of the prepattern.

A variant of this idea, the "pattern field" hypothesis, proposed by Waddington (1973), allows for the possibility of a smaller number of pattern-evoking signal distributions than of observed patterns. The central tenet is that different patterns may be produced by differences in the threshold of the cellular response to the signal rather than by the distribution of the signal itself (Fig. 2.6b). Although the pattern field idea allows a smaller number of prepatterns for a given number of patterns, both hypotheses entail fairly direct relationships between the singularities of the underlying pattern-evoking signal distribution and the final observed pattern. (It should be mentioned that there are mechanisms for generating unusual distributions of a signal from a previously uniform dispersion; the first was proposed by Turing [1952] and other models have been described since.)

The third idea, and the most influential in the last two decades, completely uncouples the shape of the final pattern from the underlying signal. This is the concept of "positional information," proposed by Wolpert (1969, 1971); here there need be

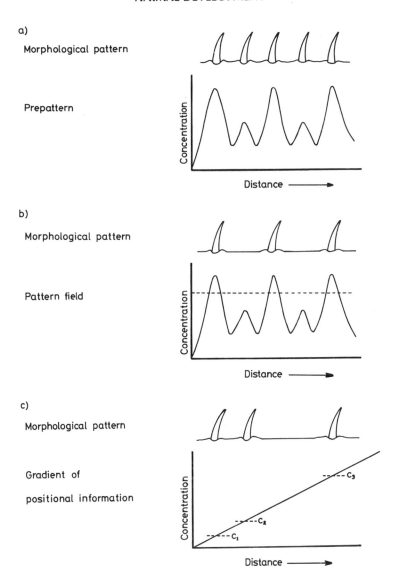

Fig. 2.6. Three mechanisms of pattern formation. See text for discussion. (Adapted from Wolpert, 1971.)

no obvious relationship between the distribution of the signal and the form of the developed pattern. Wolpert suggested the existence of a common, perhaps universally shared, mechanism by which cells ascertain their relative positions, the positional information, within an embryonic field; the cell then "reads" and interprets its position to acquire a position-specific state, the "positional value," which then sets the future pattern of differentiation at that cellular locus. In this scheme, the similarities between antenna and leg, for instance, lie in their overall similar positional information, while their differences stem from differences in cellular "interpretation" of the positional information, which results in different patterns of gene expression.

Fig. 2.7. The bristle pattern of the *Drosophila* thorax. The larger bristles, the macrochaetes, are each situated in stereotypic positions, with the pattern symmetrical about the midline; the smaller bristles, microchaetes, are in rows. (Reproduced from Richelle and Ghysen, 1979, with permission of the publisher.)

The positional information itself may be a simple monotonic concentration gradient of a morphogen across the field or may be produced by a more complicated signal-generating scheme; positional information by gradient is diagramed in Figure 2.6c and a current view is presented in Wolpert (1989).

The hypothesis of positional information has been been highly useful in explaining various observations and in thinking about pattern formation; it has, for instance, provided the stimulus for much of the current work on limb development, in particular the successful identification of the retinoic acid gradient in limb bud development (reviewed in Tickle, 1991). Nevertheless, it has not provided a clear guide for genetic thinking or experiments about the immediate responses to the positional information; too many events lie between the putative instructional events and the final outcome. This lacuna has, in effect, been the weakness of all of the global mechanisms of pattern formation depicted in Figure 2.6.

The other class of explanations, which emphasizes sequences of local events, more readily lends itself to experimental tests, including genetic ones, because of the focus on immediate events. When a mutation is found that affects a particular aspect of a pattern, one can ask whether the mutation exerts its effect just within the cell of that genotype (so-called "autonomous effects") or on neighboring cells, and, if the latter, what the nature of such "nonautonomous" effects is. In such work, one is looking at the fine-grained pattern of events rather than attempting to infer early events from late outcomes.

Though not framed originally in genetic terms, one of the more compelling hypotheses of pattern formation that relies on local cell interactions with a schematized network of such effects is the "polar coordinate model" (Bryant et al., 1981), based on experiments dealing with the regeneration of appendages and appendage primordia. The collective set of findings from much work on the regeneration of appendages as different as cockroach legs, fruit fly wings, and salamander legs suggests that a common pattern-forming mechanism is involved, which is described by the model (French et al., 1976; Bryant et al., 1981).

The central finding, on which the model is based, is that if one cuts the primordium of any regenerating appendage, each fragment will either duplicate those pattern elements it would have given rise to without growth or regenerate the missing structures, with the specific outcome depending on the fragment's initial position and size. The either/or duplication/regeneration choice is fundamental and has predictive value. Furthermore, fragments comprising less than half a disc tend to duplicate, while fragments larger than a half a disc regenerate. Both responses depend on closure of the wound, with cell contact between the cut edges, and a subsequent localized stimulation of DNA synthesis and cell proliferation at the edges (Bryant and Fraser, 1988).

The polar coordinate model consists of one postulate concerning the positional specification of cells within a field and two rules of regenerative behavior when part of that field is removed. The postulate is that the position of each cell on a collapsed cone (the idealized form of an appendage primordium) is specified uniquely by two coordinates with respect to a central point, the tip of the cone (E): The first coordinate is a radial distance from that central point, and the second is a circumferential position on the circle delimited by that radius. The coordinate system is shown in Figure 2.8.

The first rule of regenerative behavior in the polar coordinate model is termed the "shortest intercalation rule": When cells with nonadjacent positional values are brought together, the process of growth restores all intermediate positional values by the shortest numerical route. The shortest intercalation rule, applied to circumferential values, explains why small fragments undergo duplication and large fragments undergo regeneration (French et al., 1976).

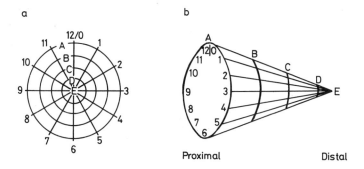

Fig. 2.8. The polar coordinate model. Each position on the appendage or appendage primordium is given by two coordinates, a circumferential position (**a**) and a radial position (**b**) along the p-d axis. Two rules, described in the text, describe pattern formation during regeneration and normal growth. (Redrawn with permission from Bryant et al., 1981, *Science* **212**, 993–1002. © AAAS.)

The second rule is the so-called "distalization" rule, that when new cells are generated by a preexisting adjacent cell, the new cells, if they possess circumferential values already present among their neighbors, will adopt a radial positional value that is one degree more distal than that of the displacing cell(s) (Bryant et al., 1981). Each round of intercalary cell division thus produces progressively more distal cells, until the process of circumferential filling of missing values is complete (Fig. 2.9).

The central feature of the model is that there are distinct positional values that are equivalent between different primordia and, by implication, comparable genetic properties underlying those positional values. Regeneration experiments involving apposition of different fragments from different discs provide some support for comparable positional value systems, and interactions, between these discs (Adler, 1979; Haynie, 1982), although not all of these experiments have demonstrated such relationships (Wilcox and Smith, 1977).

Though no explicit predictions were originally made about which gene activities are involved, a growing number of genes that seem to affect pattern in comparable geographical regions within the fruit fly have been identified and, in certain important respects, seem to be candidates for specifiers of positional values (Wilkins and Gubb, 1991). Many of these genes were first identified as playing roles in the development of embryonic segmental patterning, and it seems increasingly likely that despite the outward differences of the embryo and imaginal discs of *Drosophila*, a number of important similarities in pattern formation obtain. Some of the correlations and tests will be described in chapter 7.

Despite the successes of the polar coordinate model, there are observations that can only be accommodated by additional assumptions (Karlsson, 1981a,b; Kauffman

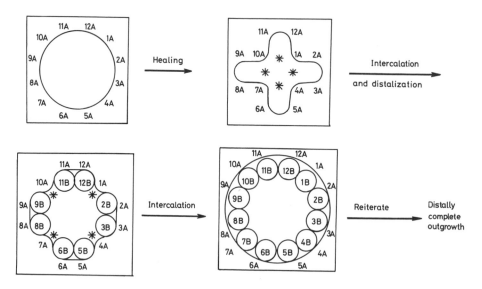

Fig. 2.9. The distalization rule of the polar coordinate model; completion of distal positional value set by outgrowth. With each round of filling in of circumferential filling of positional values, cells with values possessed by their neighbors are forced to adopt the next most distal positional values. (Redrawn with permission from Bryant et al., 1981, *Science* **212**, 993–1002. © AAAS.)

and Ling, 1981; Kirby et al., 1982) or that are difficult to reconcile with it (Meinhardt, 1983). Other models based on localized interactions have also been formulated and these too make predictions that can be tested.

One such model has been proposed by Meinhardt (1983, 1986), and involves a localized interaction between different regions or "compartments" at their boundaries. In the Meinhardt scheme, when two such boundaries cross, a singularity is generated at the intersection. This singularity becomes the source point for the elaboration of a morphogen which develops a cone-shaped distribution there; the point of intersection itself will become the distal tip of the appendage (point E in Fig. 2.8) that arises from the imaginal disc while lower concentrations of morphogen specify intermediate p-d positions. Genetic data have been obtained that bear on this model, and these too will be discussed in chapter 7. In chapter 9, we will return to some of the general phenomena and genetic questions about pattern formation, in connection with the development of two-dimensional integumentary patterns.

DOMAINS OF GENE EXPRESSION AS CLUES TO FUNCTIONAL SIGNIFICANCE AND SOME THOUGHTS ABOUT GENETIC "CONTROL"

The essence of genetic analysis in development is inference about gene function from alterations of development in mutants. Despite all the new ways for making and analyzing mutants and their phenotype effects, that strategy remains essential. Nevertheless, it is not the only approach to elucidating possible gene function. Even in the absence of mutants, inferences about the function of a gene, and the cells and tissues that require it, are often made from the observed expression patterns of the gene, deduced from the measured distributions of the gene's transcript or protein product. The assumption is that if the gene is expressed in particular cells at particular times, it must serve a function there. Several words of caution are required with respect to this assumption.

The premise behind this belief is that natural selection has optimized the expression of genes so that particular gene products are synthesized only in the tissues in which they are needed. Yet there are good reasons for doubting the premise. If the activation of transcription of genes in eukaryotes involves, not uncommonly, interactions between two or more regulator proteins which act at discrete neighboring sites in the promoter or enhancer regions (reviewed in Mitchell and Tijian, 1989; Martin, 1991), then altering one element in the regulatory network will, in principle, affect the regulation of many genes. The corollary of this idea is that for such a *combinatorial* system of gene regulation, natural selection is constrained in the degree to which it can alter the regulation of individual genes one at a time (Kauffman, 1987; Dickinson, 1988).

In consequence, it must be the case that aspects of certain gene expression patterns do *not* serve an adaptive function but are simply adventitious consequences of other aspects of the pattern which have been selected. Nor need such nonadaptive expression of genes, even of regulator genes, have deleterious consequences. The effect of a particular gene product is, almost invariably, achieved within the "context" of the other gene products of the cell. A regulatory gene product that has, for instance, a powerful effect on cell properties in one context may be entirely lacking in effect in other cells, as for instance when one or more other specific regulatory gene products with which it interacts are missing. Indeed, one might say that no regulatory gene product is an island; combinatorial "control" bears an analogous relationship to the individual regulatory gene to that of social constraints in the behavior of individual human beings.

The net consequence is that combinatorial systems lock in certain aspects of gene expression. First, one may select for certain changes but only to maximize expression for certain genes; other genes in the network may have little adaptive significance but cannot, in any case, be simultaneously selected for optimality (Kauffman, 1987). Second, the constraints on regulating expression become more severe, as one adds more regulatory factors and binding sites and rules for the interactions of the products with the sites; hence, the more complicated the regulation of a set of genes which respond to common factors in this manner, the less freedom there is to alter the regulation of single members of the set (Dickinson, 1988).

There is a terminological consequence of these constraints which should be kept in mind. Although it is not uncommon for developmental geneticists to speak of a particular regulatory gene "controlling" a process, such "control" is, in fact, always context dependent, being a function of the other regulatory factors present and required and the particular cytoplasmic inputs that the nuclei of the responding cells are experiencing. Furthermore, to speak of a gene that has not yet been characterized in molecular terms as controlling a process, simply on the basis of its mutant pheno-type, is even more mistaken: Deficiencies in many structural genes will cause a developmental process to stop. All that the word "control" signifies for such a process is that the gene product can be made rate-limiting, at some concentration.

Combinatorial regulatory systems thus have two opposing effects. With a relatively small number of factors, they allow, in principle, a large number of different regulatory networks, involving different gene sets, to emerge. At the same time, they reduce the freedom of the system to alter the expression of individual members of regulatory networks. How might eukaryotes cope with such restrictions? One possibility is the loosening of regulation such that a large degree of "sloppiness" exists in the system, with most genes "on" most of the time and many not exerting particular effects except within certain contexts. Though such a possibility may appear unlikely, it does fit some of the genetic data and molecular surveys, to be discussed in chapters 6, 7, and 8.

Another, more attractive possibility, not excluded by the first, is the existence of large degrees of *functional redundancy*, in which genes that specify products with highly similar activities are part of separate regulatory networks. Gene products activated by one set of activators whose other activities might be harmful to or unnecessary for particular cell types might be substituted by similar-acting gene products called forth by a different set of regulators. Some of these partially inter-changeable activities would involve members of the same gene family that are embedded within different regulatory networks; such a possibility may help to explain the existence of gene families whose members appear to be functionally equivalent. Other instances of quasi-interchangeable activities could involve groups of elements that are not homologous, but which happen to have similar functions.

This hypothesis that functional redundancy can act as an antidote to the inherent rigidities of combinatorial control is, of course, just a speculation. However, as we shall see, the existence of functional redundancy, a phenomenon not given much attention until recent years, is considerably more widespread than previously believed (reviewed in Tautz, 1992). In consequence, it is becoming increasingly important to understand the selective pressures that have favored the evolution of such "backup" systems. Instances of functional redundancy in the nematode, the fruit fly, and the mouse will be given in the chapters that follow, and we will return to the general questions of regulatory system properties and functional redundancy in chapter 10.

EMBRYOLOGY AND GENETICS

When T.H. Morgan published his book *Embryology and Genetics* in 1934, the two fields designated in the title were completely separate entities. It was Morgan's hope that the book would help to provide a bridge between them. Instead, although it does contain Morgan's prescient insight that development must involve the activation of different "batteries" of genes in different cells and tissues, his treatment confirmed the gulf between the two disciplines. As discussed in the first chapter, that gulf essentially persisted for more than two decades, despite the efforts of a few pioneers in the late 1930s and throughout the 1940s. To a large extent, embryologists continued to concentrate on events in the cytoplasm, while geneticists, for the most part, emphasized the importance of the nucleus (Gilbert, 1991). In the 1960s, the general conceptual gap between the two fields was at last bridged by means of the models of prokaryotic gene regulation, inaugurated by Jacob and Monod, although the wholesale fusion of approaches and techniques between developmental biology and genetics would not occur until the 1980s.

The aim of this chapter has been to illustrate the fact that all the central phenomena of classical embryology can now be thought of explicitly in genetic terms, and for certain organisms, in terms of particular genes. In addition, a different form of interdisciplinary connection is being made these days: In certain instances, it is becoming impossible to ask why certain things happen in development without inquiring into their evolutionary origins. In effect, evolutionary concerns, long banished from developmental studies in a reaction to the Haeckelian approach (Gould, 1977), are beginning to return to developmental biology. Examples will be discussed in this book as we come to them and we will return to this subject in chapter 10. Indeed, the present period in developmental biology is marked not only by the increasing precision of analysis of biological phenomena, but also by the increasing synthesis between lines of inquiry long treated as essentially unconnected.

In the next chapter we will examine how genetic analysis is being applied to the early embryogenesis of *Caenorhabditis elegans*, a member of the tribe of nematodes, one of the first invertebrate groups investigated in depth by 19th-century embryologists. The work illustrates the ways in which genetics can and does enrich embryology.

THREE

CAENORHABDITIS ELEGANS

Early Embryogenesis

The study of cell-lineage has thus given us what is practically a new method of embryological research. The value and limitations of this method are, however, still under discussion..Like other embryological methods, it has already encountered contradictions and difficulties so serious as to show that it is no open sesame.

E.B. Wilson (1898)

Thus we want a multicellular organism which has a short life cycle, can be easily cultivated, and is small enough to be handled in large numbers, like a microorganism. It should have relatively few cells, so that exhaustive studies of lineage and patterns can be made, and should be amenable to genetic analysis. We think we have a good candidate in the form of a small nematode worm, *Caenorhabditis....*

From a proposal by Sydney Brenner to the
Medical Research Council, October 1963;
quoted in *The Nematode Caenorhabditis elegans,*
edited by W.B. Wood

INTRODUCTION: *CAENORHABDITIS ELEGANS* AS A MODEL SYSTEM

Caenorhabditis elegans, a small, free-living soil nematode, occupies a special place in the bestiary of developmental genetics. Among animals complex enough to have a true neuromuscular system and a corresponding set of behaviors, it is one of the simplest and, after three decades of research, the best characterized in terms of its cellular makeup. For wild-type animals of both sexes, the somatic cells arise in a constant and stereotypic sequence of cell divisions, generating a pattern of fixed cell lineages (Sulston and Horvitz, 1977; Sulston et al., 1983). The end result is that every adult animal has a precise number and spatial arrangement of its somatic cells, which is completely characteristic of its sex.

Just as importantly, this organism has proved to be readily accessible to genetic and molecular analysis. Developmental and behavioral mutants can be obtained with ease, and once isolated, such mutants can be rapidly mapped and classified by a variety of genetic techniques. Approximately 1000 genes, perhaps 10% of the entire gene complement, have already been mapped and subjected to, at least, preliminary genetic analysis. In addition, the genome is sufficiently simple to lend itself to a thorough molecular analysis by recombinant DNA techniques. Most of the genome has now been cloned in such a way that most new gene clones can be placed within a particular region of DNA sequence (reviewed in Coulson et al., 1991), facilitating their further characterization. Furthermore, the cellular and genetic characterizations have proved to be tremendously valuable in combination. Given the complete wild-type cell lineage, one can, with contemporary microscopical techniques, pinpoint the initial cellular population affected by a given genetic defect and identify the entire sequence of events that produces the developmental derangement.

In this chapter, the initial phase of embryogenesis of *C. elegans* is covered, up to approximately the 150-cell stage of cleavage. During much of this period, there is only a low, but nonzero, level of zygotic gene transcription (Schauer and Wood, 1990). In consequence, the gene product composition of these early embryonic cells is predominantly that of the oocyte. Since maternally encoded constituents provide the bulk of the cellular material, it is not unreasonable to believe that they are a direct source of the properties of the early embryo.

The central questions about the early nematode embryo concern the relationships between these maternal constituents and the formation and properties of the early cell lineage. Are these first embryonic cells directly "instructed" by maternal constituents in their different fates? If so, are the instructions a set of qualitatively different "determinants" (Fig. 2.3a), or do they involve a quantitative or gradient-based process (Fig. 2.3b)? In either case, does the fixity of the cell lineage directly mirror, in some fashion, the architecture of the egg? The long-held and traditional view has been that the nematode egg is, indeed, a "mosaic" one, in which different regions of the oocyte stamp their character on different early blastomeres (see discussion below).

In this chapter we look at the findings that have modified, and even undermined, this view and the kinds of genetic analysis that have been used to investigate early embryogenesis in this animal. An important conclusion is that while the early embryo is in compositional terms largely a maternal construct, much of the early determination of cell fate is dynamic and connected to specifically timed events related to the cell cycle and to particular cellular interactions. A second general finding is that there are few genes expressed exclusively in oocyte development; the oocyte is an unusual cell but it is constructed primarily from gene products employed in other tissues. In chapter 6 we will take up the themes of cell lineage and cell interaction, begun here, and see how they apply to later events of embryogenesis and, in particular, to postembryonic development.

THE LIFE CYCLE AND GENETICS OF *CAENORHABDITIS ELEGANS*

Stages of Development

In any wild-type population of *C. elegans*, the great majority of the individuals are self-fertilizing hermaphrodites. These are closely related in form and function to true females of certain closely related species but, unlike the latter, are equipped to produce spermatocytes for a relatively short time and to fertilize their own oocytes. The *C.*

elegans hermaphrodite is, in effect, a modified female. In addition to this majority sex, however, there exist within each population rare males (readily distinguishable from the hermaphrodites, under a dissecting microscope, by their different, fan-shaped tail morphology). Males can mate with the hermaphrodites, and the cross progeny include equal numbers of hermaphrodites and males. As will be described below, the existence of males permits conventional genetic analysis to be carried out in this organism.

The life cycle for the hermaphrodite is summarized in Figure 3.1. The complete sequence, from fertilized egg to emergence of the adult, includes five distinct phases of approximately equal duration. These periods are, respectively, embryogenesis, which takes place inside the eggshell, and four successive larval stages. Each larval stage transition is preceded by a 2 hr period of reduced activity, termed the lethargus, which is followed by a molt, the shedding of the larval cuticle.

Oocyte formation takes place in the ovary of the hermaphrodite, and fertilization takes place as the oocytes pass through a special sperm receptacle, the spermatheca. The first cleavage divisions of the embryo take place as the zygote moves out of the spermatheca; the embryo then moves down the uterus toward the external opening, the vulva. The time of egg laying is variable, ranging from about the 2-cell to 32-cell stage. Cleavage continues and is accompanied by a progressive reduction of cell size (in the absence of cell growth during this period) and the beginnings of detectable cellular differentiation, until the embryo has reached approximately 550 cells, about 7 hr after fertilization.

The end of cleavage is followed by the second phase of embryogenesis, lasting 6 to 7 hr, which consists of a sequence of marked cellular differentiations and morphogenetic movements that transform the embryo from a round ball of cells into an elongated, twisted early larval form (the "pretzel" stage). Some of the intermediary morphogenetic stages during this transformation have been descriptively labeled the "comma," "tadpole," "plum," and "loop" (Krieg et al., 1978); a few of the stages of embryogenesis are shown in Figure 3.2.

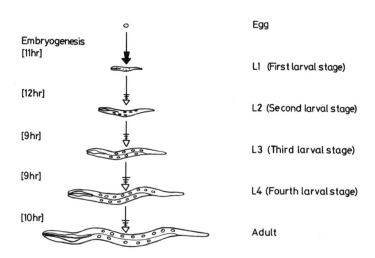

Fig. 3.1. Life cycle of *C. elegans* at 25°C. The crossed lines on the arrows between stages indicate the periods of lethargus.

The newly hatched first-stage larva (L1) is very similar in external morphology to the adult but, at 0.2 mm, is only one-fifth its length. In the ensuing four stages of larval development, two principal changes take place: There is a substantial increase in overall size (through cellular growth and division) and sexual development and maturation occur, these last processes involving the development of the gonads and the secondary sexual characteristics.

In addition to the sequence of events in the main life cycle, there is an alternative pathway that the organism can take, namely dauer larva formation. Development of the dauer larva, a nearly dormant larval form, takes place at the second molt if the animals have been subjected to starvation or overcrowding. The dauer larva does not feed and is inactive except for a touch stimulus response. Dauer larvae are slightly longer but only half as wide as normal second-stage larvae (Fig. 3.3).

Genetics and Genome Structure

The diploid chromosomal constitution of the *C. elegans* hermaphrodite consists of 12 chromosomes: 5 pairs of autosomes and 2 X chromosomes. The genotype is symbolized as 5AA + XX. The rare males have 11 chromosomes, its set consisting of 5 pairs of autosomes and a single X; its genotype is therefore symbolized as 5AA + XO. Males arise from occasional errors in chromosome pairing and separation that occur during meiosis in gamete formation, these nondisjunction events producing gametes that lack X chromosomes. When such nullo-X gametes are fertilized by normal single X-bearing gametes, the result is an XO individual, a male. In a normal population, the frequency of males is approximately 0.2%. The frequency with which males are obtained can be considerably enhanced, however, by the use of certain mutations that cause an elevated rate of nondisjunction (Hodgkin et al., 1979). Such *him* (for *high incidence of males*) mutants are very useful in certain genetic experiments, when one needs an enrichment of males for further crosses.

Self-fertilizing hermaphroditic organisms possess some advantages for genetic analysis, but pose a potential problem. The principal advantage is that mutants can be readily isolated as homozygotes, without extensive backcrossing. For any hermaphrodite that is heterozygous for a new mutation, one-fourth of its progeny will be homozygous for the mutation. In contrast, organisms with a conventional sexual system must be put through two crosses—an initial outcross and then a cross between siblings—to isolate homozygotes of new recessive mutations. This difference between the two sexual systems is illustrated in Figure 3.4. Furthermore, while morphological abnormalities in mutants of standard outcrossing animals, such as *Drosophila*, often cause difficulties in mating mutant homozygotes, the existence of internal self-fertilization obviates this difficulty in perpetuating homozygotes in *C. elegans*.

The potential disadvantage of the self-fertilizing hermaphrodite system is the obverse side of its strength: If the mutant hermaphroditic individuals were altogether incapable of outcrossing, one could not perform any of the standard genetic tests on them, since all of these tests require crosses of one sort or another. Fortunately, the existence of males which can mate with hermaphrodites circumvents this problem in *Caenorhabditis*. Since half of the sperm of the male do not carry an X chromosome, approximately half of the progeny of outcrosses are males, and these male progeny will either be heterozygous for the mutation (if the mutation is autosomal) or all

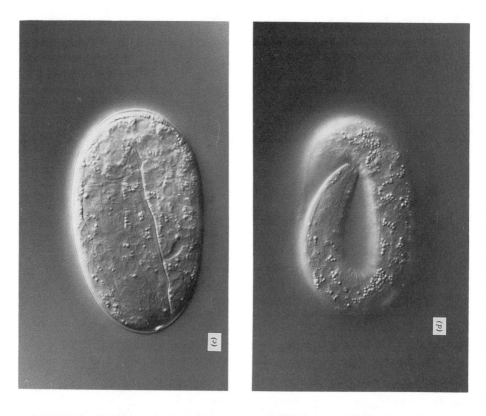

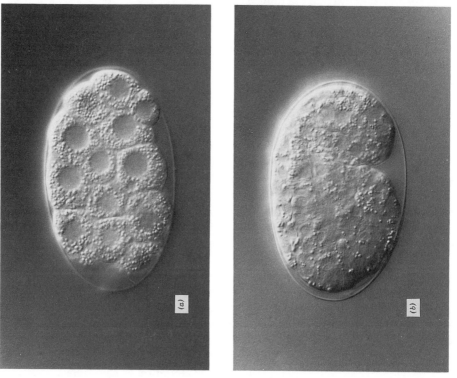

Fig. 3.2. Stages of embryogenesis in *C. elegans*. **a:** 28-cell stage, the beginning of gastrulation. **b:** Comma. **c:** Plum. **d:** Late pretzel. (Reproduced from Sulston et al., 1983, with permission of the publisher.)

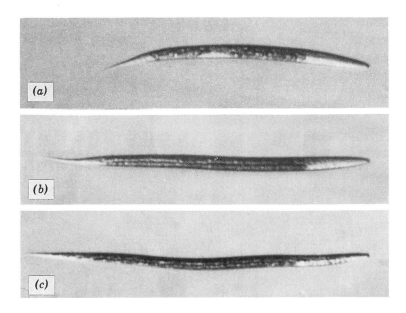

Fig. 3.3. Comparison of L2 and dauer larva. **a:** L2 larva beginning second molt. **b:** L2 larva entering the dauer stage. **c:** Fully developed dauer larva. (Reproduced from Golden and Riddle, 1984, with permission of the publisher.)

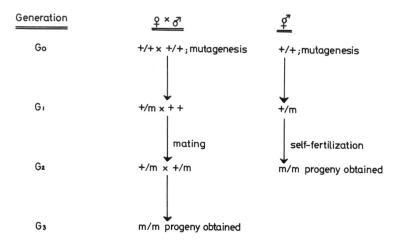

Fig. 3.4. Isolation of mutant homozygotes compared in hermaphroditic (right) and standard sexual systems (left).

mutant (if the mutation was on the X chromosome in the homozygous mutant her-maphroditic parent). In such crosses, however, care must be taken to ensure correct discrimination of the outcrossed progeny from any progeny produced by self-fertiliza-tion. This can be accomplished by the use of a second distinguishing marker in the original hermaphrodite stock. In outcrossings from this stock, none of the outcross progeny will display this second recessive marker. In *C. elegans*, this complication is further minimized because mating delays self-fertilization in the hermaphrodite through preferential utilization of the male's sperm (Ward and Carrel, 1979).

The standard procedures for genetic analysis in *C. elegans* were developed by Brenner (1974). A more recent overview of the genetics of *C. elegans* can be found in Herman (1988) and a discussion of the basic methods in Sulston and Hodgkin (1988). Although the organism lacks the impressive array of external structures that can be altered by mutation in an organism such as *Drosophila*, mutation can give rise to several classes of obvious morphological variants, which can be observed with a dissecting microscope. A few of the more common morphological variations are "dumpy," "small," and "long" phenotypes, symbolized respectively as Dpy, Sma, and Lon. Another class is "blister" (Bli), which is distinguished by fluid-filled blisters in the cuticle. A few mutant types are shown in Figure 3.5.

Equally useful are certain mutants that can be spotted on the basis of their abnormal movements. The wild-type adult moves on one side with a regular, sinusoidal move-ment in the dorsoventral (d-v) plane. The movements can be either forward or reverse. Following mutagenesis and segregation of the homozygotes, a relatively common observed class of mutants is those that have less regular movements. These have been categorized broadly as "uncoordinated," or Unc mutants, and are affected in either the neural or muscular systems but possess normal pharyngeal muscle movements and hence can feed and therefore survive. A second class with altered movement is the "roller" (Rol) mutants. These mutants rotate around the long axis of the body, produc-ing a circular motion. When cultured on a lawn of bacteria, the standard food, Rol mutants can be distinguished as those which make craters in the lawn.

In addition to phenotypic characterization, a detailed genetic characterization of any new mutant is required. The particular characteristics of the mutation that must be determined are its recessivity/dominance, the chromosome it is located on, the position (its "map location") on that chromosome, and the identity of the affected gene with respect to other mutations that produce similar phenotypes and which map in nearly the same position. These tests are essentially similar to those in other animals, but involve slight modifications because of the hermaphroditic nature of the sexual system in this organism.

To determine whether a mutation, *m*, is recessive to wild-type, homozygous mutant hermaphrodites, *m/m*, are crossed to wild-type males, symbolized +/+ (or + if X-linked), and the cross progeny, +/m, are scored for the mutant phenotype. Most mutations are found to be recessive to wild-type or weakly semidominant.

This initial cross can also give some preliminary information about the linkage group of the mutation. If only mutant sons and wild-type daughters are obtained in such crosses, then the mutation is identified as a sex-linked one, namely one on the X chromosome. If the mutation is not on the X, then it must be on one of the five autosomes. Each chromosome comprises a single linkage group; therefore if a new mutation is shown to be linked to a marker on a given linkage group, its linkage group is thereby identified. To test for linkage of this mutation, call it *a*, to a second marker,

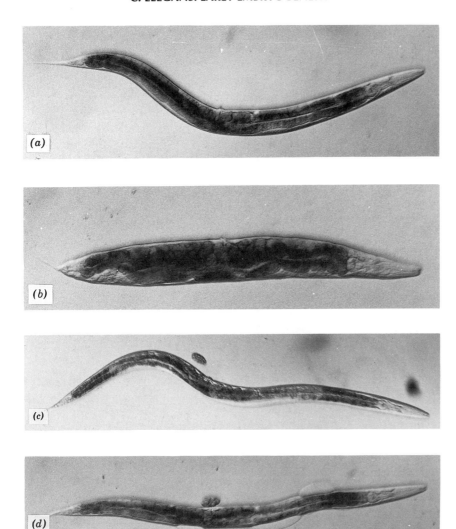

Fig. 3.5. Some morphological mutants of *C. elegans.* **a:** Wild-type. **b:** A *dpy* mutant. **c:** A *sma* mutant. **d:** A *bli* mutant. Magnification: × 133. (Photographs courtesy of Dr. J. Hodgkin.)

b, the *trans* double heterozygote, *a* +/+ *b*, is constructed, where *b* is a mutant gene that produces a different phenotype and is on a known linkage group. If segregation from the hermaphrodite produces the classical dihybrid 9:3:3:1 segregation ratio, the genes are unlinked and the new mutation must be on one of the other chromosomes. If an excess of wild-type progeny and a deficit of the double mutant phenotype are found, then linkage of the two genes is indicated. Once a gene is identified as being on a particular chromosome, its genetic distance with respect to other genes on that chromosome is established by genetic mapping procedures. However, genetic mapping in a self-fertilizing hermaphrodite involves some aspects not seen in standard sexual systems; the standard mapping procedure for estimating recombination distance between two markers on the same chromosome is described in the appendix.

Mutations producing similar phenotypes that are found to map at or near the same position on a linkage group may be either in the same gene or in neighboring genes of similar function. To determine whether two mutations are in the same gene or not, a standard "complementation test" is performed. The rationale of the complementation test is that if the mutations are in different, though neighboring genes, the double heterozygote will have one functional dose of each gene and will be wild-type, whereas if the mutations are both in the same genetic unit, the double heterozygote will have no wild-type copies of the gene and will be phenotypically mutant (Fig. 3.6).

The rules of nomenclature for *C. elegans* can be found in Horvitz et al. (1979). Briefly, phenotype classes are given by three-letter designations, beginning with a capital letter, e.g., Dpy, Sma, etc., and gene names are given by three-letter, italicized designations, all in lowercase, followed by a number indicating the particular gene in question (since many mutations in different genes produce comparable phenotypes). Thus, examples of gene names would be *dpy-1*, *lon-2*, and *unc-22*. For ease of reference, the linkage group number (I to V, and X) can be appended, e.g., *dpy-1 III*. Allele designations are italicized and given in parentheses by a one- or two-letter initial, signifying the location of the laboratory where the mutant was isolated, followed by a number; thus, for instance, *rol-8 (sc15)* would indicate a mutation isolated in Santa Cruz (*sc*) in the *rol-8* gene.

One genetic property of interest, for any organism, is its gene number. The first estimate of total gene number in *C. elegans* was provided by Brenner (1974). By measuring the overall rate of occurrence of new lethal mutations on the X chromosome (measured as the rate of newly occurring sex-linked lethals) and dividing by the average rate of mutation per gene (measured for a sample of genes that give a visible phenotype), he obtained an estimate of 300 "vital" genes (each irreplaceable for viability of the organism) on the X chromosome. Since the X chromosome represents approximately one-sixth of the total genome (measured in terms of total map length),

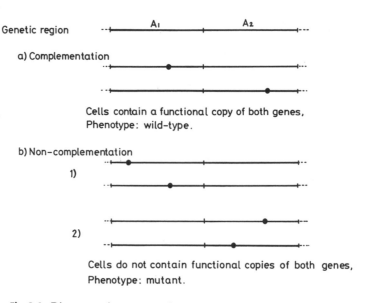

Fig. 3.6. Diagrammatic representation of a genetic complementation test.

this figure extrapolates to 1800, or approximately 2000, essential gene functions for the genome as a whole.

This must, however, be regarded as an underestimate of total gene number, since there are known cases of genes whose complete mutational inactivation does not produce lethality for the individual organism but rather other phenotypes (for instance, see Meneely and Herman, 1979; Greenwald and Horvitz, 1980). Herman (1988), reviewing the genetic evidence to date, has concluded that the total gene number (both essential and nonessential genes) is on the order of 4000. If one makes the assumption that the gene number may be three times as great as the number of indispensable vital genetic functions, however, as appears to be the case in *Drosophila* (chapter 4), then there may be 5000 to 6000 genes in total. The number may, indeed, be greater, and full molecular characterization of the genome should resolve the issue. Coulson et al. (1991) conjecture that the real number, each gene being defined as an expressible coding sequence ("open reading frame" or ORF), may be as high as 10,000.

Another feature of interest about the *C. elegans* genome, and one which is not in doubt, is its very small size, the genome being only 8×10^7 bp per haploid complement (Sulston and Brenner, 1974). Indeed, this organism has one of the smallest genomes, yet determined, of any animal species. (One other nematode species, *Panagrellus redivivus*, has a still smaller one.) The genes themselves are also unusual (with respect to most animal genomes) in possessing small introns, generally, and unusual splice junctions (Emmons, 1988). A further instance of genetic "compactness" is gene size: At 5 kb on average per gene, and with an average spacing of genes every 20 kb (Emmons, 1988) the *C. elegans* genome is somewhat more streamlined than typical higher eukaryotes.

A third distinctive feature of the genome is its complete deficiency for the highly repeated, short sequences, termed "satellite" sequences (named for their behavior in density gradients, where they usually band outside of the main nuclear DNA peak). The *C. elegans* genome appears to be an unusually stripped-down one. (It may be of interest in this connection that satellite sequences for the most part surround centromeres in eukaryotic chromosomes and *C. elegans* lacks typical centromeres; its chromosomes show attachment to spindle fibers throughout, a condition said to be "holocentric.")

However, in possessing interspersed copies of mid-repetitive sequences (as defined by renaturation studies), this animal falls into the eukaryotic mainstream. They average about 660 bp in length and are interspersed approximately once every several thousand base pairs of single-copy DNA (Emmons et al., 1980). Studies of cloned DNA sequences suggest that each family of related interspersed sequences may be present on average only about 10 times per haploid genome (Emmons et al., 1979), while renaturation studies of DNA suggest a somewhat higher figure (Emmons et al., 1980). As in other eukaryotes, the majority of these sequences probably have little direct effect on expression of neighboring genes.

There is, however, one class of mid-repetitive sequence that has mutator activity and that is proving significant, for the investigator, in the molecular and genetic analysis of genes in *C. elegans*. This is the group of transposable elements, of which the best characterized to date is Tc1, a 1610 bp element. First identified as a mid-repetitive sequence present at a low copy number in the standard, or Bristol, strain and at a high copy number in the Bergerac strain (Emmons et al., 1979; Files et al., 1983), this element is a major source of mutations (insertion mutations) in

C. elegans through its transposability. Although in Bergerac the element is considerably more unstable in somatic tissues than in the germ line, these relative rates are under genetic control; strains have been isolated that exhibit enhanced transposition in the germ line without a comparable increase in the soma (Collins et al., 1987). The types and behavior of transposons are reviewed, along with the other facts of genome structure, in Emmons (1988), while the use of transposons for gene cloning is described in chapter 6.

OOGENESIS, MATERNAL EFFECT MUTANTS, AND EARLY DEVELOPMENT

Oogenesis

Caenorhabditis has one of the smallest animal eggs that have been studied, measuring 60 × 40 μ It is, nevertheless, like all egg types, a cell of considerable complexity and one whose construction involves an intricate assembly process involving synthetic inputs from several cell types. The process of oogenesis in *C. elegans* has been described by Hirsh et al. (1976) and is summarized below.

Oogenesis takes place in the ovaries of the hermaphrodite, each ovary occupying the dorsal and distal arm of each U-shaped gonad (shown in Fig. 6.14). (The terms distal and proximal are with reference to the position of the vulva.) The ovary forms a tapered cylinder of densely packed peripheral nuclei surrounding a central core of cytoplasm. The approximately 1300 nuclei coating the inner surface of the cylinder are partially separated from each other by cell membranes but are open to the syncytial core. At the loop region of the gonad, the nuclei begin to move into the proximal arm and each, surrounded by cytoplasm, becomes enclosed within a cell membrane. As the dozen or fewer oocytes move down the proximal arm in single file toward the spermatheca, they increase in size. Figure 3.7 shows a cross section of the distal arm and loop region.

The formation of the oocyte is accompanied by the first stages of meiosis, the nuclear division process that reduces the diploid state of the gamete precursor cells to the haploid condition of the completed gamete nucleus. In the nematode gonad, the stages of meiosis exhibit a progression down the length of each gonad arm, from the distal tip to the central proximal region. At the extreme distal tip of the gonad, where oogenesis begins, the nuclei are predominantly in mitotic interphase. Meiotic figures make their appearance as the nuclei progress through the gonad. In the distal quarter of the ovary, the nuclei are predominantly in interphase; in the remainder of the distal arm, they are in pachytene (the stage of closest apposition of homologous chromosomes); and entering the loop, the nuclei enter diakinesis (in which the homologous chromosomes are still paired but less tightly than in pachytene), remaining in that phase until fertilization. (The genetic requirements for and controls on this meiotic progression will be described in chapter 6.) Accompanying this meiotic sequence is a series of processes that complete and fill the developing oocyte with the requisite cytoplasmic materials; the syncytial arrangement facilitates rapid oocyte assembly. (Syncytial arrangements of primordial germ cells, as seen here in the nematode hermaphrodite gonad, are very common in the animal kingdom, in both oogenesis and spermatogenesis [King, 1970].)

Fig. 3.7. Ovary and loop region in adult hermaphrodite of *C. elegans.* (Reproduced from Hirsh et al., 1976, with permission of the publisher.)

At 20°C, one oocyte is developed, on average, from its precursor nucleus every 40 min. The accumulation of oocytic stores that accompanies and drives oocyte growth shows a distinct regional specialization. At the distal tip of the ovary, the cytoplasm is filled with ribosomes and mitochondria, but toward the loop region there is an increasing concentration of yolk bodies and lipid droplets, the kinds of cytoplasmic inclusions that are typical of oocytes.

The yolk proteins comprise a small group of glycoproteins (Sharrock, 1983) that are encoded by a gene family, consisting of five X-linked genes and one autosomal gene (Heine and Blumenthal, 1986). They are synthesized not in the ovary itself but in the intestine, from which they are first secreted into the body cavity and then taken up by the ovary (Kimble and Sharrock, 1983). (Such production of yolk outside the ovary followed by its transport to the ovary is, like syncytial arrangements for germ cells, common throughout the animal kingdom.) The role of the thin epithelial sheath cells that enclose the ovary is unknown, but these cells may be active in the synthesis of nonyolk materials or as conduits of nutrients for biosynthesis within the gonad.

Fertilization and Early Cleavage: Origins of the Embryonic Founder Cells

In the *C. elegans* hermaphrodite, fertilization takes place in the spermatheca, as oocytes are moved into it one at a time, and involves entry of a single sperm at the pole toward the spermatheca, the future posterior pole of the embryo. Entry of the sperm triggers numerous cytoplasmic and nuclear changes. The fertilized egg rapidly develops a vitelline membrane and a hard, transparent chitinous shell, while the oocyte nucleus now completes meiosis, yielding two polar bodies and the haploid female pronucleus. The two pronuclei are situated at opposite poles (Fig. 3.8a) and the maternal nucleus migrates toward the paternal one (Fig. 3.8b,c), meeting and fusing at the center (Fig. 3.8d,e). (Early in the pronuclear migration, there is a transient "pseudocleavage" event, eccentrically placed [Fig. 3.8b].) As the zygote nucleus begins cleavage, the cleavage spindle moves slightly posteriorly and the resulting first cleavage division, which bisects the spindle, results in a relatively large anterior cell and a smaller posterior cell.

Several kinds of evidence indicate that the movements of the female and male pronuclei are principally dependent on one element of the cytoskeleton of the egg, the microtubule system (Strome and Wood, 1983; Albertson, 1984). At 25 nm in diameter, the microtubules are the largest filaments of the cytoskeleton, the elaborate set of protein fibers and filaments that crisscross the cytoplasm of eukaryotic cells, and function during mitotic and meiotic divisions as the principal component of the division spindle. The movement of the maternal pronucleus is but one manifestation of an early critical change in the egg that, to some degree, prefigures the course of embryogenesis. This change is the establishment of a manifest anteroposterior (a-p) polarity, which sets the stage for the definitive a-p polarity of the embryo.

Prior to fertilization, there are no signs of asymmetric distribution of any of the readily detectable components of the egg such as yolk, microtubules, mitochondria, and various surface antigens (Strome, 1986). Two other components that show a uniform distribution are actin microfilaments (the thinnest fibrillar component of the cytoskeleton) and an array of small inclusion bodies termed P granules. However,

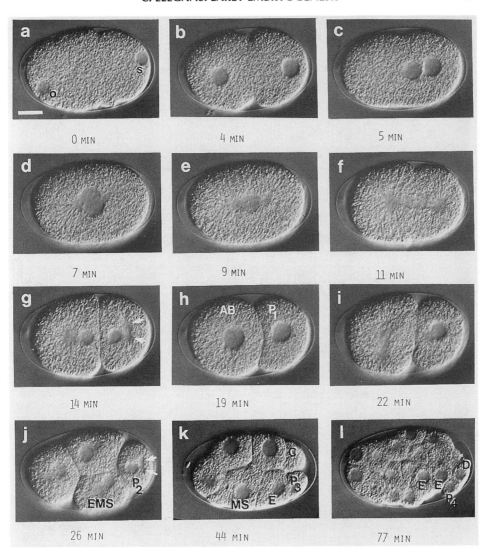

Fig. 3.8. Stages of early embryogenesis in *C. elegans* (at 23°C). **a:** Fertilized egg, approximately 45 min after fertilization with oocyte (o) and sperm (s) pronuclei shown. **b:** Beginning of pronuclear movement and pseudocleavage. **c–e:** Approach of pronuclei, rotation, and fusion. **f:** First cleavage. **g, h:** Early and late 2-cell stage. **i:** AB has begun mitosis. **j:** The 4-cell stage, arrows showing flattened centriolar region. **k:** The 8-cell stage. **l:** The 26-cell stage, start of gastrulation with inward movement of two E cells. (Reproduced from Schierenberg, 1988, with permission of the publisher.)

fertilization prompts a swift redistribution of several of these components along the a-p axis. In particular, the actin microfilaments become concentrated at the anterior pole (Fig. 3.9) and the P granules become clustered at the posterior pole (Fig. 3.10) (Strome, 1986; reviewed in Strome and Hill, 1988). Together with the posterior migration of the first cleavage spindle, these cytoplasmic events signal and, in part, comprise the first regional axial organization of the embryo. The cleavage divisions further realize and amplify this process.

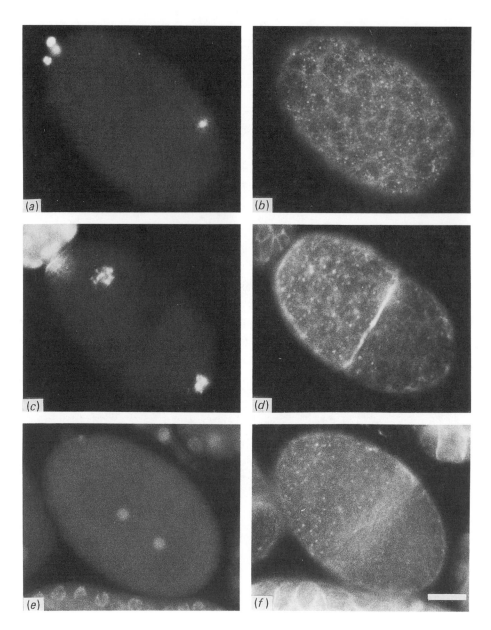

Fig. 3.9. Polarization of actin distribution in *C. elegans* embryo following fertilization. a, c, e, DAPI staining of DNA; b, d, f, immunostained for actin. **a, b:** Meiosis II, shortly after fertilization—the actin filaments are homogeneously distributed. **c, d:** Pronuclear migration stage; the microfilaments have been concentrated to the anterior half and the pseudocleavage furrow is stained. **e, f:** Telophase of first cleavage—actin filaments are concentrated predominantly in the anterior part of the embryo and in cleavage furrow. (Reproduced from Strome and Hill, 1988, with permission of the publisher. © ICSU Press.)

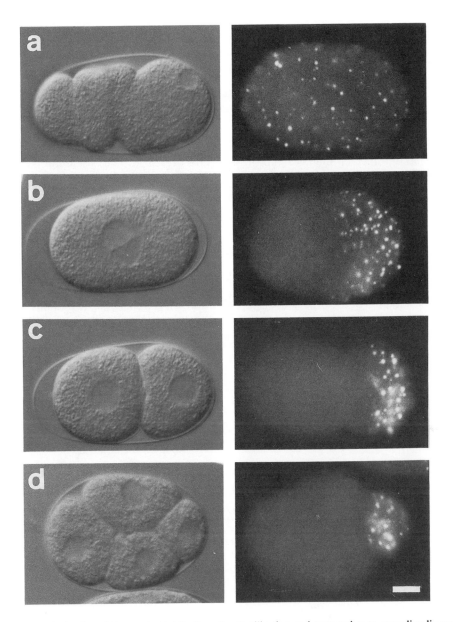

Fig. 3.10. Polarization of P granule distribution after fertilization and secregation to germ line lineage. Nomarski pictures of early cleavage stages on left; immunostaining for P granules on right. **a:** Pseudocleavage; uniform P granule distribution. **b:** Polarization of granules just prior to first cleavage. **c, d:** Segregation of P granules to P_1 and P_2 respectively. (Photograph courtesy of Dr. S.E. Strome.)

The development of a-p polarity is of crucial importance in the early structuring of the embryo but its ultimate sources are unknown. It could reflect a subtle, as yet undetected biochemical polarity in the oocyte itself, perhaps imposed during the final stages of oocyte development, or it may be that the oocyte is essentially isotropic but becomes anisotropic upon entry of the sperm. The anterior position of the egg nucleus indicates the existence of some prior a-p polarity.

The cleavage divisions begin directly after fusion of the pronuclei and show a highly invariant spatial pattern and tempo. The first four to six division rounds normally occur within the uterus, but eggs can be physically released at the time of pronuclear appearance or at the one-cell stage, and the complete sequence of divisions then observed *in vitro* (Hirsh et al., 1976; Deppe et al., 1978).

In nematodes, the first cleavage divisions of the zygote lay the foundations of the major embryonic cell lineages and of the future organization of the embryo. These divisions sequentially create or reveal the basis axes and polarities of the embryo and generate a special set of six blastomeres, each of which gives rise to one of the six major cellular lineages of the larva. These key lineage precursor cells have been termed "embryonic blast cells" (Laufer et al., 1980) or "embryonic founder cells" (Sulston et al., 1983). The sequence is schematized in Figure 3.11, and it can be seen that while three of the lineages are "pure" as to cell type (D, muscle cells; E, intestinal or endodermal cells; P_4, germ line cells), the other three give rise to both ectodermal and mesodermal cells.

The sequence of the first cleavage divisions, which produce the six founder cells, is shown in Figure 3.8g–i. The very first division sets apart the first lineage, that of the AB line, and fixes the a-p axis and polarity of the embryo and of the future adult. The

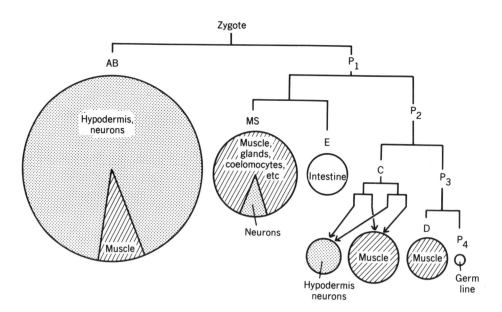

Fig. 3.11. Generation of the embryonic founder cells in the first cleavage divisions of the embryo. Ectodermal derivatives indicated by stippling, mesodermal tissues by striping. (Reproduced from Sulston et al., 1983, with permission of the publisher.)

sister cell from this first division, P_1, is always smaller than the AB cell in all wild-type embryos. Its division, in turn, is also asymmetric, is along the a-p axis, and generates a large embryonic blast cell, EMS, and a smaller cell, P_2. The EMS cell divides to give two blast cells, the E cell and the MS cell, while P_2 divides asymmetrically to give the C blast cell and the smaller P_3 cell. It is also in the third division that the beginnings of left–right (l-r) asymmetry take place by transverse divisions that take place within the AB lineage. Finally, P_3 divides into the D blast cell and its smaller sister, P_4. During the course of the divisions that give rise to the embryonic founder cells, the previously formed founder cells are also undergoing cell division. By the 16-cell stage, all five major somatic cell lineages have been established.

The sequence of P cells, beginning with P_0 and ending with P_4, are regarded as the germ cell lineage or the germ line (Boveri, 1910; Schierenberg, 1985). The first four divisions of the germ line result in a relatively small germ line cell and a larger somatic sister. In effect, each of the P_n cell divisions (until P_4 divides) is a stem cell-like division, in that one daughter retains the mother cell potential (the smaller one) while the other daughter (the larger one) acquires a new, somatic developmental fate.

The orientation and timing of the ensuing divisions similarly occur in a highly stereotyped fashion (Deppe et al., 1978; Sulston et al., 1983). Each cell in every lineage shows a fixed orientation of division, either a-p, d-v, l-r, or oblique. The cell lineage program is thus strictly constant until the end of cleavage, when the hermaphrodite embryo consists of 558 viable cells. (A total of 671 cells are produced during the embryogenesis of the hermaphrodite, but of these, 113 are eliminated by an invariant pattern of cell deaths, "programmed cell death.") Given the constancy of ancestry, embryonic cells are designated by their embryonic cell founder followed by the sequence of division positions of the cells from which they derive, e.g., ABal is the left-hand daughter of the anterior daughter of the AB founder cell.

Morphogenetic cell movements begin with a sequence of inward cellular migrations at the 28-cell stage, when the 2 cells of the E lineage (the precursor cells of the intestine) migrate internally. This movement is the first step in nematode gastrulation. Throughout cleavage, there are few dramatic signs of cellular differentiation, with the notable exception of the E lineage, whose 8 cells, when the embryo contains more than 200 cells, can be recognized by the presence of fluorescent rhabditin granules (particles composed of tryptophan breakdown products) (Babu, 1974).

The precise, stereotyped nature of the cleavage divisions raises the question as to how it is achieved, and, in particular, whether the observed regularities are produced by intrinsic (autonomous) cell division properties or through cell–cell or cell–eggshell interactions. The answer appears to be that all three factors are important.

One approach to investigating cleavage mechanisms involves selectively removing either specific cells or parts of cells, which thereby also reduces the physical pressure of the eggshell. The first analysis of this kind in *C. elegans* was reported by Laufer et al. (1980). Pressure was applied through a coverslip to eggs, bursting one or more of the embryonic blast cells; such partial embryos, because they lack some of the original embryonic material, are free of the pressure of the shell. Cleavage divisions will continue in these embryos, for many further rounds of division, if the embryos are placed in a medium containing nematode coelomic fluid (Laufer et al., 1980) or cell culture medium (Edgar and McGhee, 1986).

Observation of the partial embryos reveals two characteristic division patterns. Partial embryos derived from AB or E embryonic blast cells show successive divisions at a constant oblique angle; the result is initially a helical cell array and ultimately a ball of cells. Partial embryos derived from the P lineage, however, show a predominantly linear division pattern. The data suggest that the cleavage program of the intact embryo reflects the combined action of these two intrinsic division programs, in conjunction with interactions between cells of different lineages and possibly the eggshell. In addition, some of the ensuing cellular movements seen in these embryos appear to result from purely spatial constraints—the C blast cell, for instance, seemingly pushed to one side by the cells of the AB lineage—while others may involve selective migrations of one cell type over or around cells of other lineages, such as the migration of AB cells over and around E and MS cells (Laufer et al., 1980).

 Another interesting fact suggested by this first study of partial embryos is that a *reversal of polarity* occurs in the production of the germ line lineage, beginning with the division of P_2. In several partial embryos derived from P_1, the divisions of P_2 and P_3 produce somatic cells that are *posterior* rather than anterior to their germ line sisters. This finding has been confirmed by Schierenberg (1985, 1986, 1987), who produced partial embryos by a gentler method: laser-induced partial ablation or extrusion of nucleated partial blastomeres through laser-produced holes in the eggshell. In many of the partial P lineage embryos produced by these methods, this reversal in polarity of the somatic sister, relative to the germ line sibling cell, was seen in both the division of P_2 and P_3. A revised schematic depiction of the first cleavage events, indicating this reversal of polarity, is shown in Figure 3.12b. The partial

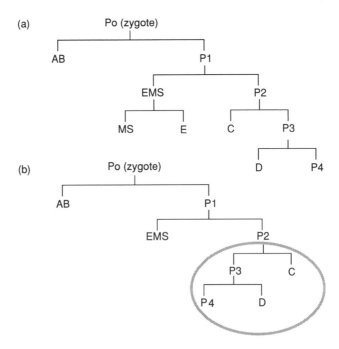

Fig. 3.12. a: Lineage diagram of founder cells, as traditionally drawn, with P_4 posterior to all somatic embryonic founder cells. **b:** Reversal of polarity in the founding of C and D lineages, as found by Laufer et al. (1980) and Schierenberg (1985, 1986).

embryos thus reveal a hidden complexity in the orchestration of the early cleavage divisions that is not readily apparent within the compressed space of the normal embryo inside its eggshell.

The Nematode Egg: Is It a "Mosaic" System?

The central question about early embryogenesis in *C. elegans* is how the first decisions of cellular fate relate to the molecular and cytological structure of the egg. The traditional view has been that the nematode egg is a typical mosaic system (Fig. 2.3a), one in which early blastomeres are assigned irrevocable and restricted fates by qualitatively unique cytoplasmic constituents derived from the oocyte.

The essential prerequisite for assigning an egg to this class is demonstrating that the inheritance of qualitatively distinct cytoplasmic fractions is invariably associated with a particular developmental fate. More than 80 years ago, Theodor Boveri discovered a unique cytoplasmic fraction of this kind in the eggs of one of the larger nematodes, *Ascaris*. The cytoplasmic fraction that Boveri found is required for both founding the germ line in *Ascaris* and for sparing the germ line cells from "chromatin diminution," a phenomenon which Boveri had discovered earlier, in 1887. Chromatin diminution is found in some genera of nematodes (as well as in several unrelated invertebrates) and consists of the selective elimination of chromosomal fragments in the somatic cells of the embryo; it is only in the germ line of these species that the chromosomes remain fully intact. (We now know that during chromatin diminution in nematodes, the chromosomal material lost is largely, and probably solely, a repetitive noncoding set of sequences and not a specific set of genes whose expression is required in the germ line [Tobler, 1986].)

Boveri found that, in *Ascaris*, the cytoplasm that the germ line inherits segregates specifically with the successively formed germ line blastomeres and that redistribution of this cytoplasm by appropriate experimental manipulation spares other blastomeres, those not destined to contribute to the germ line, from chromatin diminution (Boveri, 1910). The behavior of the germ line-specific cytoplasm fits the traditional picture of a qualitatively unique "morphogenetic determinant," and this is how it has usually been regarded, though Boveri himself recognized that it might involve a graded property and thresholds of response (discussed in Schierenberg, 1986).

In contrast to *Ascaris, C. elegans* shows neither visible signs of chromatin diminution in the somatic cell lineages nor evidence of selective DNA loss in these cell lines when tested by DNA analytical methods (Sulston and Brenner, 1974; Emmons et al., 1979). However, the early germ line cells do show characteristic cytoplasmic granules, the P granules mentioned earlier, which segregate specifically to the germ line (Fig. 3.10), the kind of behavior one would predict for a unique morphogenetic determinant. This segregation of the P granules to the posterior pole is mediated by the actin filaments in the cell, the major contractile component of the cytoskeleton of the egg, rather than by the microtubules (Strome and Wood, 1983).

Despite the interest that the P granules have excited, it is still unsettled whether they cause germ line determination or are merely associated with the segregation of the germ line. They are not necessary, however, for the distinctive stem cell-like divisions of the germ line because posterior cytoplasm can be extruded by physical means before it has received the P granules and yet exhibit asymmetric, P-line type cell divisions (Schierenberg, 1988). Since much circumstantial evidence links this kind of division with germ line fate, it seems probable that the P granules are not the germ line determinants themselves.

Do the somatic cell lineages of the embryo show any evidence for comparable qualitatively unique cytoplasmic components? One approach that has been used would seem to support the idea. These experiments involve blocking cell division in the early blastomeres in either intact or partial embryos, which have been made accessible to the cell division inhibitor cytochalasin D, and then asking if molecules specific to particular lineages appear in the blocked cells and, if so, whether they appear only in the appropriate blastomeres.

In the first such study of this kind on nematode embryos, Laufer et al. (1980) followed the appearance of the rhabditin granules, as a marker of the E lineage, in early cleavage-blocked blastomeres of intact and partial embryos. These granules, which are breakdown products of tryptophan metabolism (Babu, 1974) and are detectable by their characteristic fluorescence under UV illumination, were indeed found only in the appropriate cells. Thus, in embryos blocked at the two-cell stage, rhabditin granules appeared on schedule only in the P_1 cell, never the AB cell, and in embryos blocked at the four-cell stage, rhabditin granules were formed only in the blocked EMS cell, not the P_2 cell or daughters of the AB cell. Similar findings of lineage specificity were made in partial embryos in which the AB cell had been destroyed. In the second set of experiments of this type, Cowan and McIntosh (1985) reported comparable evidence for lineage specificity of characteristic differentiation markers in cleavage-blocked embryos, using markers for the E, MS, and AB lineages (respectively, rhabditin granules, a paramyosin antigenic determinant, and a hypodermal cuticular antigen). They also found, however, that inhibiting cleavage in certain early precursors of a lineage could inhibit expression; thus, blocking division of the EMS cell prevented expression of a muscle-specific marker, while blocking its daughter, MS, permitted expression (Cowan and McIntosh, 1985).

Such results indicate that there is a large degree of autonomy in the segregation and expression of particular differentiation potentials within the somatic lineages; evidently, *something* is segregated early in these lineages and its expression is not dependent, to a first approximation, on there being the characteristic array of cell types. This does not mean, however, that that something is necessarily a qualitatively unique determinant for each lineage. In principle, differential assignments of fate could be a product of differential partitioning of one or more substances, whose quantity determines fate assignment. Alternatively, the ability of late cleavage-blocked cells may even be a consequence of cell–cell interactions that take place between cells of different lineages, whether or not they have been blocked early in cleavage.

Although inhibiting the process of cell division need not block expression of differentiation potential, the occurrence of certain DNA replication cycles does appear to be essential for such expression. Experiments with aphidocolin, the DNA synthesis inhibitor, show that there is a critical round of DNA replication immediately after the founding of the E lineage, which is required for the subsequent expression of differentiation markers (many cell divisions later) within that lineage (Edgar and McGhee, 1988). There are some indications that the other somatic lineages may show comparable requirements for DNA replication at the time their founder cells appear, but the evidence is less strong than for the E lineage (Edgar and McGhee, 1988).

Other experiments have involved altering cytoplasmic contents or mixing cytoplas-

mic contents between different lineages. One set of experiments suggests that whatever the determinative substances might be, their activities are not sensitively keyed
to the amount of cytoplasm present in each cell. When small holes are punched in the
eggshell with a laser microbeam, and cell fragments are extruded, a large percentage
of such blastomeres can heal, followed by the resumption of normal development
(Laufer and von Ehrenstein, 1981). Amounts of cytoplasm up to 15% or so can be
removed, yet many embryos survive and develop into normal fertile adults; individual
blastomeres apparently have a fairly robust regulative capacity.

Cells can also be fused with this procedure, thereby transferring cytoplasmic
material from blastomeres of one somatic lineage to another, and one can ask if there
has been a corresponding transfer of developmental potential, as assayed by the
appearance of a characteristic differentiation marker. In one experiment of this kind,
the EMS cell was fused to AB.a (the anterior daughter of AB), and following the
tetrapolar cell division in these binucleate cells, the blastomeres, identified by their
respective positions, were scored for the ability to produce rhabditin granules; in no
case out of 17 was this ability detected in cells derived from the AB lineage-derived
cell (Wood et al., 1983). These results indicate, at least, that any E-specific determinant cannot be readily transferred between lineages; if they exist, they might be either
nucleus or cell membrane bound.

In contrast, Schierenberg (1985) has reported that if the nucleus of P_1 is extruded
and the remaining cytoplast of P_1 is fused to one of the granddaughters of AB, one can
obtain embryos that show rhabditin granules in the descendants of the recipients. This
result shows that, in certain instances, developmental potential *can* be transferred
between lineages, though the result does not reveal whether the material basis of such
potential is cytoplasm or membrane based. The basis of the discrepancy between the
results of Wood et al. (1983) and of Schierenberg (1985) has not yet been resolved; it is
not clear whether differences in the cells examined or in the techniques account for it.

Altogether, these experimental manipulations have not answered the question of
whether there are unique cytoplasmic "determinants" for the different lineages or
whether fate specification involves different quantities of one (or more) substances.
What the experiments, described above, have revealed is a large measure of developmental autonomy in these lineages from the time they are established. In contrast,
certain other recent findings have demonstrated that at least some of the defining
properties of the somatic lineages require cell–cell interactions.

One of these cases involves the E lineage: If the P_2 cell is extruded from the four-cell
embryo, the EMS cell undergoes its typical asymmetric cleavage and the E daughter
undergoes further cleavages but neither the normal gastrulation movements of the E
daughter cells nor the characteristic differentiation of intestinal cells takes place
(Schierenberg, 1987). Evidently, one of the six major lineages in the embryo, the E
lineage, requires a particular kind of cell contact, with the germ line lineage, in order
to develop its full potential. (In the cleavage-inhibition experiments, the blocked E
cell remains in contact with P lineage cells.)

The second instance of essential cell–cell interaction in early fate assignment is that
of the AB daughter cell which gives rise to the pharynx. In normal development, the
pharynx is derived ultimately from two cells of the early embryo exclusively, namely
from P_1 and the anterior daughter of AB only (AB.a); the AB contribution is quite
specific in that the posterior daughter of AB, AB.p, never contributes pharyngeal cells
(Fig. 3.13). Yet by a manipulated switching of AB.a and AB.p, Priess and Thomson

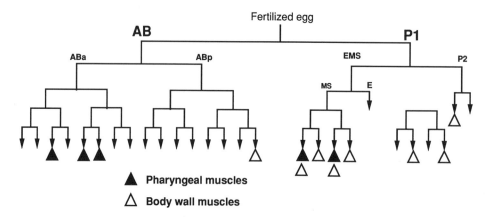

Fig. 3.13. Lineage of cells giving rise to pharyngeal muscles (solid triangles) and body muscles (open triangles). Within the AB lineage, in normal development, only the AB.a sublineage gives rise to pharyngeal muscles; the AB.p sublineage does not. (Adapted from Priess and Thomson, 1987.)

(1987) showed that when these two cells are reversed in the embryo, AB.p takes on the role of AB.a, contributing descendants to the pharynx, while the initial AB.a cell takes on the developmental options of AB.p. Clearly, cellular positioning and cell–cell interactions can play an important part in fate assignment, within the basic lineages, as well as between them.

One of the more suggestive clues to the nature of the morphogenetic system is the apparent relationship between relative blastomere size and fate assignments. The possible importance of size is indicated by the asymmetry of daughter cell sizes in every P cell division, in which the P daughter is always smaller than its founder cell sib. In contrast, initial divisions within every blast cell lineage are symmetric with respect to daughter cell size. It is difficult to escape the impression that the geometry of division or size of the daughter cells is connected in some way to the establishment of differences in developmental potential, at least between the P lineage and the five somatic lineages.

There are also some correlations between relative cell division rates and the fates of the major embryonic cell lineages. In every P cell division that generates a somatic embryonic blast cell, the P cell daughter exhibits a different cell division rate from that of the sibling blast cell, whose progeny in turn show initially synchronous and characteristic division rates (Deppe et al., 1978). These rates were, at first, correlated with the presumed points of origin of the founder cells along the a-p axis, with AB lineage cells cycling the fastest and D cells the slowest (Deppe et al., 1978). This correlation appears less significant today, however, in light of the subsequent discovery of the reversal of polarity and the corresponding reversal in position of the C and D founder cells (Figure. 3.12).

Onset of Zygotic Transcription

An important genetic consideration in evaluating models of fate assignment is whether the initial events involve maternal components exclusively or whether genes expressed from the zygotic genome play a part. The long-standing view has been that the processes are under strictly maternal control. Early molecular observations reinforced this supposition. In an analysis of transcripts by *in situ* hybridization of poly A$^+$ mRNAs, Hecht et al. (1981) concluded that there was virtually no zygotic transcription until approximately the 100-cell stage. Furthermore, the expression requirements of two E lineage markers, a gut esterase and the rhabditin granules (which are among the first biochemical differentiation markers to be expressed), were found to fit this pattern: Expression of these markers becomes resistant to α-amanitin, an inhibitor of the mRNA-synthesizing polymerase, RNA polymerase II, when the embryo consists of approximately 100 cells, with the markers themselves appearing about 30 min later (Edgar and McGhee, 1988).

Recent observations, however, suggest that there may be zygotic transcription considerably earlier. These experiments have involved the technique of measuring "run-on" transcription, in which transcripts initiated *in vivo* are continued *in vitro* and detected by incorporation of labeled precursor. (The conditions used do not allow new initiation, hence the experiments provide a snapshot of transcription in progress, at the point the cells are disrupted.) In the first such study, Cleavinger et al. (1989) reported that four to eight cell embryos of *Ascaris* (a large relative of *C. elegans* and the first nematode experimental system) showed considerable RNA polymerase II mediated transcription.

In a more extensive series of studies, Schauer and Wood (1990) showed that such transcriptional activity can similarly be detected in embryos with as few as eight or nine nuclei, that the transcripts are the typical size of *C. elegans* primary RNA polymerase II transcripts, and that the rate of zygotic transcription is approximately constant from this early stage to late embryos (containing several hundred cells). That the results are not artifactual is shown by results with particular cloned genes. While transcripts of various "housekeeping functions" are detectable both in early and late embryos, transcripts for genes known to be expressed late are found only in extracts from late embryos, while adult-specific genes show no transcripts from either early or late embryos (Fig. 3.14).

Furthermore, while the bulk of the early embryonic mRNA population is found to be maternal, the average number of genes being transcribed from the eight or nine nuclei stage onward is estimated to be substantial, between 1000 and 3000, and probably closer to the higher number. Comparisons of the number of random clones from a genomic library that hybridize to early and late run-on transcripts suggest that nearly all genes are transcribed in late embryos, while the fraction of these late genes that are also transcribed early is 70% (Schauer and Wood, 1990). If the total gene number is taken as 4000 (Herman, 1988), this would correspond to 2800 genes transcribed in early embryos; if the total number of genes is closer to 10,000, the number could be as high as 7000.

Finally, differential screening with early and late embryo transcript populations indicates that there may be as many as 20–30 zygotic genes transcribed early and specifically from the zygotic genome (Schauer and Wood, 1990). The implications of this last finding are significant. While the bulk of the early embryo's constituents, including mRNA populations, is of maternal provenance, there is a distinct zygotic genome contribution, which even includes a small number of genes uniquely expressed in the early embryo. While maternally specified components undoubtedly play the major part in orchestrating events in the early embryo, there is the possibility that some specific zygotic genes make an important contribution.

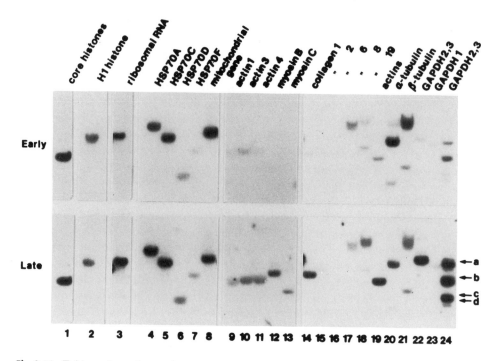

Fig. 3.14. Evidence for early zygotic transcription capacity in *C. elegans*. Shown are run-on transcription assays for various genes from early (96% pregastrulation embryos) and late (<5% pregastrulation) embryos. Various "housekeeping" genes, such as the histones, ribosomal RNA, and several of the constitutively expressed *hsp* genes, show transcriptional activity both early and late, while several genes associated with differentiation of specialized tissues, such as the myosins and collagen 1, show transcriptional activity only from late embryos, as expected. (Reproduced from Schauer and Wood, 1990, with permission of the Company of Biologists, Ltd.)

Potentialities of Genetic Analysis of Early Development

Although the various cellular manipulations and observations have revealed the complexity of the fate-specifying system in the early embryo of *C. elegans*, they have not resolved its nature. In principle, a genetic approach, the analysis of maternal effect mutants, altered in this fate-specifying system, can help to illuminate this matter. For instance, any maternal effect mutants whose progeny show altered relationships between blastomere size and/or division rate and subsequent fate assignment can help to elucidate which relationships are essential and which merely involve correlations.

Furthermore, in principle, informative mutants should be able to help one decide if there is any validity in the conventional models that relate blastomere fates to the initial organization of the oocyte (Fig. 2.3). For the simplest model, that of mosaic determinants laid down in the oocyte (Fig. 2.3a), there is one clear genetic prediction: that there should be one group of maternal effect mutants whose initial primary defects are solely within specific embryonic blast cell lineages. Hermaphrodites homozygous for such mutations, if the genes are exclusive maternals and are not required for hermaphrodite viability, should produce embryos altered only in specific blast cells, the other lineages developing independently.

In contrast, gradient models, involving preset gradients in the oocyte (Fig. 2.3b), predict that *no* maternal effect mutants will be found that affect only one somatic lineage and that any and all which are affected in the primary morphogenetic system will be affected in more than one, and perhaps all, somatic lineages. This prediction, indeed, should hold even for dynamic gradient systems, in which the gradient is not fixed within the egg but comes into being during early cleavage. As seen earlier, the phenomenon of polarity reversal (Fig. 3.12) suggests that there *is* an important element of dynamism in the specification system. In addition, mutations that shift the concentrations of the putative gradient substance should cause transformations of certain lineages into others. Such transformations have been seen in the *Drosophila* embryo, which is now known to be partly gradient based (see chapter 4).

Finally, if the system involves only partial direct spatial specification of the embryo by maternal components (Fig. 2.3c), with the rest being carried out by ensuing processes (perhaps involving zygotic functions), the genetic prediction is that only one or a few blastomeres would be directly affected by maternal effect mutants, with effects on other parts of the embryo being indirect and involving expression of zygotic functions. As we have noted, there is some degree of early zygotic genome transcription in the *C. elegans* embryo. Though the emphasis in the rest of this chapter will be on maternal "control" of early embryogenesis, we will return, at the end, to discuss the possible zygotic contribution.

Maternal Effect Mutants: The Uses and Limitations of Temperature-Sensitive Mutants

Maternal effect mutants are an essential resource for analyzing the maternal contribution to early embryonic development, as discussed in chapter 2. Obtaining such mutants in a hermaphroditic species, however, involves some additional complexities not encountered in standard sexual systems. One might begin by screening for recessive mutants that make sterile though fully viable hermaphrodites, but within this group could be those that are sterile because of sperm defects (the sperm also being supplied by the hermaphrodite). The mutants of interest therefore will be among those that lay fertilized eggs (and hence make functional sperm) but whose embryos experience aberrant development. Even among these embryogenesis-defective mutants, however, only some are expected to be maternal effect mutants; the rest will be mutant in genes expressed and required by the embryo itself, zygotic genes. Identification of the embryogenesis-defective mutants, among the set of hermaphrodite steriles, must therefore be followed by a further screening to identify the maternal effect mutants.

A further potential difficulty involves the degree of residual expression, or "leakiness," in the isolated mutants. In most screens, a substantial proportion of the mutants that are isolated turn out to be hypomorphs. Because hypomorphic alleles frequently give variable or weak developmental aberrations (Hadorn, 1961), one might be tempted to discard these mutants and save only those with little or no residual gene activity, the amorphs. Yet, in so doing, one would be eliminating one major source of potentially informative mutants. Furthermore, if applied to the isolation of maternal effect mutants, this elimination of the hypomorphs would produce the following problem: Any individual homozygous for an amorphic maternal effect mutation would be altogether incapable of producing progeny; therefore, if that individual were the sole carrier of the mutation, the mutation would itself be lost. Only if heterozygous sibs had been saved could the mutation be preserved in a stock. Unfortunately, the need to save sibs of all tested individuals substantially increases the work; the propagation of extreme defective mutations, in fact, does involve constant perpetuation of the mutations in heterozygous carriers. Yet, amorphs, totally lacking the gene function, are usually the most informative about the role of the wild-type gene (Hadorn, 1961).

In effect, for a thorough analysis of the maternal genetic contribution to embryogenesis, one needs to cast one's net as widely as possible and obtain both amorphic and hypomorphic mutations. Special strategies are needed for the former (and these are discussed in the next section) and involve more effort, but the results can be directly informative about the roles of particular genes. In contrast, hypomorphic mutants are easier to collect and, in general, to maintain; although their phenotypes must be interpreted with caution, they frequently provide valuable clues to the expression and function of the corresponding wild-type genes.

One broad class of mutants, which for the most part prove to be hypomorphic, are temperature-sensitive (or ts) mutants. Such mutants comprise a subcategory of the general group of *conditional* mutants—those that produce lethality under one standard environmental condition, the *restrictive* state, but which permit survival under an alternative condition, the *permissive* state. The major advantage of conditional lethals is that stocks can be grown at the permissive condition but then studied at the restrictive one. Of the various kinds of conditional lethals, ts mutants are among the easiest to isolate and the most versatile for subsequent analyses. As their name implies, mutants are isolated that survive at one temperature, but which die or manifest an abnormality at a second temperature. It is customary to employ the opposite extremes of the temperature range for growth of the wild-type at the restrictive and permissive temperatures. The first wave of genetic studies of early embryogenesis in *C. elegans* involved large-scale isolations of ts maternal effect mutants and, in this section, we will review this work and the kinds of information it produced.

A primary reason for choosing temperature sensitivity as the form of conditional lethality is that the great majority of protein-coding genes can mutate to give ts alleles. In principle, therefore, the selection of ts mutants can allow the identification of mutations in most genes for a given developmental process, in effect allowing saturation mutagenesis for that process. The reason for the ease of finding ts mutants is that every wild-type protein has an optimal thermal stability within the normal temperature growth range. Alterations in the amino acid sequence of the protein, produced by point mutations, can lead to an altered stability of the polypeptide chain in its active conformation; at the restrictive temperature, the mutant protein becomes denatured.

This susceptibility to denaturation may occur preferentially during synthesis of the protein chain, in which case the mutant is said to be "temperature-sensitive in synthesis" or *tss* (Sadler and Novick, 1965), or the temperature sensitivity may result from susceptibility of the correctly folded protein chain, in which case the protein is said to be "thermolabile" or *tl*. Unfortunately, it is not always possible to distinguish between these alternatives (unless one can isolate the protein itself), and this can lead to ambiguities in interpretation, as discussed later.

Despite some of the potential pitfalls in the analysis of ts mutants, the collective work on maternal effect ts mutants in *C. elegans* has proved useful in three respects. First, the relationship between the gene set expressed in oogenesis and those gene sets used in later stages of development has been elucidated; the results show that the specialness of the oocyte as a cell does not result from large numbers of oogenesis-specific genes. Second, some of the mutants have provided suggestive information about the relationship of cell cycle events to fate allocation in the early embryo. Third, some of the ts mutants have been used to isolate nonconditional mutants in the same genes, facilitating further characterization of those genes. The discussion of ts mutants that follows is intended both as an explication of what these mutants have told us about *C. elegans* development and to illustrate the versatility of ts mutants, in general, for exploring gene function.

A general procedure for obtaining large numbers of ts mutants of *C. elegans* was described by Vanderslice and Hirsh (1976) and is outlined in Figure 3.15. Late larvae are exposed to the mutagen ethyl methane sulfonate (EMS), which induces a high frequency of point mutations, and their progeny are allowed to self-fertilize to produce homozygotes at the permissive temperature. These progeny will include a fraction of induced ts homozygotes. Individual larvae are then allowed to form isolated larval broods, which are tested at the permissive and restrictive temperatures. Putative ts mutants are identified as those that accumulate one predominant developmental stage at the restrictive temperature but which show all stages at the permissive temperature.

In the particular scheme shown, 16°C was chosen as the permissive temperature and 25°C as the restrictive temperature. A representative sample of mutant phenotypes is shown in Table 3.1. The classes of particular interest for analyzing oogenesis and embryogenesis are those that produce fertilized eggs that develop poorly or not at all. These are defective in embryogenesis, either because of maternal or zygotic defects, and are designated either *zyg* (for zygote defective) or *emb*.

To date, three extensive searches for ts developmental mutants and specifically embryogenesis-defective mutants have been made (Hirsh and Vanderslice, 1976; Miwa et al., 1980; Cassada et al., 1981), and most of the mutants from these searches

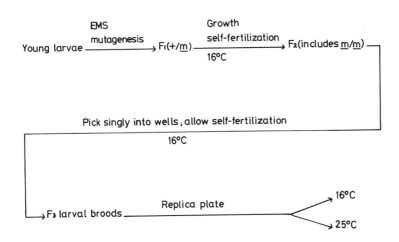

Fig. 3.15. Mutagenesis and screening for ts mutants of *C. elegans*.

Table 3.1. Classes of *Caenorhabditis elegans* ts Mutants From One Screening[a]

Phenotype at restrictive temperature	Frequency	Designation	Presumptive defect
ts mutants			
No offspring or eggs	112 (50%)	*gon*	Gonad formation or function
Unfertilized eggs only	24 (11%)	*sp*	Spermatogenesis
Fertilized eggs only	21 (9%)	*zyg = emb*	Embryogenesis
Intermediate growth stage	48 (22%)	*acc*	Larval essential function
Gross morphological defects	8 (4%)	*mor*	Larval or general function
Sterile but morphologically normal progeny	10 (4%)	*abF*₁	Early germ line function
Total	223 (100%)		

Source. Data from Hirsh and Vanderslice (1976).

[a]Wild-type, all stages present.

have been characterized genetically and developmentally (Schierenberg et al., 1980; Wood et al., 1980; Isnenghi et al., 1983; Denich et al., 1984; reviewed in Wood, 1988). These mutants have been mapped and extensive complementation tests between the three sets have been carried out. In these early studies, 54 genes required for embryogenesis, and designated as either *emb* or *zyg* loci, were identified and found to be scattered around the genome, with all chromosomes represented. Many of these *emb* genes also appeared to be clustered in certain regions, as judged from the mapping, but this phenomenon appears to reflect the existence of relative recombinational "cold spots" rather than genuine clustering (discussed in Emmons, 1988).

The set of *emb-zyg* mutants provides the source material for approaching one of the key questions about oogenesis: How distinctive is the gene set expressed in oogenesis compared to gene sets expressed in later stages? To answer this question, each mutant must first be individually analyzed to assess the time of action of its wild-type allele during development; the collective pattern can then provide a provisional answer.

This categorization of time of wild-type gene action is provisionally provided by means of three simple genetic tests, each involving the scoring of progeny phenotype as a function of the input parental and output zygotic genotypes (Wood et al., 1980). These tests are comparable to the standard crosses that define maternal effect mutants in conventional sexual systems but are adapted to the genetic peculiarities of the self-fertilizing hermaphrodite, and are described in Table 3.2. The results of the crosses establish whether the gene is a maternal or zygotic genome function, and if maternal, whether an exclusive or partial maternal. Since a "maternal" function in a self-fertilizing hermaphrodite might, in principle, involve a component contributed by the sperm, the crosses are designed to determine, in addition, whether a given gene is required in oogenesis or in the sperm's contribution to the zygote.

The three crosses generate five possible categories of embryonic-essential function, defined in terms of the gametic and zygotic patterns and requirements of gene expression. Table 3.3 presents a tabulated summary of 47 of these genes. The striking

Table 3.2. Genetic Tests of emb Mutants to Establish the Time of Action[a]

Selfing or S-test: To determine whether the function must be expressed by the zygotic genome.

$+/m$ 25°C, self-fertilization   ⟶ $+/+$ $+/m$ m/m

                                           1   :   2   :   1

Results and interpretation: If all genotypic classes survive, then expression of the wild-type gene function by the maternal genome is sufficient. If the m/m class dies during embryogenesis, then expression of the wild-type function by the zygotic genome is necessary.

Rescue or R-test: To determine which maternal effect functions are also zygotically active and capable of rescuing progeny from maternally defective hermaphrodites.

$$m/m \times +/+ \longrightarrow m/+$$
25°C cross

Results and interpretation: If all $m/+$ progeny are inviable, then the function is a strict maternal. If some cross progeny are viable, then the maternal effect function is provisionally classified as one that is expressed by the zygotic genome as well. Definitive classification is provided by the H-test.

Heterozygous rescue or H-test: To determine whether rescue of a maternal effect mutation is by the wild-type gene itself or via the function of this gene during sperm production.

$$m/m \times +/m \longrightarrow m/m \qquad +/m$$
25°C cross   1   :   1

Results and interpretation: If only the $+/m$ class is rescued, then the wild-type allele is required in the zygote itself for rescue. If both classes survive, then wild-type function is necessary for sperm production to supply a needed sperm component for the zygote.

---

[a]For details, see Wood et al. (1980).

fact is that for this set of genes, identified by the mutant screening procedure described, there is an overwhelming preponderance of those that show maternal expression; the largest class (class 1) are those identified as exclusive maternals, and the second largest class (class 2) are the partial maternals. Only one gene in this set has been clearly identified as specifying an essential paternal (sperm-donated) gene product (class 4), while the set of strict zygotic genome genes for embryogenesis consists of only 3 of the 47 genes.

Although the genes listed comprise only 1% or fewer of the total gene set of *C. elegans*, the sample is large enough to justify the conclusion that genes defined as essential for embryogenesis by the screening criteria used are for the most part expressed in oogenesis, with a large fraction of these also expressed by the zygotic genome during early development. There is, however, one qualification to be noted in evaluating these results. Table 3.3 classifies genes on the basis of their mutant phenotypes, and in many cases in terms of the behavior of only one or two alleles of the gene. Nevertheless, just as in *Drosophila*, where the properties of the allele can influence the classification of a gene as a maternal effect function or a more generally essential function (Bischoff and Lucchesi, 1971; Perrimon et al., 1986; chapter 4, this volume), allele-specific properties can influence the classification of a gene. An

Table 3.3. Functional Classes of 47 _emb_ Genes, Classified by Parental Effect Tests

Class	Description	Number
1	Exclusive maternal; maternal expression is both necessary and sufficient	28
2	Partial maternal; zygotic expression rescues embryos derived from _m/m_ hermaphrodites	11[a]
3	Zygotic expression is essential	3[a]
4	Paternal expression is sufficient to rescue	1
5	Maternal _and_ zygotic expression is essential	5

Source. Data from Wood (1988).

[a]_let-2_, depending upon the degree of residual expression of the particular allele, can be classified as either a class 2 or class 3 gene function.

instance is the _let-2_ gene (_lethal-2_, initially identified as a general lethal), which has been placed in both classes 2 and 3 on the basis of the disjoint behavior of two of its alleles. This difference probably reflects differential "leakiness" of the alleles, a point discussed below.

This difficulty aside, the cross tests reveal that most of the _emb_ genes are expressed maternally and hence during oogenesis, with such expression being sufficient in the majority of cases to allow mutant embryos to hatch. One may ask, however, whether these genes are restricted in their expression to oogenesis/early embryogenesis or whether they are expressed at still later times in development as well. To determine this, one allows mutant embryos to hatch at the permissive temperature, then shifts them to the restrictive temperature at successive stages after embryogenesis and examines the animals to see if any additional mutant phenotypes have appeared.

When this test is carried out, most of the mutants exhibit secondary postembryonic defects (Vanderslice and Hirsh, 1976; Wood et al., 1980). These secondary phenes consist of defects in gonad formation during larval development (the Gon phenotype), cessation of growth and accumulation at the L2, L3, or L4 stages (the Acc phenotype), and various morphological defects arising during larval development (the Mor phenotype). Thus, of the 21 ts mutants of the _emb/zyg_ set isolated by Wood et al. (1980), only 2 display the Emb phenotype solely. Of the remaining 19, 8 show both Gon and Acc phenotypes, 9 show the Gon phenotye without other defects, and 2 have Mor defects. Similarly, in the _emb_ mutant set described by Cassada et al. (1981), 30 of the 35 maternal effect mutants showed secondary mutant phenes. In few of these cases is the presence of secondary mutant phenes likely to be caused by the presence of secondary mutations in the stocks. All of the strains described by Wood et al. (1980) were purified of potential second-site mutations by several rounds of backcrossing and reisolation. In the few instances where reversions of the Emb phenotype were selected, the secondary phenes were found to be simultaneously reverted. The various developmental defects produced by temperature shift in most mutants must therefore reflect the requirement at later times of these maternally expressed genes.

Indeed, the genetic classification scheme, based on the tests shown in Table 3.2, only scores development as a function of expression up to the stage of hatching, and therefore is a necessarily incomplete classification of expression pattern. Furthermore, it gives only a relative quantitative measure of gene expression during early development. It is easy to imagine, for instance, a class 3 function (zygotic expression required) being classified as a class 2 allele (a partial maternal) on the basis of a leaky mutation; the degree of hypomorphic expression during either oogenesis or embryogenesis might be sufficient for rescue in such a case. This is undoubtedly the source of the discrepancy for the *let-2* alleles mentioned above, and for two mutant alleles of the *emb-9* gene, *B189* (class 2) and *B117* (class 3) (Wood et al., 1980).

Evidently, the genetic classification of patterns of gene activity is both incomplete and subject to error. A more precise means of delineating either the time of gene expression or of activity of the gene product is required. Indeed, for ts mutants, the very property of temperature sensitivity allows approximate determinations of this kind (Suzuki, 1970). As noted above, the additional periods in which the *emb* mutant gene products are required were determined by phased exposures of mutant animals to the restrictive temperature. Such experiments can establish the specific period(s) during which temperature inactivation of the gene product produces the mutant phenotype. For all *tl* mutants, this temperature-sensitive period or TSP is the approximate period in which the gene product is active and required in the cell (Suzuki, 1970)—unless the product is synthesized long before its use, in which case the TSP measures the period from the beginning of its synthesis to the end of its required use. For *tss* mutants, the TSP measures the period of synthesis of the gene product. Although the TSP is therefore not an unambiguous property, it is often surprisingly informative.

An alternative method for determining the TSP is the *reciprocal temperature-shift experiment*. Two sets of synchronously developing animals are placed at the restrictive and permissive temperatures respectively at the beginning of development. At fixed intervals, one batch of each is shifted quickly to the other temperature. The fraction of normal or surviving individuals following each timed shift is scored at the end of the experiment. The reciprocal temperature shifts generate reciprocal survival curves that together define the TSP.

The logic of the experiment can be illustrated with the simplest case, that of a mutant that synthesizes a TL protein during the same period in which the protein is required to act. The beginning of the TSP is defined by a series of downshifts from restrictive to permissive temperature. Imagine a group of embryos or eggs beginning development at the restrictive temperature, with batches being shifted down to the permissive temperature at successive periods. Those that are downshifted before the critical developmental period will all survive, while those maintained at the restrictive temperature during this period and then downshifted will fail to execute the necessary cellular function and die. (In the latter case, death may not follow instantaneously but may occur at a later stage.) In the downshift series, therefore, the developmental period in which the survival rate decreases from 100% to 0% defines the period of irreversible developmental defects, an interval denoted as the *defective execution stage* (Hartwell et al., 1970). The midpoint of the curve, 50% survival, can be taken as the average time of defective execution, and the point of inflection of the curve as the *beginning of the TSP*.

The temperature-upshift curve gives the converse result and defines the end of the TSP. In this series, mutant embryos shifted to the restrictive temperature before the protein is synthesized will all have defective protein during the critical period and suffer a lethal defect. In contrast, all embryos shifted after this period will survive. Again, shift-up during the critical period gives fractional survival, the proportion depending upon the time of upshift, the synchrony of the animals, and the fraction of expressed activity necessary to ensure survival. In upshift then, the phase in which survival begins to increase and goes to 100% is the *normal execution stage* (Hartwell et al., 1970) during which the gene activity is required. The midpoint of the upshift survival curve marks the midpoint of the normal execution stage, and the beginning of the rise in survival represents the *end of the TSP*. The two survival curves from the reciprocal shift experiments together delimit the TSP as the region of overlap, where survival is changing in both curves.

A hypothetical example, for the situation described, is shown in Figure 3.16a. In this somewhat idealized situation, the time of occurrence of the defective and normal execution stages are nearly coincident, with the time of completion of normal execution occurring slightly later than the occurrence of defective execution in the absence of the wild-type gene activity. For both curves, survival is plotted as a function of developmental stage rather than time. This standardization is necessary because the rate of development is a function of temperature, whether for wild-type or mutant strains. Thus, in wild-type *Caenorhabditis*, development at 25°C is twice as rapid as at 16°C (Byerly et al., 1976). The most reliable method for determining stage of development is by direct observation of wild-type and mutants, since the mutants' growth rates even at the permissive temperature are often somewhat slower than that of the wild-type.

Besides the simplest situation described above, other possibilities exist; these are shown in Figure 3.16b,c. In Figure 3.16b, the defective execution stage comes well after the normal execution stage. In terms of the formal logic of the experiment, the result seems paradoxical, since the *beginning* of the TSP, defined as the beginning of the downshift curve, seems to come *after* the presumed *end point* of the TSP, defined as the end of the rise in the upshift curve. However, assuming that the staging at the two temperatures has been performed correctly, there is a simple biological explanation: The wild-type function is normally carried out at a comparatively early point in development, but the protein is synthesized over a much longer period and can carry out the correct function later than the normal time. In this situation, defective execution—the point of irreversible damage, as measured by the downshift curve— would, indeed, follow completion of gene product action at the normal, early time. The diagram in Figure 3.16c shows the converse result, with defective execution substantially preceding the normal execution point. This result also presents a seeming puzzle. How can the failure to perform a cellular function produce a deleterious effect before that process is normally completed? One possibility is that the wild-type protein serves as a check on some process, extended in time, which if completed prematurely causes aberrant development.

When one measures the TSPs for a large set of maternals, the results bear out the general expectations for time of action: Exclusive maternals, expressed by the maternal genome and whose products are stored in the egg, generally have their first TSPs during oogenesis or early cleavage, while partial maternals show later and often prolonged TSPs, and strict zygotic functions have TSPs located distinctly within the period of embryogenesis and often late (Schierenberg et al., 1980; Wood et al., 1980; Isnenghi et al., 1983).

Somewhat discouragingly, however, the phenotypes for most of these early-acting maternals indicate a high degree of pleiotropy, involving multiple defects in zygote formation or early cleavage. These defects affect pronuclear fusion or polar body

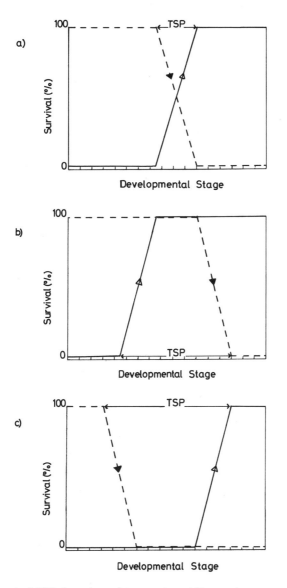

Fig. 3.16. Measurement of TSPs in reciprocal temperature-shift experiment: some patterns. Details described in the text. Dashed line, downshift survival curve; solid arrowhead, defective execution point; solid line, upshift survival curve; open arrowhead, normal execution point. (Adapted from Miwa et al., 1980.)

formation, placement of the first or subsequent cleavage planes, abnormal cytoplasmic streaming, timing defects in the early cleavages, and a variety of other nuclear and cytoplasmic abnormalities (Denich et al., 1984).

Nevertheless, even pleiotropic defects can be informative, when exhibited by a group of mutants. An early speculation, based on the relative rates of cell division of the major somatic cell lineages, which are characteristic for those lineages, was that these division rates might be causally linked to the particular fates that are allocated (Deppe et al., 1978). Yet many of the mutants experience slower rates of cell division but exhibit normal differentiation within somatic lineages (Schierenberg et al., 1980; Denich et al., 1984).

Furthermore, the detailed analyses of several of the mutants have yielded information of interest, despite the pleiotropy of the mutants. Some examples are shown in Figure 3.17, which illustrates the measurement of TSPs for three *emb* mutants, two alleles of a class 1 (exclusive maternal) function (*emb-5*) and a single mutant of a class 3 (strict zygotic) function (*emb-9*). The combined results indicate that both wild-type genes are required for particular morphogenetic processes. The timing of the TSPs provides one indication: For both alleles of *emb-5*, the TSP occurs early, roughly between the 8- and 30-cell stage of development, the period that includes the beginning of gastrulation, while the TSP for *emb-9* occurs during the later period of dramatic morphogenesis (between the comma and pretzel stages of development). Different alleles of the same gene, however, can have slightly different TSPs; thus while the respective TSPs for the two *emb-5* mutants (Fig. 3.17) occur at nearly the same time, when the embryo consists of 24–28 cells, the *hc61* shows defective execution (the downshift curve) slightly preceding normal execution. Microscopic examination of the living embryos of these two mutants pinpoints the developmental defect(s), in both mutants, as occurring in the endodermal (E) cell lineage; it has also explained the difference in their TSPs. In normal embryos, two E cells, consisting of the anterior division product of the E cell, E.a, and the posterior division product, E.p, migrate internally from their initial posteroventral location at about the 28-cell stage. Following migration, the two cells divide to yield four E lineage cells, an anterior dorsal and ventral pair (E.ad and E.av) and a parallel posterior dorsal and ventral pair (E.pd and E.pv). These cells give rise to the corresponding four sections of the adult intestine, e.g., E.ad gives rise to the anterior dorsal row of intestinal cells. In contrast, in both of the *emb-5* mutants shown, *hc61* and *hc67*, the E cell divisions occur precociously relative to wild-type (Schierenberg et al., 1980). The premature division is followed by a prolonged migration phase, at the end of which the four E cells find themselves in an abnormal anatomical environment, apposed to an additional D cell and four extra MS cells (generated by divisions in those lineages during the period of delayed E cell migration). Furthermore, the E.ad and E.av cells are now in reversed positions (Fig. 3.18). Inevitably, the end result is aberrant intestinal development.

The subtle difference between the two mutants concerns the relatively accelerated time of defective execution in *hc61*. In the latter, E.a and E.p divide before migration begins, while in *hc67*, division occurs during migration. The relative difference in time of defective execution therefore signals a genuine difference in the behavior of the two mutants. In several respects, *hc61* seems to be the more severely defective allele, showing more extreme secondary phenes; the accelerated inward movement of its cells is one example.

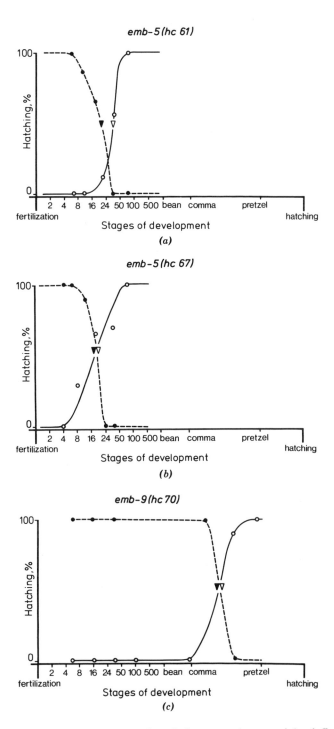

Fig. 3.17. Comparison of two TSPs for two allelic exclusive maternal mutants (of *emb-5*) and for that of a zygotic essential function (*emb-9*). Symbols as in Figure 3.16. Each point represents at least 10 eggs. Results are discussed in the text. (Reproduced from Miwa et al., 1980, with permission of the publisher.)

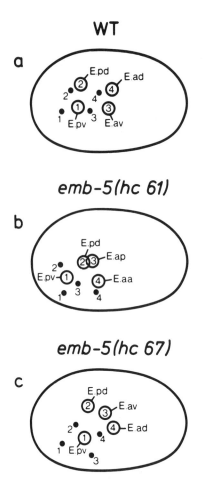

Fig. 3.18. Alteration of E lineage cell behavior in *emb-5* mutants. The solid dots indicate the positions of the four E lineage cells just after the division of E.a and E.p; the open circles, their positions just before the next division. Orientation: Anterior to right and dorsal to top. In *hc61,* the division of E.a is a-p rather than d-v; E.ap therefore corresponds to E.av and E.aa to E.ad. (Reproduced from Schierenberg et al., 1980, with permission of the publisher.)

The results, in sum, indicate that the wild-type *emb-5* gene product influences the gastrulation process directly and acts specifically to delay the second division in the E cells relative to the migration process. One cannot conclude, however, that this is *the* function of the *emb-5* gene product in development because other *emb-5* mutants, isolated subsequently, show additional and apparently independent effects, including an altered division plane in the first cleavage and precocious divisions in P_4 (Denich et al., 1984). These results confirm a role for the *emb-5* gene product in the cleavage division process but suggest that it is more general than the E lineage cell division results shown in Figure 3.18. The *apparent* specificity of the latter presumably reflects either a degree of special sensitivity to reduced dosage of the *emb-5* gene product in the E lineage or particular characteristics of the protein products in *hc61* and *hc67*.

It will be apparent from this example, and many others that could be cited, that the use of TSPs to infer the time and mode of action of the wild-type gene product is not free from ambiguity. The two principal sources of ambiguity flow from the degree of residual expression, which can alter the length or position of the TSP, and the nature of the mutant defect, whether it is of the TL or TSS kind. When there is a large temporal difference between synthesis of a protein and its utilization, the basis of a ts defect in that protein becomes an important factor in interpreting the nature of the TSP. Thorough discussions of the technical and analytical problems in studying ts mutants can be found in Hirsh (1979) and Wood et al. (1980).

While the genetic and developmental characteristics can provide strong clues as to time and manner of gene function, one ultimately needs direct biochemical or molecular characterization of the gene product for rigorous definition of function. That information is not available yet for *emb-5* but it is for *emb-9,* whose TSP is during the comma-to-pretzel morphogenesis stage (Fig. 3.17c). The gene has now been cloned and shown to encode one of the collagen basement membrane molecules (Guo et al., 1991). Collagen is a rigid structural protein, typically forming trimers, and the mutant alleles of three ts mutants have been cloned, including *hc70* (Fig. 3.17c), and shown to cause amino acid substitutions in the triple helical domain of the molecule. This result helps to explain the developmental phenotype, an arrest of morphogenesis during the transition to the pretzel stage (Fig. 3.2b–d). The TSP indicates that the gene product acts (or is synthesized) when the cells of the hypodermis are changing shape, elongating the embryo into the pretzel form (see chapter 6). The basement membrane, underlying the hypodermal cells, undoubtedly provides a measure of traction for this shape change to be accomplished. The identification of the *emb-9* gene as a collagen basement membrane gene also helps to explain a previously puzzling characteristic of many of its ts mutants, namely that they are semidominant (Miwa et al., 1980; Wood et al., 1980, where *emb-9* is designated *zyg-6*). Mutant versions of gene products that participate in complex assembly processes will often exhibit such semidominance because the altered gene product interferes with the assembly reactions involving the product of the wild-type allele. We will see more examples of this phenomenon when we look at the construction of the nematode cuticle in chapter 9.

Nonconditional, Amorphic Maternal Effect Mutants

As we have seen, any residual mutant "leakiness" can affect the final mutant phenotype and, hence, alter the interpretation of the role of the wild-type gene. Indeed, the only situation in which one can make strong inferences about wild-type gene product function(s), from mutant phenotypes, is when one is studying known amorphs or null alleles for the gene in question. Because most ts mutants retain some degree of wild-type function, it often happens that the principal value of many of the more interesting ts mutants—the ones that appear the least pleiotropic and which create sharp, specific aberrancies in development—is to pinpoint the genes for which one would like to obtain amorphs.

In recent years, two comparatively simple methods for obtaining nonconditional, amorphic mutations in *emb* genes have been devised. The first is of general utility whenever one has as starting material a conditional lethal mutation, such as a ts mutant, in a potentially interesting gene; it relies on the principle that mutations which fail to complement a known mutation in a given gene (Fig. 3.6) are presumptive new mutations in that gene. Applied to *C. elegans,* this method involves mating males that are homozygous for a ts mutation to mutagenized hermaphrodites carrying a marker mutation (this permits cross progeny to be distinguished from self-progeny) and then allowing progeny production to take place at the permissive temperature; wild-type progeny (those heterozygous for the marker mutation) are then isolated and tested under the restrictive condition for defectiveness. F_1 isolates showing the mutant phenotype must carry a new mutation that fails to complement the original mutation and hence one presumed to lie in the same gene. The method, in principle, can yield many new mutant alleles, some of which may be amorphs and which, altogether, can be used to construct a so-called "phenotypic series" (of mutant severities). Even if no amorphs are obtained, the latter can provide information for predicting the likely amorphic phenotype.

In the experiments reported by Kemphues et al. (1986), new mutations were sought in the *zyg-11* gene, a strict maternal previously identified by the *b2* allele of Wood et al. (1980). Embryos from homozygous *b2* hermaphrodites show an early and interesting mutant phenotype, the replacement of the asymmetric first cleavage division with a medial cell division (which effectively gives two AB-type cells). To determine if this is the amorphic phenotype, males homozygous for *b2* were mated to *dpy-10* hermaphrodites that had been mutagenized with EMS, and individual F_1 animals were plated on separate plates at 16°C; some wild-type adults from each plate were then transferred to 25°C and after 2 days these were screened for the presence of unhatched eggs. Candidate mutants were then reisolated from the original plate and retested.

By this method, three new mutant alleles of *zyg-11* were isolated, and in addition, two more were found in complementation tests of mutants isolated in other screens. Two of the new mutants show some temperature sensitivity, and three were absolutely defective, allowing no progeny production from homozygous hermaphrodites at any temperature. All mutants are, like the original mutation *b2,* strict maternal effect mutants, and the ts mutants show a TSP around the time of fertilization. Phenotypic examination of the various ts mutants, and comparison to the absolute defectives, indicates that the primary defect in embryos derived from mutant mothers is a greatly prolonged meiosis II division, which leads to the subsequent cleavage and cytoplasmic aberrancies.

The absolute phenotypic defectiveness of three of the mutants justifies their provisional classification as amorphs, but it is desirable to be certain that the observed phenotype is an amorphic one and not influenced by even a small degree of leakiness. In *Drosophila,* as we shall see in the next chapter, the Mullerian tests for classification can be applied, but this requires a genetic deficiency (a small deletion) of the chromosomal region containing the gene under test and in *C. elegans* many genes are not "uncovered" by such deficiencies.

There is, however, a test borrowed from prokaryotic genetics that can be used to determine if one is dealing with a particular category of amorphs, namely those produced by "nonsense" or terminator codons within the mutant gene. Some background explanation might be useful here. The great majority of mutants are "missense" mutants, which cause the production of defective proteins through the insertion of an inappropriate amino acid at a critical site. Nonsense mutants, in contrast, cause premature termination of translation of the mutant protein. The consequence is a shortened and generally nonfunctional gene product, which, in most cases, is degraded by the cell. There exist, however, special suppressor mutations for chain terminator mutations that can prevent this chain termination. These suppressor strains possess an altered cellular activity that allows the placement of an amino acid at the position of the stop codon, thereby permitting completion of synthesis of the polypeptide chain. Although the inserted amino acid is generally not the one present at the same position in the wild-type protein, the completed protein chain usually possesses much of the wild-type activity. In bacteria, yeast, and nematodes (Wills et al., 1983), the nonsense suppressor mutation is altered in a tRNA gene, and specifically in the anticodon region, the triplet of nucleotides that recognizes and binds to the codon of the mRNA. The mutant tRNA pairs with the given nonsense codon and, in so doing, inserts the amino acid that it carries into the growing polypeptide chain. (The normal tRNA function is not lost, however, because every organism contains several copies of each tRNA gene in its genome, of which only one is altered in the suppressor gene.)

When a particular nonsense suppressor, *sup-7,* was tested for its ability to alleviate the mutant defects in the *zyg-11* strains, one mutant, *mn40,* was found to be suppressible as was one allele of another strict maternal effect gene, *zyg-9,* whose mutants also show defects in meiosis and nuclear and cytoplasmic movements in the fertilized egg (Hirsh et al., 1985; Kemphues et al., 1986). The demonstration of suppression indicates that the mutant allele is a chain terminator mutant and hence a probable null allele, while the similarity of developmental phenotype between this mutant and the other two strong *zyg-11* mutants strengthens the case that these two are also amorphs. (The inference that a chain terminator mutant is a null is reasonable, and holds in most instances, but a few cases are known in *Escherichia coli* and yeast, where the incomplete polypeptide retains some activity.)

The second method for obtaining absolute defectives is potentially more useful because it doesn't require an initial mutation; in principle, it allows the isolation of amorphic mutations in any exclusive maternal effect gene. The method makes use of the fact that certain mutants affected in particular postembryonic (larval stage) cell division sequences make defective vulvas and are consequently incapable of egg laying. Since fertilization in such hermaphrodites still takes place, the fertilized eggs give rise to larvae inside the body of the hermaphrodite (the "bag of worms" phenotype), which eventually destroy the parent. Therefore, in egg laying deficient strains, the only individuals that will be spared from being devoured by their offspring are those that are simultaneously homozygous for a second mutation which prevents either fertilization or embryogenesis.

To isolate new strict maternal effect mutants, including absolute defectives, Kemphues et al. (1988a) employed the following strategy: Strains carrying either of two mutations that block egg laying were mutagenized with EMS, and the F_1 progeny were picked to individual plates and cloned, with subsequent incubation to allow the production of an F_3 generation; those plates showing surviving F_2 possess mutational blocks in either gonadogenesis, fertilization, or in maternal effect functions (Fig. 3.19). The existence of refractile (fertilized) eggs indicates the class of choice, those carrying new maternal effect mutations. The mutations were then reisolated from heterozygous sibs from the original plate and then saved for further genetic and phenotypic analysis.

While, as noted earlier, many exclusive maternal effect mutants show a variety of meiotic, cytoplasmic, or nuclear structure abnormalities, a group of eight, defining four new genes, were identified that showed abnormalities in the cleavage process

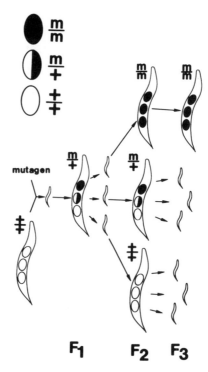

Fig. 3.19. A scheme for selecting nonconditional maternal effect mutants. An Egl (egg-laying defective hermaphrodite) is mutagenized and individual F_1 progeny are cloned. Those clones yielding one-quarter F_2 survivors carry a new maternal effect mutation. See text for further discussion. (Reproduced from Kemphues et al., 1988a, with permission of the publisher. © Cell Press.)

without any obvious prior defects in meiosis or nuclear structure (Kemphues et al., 1988a). Mutants defective in these four genes are characterized by symmetric cleavage divisions, beginning at the first (AB-P_1 division) or the second cleavage division, failures in P granule segregation, and the subsequent production of disordered embryos with higher-than-normal cell numbers and the occurrence of some cell differentiation (as shown by the production of muscle and pharyngeal cells). Because of their defectiveness in early cleavage and segregation of developmental potential, they have been designated *partitioning defective (par)* mutants and the four genes named *par-1, par-2, par-3,* and *par-4.* In mutants of the first three, the posteriorward movement of the zygote nucleus fails to occur, with a resultant positioning of the spindle in the center of the cell; in *par-4* mutants, the position of the first spindle is normal, but in all mutants, the second division is aberrant (with a failure of the normally occurring rotation of the P_1 nucleus). The immediate developmental consequence of these defects is the abolition of the polarized germ line cleavages. Indeed, the fundamental defect seems to involve a failure of polarization, as shown most vividly by their failure to segregate P granules normally to the posterior pole. Though P granule segregation requires the operation of the actin filament system, none of the *par* genes, however, maps to any of the known actin loci (cited in Kemphues et al., 1988a).

Further analysis of the *par* mutants should help to elucidate the relationship between asymmetric cleavage divisions in the early embryo and the early differential allocation of cell fates. Characterization of mutations in yet another gene, *cib-1* (changed *i*dentity of *b*lastomeres *1*), may help to illuminate the equally important matter of the relationship between cell cycle timing and embryonic blast cell fate. The first *cib-1* mutant was isolated fortuitously, as a segregant from a ts stock, and proved to be an amorph while subsequent alleles, the majority being ts, were isolated on the basis of noncomplementation with the first allele (Schnabel and Schnabel, 1990).

The *cib-1* gene is an exclusive maternal effect gene, whose action(s), as judged from TSPs, take place soley during the first cleavage divisions. The mutants have the interesting property of causing delays in the second, third, or fourth cleavage division of the P lineage specifically, with the more strongly defective alleles causing earlier and longer pauses than the leakier mutants. The consequence of a prolonged pause is that the affected P cell skips a cell cycle, and when it next divides it has apparently acquired the fate of its normal somatic cell daughter. Thus, a sufficiently paused P_1 cell, when it does divide, does so as an EMS cell (Schnabel and Schnabel, 1990). A possible clue to the nature of the *cib-1* defect is that long pauses are accompanied by a suppression of the S phase that would normally occur prior to that division.

Comparable results can be achieved by suppression of the first cleavage division through various physical treatments (cold shock, pressure, or centrifuguation). Schlicht and Schierenberg (1991) have found that embryos recovering from such treatments suffer a net delay of one cleavage division and never give rise to the full complement of six basic lineages, producing from one to five before the germ line lineage ceases its asymmetric divisions. (It is not known whether these delays are followed by a suppressed S phase, as in the *cib-1* mutants.) They have interpreted these results in terms of the following hypothesis: that there is a cytoplasmic factor required for each germ line division, whose amount decreases both with time and each germ line division, that each embryonic founder cell is stamped with special characteristics by the germ line division in which it arises (see also Schierenberg, 1985, 1986, 1987), and that delays of cleavage by one division leave less of this factor for subsequent germ line divisions. The potential significance of these results and those concerning the *cib-1* mutants will be discussed in the penultimate section of this chapter.

Estimating Total and Uniquely Expressed Maternal Genes; Implications of the Estimates

One of the central questions about the oocyte concerns the molecular basis of its distinctiveness. Does it have the capacity to generate the animal because it is a uniquely complex cell, with many distinctive gene products, or should it be viewed rather as perhaps only a slightly more complex than average stem cell?

One general measure of cellular distinctiveness is the set of genes expressed by a particular cell relative to other cell types. By this measure, the genetic characterization of the nematode oocyte would seem to indicate that it is not a highly unusual cell. The data on ts mutants indicate that the great majority of genes required for early embryogenesis are expressed maternally and that the largest part of these are expressed at other developmental stages and by other cell types. Even among those classified as exclusive maternals, on the basis of the failure of zygotic expression to rescue them, a substantial fraction may be expressed by the zygotic genome but at levels insufficient for rescue.

This genetic characterization of oogenesis gene functions in *C. elegans* is very reminiscent of the molecular data on expressed RNA sets in sea urchin and amphibian embryos (reviewed in Davidson, 1986); these embryo types also show a broad overlap between the gene set expressed in oogenesis and those expressed in other cell types. Nevertheless, in thinking about the nematode oocyte, one needs to go beyond the qualitative impressions garnered from the small set of identified oogenesis genes to quantitative estimates of the numbers of genes expressed in oogenesis.

Although the existing set of ts mutants can be used to estimate in various ways both 1) the total number of genes expressed in oogenesis and 2) the number of uniquely expressed oogenesis genes (see Wilkins, 1986, pp. 71–72), all such estimates are compromised by the various assumptions one makes about the representativeness of the mutant alleles obtained and by the residual leakiness of many of the mutants. A different approach to identifying unique oogenesis genes would involve scoring the frequency of mutation to exclusive maternal effect lethality (class 1), relative to the average loss-of-function mutation rate for vital loci, for a large group of genes. Genes that mutate to the strict maternal lethal phenotype at frequencies comparable to those for loss of essential functions are likely to be "pure" oogenesis genes, while those that give the maternal lethal phenotype relatively infrequently are likely to be genes expressed both during oogenesis and at other stages, the maternal effect alleles being atypical ones (Perrimon et al., 1986).

Kemphues et al. (1988b) have carried out such an analysis for LGII (the second linkage group), using the screening method shown in Figure 3.19 to isolate the mutant set. From a large screening of more than 12,000 mutagenized chromosomes, they were able to identify only four genes on this linkage group that mutate consistently to strict maternal effect lethality. Of these only two are required for embryogenesis, one is required for meiosis, and one shows a complex expression pattern not restricted to oogenesis. Thus, for approximately one-sixth of the genome (since there are six lineage groups altogether and all are approximately the same size) only two or three genes seem to be expressed uniquely in oogenesis. If one extrapolates to the whole genome, this would come to only a total of 12–18 expressed solely in the oocyte.

Nevertheless, while the oocyte may not possess numerous unique gene products, the comparatively few that it does have may make a significant contribution to the oocyte's capacity as progenitor cell. Furthermore, the uniqueness of an oocyte can derive from other features beyond uniqueness of its gene products. The most import-ant of these aspects, undoubtedly, is the particular *combinations* of gene products that it possesses. This is particularly easy to envisage for transcriptional properties: Given the multiple factors of various sorts required for the transcription of individual genes (reviewed in Martin, 1991), it is apparent that special combinations of such factors, where each is found in some other cell type, can generate cell-specific transcriptional capacities. The importance of combinatorial control has been referred to earlier (chapter 2, "Domains of gene expression as clues to functional significance and some thoughts about genetic 'control' ") and is discussed again in later chapters.

In addition to special combinatorial features, some part of the uniqueness of the oocyte, and its long-term influence on the embryo, may involve differences in certain general cellular properties, relative to somatic cells. Thus, two special features of many oocytes are the abundance of their transcripts and the fact that many of these may be long lived and exert their effects only later in embryogenesis. This "strategy" of expression seems to be employed, for instance, in the sea urchin, where many genes are continuously or repeatedly transcribed from early development onward to permit the requisite accumulation of their protein products for later stages of development (Hough-Evans et al., 1977; Davidson, 1986, pp. 172–173). In the sea urchin and in *Xenopus*, many of the stored transcripts may require additional processing (Davidson et al., 1983), but this would not be inconsistent with the idea of delayed utilization.

A genetic test of such maternal expression with delayed utilization is, indeed, possible in *C. elegans*, involving genes expressed in and required in the larval stages, as recognized by their zygotic Acc or Mor phenotypes (Table 3.1). In a test of this kind, involving homozygotes of 12 different acc_2 mutants (those that accumulate in the L2 stage), 4 were found to be rescued by maternal + allele expression (in heterozygous mothers) (Wood et al., 1980). The results do not identify the kind of components that are being stored in the eggs—whether mRNAs, proteins, or final metabolic products—but they show that there is maternal expression (transcription) of genes whose action is apparently not required until the first or second larval stage. The proportion of maternally rescuable larval-stage genes in any such test is, in fact, likely to be an underestimate of the true level of expression because these tests score only rescue and might not detect lower levels of expression insufficient to rescue development. Molecular experiments provide some support for the notion of long-lived maternal products and, specifically, long-lived maternal RNAs: DNA probes for 4 of 14 cloned genes identified by the run-on transcription test as being preferentially expressed early in embryogenesis detect abundant transcript amounts for these genes late in embryogenesis as well, when they are not being expressed at high levels by the zygotic genome (Schauer and Wood, 1990).

WHERE IS THE "MOSAIC EGG" TODAY?

What is the status today of the traditional idea that the nematode egg is a "mosaic" one? It has essentially vanished, in fact. From experimental physical manipulations of the embryo, there is little evidence for either transferable cytoplasmic "morphogenetic determinants" or for individual determinants for the individual six major lineages. Although the posterior polar plasm has distinctive properties relative to the anterior cytoplasm of the egg (Schierenberg, 1988), this is consistent with a large number of alternative models of egg organization. Nor have genetic approaches produced such evidence: None of the maternal effect mutants neatly eliminates one of the lineages while leaving the cell division properties and fates of the others unchanged. On the other hand, the existing mutants do not support the other classical alternative, the static gradient model, in which a gradient along the a-p axis allocates founder cell fates as a function of the concentration of the active substance.

The negative verdict of the genetic data on both models may, however, simply reflect a rather trivial fact: that too few mutants have yet been collected. If, after all, there are only about 20 genes uniquely expressed in oogenesis, and some of these 20 encode critical "determinants," then they might easily have escaped detection. Another equally possible alternative is that the fate-setting system, whatever it is, involves genes whose products are used at multiple stages in development and that mutants of these genes might fall into the category of general lethals rather than maternal effect lethals.

Nevertheless, the information that does exist points in a rather different direction and, in particular, to connections, still poorly understood, between fate setting and certain basic cell division and cell cycle properties. The *par* mutants suggest that any interference with the setting up of a-p polarity or the occurrence of asymmetric first or second cleavage division prevents the development of the germ line lineage (P_1 through P_4) and implicate the cytoskeleton as essential for correct fate assignment in the first two critical cleavage divisions (Kemphues et al., 1988a). An inference one can draw from the *par* mutant phenotypes is that it is the establishment of a-p polarity that is fundamental to the differences between the AB and P (germ line) lineages rather than the allocation of AB and P "determinants" per se.

Furthermore, the reversal of polarity that takes place in early cleavage (Fig. 3.12) indicates that the polarized cell divisions of the early embryo are produced by a dynamic mechanism in which the allocation of blast cell fates is a *sequential* process rather than something that is prefigured in the oocyte or fertilized egg. The dynamic nature of the process and its relationship to cell cycle events are further indicated by the *cib-1* mutants (Schnabel and Schnabel, 1990). In these mutants, blastomere "pausing," with skipping of particular S phases, leads to subsequent cell fate changes, showing that while early fate assignments are connected to the normal temporal execution of certain cell cycle events, a blastomere may "default" to the state of one of its progeny if cell cycle progression is interrupted. The cleavage suppression results of Schlicht and Schierenberg (1991) also fit this picture.

It would appear that the process of fate allocation may be both more progressive and subtler than traditionally conceived. The initial compositional and/or gene expression differences between the lineages as a whole may initially be few in number or slight in character and only emerge slowly through numerous cell interactions as development proceeds. This view of fate assignment in the nematode egg is dramatically different from the classical picture of the mosaic egg and, indeed, is much closer to an embryo type considered at the opposite end of the mosaic-regulative range, the mouse embryo (chapter 5).

In considering what the critical cell cycle connection to fate assignments might be, one must conclude that it cannot be the occurrence of cell division itself, since, as we have seen, lineage-specific potentialities can appear in the absence of cell division (Laufer et al., 1980; Cowan and McIntosh, 1985). Furthermore, it cannot be a process obligately coupled to cell division rates, despite the differences in these rates between cells of the different lineages (Deppe et al., 1978); too many mutants show significantly slower cell division rates yet express markers characteristic of the major lineages (Schierenberg et al., 1980; Denich et al., 1984).

Finally, the results with the *cib-1* mutants show that cell fate changes occur in the absence of cell division and may involve a "skipped" round of DNA replication. These findings rule out a counting of rounds of cell division and of DNA replication as necessary for fate allocation; they also eliminate a critical role for the DNA:cytoplasm ratio as the allocating device (Schabel and Schnabel, 1990).

The results are more in accord with the idea that there is a "developmental clock" that assigns fate as a function of time or some process normally linked to timing but independently of cell cycle events such as the number of cell divisions (Schnabel and Schnabel, 1990). The hypothesis of Schlicht and Schierenberg (1991) is comparable, positing a substance in the early embryo that governs the soma–germ line divisions, which declines in concentration (with time) and whose concentration determines embryonic founder cell fates at the time of their formation. The decrease in the critical substance, as a function of time, would effectively provide the "developmental clock" proposed by Schnabel and Schnabel (1990).

The important strategic question for the investigator concerns how best to explore further the nature of the morphogenetic system. Additional mutant hunting will play a role in this further characterization. Other genes that are like *cib-1* in their mutant behavior have been identified (cited in Schnabel and Schnabel, 1990), and their analysis should prove helpful. In addition, it should be possible to isolate maternal effect *emb* mutants, by the method of Kemphues et al. (1988a), that show the asymmetric cleavages of normal development (in contrast to the *par* mutants) but which exhibit altered distribution of differentiation potentials. Such mutants would be expected to produce inviable eggs. Any mutants of this general type may be either altered in the processes that follow establishment of a-p polarity, after fertilization, or defective in substances that execute these processes. Finding mutants of this type may be difficult, but they would be invaluable in exploring fate allocation in the first cleavage divisions.

An entirely different approach might be to use reverse genetics, to first search out gene products that are asymmetrically distributed at the successive early stages and then to identify the genes that encode them. In principle, one approach would be to do a thorough screening for asymmetrically distributed RNAs by *in situ* hybridization using the genomic clones generated by the total genomic mapping now in progress (Coulson et al., 1991). This would also be labor intensive, but improvements in the methods may make it less so with time. Furthermore, reverse genetic techniques should be applied to the identification of those 20–30 genes expressed uniquely from the zygotic genome in early embryogenesis (Schauer and Wood, 1990). Mutant analyses of these genes will determine whether or not the early fate-assignment processes do involve a zygotic genome contribution.

Another approach might be to make monoclonal antibodies to early embryos of the strain of *C. elegans* that does not make the highly antigenic P granules (see Wood, 1988) and use these antibodies to screen for asymmetrically distributed protein products. Identification of any such proteins could be followed by their sequencing, and then the use of oligonucleotide probes to clone the encoding genes. By screening against the bank of genomic clones, it should be possible to identify the DNA region in which the gene(s) reside and, then, to isolate mutants.

CELL LINEAGE IN THE EARLY EMBRYO: A POSTSCRIPT

The determinate cell lineage seen in the embryos of nematodes and various other animal species was a source of fascination for many of the pioneering embryologists at the turn of the century, men such as E.G. Conklin, F.R. Lillie, C.O. Whitman, and E.B. Wilson. Yet, as Wilson recognized, the phenomenon in itself was no "open sesame" to an understanding of early development. Despite a wealth of comparative and experimental studies, carried out from the late 1870s onward, questions about both the origins of fixed cell lineages, in the structure of the oocyte, and their significance for early embryonic development continued to persist.

A major contribution to understanding cell lineage in the nematode was made by Boveri (1910), who discovered that in rare, doubly fertilized embryos, one could get altered early cleavage divisions, resulting in various ratios of AB and P-type cells in four-cell embryos. Boveri recognized that such results bespoke a certain dynamism both in the cleavage mechanism and in the fate-setting process itself and argued that each determinative event might involve a binary decision, linked to but potentially dissociable from the cell division event itself. Despite Boveri's results, and his undoubted influence in the field, the image of the nematode embryo as a mosaic system of independent cells, allocated fates from a presumably mosaically organized egg, persisted for most of this century, not least because of Boveri's other discovery in the nematode, that of the germ line "determinant" cytoplasm.

In the last few years, there has been a return to Boveri's view of the early cell lineage as a dynamic process, involving cell interactions in the setting of cell fates and capacities. Some of these results have been reviewed in this chapter and others will be described in chapter 6, where we will return to the questions posed by fixed cell lineages. All of the available evidence, both from experiments involving surgical intervention and from analysis of maternal effect mutants, has failed to support either of the conventional static models of fate allocation in the early embryo, involving either a preset mosaic of determinants in the oocyte or a fixed oocyte gradient. Instead, the processes appear to involve binary soma–germ line cell divisions, each one setting the stage for the next, and some form of coupled process allocating fates and capacities at each division (Schnabel and Schnabel, 1990; Schlicht and Schierenberg, 1991). The early events probably involve solely maternal components, but the participation of some zygotic genome products in these interactions is a possibility. The precise molecular basis of these "decisions" and the nature of the coupling to the cell division process still remain obscure, but further mutant hunts and the utilization of "reverse genetic" approaches should assist in revealing their nature.

FOUR

DROSOPHILA MELANOGASTER
From Oocyte to Blastoderm

When Thomas Hunt Morgan left embryology for genetics he deserted *Hydra*, planaria, see urchins and amphibia for *Drosophila*. The fruit fly was recognized, and still is recognized, as an ideal organism for the study of eukaryote genetics, but it now seems that it will turn out to be equally suitable for answering many questions concerned with development.

H. Schneiderman and P.J. Bryant (1971)

INTRODUCTION: THE COMING OF AGE OF *DROSOPHILA*

At the start of the 1970s, the potential of the fruit fly as a model system for investigating animal development was sensed by few outside the then relatively small community of Drosophilists. Yet today, the prediction of Schneiderman and Bryant appears, if anything, too modest. *Drosophila* is now the animal whose development we know the most about, and ideas derived from *Drosophila* work are often the starting point for discussions about development in other systems.

The coming of age of *Drosophila,* from the early 1970s, when the fruit fly work was mostly a self-referential world of interesting but pleiotropic and largely mysterious mutants, to its preeminence today, can be traced to the confluence of two principal streams of research. The first was that initiated by T.H. Morgan and his colleagues, the detailed analysis of the genetics and cytogenetics of the fruit fly, which made it the most completely genetically characterized animal system. To this main stream was added the important tributary, in the late 1970s and 1980s, of saturation mutagenesis for obtaining mutants in every step of a genetic pathway; the seminal contribution in this regard was made by Christianne Nüsslein-Volhard and Eric Wieschaus. The coincidental revival of techniques of clonal analysis, by, among others, Peter Bryant, Antonio Garcia-Bellido, and John Merriam, also made a significant contribution to the genetic renaissance of *Drosophila*.

The second critical input was that of molecular biology, in particular, the development of techniques for cloning genes, which, by the early 1980s, made it possible, in principle, to clone any *Drosophila* gene, facilitated both by the relatively small genome size of the fruit fly and the extensive cytogenetic characterization of the giant chromosomes of the larval salivary gland. The latter made possible both the techniques of chromosome "walking" and "jumping" and the subsequent mapping of cloned sequences to specific chromosomal regions (Bender et al., 1983a).

It was the conjunction of these various lines of research that allowed the rapid development of the fruit fly as a system for investigating developmental mechanisms, a process that continues unabated today. In this chapter we concentrate on one aspect of this research, the analysis of the maternal genetic factors that provide the initial cues for the patterning of the *Drosophila* embryo. A central conclusion of this work is that there are four different maternal systems, involving essentially nonoverlapping gene sets, that govern this initial patterning; three are involved in establishing pattern along the anteroposterior (a-p) axis and one along the dorsoventral (d-v) axis. In chapter 7, we will look at the next phase of the process, involving the differential activation of various zygotic genes by these maternal gene patterning systems, and examine how the combined maternal-zygotic genetic hierarchy establishes the initial visible pattern of the embryo. Before turning to the first aspects of this material, however, we will briefly review some of the basic biology and genetics of the fruit fly.

THE LIFE CYCLE AND GENETICS OF *DROSOPHILA MELANOGASTER*

Stages of Development

The life cycle of *Drosophila* involves a period of embryogenesis, completed within the eggshell, an ensuing sequence of three larval growth stages or "instars," and a final stage of metamorphosis within the pupal case to produce the adult or "imago." Although the sequence of embryogenesis and larval growth is superficially comparable to that of *Caenorhabditis*, the final stage is considerably more complex than the larval–adult transition of the nematode and results in an animal dramatically different from the final larval form. The difference in complexity between the nematode and the fruit fly is further reflected in their constituent cell numbers. At hatching, the first larval stage of *Drosophila* consists of approximately 10,000 cells (Madhavan and Schneiderman, 1977), nearly 20 times that of the *Caenorhabditis elegans* L1 form. The adult fruit fly consists of over a million cells, while the adult *Caenorhabiditis* hermaphrodite has only 2500–3500 cells and nuclei (somatic plus germ line) in all. The life cycle in *Drosophila* wild-type strains takes 10–12 days at 25°C; it is diagramed in Figure 4.1.

The first phase of embryogenesis in *Drosophila* consists of a sequence of syncytial nuclear cleavage divisions, which convert the fertilized egg into an ellipsoid monolayer syncytium of approximately 6000 nuclei, surrounding a yolk mass. The first nine cycles of nuclear division are accompanied by migration of most of the daughter nuclei toward the periphery of the egg cell, and the final four divisions take place in the outer cytoplasm or periplasm (Turner and Mahowald, 1976; Zalokar and Erk, 1976). At the 10th division, approximately 20 nuclei at the posterior end of the embryo become enclosed along with the polar cytoplasm by cell membranes to become the first formed cells of the embryo. These "pole cells" later give rise to the germ line of the animal. (Thus, like *C. elegans*, the germ line originates from cells at the posterior end of the embryo, cells that originate with the posterior plasm of the egg.) At the end of cleavage the syncytial monolayer is converted into a cellular monolayer as cell

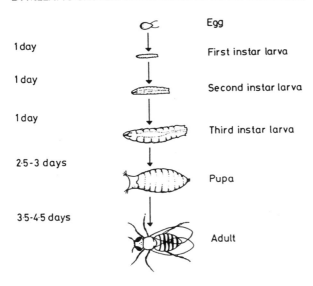

Fig. 4.1. Life cycle of *D. melanogaster* at 25°C.

membranes grow down and around the nuclei. The embryo is now at the cellular blastoderm stage. Blastoderm is reached approximately 3.5 hr after fertilization and marks the point of large-scale zygotic genome activation. The blastoderm stage is rapidly succeeded by a series of cellular invaginations and movements of cell sheets that constitute gastrulation. Further events of morphogenesis and differentiation, over a period of more than 17 hr, create the first instar larval form, which hatches from the egg case about 22–24 hr after fertilization.

The three stages of larval development are outwardly devoted to larval growth. However, the larva contains cells that later give rise to the structures of the adult fly; the larval growth period is also one of rapid multiplication of most of these cells. The presumptive imaginal cells are arranged in clusters in specific sites on the surface of the larval body in two kinds of cell groups, *imaginal discs* and *abdominal histoblast nests.* The imaginal discs, 17 in total, give rise to all the external head and thorax structures, the genital disc, and much of the musculature of the imago. The abdominal histoblast nests generate the outer surface of the abdominal segments of the adult (except for the eighth segment which is formed in part by the genital disc). The positions of the various imaginal cells groups and the structures they give rise to are shown in Figure 4.2.

The imaginal cell groups are set apart from the surrounding larval cells during the first half of embryogenesis, becoming detectable as small invaginated clusters at 9–10 hr by antibody staining specific for diploid epithelial cells (Bate and Martinez-Arias, 1991). By hatching, the disc and histoblast cells are visibly different from the surrounding larval cells in having smaller nuclei and an undifferentiated epithelial appearance. A fundamental cytological difference between imaginal and larval cells is that the former are diploid cells capable of repeated rounds of cell division, while the majority of larval cells grow without mitosis or cell division. The chromosomes of the larval cells continue to undergo rounds of replication, but these chromosomes remain undivided after each round of DNA replication, becoming multistranded or "poly-

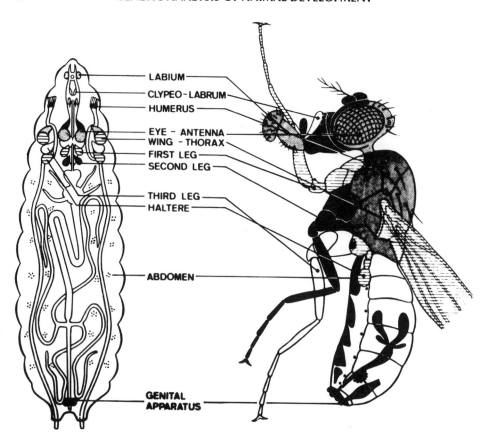

LABIUM
CLYPEO-LABRUM
HUMERUS
EYE - ANTENNA
WING - THORAX
FIRST LEG
SECOND LEG
THIRD LEG
HALTERE
ABDOMEN
GENITAL APPARATUS

Fig. 4.2. The imaginal discs and histoblast nests of *Drosophila*. The locations of the discs and histoblast nests in the third instar larva are shown on the left and their derivatives in the adult on the right. (Reproduced from Nöthiger, 1972, with permission of the publisher).

tene." Among the imaginal cells, the disc cells differ from the abdominal histoblast cells in that they divide during the larval stages; the histoblast cells only commence their cycles of cell division during the pupal stage.

During the pupal period, most of the larval cells undergo dissolution (histolysis). The form of the imago gradually takes shape as each imaginal disc progressively unfolds and differentiates. During the final hours of the pupal period, the various disc structures (eyes, legs, wings, etc.) knit together to produce the adult. The sequence of differentiative and morphogenetic changes is triggered at the end of the larval period by the secretion of the steroid hormone ecdysone from the larval ring gland.

Genetics and Genome Structure

The genetic system of *D. melanogaster* is conventional for an outbreeding animal. Its genome is organized into three pairs of autosomes and two sex chromosomes, females having the constitution 3AA + XX and males, 3AA + XY. The X chromosome is designated as the first chromosome, the two large metacentric chromosomes as the second and third chromosomes, and the comparatively tiny autosome as the fourth chromosome. The sex chromosomes have certain structural peculiarities. The X chromosome material to the right of its centromere (as the chromosome is conventionally drawn) is largely densely staining or heterochromatic and devoid of essential genetic functions. The Y chromosome appears wholly heterochromatic in all cells except spermatocytes but contains at least six loci that are essential for male fertility (Gatti and Pimpinelli, 1983). Apart from their chromosomal difference, male and female *D. melanogaster* differ in their capacity for recombination. In standard laboratory strains, males completely lack recombination, which takes place only in females.

More than eight decades of *Drosophila* work has produced a wealth of mutations affecting visible traits. Many of these mutant lines are of direct interest for developmental studies. These mutations range from those producing small differences in eye or body pigmentation, or in bristle form or pattern, to those producing profound developmental changes. Particularly interesting are the homeotics, which substitute one imaginal structure for another, e.g., a leg for an antenna. Thousands of mutants of *Drosophila* have been discovered and characterized; source books describing many are Lindsley and Grell (1968) and Ashburner (1989). As more and more genes are cloned and mapped, an increasing number of molecular markers are being added to the conventional genetic map, a situation that facilitates further characterization of the genome itself and of new genes (reviewed in Merriam et al., 1991).

The genetic nomenclature for *Drosophila* is more complex than that of *Caenorhabditis,* reflecting its longer history. Gene names are italicized, and recessive visible mutants are usually given a lowercase one-, two-, or three-letter notation that is an abbreviation of the phenotype description (e.g., r for *rudimentary*) and dominant mutations are given designations with an initial uppercase letter (e.g., *Cy* for *Curly wings*). The wild-type allele is given by a + superscript (e.g., r^+, Cy^+), and particular mutant alleles are also given by a superscript (e.g., r^{49}). In the past, lethals were generally denoted by an *l,* followed by the number of the chromosome bearing the lethal, and an individual allele designation. A recently introduced system for designating genes as a function of their position on the cytogenetic map (see below) is given in the appendix, along with the nomenclatural rules for the various kinds of chromosome aberrations and rearrangements. Shorthand designations for phenotypes and the proteins encoded by genes are not fully standardized but usually are given by the nonitalicized gene name, with the first letter capitalized (e.g., the *hunchback* [hb] protein is designated Hb, as is the phenotype). Enzyme names are given a three-letter designation in caps (e.g., the *Adh* gene encodes the ADH enzyme).

There are two reference systems for describing gene locations in *D. melanogaster.* The first system is that of genetic map distance, calculated from meiotic recombination frequencies, with respect to the ends of each chromosome. For example, the location of the lethal complementation group *l(3)S12* is 3–52 or about 52 map units from the left end of chromosome 3. The second system used to designate gene locations is cytogenetic. It is based on the distinctive appearance of the polytene chromosomes of the larval salivary gland cells. When these chromosomes are spread and stained, they are readily distinguishable by their large size and characteristic banded pattern (Fig. 4.3). The width of these chromosomes reflects the numerous side-by-side chromatids of which they are composed—up to 1024 chromatids by late third instar—and their length results from their relative absence of condensation. The

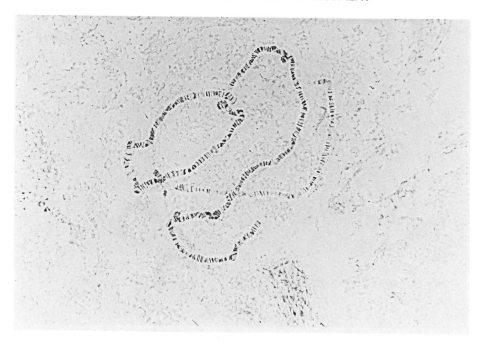

Fig. 4.3. Polytene chromosomes from the salivary gland of the third instar larva. (Photograph courtesy of Dr. M. Ashburner).

source of the banding pattern is unknown, but each chromosome region has a *distinctive banding pattern.* Furthermore, by using a variety of genetic strategems, it has been shown that each genetic function has a fixed location in a particular band or (less darkly stained) interband region. Each gene can be described by its site on the salivary gland chromosomes, and each salivary chromosome represents a magnified version of the euchromatic portions of a standard chromosome. (The heterochromatic portions of chromosomes, located around the centromeres, are substantially un-derreplicated in third instar polytene chromosomes.) In the nomenclatural system of Calvin Bridges (1935), the genome is divided into 102 sections. Each section consists of six subsections, labeled A to F, and each band within a subsection is given a number; gene locations are given in terms of band number. Thus, the gene for the enzyme alcohol dehydrogenase, *Adh,* maps to 35B2.3. The designation 2.3 indicates the resolution of the cytogenetic gene localization, in this case to within band 2 or 3 in section 35B. The relationships between the genetic and cytogenetic maps are summarized in Figure 4.4.

When one turns from the characterization of the genome by genetic or cytogenetic means to the exploration of its molecular organization, many questions arise. The fundamental problems are those encountered in the exploration of any eukaryotic genome—we have touched on some of them in considering that of *C. elegans.* They concern the number of genetic functional units in the genome, the nature of regional organization of gene groups within chromosomes, and the role of the various repetitive sequences that are scattered around the genome. Although much remains unknown, there is a substantial data base for approaching these questions in *Drosophila.* A good general, though slightly dated review of *Drosophila* genome structure can be

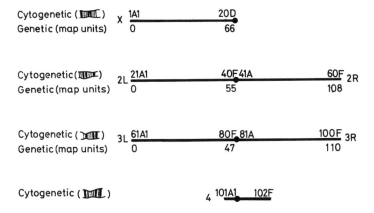

Fig. 4.4. The cytogenetic (top) and genetic (bottom) maps of the *Drosophila* chromosomes. See text for details.

found in Spradling and Rubin (1981), while a more recent overview of the major classes of sequence is given in Ashburner (1989).

Total DNA content and gross sequence composition have been determined by DNA renaturation studies. The haploid genome content is 1.7×10^8 bp and thus about twice that of *C. elegans.* Of this, approximately 74% renatures as single-copy sequences, 12% as mid-repetitive sequences, and 14% as highly repeated or "satellite" DNA. The last group of sequences comprises four major species, each consisting of a small oligonucleotide repeat present thousands of times. They are located predominantly or exclusively in the centromeric heterochromatin. The mid-repetitive sequences, in contrast, are scattered around the genome, although about 10% appear to be localized in the centromeric heterochromatin. About three-quarters of the mid-repetitive sequences are in the category of mobile or transposable elements, of which more than 40 families have been identified to date (Finnegan and Fawcett, 1986).

Much debate has centered on the question of how many genes exist in the *Drosophila* genome. The controversy stems from a disparity in the estimates produced by genetic means and those produced by molecular analyses. The genetic tests suggest that there is one genetic function, defined as a genetic complementation group, per (polytene chromosome) band (see Judd et al., 1972; Lefevre, 1974). Since there are approximately 5000 total bands in a polytene chromosome complement, this yields an estimate of about 5000 genes for the total genome. In fact, several bands contain more than one complementation group (Young and Judd, 1978), and a more realistic estimate of gene number, defined as the number of complementation groups, is probably 6000 to 10,000 (Judd, 1977).

However, the genetic tests detect only those genes whose products are required either for viability or fertility, or whose absence causes a visibly mutant phenotype. There are several reasons for suspecting the presence of genes whose presence would not be detected in these ways. In the first place, null alleles of a large proportion of enzyme-coding loci produce no phenotypic change (O'Brien, 1973). Flies that are completely deficient in these gene activities may be perfectly viable and fertile under standard laboratory conditions and show no visible effects. (Some of these genes might be essential for viability in the wild.) Secondly, saturation hybridization studies, designed to detect all mRNAs complementary to single-copy DNA, have revealed considerably more than 5000 distinct mRNA sequences; the numbers are closer to 15,000 for some stages and even tissues. These analyses will be discussed in more detail in chapter 7. The molecular evidence thus supports a total gene number considerably in excess of 5000, perhaps on the order of 10,000–15,000. A gene number of 15,000 or so would imply that the average gene size in *Drosophila* is approximately 10,000 bp, a fairly typical size for a eukaryotic gene. The molecular estimates are probably the more reliable ones of total gene number than the genetic ones, given the evidence that not every gene is strictly essential for viability.

A second question about genome organization concerns the placement with respect to one another of genes of related cellular function. A variety of different patterns have been observed. Some gene groups, such as the genes for the histone proteins and the ribosomal RNAs, exist in single clusters. Other gene groups are widely dispersed, and still others, such as the tRNAs, are partially dispersed and partially clustered. For the groups of genes that specify enzymes of single metabolic pathways, there is apparently no clustering. On the other hand, there are clusters of homeotic genes within a relatively small portion of chromosome 3; these genes will be described in detail in chapter 7.

The third general question about genome organization concerns the biological function, if any, of the various repeated, noncoding sequences. The highly repeated satellite sequences are almost certainly nonessential as they can be deleted in large blocks without major effects on viability, although they may have some role in promoting chromosomal pairing during meiosis. The importance, if any, of the mid-repetitive sequences for gene expression remains, to some extent, an open question but, as noted above, the great majority appear to be transposable elements and fully dispensable individually. (They have evolutionary significance, however, in that they provide a major source of spontaneously occurring mutants.)

OOGENESIS IN *DROSOPHILA*

Oocyte formation is a considerably more complex process in *Drosophila* than it is in the nematode, involving more cell and tissue types and a correspondingly more complex set of cellular interactions to produce the oocyte. The ovaries, which are connected via separate oviducts to the common oviduct, and the accessory organs of the female reproductive system in *Drosophila* are shown in Figure 4.5.

The ovaries have a composite cellular origin. The germ line cells, which include the oocytes, are formed from the pole cells, which are set aside during late cleavage and which then migrate to the rudimentary gonad. The cells that surround each developing egg chamber and make up the walls of the egg chambers are derived from somatic mesoderm. The tapered structures that encase the egg chambers are termed ovarioles and each ovary consists of a cluster of 10–17 ovarioles. Mature eggs are released from the posterior ends of the ovarioles into the oviducts, and fertilization occurs in the uterus. Because sperm can be stored in the spermatheca and seminal receptacles, fertilization of the egg may take place sometime after mating.

The tapered anterior tip of the ovariole is called the germarium and is the site of origin of the pro-oocyte. The pro-oocyte arises in a characteristic and complex set of

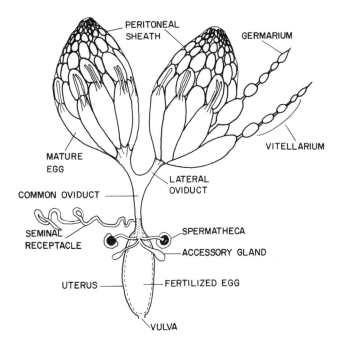

Fig. 4.5. Structure of the female gonads in *Drosophila*. The gonads are shown from the dorsal side; two ovarioles are shown on the right, plucked loose from the mass, and the uterus is shown in its distended, egg-carrying form. (Reproduced from Mahowald and Kambysellis, 1980, with permission of the publisher).

cell divisions that begins with an oogonial stem cell. Each stem cell first divides to give a daughter stem cell and a so-called "cystoblast" cell. Each cystoblast in turn gives rise to a set of 16 sister cells in four rapid sequential mitotic divisions, the set of 16 comprising a "cyst." One of these 16 cells becomes the pro-oocyte and eventually the oocyte. Its 15 sister germ line cells become "nurse cells," whose function is to synthesize materials to supply the growing oocyte. The 16-cell cyst, surrounded by a layer of somatic cells, is termed the egg chamber. The sequence of divisions, from stem cell to early cyst, is summarized in Figure 4.6.

The process by which one of the 16 cystocytes is singled out to become the pro-oocyte provides an illustration in *Drosophila* of a direct relationship between cell lineage and cell fate. The key to this relationship is in the pattern of intercellular "bridges," or channels, that connect the 16 cells of the cyst. These intercellular bridges are vestiges of incomplete cell division, each marking the site of an earlier division. Examination of the pattern and number of these bridges shows that eight cells within each cyst are connected to only one other cell, four cells are connected to two cells each, two cells are connected to three other cells, and finally, two cells have four separate bridges and are connected to four cells each. It is invariably one of the two four-bridge cells that becomes the pro-oocyte. Since each bridge represents the occurrence of an earlier division, it follows that the two cystocytes with four bridges were the first to be generated from the cystoblast. Both of these cells may have an initial capacity to form an oocyte, since both possess synaptonemal complexes (an essential structure for chromosome pairing) and commence meiosis. However, by the

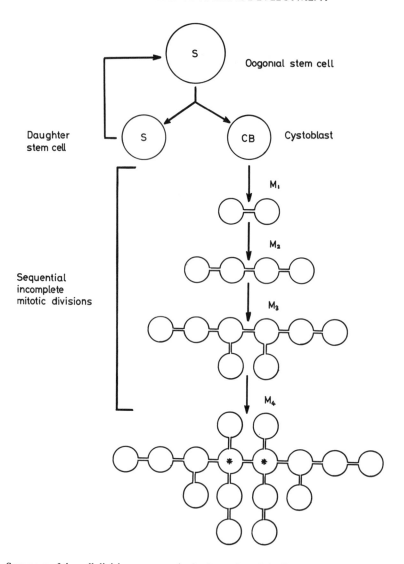

Fig. 4.6. Summary of the cell division sequence in the formation of the *Drosophila* egg chamber. The two four-canal cells are starred. Details in text. (Adapted from King et al., 1982).

time the 16-cell cyst has passed the middle of the germarium, one cell has aborted its meiotic progression, leaving a single pro-oocyte (Carpenter, 1975).

It is not known what factors determine the selection of the definitive pro-oocyte. However, the chosen cell is usually the most posterior. This might suggest that an oocyte "determinant," initially attached to the posterior cortex of the cystoblast parent cell, determines the selection (Brown and King, 1972). Alternatively, a determinative influence of the posterior position itself might somehow select one of the two four-bridge cells. In either event, the posterior positioning of the oocyte with respect to the nurse cells defines the a-p polarity of the egg chamber and later of the egg itself and of the embryo.

Each ovariole consists of two distinct regions, the germarium, in which the egg chambers are first formed, and the vitellarium, in which the major phase of oocyte growth and development takes place. The final stages of egg chamber development in the germarium involve the covering of the cyst by a monolayer of somatic "pre-follicle" cells. Within the first 30 hr of entering the vitellarium, these 80 somatic cells divide a further four times to give 1200 follicle cells, which constitute the definitive somatic cell covering of each cyst.

The development of the *Drosophila* oocyte is conventionally divided into 14 stages (King, 1970) (Fig. 4.7). During the first six stages, oocytes and nurse cells remain roughly the same size as each other, while increasing in volume by approximately 40-fold. Each volume doubling takes about 9 hr. During stage 7, the egg chamber elongates and growth temporarily slows. Beginning in stage 8, the process of vitellogenin (yolk protein) accumulation begins. Some vitellogenin synthesis takes place in the ovary (in the follicle cells) (Brennan et al., 1982), but, as in *C. elegans* and vertebrates, most of the yolk synthesis takes place outside the ovary. For *Drosophila,* the principal site of yolk protein synthesis is the fat body; the yolk polypeptides are released into the hemolymph and taken up by the developing oocyte.

From stage 8 onward, growth of the oocyte is rapid with respect to the nurse cells. Between stages 8 and 12, the oocyte increases in volume 1500-fold, doubling in size every 2 hr. Part of this increase is at the expense of the nurse cells, which, beginning in stage 11, empty their contents into the oocyte. As a result, by the end of oogenesis the entire interior of the egg chamber is oocyte. Stages 13 and 14 involve the secretion of the outer protective coverings of the egg—the vitelline membrane and the chorion—and the progression of the oocyte into metaphase of meiosis I. The oocyte remains in this stage until fertilization. The total increase in oocyte volume during development is approximately 90,000-fold. The mature oocyte, shown in stage 14, is considerably larger than the *Caenorhabditis* egg, measuring 140 μ × 500 μ.

The development of the oocyte involves two kinds of cellular interaction. The first set of interactions occurs between the oocyte and the nurse cells. In more primitive insects, the oocyte synthesizes a large part of its final mass. However, the fruit fly oocyte is largely the passive recipient of the biosynthetic activities of the nurse cells. The nurse cells themselves are polyploid, reaching 512 and 1024 times the haploid DNA content. This degree of polyploidy gives the nurse cells several thousand genomes for synthesizing needed gene products (proteins, ribosomes, mRNAs, etc.) for the oocyte. The oocyte nucleus is not completely inactive during the developmental sequence; it shows a brief period of transcriptional activity during stages 9 and 10 (Mahowald and Kambysellis, 1980), but the nurse cells provide by far the greater part of the store of gene products. Transfer of materials from nurse cells to oocyte is effected by means of intercellular channels, which are large enough to transmit mitochondria, but, in addition, there is some selectivity in the molecular species that are transported (Gutzheit and Gehring, 1979) and, correspondingly, some selective anchoring of substances in the anterior region (Berleth et al., 1988).

The second class of cellular interactions within the egg chamber are those that take place between the germ line cells—the oocyte and nurse cells—and the somatic follicle cells. These follicle cells also become polyploid, between stages 6 and 10, although to a smaller degree than the nurse cells. The requirement for polyploidy in

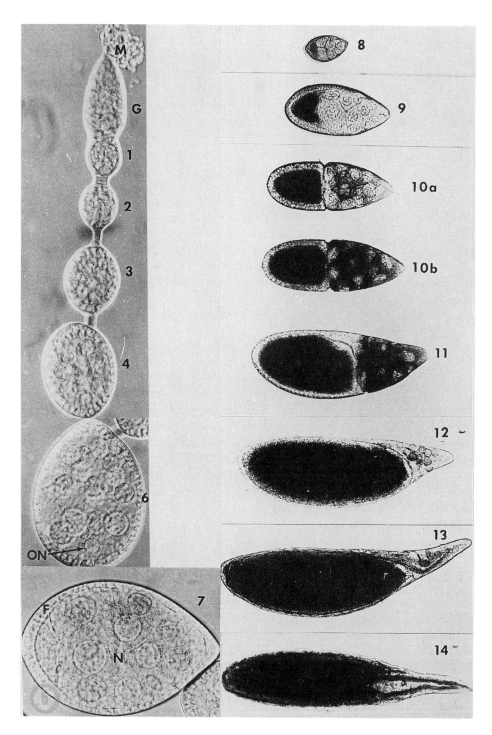

Fig. 4.7. Stages of development of the *Drosophila* egg chamber; the classificatory scheme is by King (1970). Stages shown on the left with Nomarski optics (×380); stages shown on the right are the vitellogenic stages, shown by bright-field microscopy (×100). The oocyte is on the left in the vitellogenic stages depicted. F, follicle cell layer; G, germarium; M, muscle sheath (stripped away); N, nurse cells; ON, oocyte nucleus. (Reproduced from Mahowald and Kambysellis, 1980, with permission of the publisher).

these somatic cells is presumably also related to biosynthetic needs: first, the secretion of the vitelline membrane and then, the layers of the chorion (the eggshell) by this enveloping cellular monolayer.

The follicle cells undergo a characteristic set of migrations between stages 7 and 13, which are significant for the final morphogenetic stages of egg development. Between stages 7 and 10, most of the follicle cells migrate posteriorly, coming to surround the oocyte within the egg chamber and leaving only 80 or so thin epithelial cells covering the nurse cells. During stages 9 and 10, a small group of follicle cells, the "border cells," move from the anterior end of the egg chamber *through* the cyst and come to lie just anterior to the oocyte. These cells later move back through the nurse cells to the anterior end. They secrete the anterior portion of the vitelline membrane of the oocyte and perhaps also the micropylar opening, through which the sperm later enter. The main covering layer of follicle cells first secretes the posterior vitelline membrane during stages 9 through 11. It then switches to secreting the successive layers of the chorion, the shell of the *Drosophila* egg, between stages 11 and 14. The follicle cells are also responsible for forming several distinctive chorionic structures, the two long appendages of the dorsal surface (see Fig. 4.26) and a chorionic plaque at the posterior end of the egg (Turner and Mahowald, 1976).

The chorion as a whole consists of more than 20 proteins (Waring and Mahowald, 1979), most or all being encoded in two gene clusters, one on the X and one on the third chromosome. Their synthesis in bulk during the relatively brief final stages of oogenesis is ensured by a process of preferential DNA amplication of these two gene clusters (a process termed "DNA amplification") within the follicle cells (Spradling and Mahowald, 1981; reviewed in Orr-Weaver, 1991). The completion of chorion synthesis endows the egg with a distinct a-p and d-v polarity (Fig. 4.26). These polarities foreshadow those of the embryo itself.

Viewed as a developmental pathway, oocyte formation is highly complex, involving numerous cellular interactions. The sequence of events is schematized in Figure 4.8. When one considers that the completion of most of the individual steps, each symbolized by a single line, involves and requires the activities of numerous genes, the extent of the underlying genetic complexity is apparent.

Indeed, much evidence indicates that the successful completion of oogenesis requires the expression on the order of 70–80% or more of all genes in the *Drosophila* genome. These estimates are derived from several studies in which female germ line clones are made homozygous for either deletions or lethal point mutants and scored for their ability to make functional eggs. Germ line homozygosity for these deletions or mutations is produced by X-ray–induced mitotic recombination (this technique will be described in more detail later) within a dominant female-sterile background, characterized by zero egg production (Garcia-Bellido and Robbins, 1983; Perrimon et al. 1984). The resulting clones are simultaneously homozygous for the mutation under test and for the wild-type allele of the dominant *Fs;* egg production signifies that the test mutations are in a gene that is dispensable in the female germ line. In an extensive survey of this kind, Perrimon et al. (1984) screened 48 lethal point mutants on the X chromosome and found that homozygosity in the germ line for only 13 (25%) permitted normal oocyte production; for the remaining 35 (75%), oocyte production was either prevented or abnormal eggs were obtained. As a number of the tested mutations were hypomorphs, and hence might

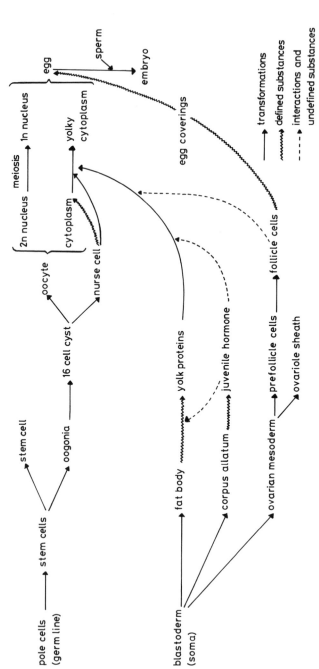

Fig. 4.8. Pathway of oocyte differentiation in *Drosophila*. (Adapted from King and Mohler, 1975).

be less severe in their effects in the germ line than amorphs would be, the estimate that 75% of all X chromosome genes are essential for normal development of the female germ line is probably an underestimate. The results apply specifically to genes on the X chromosome but there is no reason to believe that the fraction of required autosomal genes in the germ line will be lower than that for X-linked genes, particularly in the light of earlier, comparable studies which included autosomal regions (Garcia-Bellido and Moscoso del Prado, 1979; Garcia-Bellido et al., 1983). One further indication that this is so is simply that so many of the hypomorphic visible mutants of *Drosophila,* of which the majority are autosomal, are associated with some degree of infertility in females (for a discussion of this point, see Perrimon et al., 1986). The results, in sum, strongly indicate that in *Drosophila* as in *Caenorhabditis* (see chapter 3), a substantial fraction of the informational content of the genome is required for normal oogenesis.

Indeed, in general, the expressed gene set in oogenesis (for vertebrates as well as invertebrates), the "maternal gene set," is found to be a particularly large one (see Davidson, 1986, for review). It is, however, the genetic data from the nematode and the fruit fly that establish that a substantial fraction of the genome is actually *required* for oocyte development or early embryonic development or both. Furthermore, it appears that the great majority are not expressed exclusively in oogenesis but in other stages as well. From a comparison of those genes that can mutate rarely to give maternal effect lethals, as opposed to those that give a maternal effect predominantly as their mutant phenotype, Perrimon et al. (1986) estimate that the majority of genes in the genome can do so. In contrast, perhaps only 75 or so genes are expressed exclusively in oogenesis (Perrimon et al., 1986). This number is comparable to the dozen or so genes in the *C. elegans* genome estimated by Kemphues et al. (1988b) to be expressed solely in oogenesis. If the initial cues for patterning in the early *Drosophila* embryo are supplied primarily by exclusively maternally expressed genes, it would appear that their identification probably requires a saturation mutagenesis approach but that, given the small size of the set, it might not be too difficult to ascertain how they work as a whole. The results described later in this chapter bear out these suppositions.

The Drosophila Egg: Storehouse of Genetic Messages

The egg of *Drosophila,* like all mature oocytes, is rich in stored RNA molecules. The predominant RNA component is ribosomal RNA (rRNA), the bulk of which is transferred in the form of completed ribosomes to the oocyte during the period of nurse cell breakdown. However, a small but significant portion of the egg RNA is mRNA. Poly A^+ mRNA alone comprises about 2% of total egg RNA (Lovett and Goldstein, 1977; Anderson and Lengyel, 1979), and nonpolyadenylated mRNA may make a further contribution. The bulk of the maternally produced, stored mRNA is derived from transcription of nurse cell nuclei during the stages of egg chamber growth, but some fraction may be derived from the oocyte nucleus during the comparatively brief period of its activity in stages 9 and 10. The total informational complexity of the oocyte mRNA pool is equivalent to 1.2×10^7 bp of DNA, or about 10% of the single-copy DNA of the genome, and corresponds to approximately 8000 distinct protein coding sequences (Hough-Evans et al., 1980).

Some part of this mRNA pool, perhaps nearly all, is used to code for proteins that are required in early embryogenesis. A few of these mRNA species have been identified and their protein products are known to play obvious and important roles in cleavage. For example, a large supply of tubulin molecules is needed for the rapid construction of mitotic spindles during cleavage. Part of this need is met by stored supplies of tubulin protein subunits themselves, amounting to 3% of the total protein of the mature egg (Loyd et al., 1981), but these protein stores are insufficient to meet the demands for tubulin during early embryogenesis. To make up the difference, the egg also possesses a large store of tubulin mRNA, amounting to approximately 4% of the total poly A$^+$ mRNA population (Loyd et al., 1981; and see Davidson, 1986, p. 321). The histones are a second class of protein required in bulk during the rapid cycles of nuclear cleavage division; there is no evidence for stored histone proteins in the *Drosophila* egg, in contrast to the tubulins, but there is a substantial pool of histone mRNAs, amounting to approximately 2% of the total maternal mRNAs (Anderson and Lengyel, 1979).

The demand for translatable maternal mRNA of various kinds is particularly high during the cleavage divisions because of the rapidity of these divisions (see next section) and the negligible degree of zygotic genome expression during these divisions. Nevertheless, a considerable portion of the maternal poly A$^+$ mRNA remains unutilized during the first part of embryogenesis. By as late as 3.5 hr, the time of blastoderm formation, 33% of this mRNA pool is still unattached to ribosomes and found packaged in special storage ribonucleoprotein (RNP) particles, of sedimentation value 30S–50S (Lovett and Goldstein, 1977). This stored mRNA species seems to represent a surplus storage capacity for protein synthesis following zygotic genome activation.

After zygotic genome expression is activated at blastoderm formation, the mRNA pools become increasingly dominated by zygotic genome transcripts. It is of interest, nevertheless, to know whether some maternal mRNA messages continue to persist in the developing animal and make a *functional* contribution at postblastodermal stages. As we have seen, there is evidence that some maternal mRNAs in *C. elegans* are not utilized until the larval stages.

There have been relatively few explorations of this question in *Drosophila*, but one of note was performed by Gerasimova and Smirnova (1979). These investigators studied the male progeny or attached-X females; these progeny could inherit maternal X-encoded products but not a maternal X and were mutant on their paternally derived X for two X-linked enzymes, 6-phosphogluconate dehydrogenase (6-PGD) and glucose-6-phosphate dehydrogenase (G6PD). Any synthesis of either of these two enzymes in these male progeny must come, therefore, from maternally supplied templates. The time course of appearance of the two enzymes in these mutant male embryos was found to show a 25- to 30-fold increase of both enzymes. This increase took place between early embryogenesis and the late larval period, and all of the activity produced was of the maternal isozyme form. To determine whether the increases in activity might be the result of some form of progressive activation of stored inactive subunits, Gerasimova and Smirnova measured the absolute amount of G6PD protein by immunoprecipitation to determine whether traces of an inactive enzyme form could be found. The increase in enzyme activity was, in fact, however, found to be paralleled by an increase in enzyme protein, suggesting that the entire increase in activity was brought about by *de novo* synthesis. These results provide the strongest evidence to date that maternal mRNAs can persist and function until late in *Drosophila* development.

Cleavage, Blastoderm Formation, and Activation of the Zygotic Genome

Fertilization in *Drosophila* occurs as the mature oocytes are released into the oviducts. A single sperm enters the egg cytoplasm via a special channel in the dorsal anterior surface termed the "micropyle," and fertilization provokes the oocyte, suspended in metaphase of meiosis I until this point, to complete meiosis, yielding two polar body nuclei and the female pronucleus.

The haploid male and female pronuclei do not unite after completion of meiosis II but become closely apposed and undergo cleavage in parallel (Sonnenblick, 1950), this cleavage division being the first of 13 syncytial nuclear cleavage divisions that will take place (Zalokar, 1975; Turner and Mahowald, 1976). Immediately following this nuclear division, whose axis is oriented at random within the egg (Parks, 1936), nuclear fusion (syngamy) takes place between the two pairs of haploid nuclei to produce two diploid nuclei. As successive syncytial nuclear cleavage divisions occur, the daughter nuclei move outward through the yolk to the periphery, each enclosed in a jacket of cytoplasm. The propulsive force for this movement may be provided by the rhythmic contractions of the ooplasm that accompany cleavage (Fullilove et al., 1978), but the nuclear movements also require the microtubule system (Zalokar and Erk, 1976), as is the case in the early *C. elegans* embryo (Albertson, 1984). The entire sequence of cleavage divisions is diagramed in Figure 4.9.

The first nine divisions occur with great rapidity, approximately once every 8 min at 24°C (Rabinowitz, 1941). By the end of the 10th division, the nuclei have reached the surface (Foe and Alberts, 1983) and pooled their associated cytoplasmic coats to form a thickened periplasmic layer. Just prior to this, during the ninth division, a few (1–5) nuclei have reached the posterior pole, creating small bulges in the egg surface. These undergo two divisions, in synchrony with the other surface nuclei, and then become enclosed by cell membranes during the 10th division. Each of these pole cells then undergoes one or two further divisions to give a final number ranging from 23 to 52 (Turner and Mahowald, 1976; Underwood et al., 1980a; Technau and Campos-Ortega, 1986b). The pole cells are the initial germ line cells, the precursors of the gametes, and later migrate dorsally and then invaginate. (Subsequently, they become enveloped by mesodermal cells in the first stage of gonad formation.)

The surface syncytial nuclei undergo a total of four cleavage divisions in the periplasmic layer, each cycle being slightly longer than the preceding one, with the 13th division lasting approximately 21 min. A part of the increase in cycle duration reflects an increase in the length of the visible mitotic process, in which the nuclear membrane fragments and chromosomes condense and separate, but the predominant increase involves a progressive lengthening of the interphase period (Foe and Alberts, 1983). In the periplasm, each nucleus is the center of a discrete domain of microtubules and actin, as shown by immunofluorescent staining (Karr and Alberts, 1986). These final cleavage divisions at the surface of the embryo comprise the "syncytial

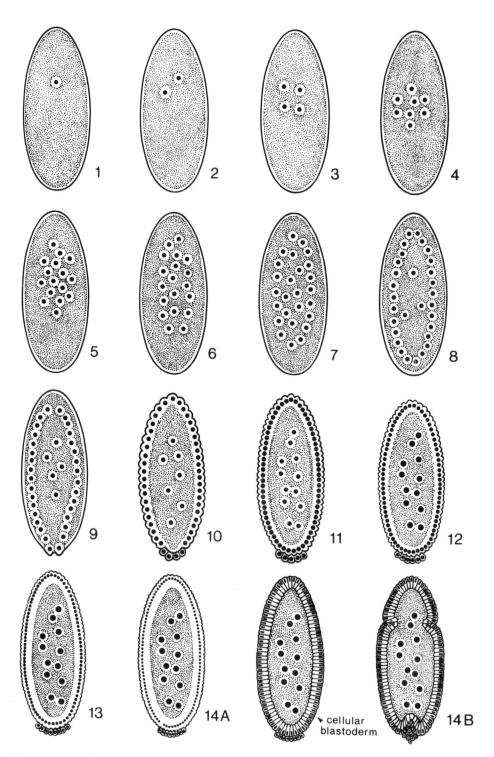

Fig. 4.9. Stages of nuclear cleavage divisions in *D. melanogaster*. The number to the right of each embryo represents its developmental stage and the total number of nuclear division cycles experienced by the dividing nuclei. Anterior at top, posterior at bottom. Solid black circles, nuclei; stippled areas, yolky cytoplasm; clear areas, yolk-free cytoplasm. See text for details. (Reproduced from Foe and Alberts, 1983, with permission of the Company of Biologists, Ltd.)

blastoderm" stage. During the 13th division, changes in chromatin structure appear, in particular the appearance of newly condensed heterochromatic regions; the number of these heterochromatic regions further increases during the 14th cycle, when cell membranes have begun to form (Foe and Alberts, 1985). Altogether, the 13 nuclear cleavage divisions, whose duration is approximately 2.5 hr at 25°C, populate the surface layer with an average of 6000 nuclei per embryo. Surprisingly, however, the final number of surface nuclei shows a wide range, from 5600 to 6500 per embryo (Turner and Mahowald, 1976). Such variability indicates that the fate assignment mechanism, discussed below, cannot act by allocating discrete numbers of cells to particular fates but must act on a regional basis.

At the end of the syncytial phase, membranes grow downward from the cell surface to enclose each nucleus within a long columnar cell, 6 μm wide and 30 μm deep (Fullilove and Jacobson, 1971). The process of cellularization is first completed in the ventral region and the developmental stage at which all surface nuclei have been enclosed is termed the "cellular blastoderm" stage. The cell division period following the 13th nuclear division is much longer than the preceding syncytial divisions, lasting a minimum of 65 min, though different regions within the embryo, as noted earlier, show characteristically different durations, forming discrete "mitotic domains" (Foe and Alberts, 1983; Foe, 1989). Shortly before cellularization is completed in the dorsal region, the embryo begins a series of cell movements and invaginations that give rise to the internal organs and the segmented form of the larva (chapter 7).

Traditionally, the cellular blastoderm stage has been regarded as a key transition point in the embryogenesis of *Drosophila*. First, and most obviously, it marks the boundary between the syncytial and cellular phases of development. Second, it marks an approximate boundary between the early, maternally directed phase of embryogenesis and the period in which the zygotic genome comes to dominate the course of development. This critical transition in fact begins during the last stages of the syncytial blastoderm and takes the form of a rapid activation of transcription in the previously quiescent embryonic nuclei. This zygotic genome activation, furthermore, is necessary for the completion of blastoderm formation: Eggs injected with the RNA synthesis inhibitor α-amanitin at the end of the ninth cleavage division fail to cellularize (Gutzheit, 1979).

Despite the impression conveyed by much of the earlier literature of a massive and complete activation of transcription just prior to cellular blastoderm, there is in reality an increasing level of zygotic genome expression during syncytial blastoderm and a small degree of some zygotic transcription in the earlier stages. Examination of spread chromatin from early embryos shows a low but significant frequency of ribonucleoprotein fibril gradients in early embryos, signifying the existence of early transcription (McKnight and Miller, 1976). These first transcriptional events can be detected as early as the 64-nuclei stage and can be identified as nonribosomal transcription units by their fibril morphologies. Measurements of newly labeled transcripts indicate that one of the first major classes of nuclear gene to be activated are the histones (Anderson and Lengyel, 1980), whose expression commences in cycle 11 (Edgar and Schubiger, 1986), while other classes show a progressively increasing transcription rate from cycle 12 onward, with maximal expression being achieved during formation of the cellular blastoderm (Edgar and Schubiger, 1986).

Given the slight extent of transcription in the early cleavage stages (divisions 1–10), it is tempting to conclude that this initial zygotic genome expression is unimportant for embryogenesis. Nevertheless, some reports in the early literature of zygotic genome effects in the first stages of cleavage (reviewed in Wright, 1970) have not been refuted and, more recently, convincing evidence of defects in early cleavage and pole cell formation in homozygotes for certain lethal mutations in the gene *engrailed* has been presented (Karr et al., 1985). (The *engrailed* gene plays a role in other developmental processes as well, as will be discussed in chapter 7.) Altogether, however, the exact number of loci whose zygotic expression is significant prior to the final cleavage divisions is unknown. The bulk of the evidence, however, suggests that the number is likely to be small and that the events of this phase are substantially under the direction of maternal genome-specified products.

An important question concerns the nature of the trigger for zygotic genome activation. The key events appear to be the achievement of a critical nucleo-cytoplasmic (n-c) ratio, which provides the signal for cellularization (Edgar et al., 1986) and the gradually lengthening interphase periods that take place during the final four syncytial cleavage divisions in the periplasm, which serve to activate zygotic genome transcription (Edgar and Schubiger, 1986). The role of the n-c ratio in determining the onset of cellular blastoderm formation was demonstrated by the discovery that in haploid embryos, whose n-c ratio is one-half that of normal diploid embryos at each cleavage division, there are 15 rather than 14 cleavage divisions. (In these experiments, haploid development was induced by using either of two mutations. Since one of these creates a paternally inherited defect in male pronuclear development, the additional cleavage division in the haploid embryos cannot be attributed to a maternal effect [Edgar et al., 1986].)

The role of interphase period length in regulating zygotic genome transcription was demonstrated by injecting cycloheximide and cytosine arabinoside, a treatment that inhibits both protein and DNA synthesis and thereby artificially prolongs the interphase periods. When this regimen is carried out after nuclear division cycle 10, general activation of transcription was found to occur (Edgar and Schubiger, 1986). However, as this treatment does not activate transcription if carried out prior to division 10, it appears that some protein synthesis is required at this division to permit transcriptional activation in the somatic nuclei.

Interestingly, the pole cell nuclei, which are physically sequestered from the somatic cells prior to cellular blastoderm, remain untouched by the wave of transcriptional activation, persisting throughout the cellular blastoderm stage in a state of transcriptional quiescence (Zalokar, 1976). The pole cells, however, are particularly active sites of protein synthesis from the time of their formation (Zalokar, 1976). Indeed, overall protein synthetic rates vary little throughout these early stages of *Drosophila* development, even taking place in unfertilized eggs (Zalokar, 1976). Thus, the shift from maternal mRNA templates to zygotic genome transcripts is not marked by any major changes in general translational rates. Nevertheless, the cytological transition from syncytial to cellular blastoderm marks a significant point in the developmental history of the embryo, the end of the period of dominance of maternal gene products. This earlier period, however, has set the stage, in an important respect, for what follows; as the evidence reviewed in the next section shows, not only are cells of the blastoderm destined for certain fates but firm developmental commitments have been impressed on the cells, and possibly the nuclei, of the embryo by this stage.

FATE MAPS AND DETERMINATIVE STATES IN THE BLASTODERM EMBRYO

Although the approximately 6000 somatic cells of the cellular blastoderm appear to comprise a uniform, homogeneous cell population, the reality is that it is a quiltwork of well-defined developmental destinies and capacities. Fate mapping procedures have been used to identify which groups of cells will give rise to particular parts of the embryo and the imago, while various transplantation experiments have been used to determine the range of capacities of cells from particular regions. We will first take a look at the fate mapping results, but a central conclusion to be noted at the beginning is that the fate maps are, to a large extent, congruent with the landscape of developmental capacities. At the *Drosophila* blastoderm stage, cells not only have particular fates but have already been substantially restricted in their potencies.

There are four general categories of the fate mapping procedure: 1) direct and careful microscopic observation of cells during development; 2) marking cells by dyes or other biochemical markers and tracing the fates of the cells so marked; 3) inducing localized cellular damage and ascertaining which structures or regions are missing (a "defect map"); 4) genetic marking, in which a scorable genetic difference is introduced into one or a few cells and their descendants detected by the marker. All four kinds of method have been used with *Drosophila*, producing a large set of complementary, though generally matching, results.

The classical fate map of the embryo was produced by Donald Poulson, one of the pioneers of *Drosophila* developmental genetics, who constructed it on the basis of direct observation of the developing embryo (Poulson, 1950). It is depicted in Figure 4.10 and shows that each region of the blastoderm cellular monolayer can be described and located by a pair of Cartesian coordinates. The longitudinal or a-p coordinate designates segmental fate, while the transverse or d-v coordinate designates germ layer (tissue type) fate.

The second method of fate mapping that has been used employs the induction of localized defects. In embryos possessing little capacity for regulatory replacement of cells, a selective ablation of early cells results in a specific deficiency or range of deficiencies in the surviving embryos. The induced-defect methods therefore can only be used to make a fate map where some localized restrictions on capacity have already been imposed. The potential drawback in induced-defect experiments is that the defects themselves may interfere with the developmental responses of the surviving cells.

One of the first comprehensive induced-defect fate mappings involved selective cell ablation with a UV microbeam. By focusing a microbeam of 15 μm diameter, sufficient to kill about 15 cells at a time, Lohs-Schardin et al. (1979b) induced localized defects in blastoderm embryos and then mapped the sites of damage in the surviving first instar larvae. The mapping of first larval thorax and abdominal precursor cells along the a-p axis of the blastoderm by this method is shown in Figure 4.11. Taking

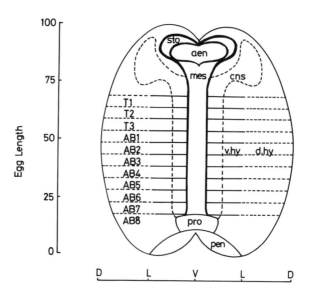

Fig. 4.10. Blastodermal fate map for the *Drosophila* embryo. Projection drawn for an egg cut down the dorsal midline and folded out. The ordinate shows the relative distance, in egg length (EL) from the posterior end (0 EL) to the anterior end (100 EL); the abscissa indicates the position along the d-v axis. AB1–AB8, abdominal segments; aen, anterior endoderm; cns, central nervous system; d.hy, dorsal hypoderm; mes, mesoderm; pen, posterior endoderm; pro, proctodaeum; sto, stomodaeum; T1–3, thoracic segments; v.hy, ventral hypoderm. (Adapted from Nüsslein-Volhard, 1979b.)

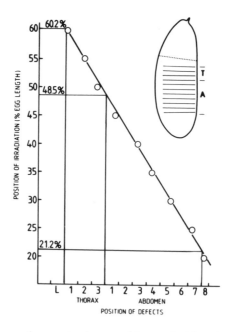

Fig. 4.11. Induced-defect map of segment anlagen position on the blastoderm-stage embryo. From the observed centers of defects, for each position of irradiation, the map (inset) was constructed. See text for discussion. (Reproduced from Lohs-Schardin et al., 1979b, with permission of the publisher.)

the posterior pole as 0% egg length (EL) and the anterior pole as 100% EL, the entire larval thoracic and abdominal segment precursor regions are found to fall between 20% and 60% EL, thus extending only 40% of the length of the blastoderm.

A second method involved selective cell removal, in which a micropipette was inserted into the embryo surface and small cell patches were extracted. Using this method, Underwood et al. (1980) deleted lateral cell groups of 15–30 cells at a time, at the cellular blastoderm stage and in slightly older embryos, scoring the sites of damage in 9 hr embryos by scanning electron microscopy. They found a similar blastodermal placement for the thoracic and abdominal regions as obtained in the laser beam ablation experiments. In addition, they localized the head segments to between 60% and 70% EL and two putative vestigial abdominal segments A9 and A10 (which are not distinguishable as separate segments in adult flies) to between 20% and 15% EL approximately. The entire larval epidermis is therefore apparently prefigured along 55% of the a-p axis and blastodermal cells outside these limits cannot replace cells lost between them. Although multiple segment defects were found to be the rule with both kinds of induced defects, the defects always involved adjacent segments. Evidently, precursor cells for consecutive segments are arrayed consecutively within this region. The estimated distance between larval segment anlagen on the blastoderm surface was three or four cell diameters.

Lohs-Schardin et al. (1979b) also measured the d-v extent of the larval epidermal precursor region on the blastoderm. The results showed that damage can be induced throughout each half circumference except for the midventral 20% and the middorsal 4%. The latter areas correspond to those occupied by the cells that give rise to the midventral (mesodermal) invagination and the amnioserosal membranes, respectively. Therefore, roughly 76% of the egg circumference, within the determined a-p limits, gives rise to the area of segmented larval epidermis. Altogether, each lateral epidermal anlagen is approximately a square of 200 µm × 200 µm on the blastoderm embryo. From cell counts at this stage, a region of this size corresponds to about 1100 cells, or 2200 cells total for both sides, this number being just slightly more than a third of the total number of blastodermal cells. Part of the remainder goes to form the epidermis of the cephalic segments and the terminal region (the putative A9 and A10 segments), but the greater part, approximately 3000 cells, goes to form internal tissue.

It is possible to do a comparable fate mapping of the blastodermal cell precursors of imaginal regions and structures using UV microbeam-induced damage, though the interpretation is somewhat complicated by the mortality that often intervenes between damage induction and eclosion (Lohs-Schardin et al., 1979a). The results indicate a high degree of congruence between the larval and imaginal fate maps: Thus, comparable thoracic primordia are damaged in the same positions, along the a-p axis and the average space between leg primordia is about three cell diameters, the same distance as between larval segment anlagen. These results, and others noted below, are compatible with the notion that not only do the imaginal precursor cells arise early and within a sea of prelarval cells but share their initial segmental or regional character with their prelarval cell neighbors.

In contrast to induced-defect mapping, genetic fate mapping permits the relative localization of different precursor cells without physical trauma to the embryo. In principle, it can define boundaries between anlagen more precisely than is possible with surgical methods. The genetic fate mapping of the larval surface has confirmed and extended the results of the induced defect methods.

To genetically fate map the structures of the larval surface in terms of their origins in the blastoderm stage embryo, one needs both a suitable genetic marker, one that can be scored throughout the larval epidermis, and a method for inducing clones in the epidermis. The fate mapping method that is suitable for the *Drosophila* embryo is that of gynandromorph analysis, originated by A.H. Sturtevant in the 1920s. Sturtevant had previously invented the basic method of gene mapping itself, from measured frequencies of recombinant progeny, while working as an undergraduate assistant in T.H. Morgan's laboratory in 1911. His development of gynandromorph fate mapping more than a decade later was, if less momentous than the discovery of genetic mapping, no less insightful.

Gynandromorphs (or gynanders) are animals that are part female and part male. The male sector, which may comprise as much as 50% of the animal but is often less, is usually a coherent patch or region and may occupy any part of the fly. When the male sectors are suitably marked (see below), this variety of patterns can be readily seen; a diagram of some gynandromorph patterns is shown in Figure 4.12. The first *Drosophila* gynandromorphs were reported by Morgan and Bridges (1919), who showed that the female sectors are XX in chromosome composition and the male sectors XO.

To understand how gynanders can be used to fate map structures or regions of the larval surface, one must understand how they arise. As Sturtevant surmised, the origin of the gynandromorphs is in the loss of one X chromosome, in an XX zygote, from one (or possibly more) of the nuclei during early cleavage. The part of the blastoderm populated by the descendants of the XO nucleus so formed will all be XO and hence chromosomally male. The fact that the male sectors comprise discrete regions indicates that the XO nuclei travel as a group, rather than dispersing, as they head toward the blastoderm surface. Furthermore, since any early nucleus serendipitously distinguished by the loss of an X can give rise to any set of structures on the surface of the adult fly, it follows that the initial orientations of the nuclei must not be fixed and that the developmental fates of these cleavage nuclei are not specified.

The adaptation of the facts of gynander formation for fate mapping follows from Sturtevant's realization that, since the dividing line between male and female tissue can fall anywhere, the line is much less likely to fall between two structures whose progenitor cells are close together at the time of cell fate specification than if they are far apart. Imagine, for instance, that one has a ball (the blastoderm embryo) that is densely decorated on its surface with a variety of distinctive elements (larval or adult structure anlage) and that one neatly slices the ball through the middle in any one of an infinite number of possible but randomly chosen places (the dividing line between XX and XO cells). Surface elements that are close together have an inherently smaller probability of being separated by the plane of separation than those that are far apart. In the absence of secondary complications, the frequency of separation of different elements should provide a *measure of distance between them*. For the *Drosophila* blastoderm embryo, the distance between anlage for different structures is therefore indicated by the *frequency with which the mosaic border falls between these different structures*. To perform the analysis, one needs only a genetic marker that unambiguously distinguishes all XO tissues from all XX tissues and a high frequency of gynandromorphs.

To fate map the portion of the blastoderm embryo that gives rise to the segmented region of the larva, and to estimate the size of individual segment anlage, Szabad et al.

Fig. 4.12. Diagrammatic representation of a sample of adult gynandromorphs generated by loss of ring-X chromosomes. Male (XO) areas indicated by shading. (Reproduced from Janning et al., 1979, with permission of the publisher.)

(1979) carried out a gynandromorph fate mapping based on the above considerations. While Sturtevant (1929) had employed the nondisjunctional property of the *claret* mutant of *D. simulans* (now termed *ca*nd for *claret nondisjunctional*) to generate gynandromorphs (this mutant causing a high rate of maternal X chromosome loss in early embryos), the method of choice today in *D. melanogaster* involves the use of an unstable X chromosome, the ring-X chromosome *In(1)w*vc. This is a fully functional X chromosome that has lost the small, genetically inert right chromosome arm and has the tip of the major (left) arm joined back to form a ring. During the first few cleavage divisions of the early embryo, the ring-X is unstable and tends to be lost, producing large XO patches. For postblastodermal development, the chromosome is apparently as mitotically stable as normal (rod-X) chromosomes, though it may be that cells with later mitotic losses are simply inviable due to the "wrong" (female) setting of dosage compensation for XO cells (see chapter 7).

The loss of the ring-X during early cleavage permits the generation of many gynanders. If, for instance, the loss occurs in the second cleavage division, as two nuclei are giving rise to four, the outcome can be diagramed as follows, where only the X chromosomes are shown:

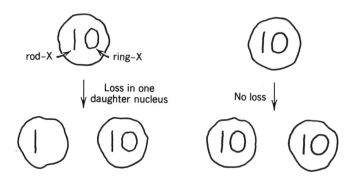

The resulting embryo will be one-quarter XO tissue, assuming that there is no preferential multiplication of male or female cells.

The ring-X system permits a fate mapping of the larval epidermal surface. If one constructs ring/rod-X heterozygotes, in which the ring-X carries a *mal+* allele and the rod-X carries a *mal−* allele, then ring-X loss produces XO *mal−* nonstaining regions on the larval surface. Szabad et al. (1979) performed such an analysis on 152 third instar larval gynandromorphs, obtained from 2231 larvae in a cross of *mal−* rod-X females by *In(1)w^vc*, *mal+* males. Of these 152 mosaics, 140 could be scored; Figure 4.13 shows the distribution of enzyme activity in the epidermis of 48 of these specimens. For the most part, there is a single male and a single female patch, but 18% of those analyzed had two male or female patches. (For a discussion of the possible significance of double losses of the ring-X, see Zalokar et al., 1980).

The frequencies with which different segments or cell groups are separated by the mosaic borders can be taken as a measure of the intervening distances of their precursor cells on the blastoderm surface, as explained above. The measure of distance is the percentage of mosaics in which the two segments (or cell groups) fall on opposite sides of the border. The unit of distance is taken as 1% separation and has been termed the "sturt," in honor of Sturtevant (Hotta and Benzer, 1973). Thus, if two cell groups fall on opposite sides of the male/female border 4% of the time, they are said to be four sturts apart.

The analysis produced several points of interest. In the first place, borders were, in general, found to separate different contiguous sections of the larval epidermis with nearly equal frequency. The measured distance between grid sections was 4.5 sturts and the probability of a single cell interface being mosaic was 2.2 sturts. Such even spacing of segment precursor regions is consistent with the results of the induced-defect experiments. Secondly, the size of the blastodermal precursor area was estimated at between 1630 and 2400 cells, which is satisfyingly close to the estimate of 2200 cells derived from the defect mapping experiments.

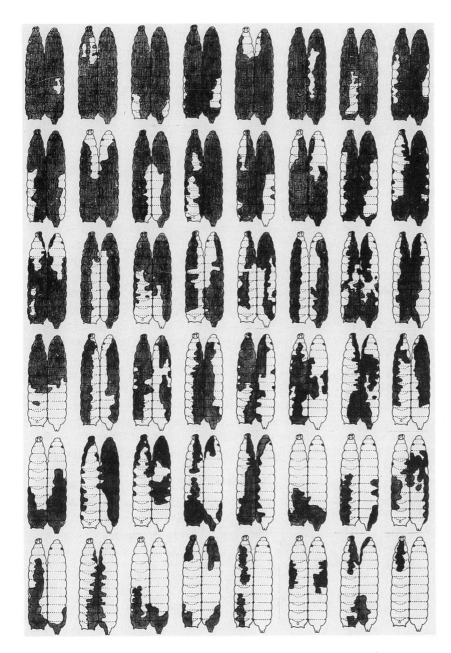

Fig. 4.13. Larval gynandromorphs. Distribution of XO (*mal⁻*; white) and XX (*mal⁺*; dark) cells in the epidermis of 48 third instar larval gynandromorphs. (Reproduced from Szabad et al., 1979, with permission of the publisher.)

However, the most interesting finding from the genetic larval fate mapping pertains to the relationship between larval and imaginal cells in the abdomen. The imaginal abdominal cells are the histoblasts, cells that are quiescent during larval development and which only begin extensive proliferation during pupation. In all the gynandromorphs scored in this experiment, the *mal* phenotype of the histoblasts was never found to be different from that of the surrounding larval cells. In those cases where the histoblast nest was cut by a mosaic border, the border was observed to run smoothly into the neighboring larval cellular region. The observations suggest that the histoblasts are singled out from a sheet of common precursor cells for both histoblast and larval cells. It appears highly probable that the other groups of imaginal cells (of the thoracic and cephalic regions) are similarly derived from a precursor cell pool within each segmental region that also produces larval cells. The nature of the assignment process that allocates larval and imaginal fates within such precursor cell pools is, at present, completely unknown.

The fate maps, in total, give a consistent picture of what the different localized regions of the blastoderm will give rise to in the embryo. By their nature, however, they do not provide direct information on states of developmental commitment. Nevertheless, one other set of direct observations and measurements indicates that neighboring regions delimited as different on the fate map are already somewhat different in their properties. Careful observation of living embryos shows that the distinctiveness of these fates is prefigured in nearly all cases by the timing of onset of mitosis in groups of contiguous cells (Foe, 1989). Altogether, during the first postblastodermal cell division, 25 "mitotic domains" can be identified, most consisting of left and right members, a few straddling the dorsal or ventral midlines (Foe, 1989). Indeed, there are more such synchronous cohorts than estimated regions of difference in the conventional fate map, suggesting that the initial array of commitments is even more diverse than suggested by the map. The precise relationships between these differential timings of mitosis and states of commitment are unknown, but embryos homozygous for the mutation *string*, which blocks all postblastodermal mitoses, nevertheless exhibit essential normal gastrulation and initial differentiative events (Edgar and O'Farrell, 1989), showing that mitosis *per se* is not essential for expression of the first commitments.

Whatever the precise significance of the mitotic domain map, one would like to know whether the regions singled out correspond to states of developmental restriction, that is of "determination." To assay states of determination, cells must first be removed from their normal locations, cultured for several cell generations and then allowed to express their developmental capacities; only if the progeny cells form the structures that they would have given rise to *in situ* can the tested cells be regarded as determined.

The first experiment to test for states of determination in blastodermal cells was carried out by Chan and Gehring (1971). They assayed anterior and posterior halves of blastoderm embryos to establish which structures could be formed by each half of the embryo, bisected across the a-p axis at approximately 50% EL. The procedure involved three steps: Blastoderm embryos from a genetically marked strain were first bisected into anterior and posterior halves; the cells of the two sets of half embryos were then cultured separately *in vivo* and exposed to conditions which provoke differentiation of imaginal structures; and finally, the tested implants were scored for their inventory of imaginal structures.

To culture blastoderm cells, the embryos were dissociated by gentle homogeniza-tion in a balanced saline solution and then injected into the abdominal cavity of wild-type female adult flies. This *in vivo* culture method allows prospective imaginal cells to grow without differentiating. To obtain subsequent differentiation of the cultured pieces, implants were removed from the adult hosts, after a growth period of 10–14 days, and placed into the body cavities of early third instar larvae. Exposure to ecdysone during pupation of these host larvae stimulates differentiation of the implants and the tested fragments are then directly scored for the imaginal structures they give rise to. To ensure that absence of a particular disc type in the cultured implants was not an artifact produced by the procedure, Chan and Gehring first combined the anterior and posterior halves with a comparable mass of whole blasto-derms of a second distinctive genotype before homogenization (Fig. 4.14). With the difference in cuticular markers of the test and control cells, any structure could be assigned unambiguously either to the tested half blastoderms or to the control embryos. (The host animals, both larvae and adults, in which the implants were grown, were of wild-type genotype, permitting clear delineation of donor and host cell contributions).

Anterior half blastoderms, with the exception of one doubtful case, yielded only head and thoracic imaginal structures (44 of 45 implants), and posterior halves produced only abdominal and thoracic derivatives (26 classifiable implants), consis-tent with the general fate map of the imago (Lohs-Schardin et al., 1979a). The Chan-Gehring experiment thus provided the first demonstration that some degree of determination, at least to "anteriorness" or "posteriorness," has occurred by blasto-derm.

Several experiments indicate that, in fact, determined states are even more pre-cisely specified at cellular blastoderm, with respect to both segmental fate and germ layer fate. Illmensee and Mahowald (1974), using genetically marked donors and

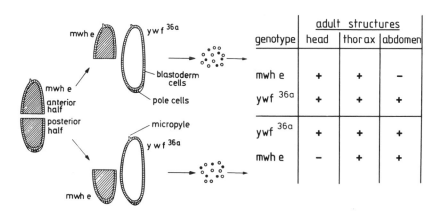

Fig. 4.14. Preparation of anterior and posterior half blastoderms for testing of determinative states. Genetically marked embryos (*mwh e*) at the blastoderm stage are bisected, and the anterior and posterior halves are separately mixed with whole carrier embryos (genotype: *ywf*^{36a}). The mixtures are dissociated and filtered to make cell suspensions, which are concentrated by centrifugation and injected into adult females for *in vivo* culture. The developmental contributions of experimental and control cells are shown on the left. (Figure courtesy of Dr. W.J. Gehring.)

hosts, found that when groups of a few anterior blastoderm cells were placed within the posterior polar region of recipient blastoderm embryos, a small number of anterior imaginal implants (labial, eye, antennal) were recovered from the body cavities of the surviving hosts following metamorphosis. Illmensee (1978) later reported comparable results with genetically marked *single* blastodermal cells, in which the markers permitted tracing of both imaginal (cuticular) derivatives and larval derivatives (the marker used for the latter being mal^+, which is detectable by histochemical staining). In these experiments, anterior and posterior cells were found to retain, respectively, anterior and posterior segmental states when placed in novel (heterotopic) locations. Interestingly, he also found that midventral cells, which normally give rise to the mesodermally derived muscle tissues, exhibited some developmental plasticity, being capable of giving rise to fat body, Malpighian tubules, or midgut cells depending on the site of implantation (Illmensee, 1978).

The most extensive tests of blastodermal cell state with respect to imaginal fates, however, were performed by Simcox and Sang (1983), who transplanted individual groups of six marked blastodermal cells to either comparable (homotopic) locations or different (heterotopic) ones along the a-p axis. In all cases where the implants could be scored unambiguously, the developed implants displayed adult cuticular structures characteristic of the site from which they came rather than the host site in which they were placed.

Most of the findings described above pertain to the acquisition of determined states along the a-p axis, and hence of segmental identity, but there are, in addition, results that suggest that somatically heritable states have also been imposed along the d-v axis, with respect to histotype. The method employed does not involve genetic marking but the use of chemical cell lineage tracers, in particular the enzyme horseradish peroxidase (HRP), which can be detected following histochemical staining in sections, or fluorescein dextran (FITC-dextran), which can be scored in living embryos by means of fluorescence. The method was developed by Technau (1986) and has been used to trace the fate of single cells from postblastodermal embryos (from the early gastrula stage, when the previously existing cytoplasmic connections between nonsister cells have been severed). Syncytial embryos are injected and perfused with the marker solutions, and then single cells are transferred to either homotopic or heterotopic recipient embryos at either the same or later stages. The results indicate that commitments to broad germ layer type (ectodermal, endodermal, mesodermal) have been imposed by the early gastrula stage (prior to the first postblastodermal cell division) and are not reversed by transplantation to novel sites (Technau and Campos-Ortega, 1986 a,b; Beer et al., 1987; and see review by Technau, 1987).

Determination, as measured by the capacity to retain a characteristic developmental fate following placement and growth in a new location, is traditionally defined (because of the nature of the experimental operations) as a cellular property: The tests are carried out on intact cells and these are found either to be restricted to certain developmental paths, as seen in their descendants, or not so limited. However, if one wants to understand the molecular basis of the phenomenon, it is crucial to know whether there is a set of nuclear properties that correlate with the cellular behavior. Presumably, determination involves some restriction in capacity of gene expression, and, indeed, as mentioned earlier, there is a progressive heterochromatinization within the nuclei, a change which is suggestive of a measure of gene inactivation, in nuclear cycles 13 and 14 (Foe and Alberts, 1985). Questions about determination may therefore be intimately connected to such changes in nuclear state and how they are created.

The classical approach to the question of nuclear states in determination has been by means of nuclear transplantation. Nuclei from somatic cells at some stage of development are placed in an unfertilized or fertilized egg and their capacity to participate in subsequent developmental events is tested, usually in place of the host oocyte or zygote nucleus. (In the syncytial embryos of insects, a variant of this approach is to add the donor nucleus to the host embryo without removing the host nucleus; successfully developing embryos are nuclear transplant chimeras.) If the transplanted nucleus can give rise to any adult structure, it is said to be "totipotent" and inferred to have experienced no irreversible genetic changes.

The principle of nuclear transplantation experiments is easy to describe, but the interpretation of the results is not always straightforward, particularly for the majority of test nuclei which do *not* successfully supplant the zygote nucleus in development. Most of these failures are attributable to the failure of the donor nuclei to "adjust" to the cytoplasmic environment of the egg, in particular to the imposed rapid rate of replication that takes place in the egg. Damage during manipulation might also be a contributory cause. However, in principle, some indeterminate proportion of the failures may be produced by irreversible genetic change or by some rigid form of chromosome modeling that cannot be erased by the egg cytoplasm.

In *Drosophila,* numerous nuclear transplantation experiments have been performed (reviewed by Illmensee, 1978). These experiments have typically involved taking genetically marked nuclei from one stage or location, placing them in novel locations in developing host embryos, with different genetic markers, of the same or different stages, and establishing, from the pattern of derived donor adult structures, whether the donor nuclei can give rise to the full range of possible structures or whether developmental restrictions in these nuclei can be detected. Significantly, it is found that nuclei from blastodermal or postblastodermal cells have the same range of developmental capabilities as nuclei from cleavage stage embryos. Since some degree of cellular determination has taken place by blastoderm, as shown by all the experiments described above, this finding of nuclear totipotency seems to suggest that the process of determination involves no irreversible alteration of nuclear state but is "carried" solely in the cytoplasm. In this view, the nuclear state that accompanies cellular determination is a highly labile one, perhaps fully comparable to the transient regulatory states that exist in prokaryotes.

This inference does not necessarily follow. Determination might well involve the tight binding of regulatory molecules and/or certain kinds of chromosomal folding patterns, affecting gene expression. In this circumstance, self-perpetuation of the determined state would involve recurrent synthesis and binding of the regulatory molecules and/or self-templating of the chromosomal folding pattern (Weintraub, 1985). If a nucleus whose chromosomes possess such properties is placed in a cytoplasm that lacks the requisite molecules for perpetuating this state, and then required to replicate, its chromosomes will lose their distinctive properties through simple dissociation of the bound molecules or by titration in successive rounds of replication. This may well be the situation experienced by a somatic nucleus placed in the very different cytoplasmic environment of the fertilized egg. However, if nuclear states are fairly rigidly set or require several rounds of replication to be erased, then a determined nucleus placed in a cytoplasm where it could not rapidly replicate would retain its state.

Kauffman (1980) performed such a test with heterotopic nuclear transplantations at the late syncytial blastoderm stage, using genetically marked nuclei to distinguish cells derived from donor nuclei, and examining imaginal implants cultured from these test embryos. He concluded that nuclear commitments *have* been imposed. However, a comparable attempt to demonstrate nuclear states in syncytial blastoderm embryos was made by Simcox and Sang (1983), though with a smaller sample and different assay conditions, and met with negative results. Thus, while Kauffman's results are provocative and provide a provisional answer to the question about the stability of nuclear states at the late syncytial blastoderm stage, a confirmation would be valuable.

Nevertheless, to sum up, the evidence shows that the first determinative events have occurred by the cellular blastoderm stage and these involve both segmental and germ layer assignments. These restrictions on cellular capacity *may* have moderately stable nuclear state correlates. Whatever the corresponding nuclear states, however, a central task for a genetic analysis of development in *Drosophila* is to identify the maternally specified "morphogenetic system(s)" that provides the initial differences in developmental assignment between cells at the cellular blastoderm stage.

Experimental Embryology and Analysis of the Morphogenetic System

In particular, one would like to know whether the morphogenetic system consists of a mosaic of "determinants" within the early embryo (either preexisting in the egg or forming after fertilization) or, alternatively, the presence of one or more morphogenetic gradients, or, even, some combination of these two mechanisms. For many years, this question could not be answered with any degree of certainty for *Drosophila*. The small size of the embryo, relative to that of other insect embryos, and the rapidity of the assignment of developmental fates to early cells, relative to most other types of animal embryo, seemed to preclude detailed investigation of the morphogenetic system of the fruit fly. The principal clues were indirect ones and were derived from experiments on the embryos of more primitive and, in general, more slowly developing insect embryos.

The evidence obtained from these embryos (reviewed in Sander, 1976) indicated two interesting properties of their morphogenetic systems. The first is that it apparently consists of some kind of information in the form of opposing gradients along the longitudinal axis; the first extends from a high point at the anterior end and declines posteriorly while the second extends from a high point at the posterior end and decreases as one moves anteriorly in the embryo. The second conclusion was that these gradients are propagated from both poles during the late cleavage divisions of the embryo rather than existing, fully formed, within the unfertilized egg; the picture is thus rather different from both of the classical alternatives—a mosaic of determinants or a gradient system envisaged as in place within the mature egg.

The experiments that yielded these findings involved constricting, or "ligating," embryos at successive stages, in different positions between the two poles, and monitoring the capacity of the ligated fragments to differentiate particular embryonic or larval structures. In these embryos, the capacity to give rise to the complete inventory of cuticular structures was found to develop only gradually during cleavage, and, in particular, the development of central (thoracic) structures requires exposure to influences emanating from both poles. Thus, while ligation in the center of early cleavage prevents normal development of the central segments, but permits that of the terminal segments, ligation at blastoderm may interfere only with the development of only one segment. The inability of early ligated embryos to specify the center, in contrast to blastodermal-stage embryos, has been classically known as the "gap phenomenon."

Furthermore, when posterior cytoplasmic material is moved anteriorly in leafhopper embryos (an embryo type in which this manipulation can be readily performed), it was found that complete but miniature embryos could be formed anterior to this material and that abdominal segments of reversed polarity formed posterior to it (Sander, 1975). These experiments show that the posterior pole material can exert a long-range influence on pattern; other experiments, involving a combination of ligature and movement of posterior pole material, indicated that an opposing influence spreads from the anterior pole. In particular, various kinds of damage inflicted at the anterior pole in *Smittia* embryos can cause the substitution of a posterior-half pattern for the normal anterior-half pattern (reviewed in Kalthoff, 1979). These findings also support the idea that in normal development there is a substance located at the anterior pole whose influence in normal development can be exerted (either directly or in a series of steps) throughout the anterior half; when this substance is missing, there is a transformation throughout the anterior half to a posterior-half pattern.

Although many of the surgical manipulations are more difficult to do in *Drosophila* than in the embryos of less advanced insects, ligation experiments subsequently carried out on the fruit fly embryo revealed a gap phenomenon and suggested that the formation of posterior and middle thoracic structures in both the embryo (Schubiger et al., 1977) and the imaginal anlagen (Schubiger, 1976) requires influences that spread from both poles (reviewed in Schubiger and Newman, 1981). However, in contrast to the experiments on *Smittia* and other embryos in which double-abdomen embryos were induced by various treatments, the first attempts to induce such defects in *Drosophila* by the same procedures were unsuccessful (Bownes and Sang, 1974; Bownes and Kalthoff, 1975). It was unknown whether these failures to manipulate the system in ways comparable to that of more primitive insect embryos reflected technical limitations or actual differences in the morphogenetic systems between *Drosophila* and these systems.

Until the mid-1980s, therefore, the overall picture was unclear; while there was some indirect evidence that gradients emanating from both poles play a part in the *Drosophila* morphogenetic system, strong confirmation had not been obtained. In addition, no experimental work in *Drosophila* had been done to establish how the d-v pattern is created.

The first decisive experimental findings on the *Drosophila* system were made when it was discovered that the key to carrying out successful operations on these embryos requires the removal of any leaked cytoplasm from the surface of the operated embryo (Frohnhöfer et al., 1986). If this residual cytoplasm on the exterior is allowed to dry, the operated embryos show a high early mortality, but if it is removed early, the experimental embryos develop well into the stages of cuticular differentiation or beyond. With this seemingly trivial finding about technique, it became possible to test for the effects of removing cytoplasm from or transplanting cytoplasm between

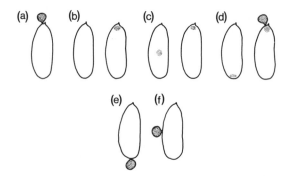

Fig. 4.15. Cytoplasmic extrusion and transplantation experiments in early embryos. **a**: Egg pricked at anterior and cytoplasm extruded. **b**: Posterior polar plasm transplanted to anterior end. **c**: Central cytoplasm (from prospective abdominal region) transplanted to anterior end. **d**: Cytoplasm extruded from anterior end and posterior polar plasm simultaneously transplanted to anterior end. **e**: Removal of cytoplasm from posterior end. **f**: Removal of cytoplasm from central region. See text for discussion.

different positions along the a-p axis on the specification of pattern elements along that axis. Some of the experiments carried out by Frohnhöfer et al. (1986) are diagramed in Fig. 4.15 and the results are given in Table 4.1.

The conclusions can be quickly summarized. Removal of cytoplasm from early cleavage embryos (experiment A) reduces head development and in some cases causes the development of extreme posterior structures (the telson). Interestingly, transfer of posterior pole material to the anterior pole, without removal of anterior cytoplasm, produces similar defects (experiment B). However, the simultaneous removal of anterior pole material and transplantation of posterior pole material (experiment D) produces an even more dramatic effect, a high percentage of double-abdomen embryos (either symmetrical, as observed in *Smittia,* or often partially duplicated and asymmetric). Since transfer of central cytoplasm (experiment C) by itself is even less effective than simple transfer of posterior pole cytoplasm in reducing head development, the induction of double-abdomen embryos requires posterior pole cytoplasm specifically. Thus, putting the findings together, one concludes that posterior material can antagonize the action of anterior pole material and exert its effects over sizable distances within the embryo. Comparably strong long-range defects can be produced within the abdominal region by removal of posterior pole material (experiment E) (even without the simultaneous transfer of anterior material). Such defects are much rarer when material from the abdomen itself is removed (experiment F).

Altogether, the results provide the best and most direct evidence that in the wild-type embryo, the initial determinative events involve the spread of effects, and presumably substances, which are initially located at the poles, along the a-p axis of the embryo; the picture is thus in agreement with the results obtained with the other insect embryo systems previously tested. One additional finding, which is intriguing, is that posterior terminal structures can be evoked in the absence of other posterior (abdominal segment) structures. This indicates that there is some latent capacity at the anterior pole for producing terminal structures and that this capacity is independent of that for forming the next most posterior regions (the posterior abdominal segments) normally physically contiguous with them.

Table 4.1. Pattern Changes Following Cytoplasmic Transfer or Removal

Changes at anterior end		Embryos with pattern (% survivors)				
					Posterior duplications	
Experiment[a]	No. of surviving embryos	Normal	Small head	Telson	Asymmetric DA[b]	Symmetric DA[b]
A	46	0	93	7	0	0
B	90	54	43	2	0	0
C	32	94	6	0	0	0
D	61	0	20	34	30	16

Changes at posterior end		Embryos with pattern (% Survivors)			
				No. of abdominal segments missing	
Experiment[a]	No. of surviving embryos	Normal	Defective telson	1 or 2	>2
E	76	67	5	15	18
F	68	90	1	10	0

Source. Data from Frohnhöfer et al. (1986). Reproduced with permission from the Company of Biologists, Ltd.

[a]See Figure 4.15 for diagram of experiment.

[b]DA, double-abdomen embryo.

These experiments delineate some of the principal characteristics of the *Drosophila* morphogenetic system and suggest, in fact, that it is composed of several dissociable subsystems. However, the experiments provide only a phenomenological characterization; they do not reveal anything about the nature of these subsystems. The specific questions that the experiments raise include the following: 1) Are the anterior and posterior pole "influences" actually molecular gradients of some sort? 2) If so, what are the molecular species of which these gradients are composed? 3) How are these products laid down in the egg? 4) How do these molecules actually provide the first developmental instructions of the blastodermal cells? and 5) While the experiments described above pertain to determinative decisions along the a-p axis, how are assignments to germ layer type, along the d-v axis, made?

Hunting for Maternal Effect Mutants

To answer such developmental questions, one must turn from the methods of experimental embryology to those of genetics and, in particular, to the isolation of maternal effect mutants. The identification and analysis of such mutants has been the focus of effort of several laboratories for nearly two decades. What has emerged from this large collective effort, and which will be described in detail below, is that the morphogenetic system of *Drosophila* consists of four virtually independent subsystems, each based on a relatively small set of strict maternal effect genes. Three subsystems instruct cells as to their regional fate along the a-p axis while the fourth creates an initial specification of germ layer fate along the d-v axis. Thus, just as the *Drosophila* blastoderm fate map can be *drawn* in terms of Cartesian coordinates (Fig. 4.10), it appears that the embryo is organized by four nearly independent gene systems, most of whose products act along one axis or the other. (That there is some overlap between these systems, involving some of the genes, will also become apparent during the recounting.)

There were several recognizable phases to this work. The first involved the analysis of a single maternal effect mutation, *bicaudal (bic)*, discovered fortuitously, which was observed to cause a profound, whole-scale transformation in embryonic pattern (Bull, 1966; Nüsslein-Volhard, 1977, 1979a,b). The phenotype of embryos from *bic* mothers is striking and is discussed below; however, the primary significance of the mutant today is that it showed that mutations affecting the morphogenetic system exist and can be isolated. The second phase, in the early 1970s, involved intensive, but, in retrospect, somewhat numerically limited, isolations of female-sterile (*fs*) mutants, from screenings of 1000 to 2000 mutagenized chromosomes, followed by a secondary screening for those that produce aberrant embryos (Bakken, 1973; Gans et al., 1975; Rice and Garen, 1975; Zalokar et al., 1975; Mohler, 1977). The third phase, beginning in 1978 and extending to the present, has involved even more intensive screenings, of up to 10,000 mutagenized chromosomes at a time, for both zygotic and maternal effect mutants that create dramatic developmental transformations in embryonic pattern (Nüsslein-Volhard and Wieschaus, 1980; Jurgens et al., 1984; Nüsslein-Volhard et al., 1984; Wieschaus et al., 1984; Schüpbach and Wieschaus, 1986).

It is the latter searches, which have employed improved screening methods for examining embryonic cuticular phenotypes from large numbers of strains simultaneously (Wieschaus and Nüsslein-Volhard, 1986), that have yielded the greatest number of maternal effect mutants of interest and which will be the principal focus in this discussion. However, the work of the second phase, the screening for, isolation of, and characterization of *fs* mutants, yielded some findings of interest which should be noted (Bakken, 1973; Gans et al., 1975; Rice and Garen, 1975; Mohler, 1975).

The kinds of screening procedures used for isolating both autosomal and sex-linked *fs* mutants are diagramed in Figure 4.16. The methods rely on the use of specially constructed chromosomes termed "balancer chromosomes" for the isolation and perpetuation of the obtained *fs* mutants that are obtained. Balancer chromosomes carry multiple inversions, which suppress recombination, and usually dominant mutations, e.g., *Curly wing (Cy)* or *Bar eye (B)*, for their ready detection in progeny. The function of balancer chromosomes is to ensure preservation of particular groups of linked alleles on the homologs, without recombinational loss of the mutations. Thus, any heterozygote for a balancer and a mutation-bearing chromosome (Bal/*m*) will produce only two kinds of gametes, balancer-bearing gametes and mutation-bearing gametes. In practice, it is convenient if balancer chromosomes carry one or more dominants and at least one homozygous lethal. Heterozygotes in which the homolog also bears one or more lethals will then form a self-perpetuating condition, a so-called "balanced lethal" stock, permitting the recessive lethal-bearing chromosome to be maintained indefinitely.

The essence of the schemes shown in Figure 4.16 is that they serve to transmit single mutagenized chromosomes into individual flies; the male and female progeny of these flies are then crossed to produce female mutant homozygotes. New chromosomal lines that produce female sterility specifically are then screened for egg production.

a) Isolating autosomal f̲s̲ mutants

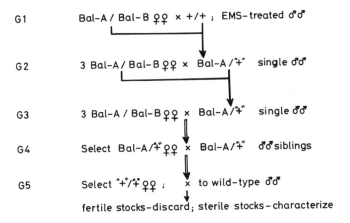

b) Isolating sex-linked f̲s̲ mutants

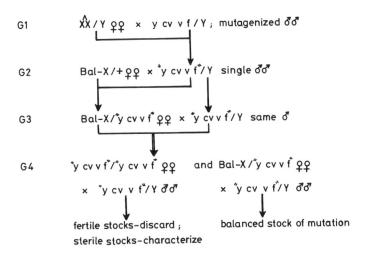

Fig. 4.16. Isolation of female-sterile (*fs*) mutants. **a**: A scheme for isolating autosomal *fs* mutants (based on that of Bakken, 1973). **b**: A procedure for isolating sex-linked *fs* mutants (from Mohler, 1977). Bal-A and Bal-B, autosomal balancer chromosomes with different dominant marker mutations; +, mutagenized wild-type chromosome; X̂X, attached-X chromosome; Bal-X, an X balancer chromosome carrying a dominant *Bar* mutation; *y cv v f*, and X chromosome with four recessive marker mutations.

119

The great majority of the *fs* mutants obtained by such schemes are defective in some aspect of oocyte formation, some obviously so, others in a more subtle fashion, while a smaller fraction of the *fs* mutants make normal eggs but are defective in either mating or egg-laying behavior. Among the remainder, the *fs* mutants that make apparently normal eggs but which cannot support normal embryogenesis were selected for further study, these being the maternal effect mutants.

Many of the primary defects observed in the embryos of the *fs* strains isolated in these mutant hunts (Bakken, 1973; Rice and Garen, 1975; Zalokar et al., 1975) seem to be in metabolic or structural components required for cleavage. Indeed, several of the mutant syndromes are similar to those observed when permeabilized *Drosophila* embryos are exposed to various drugs that interfere with macromolecular synthesis of cell division (Zalokar and Erk, 1976). For example, one X chromosomal mutant, *fs(1)1459*, forms a normal-appearing syncytial blastoderm but then proceeds to lose nuclei from the surface to the interior. This effect is mimicked by treatment with the actin microfilament antagonist cytochalasin B. This comparison suggests that the mutation might affect microfilament function, with consequent inhibition of cellularization. Other mutants show clumped chromatin in early cleavage, an effect mimicked by treatment with the transcription inhibitor actinomycin D; these genetic defects may therefore involve direct or indirect interference with RNA synthesis.

However, as seen with *C. elegans* maternal effect mutants, where different mutant alleles for the same gene (defined by their lack of complementation) are known, the developmental syndromes can differ dramatically with respect to the apparent nature of the developmental defect. For instance, a temperature-sensitive (ts) mutant designated *fs(1)1497* produces embryos with blastoderms irregularly populated with nuclei; these embryos then proceed to undergo seemingly precocious gastrulation (with respect to nuclear number) before dying. However, another ts mutant allele of the same gene produces eggs at the restrictive temperature that develop successfully to the larval stage but which lack mouth hooks, a "milder" phenotype but one with a seemingly specific morphological defect (Zalokar et al., 1975). This particular example is noted here because the gene, subsequently named *swallow* (*swa*), has been shown to play a part in the localization of a key substance in the a-p morphogenetic system, the *bcd* gene protein, discussed below.

Although the general picture presented by these mutant collections is similar to the one we have seen for the *C. elegans* maternal effect *emb* mutants, namely one of generalized embryonic arrest produced by interference with one or more aspects of general cellular function, some of the embryonic phenotypes give hints of interesting graded properties within the egg. Thus, some of the mutant defects are not spatially homogeneous but exhibit a-p polarity. For example, two ts maternal effect mutants show a polarized defective cellularization at blastoderm (Rice and Garen, 1975). Embryos produced from *mat(3)3* mothers remain acellular in the posterodorsal region while *mat(3)6* is acellular at blastoderm in the central region. A second form of a-p polarity is revealed in the occurrence of polarized patterns of mitosis or "mitotic waves" in one, and perhaps two, of the mutants (Zalokar et al., 1975). This effect is also produced by DNA synthesis inhibitors (Zalokar and Erk, 1976). In these drug-induced mitotic waves, the mitotic pattern was generally that of a more advanced mitotic stage at one or both poles, grading to earlier stages toward the middle. Such ends-to-middle mitotic waves were subsequently reported in wild-type embryos for the last four cleavage divisions, those of the syncytial blastoderm (Foe and Alberts, 1983); the mutants described above, and the DNA synthesis inhibitor treatments, seem to produce an exaggerated form of these mitotic gradients of the normal embryo. The occurrence of such polarized effects signals the existence of an underlying a-p polarity within the egg and it is this subject to which we shall now turn.

MATERNAL COMPONENTS THAT ORGANIZE THE ANTEROPOSTERIOR AXIS OF THE EMBRYO

The intensive mutant screens of Nüsslein-Volhard and Wieschaus and their colleagues have identified a set of strict maternal genes whose products play a key role in instructing cells as to their developmental fate along the a-p axis; females homozygous for these mutations produce embryos that display distinct and recognizable segmental transformations along this axis. All of the genes involved have been identified on the basis of several alleles and, for all, amorphic or nearly amorphic alleles are known. Reviews of the genetic work can be found in Nüsslein-Volhard et al. (1987) and Nüsslein-Volhard (1991).

Three general points should be made about this set of genes. The first is that it is relatively small, approximately 20 genes in all. With one exception, all these genes are strict maternals, for which the embryos of mutant mothers, who are themselves without morphological defects, develop defectively; furthermore, most of the mutations are recessive, so that the effects are seen only when the mothers are homozygous for the mutations. Despite the relatively small size of the gene set, it is probably close to complete; the mutant screens have now virtually "saturated" the chromosomes for mutations affecting the morphogenetic system and it seems probable that few new exclusively maternal genes will be discovered. (That some partial maternals play major roles cannot be excluded.)

The second point is that each gene affects fate specification along the a-p axis exclusively; fate specification along the d-v axis, affecting germ layer identity (see Fig. 4.9), is nearly or completely normal. The third point is that the set can be divided into three general classes: those affecting development of the anterior half of the embryo, those affecting development of the posterior half of the embryo, and, most surprisingly, those causing alterations at *both* the anterior and posterior tips of the embryo. Representative mutant embryos of the anterior, posterior, and terminal maternal gene systems are shown in Figure 4.17. It should be stressed at the outset, however, that this division of phenotypic effects observed is not actually as neat as the above description implies; mutations in several of the genes, both of the "anterior" and "posterior" groups produce effects in other regions, effects which will be noted in the detailed discussion of the mutants.

Maternal Setting of the Anteroposterior Pattern: The Anterior Group

The genes of the anterior group are listed in Table 4.2. We will begin with the three genes whose mutant embryo phenotypes are the most complete alterations in pattern, the first three listed in the table, and in particular *bicaudal* (bic), whose analysis and properties illustrate some basic points.

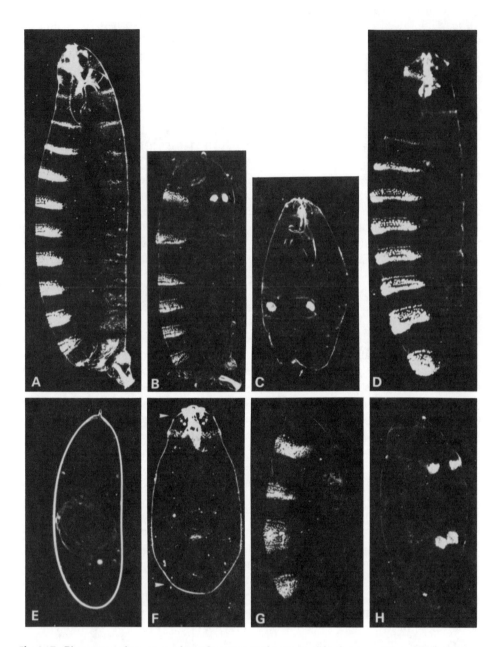

Fig. 4.17. Phenotypes of progeny embryos from maternal mutants of the three a-p pattern-forming systems. **A**: Embryo from wild-type mother. **B**: Embryo from mother deficient in the anterior system gene (*bcd*). **C**: Embryo from mother lacking posterior system activity (*osk*). **D**: Embryo from mother deficient in terminal gene system (*tor*). **E**: Embryo, still in vitelline coat, from mother deficient in all three systems. **F**: Embryo from mother lacking posterior and terminal gene activities (*osk tsl*). **G**: Phenotype of embryo from mother lacking anterior and terminal systems (*bcd tsl*). **H**: Phenotype of embryo from mother lacking anterior and posterior systems (*bcd osk*). (Reproduced from Nüsslein-Volhard, 1991, with permission of the Company of Biologists, Ltd.)

Table 4.2. Maternal Anterior Genes of *Drosophila melanogaster*

Gene	Map position	Embryo phenotype	References
bicaudal (*bic*)	2-67	Variable double abdomen (recessive)	Bull (1966); Nüsslein-Volhard (1977)
Bicaudal C (*BicC*)	2-51	Variable double abdomen (dominant)	Mohler and Wieschaus (1986)
Bicaudal D (*BicD*)	2-52.9	Variable double abdomen (dominant)	Mohler and Wieschaus (1986)
bicoid (*bcd*)	3-47	Deletion of head and thoracic segments; acron to telson transformation	Frohnhöfer and Nüsslein-Volhard (1986a,b); Nüsslein-Volhard et al. (1987)
exuperantia (*exu*)	2-93	Weak *bcd* type	Schüpbach and Wieschaus (1986b); Frohnhöfer and Nusslein-Volhard (1986b)
swallow (*swa*)	1-14	Weak *bcd* type	As for *exu*

bic, a spontaneous recessive second chromosome mutation discovered by Bull (1966), causes the production of double-abdomen embryos from homozygous mutant mothers. In such embryos, the two abdominal regions are in mirror-image symmetry, an effect that is strongly reminiscent of the induced reversed-abdomen embryos of lower insects (Kalthoff, 1979). There is, in fact, a range of defects in embryos from *bic* mothers but, even in the most completely mirror symmetric embryos, there are fewer than the normal number of abdominal segments in each half and these are broader individually than those of the wild-type (Nüsslein-Volhard, 1979a). A comparison of the integuments of first instar larvae from wild-type and *bic* mothers is given in Figure 4.18.

Although its low penetrance and variable expressivity, in homozygous mothers, initially made *bic* difficult to study, it was found that when the mutation was placed over a second chromosome bearing the deficiency vg^B, which deletes the wild-type allele of *bic,* its penetrance is increased to 100% (i.e., all females produce *bic* eggs) (Nüsslein-Volhard, 1977, 1979a,b). Even under these improved conditions, the expressivity of the mutation is not 100% but varies according to the age of the mother, with younger females showing higher expressivity, and with temperature, higher maternal growth temperatures favoring mutant expression.

A preliminary question about any mutant phenotype is whether the mutant effects stem from a qualitatively new genetic activity, a so-called "neomorphic" mutation (Muller, 1932), or simply from loss of normal gene activity. In general, recessive mutants are loss-of-function mutants while dominants are neomorphs, producing a novel function of some sort. There are, however, other classes of dominants; some dominants are such because the gene is haplo-insufficient, a condition in which loss

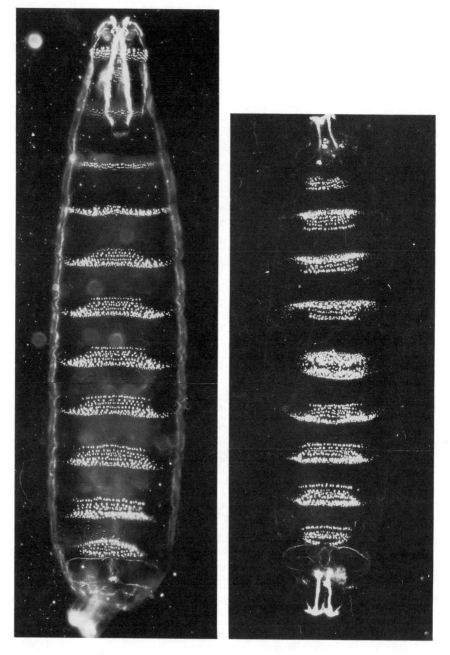

Fig. 4.18. Segmentation in a wild-type larva (left) and a larva from a homozygous *bic* mother (right). Anterior at the top, posterior at the bottom. Abdominal segment phenotype shown by the presence of a thick denticle hook band. (Photographs courtesy of Dr. C. Nüsslein-Volhard.)

of one copy produces a mutant effect, because two full two doses of the wild-type activity are normally required. Yet other dominants are hypermorphs, which create their effects through an excess production of the wild-type activity.

The *bic* mutant is recessive and its phenotype therefore presumably results from an insufficiency of maternal bic^+ activity. However, as we have seen, if the mutant is a hypomorph, possessing some residual wild-type activity, the phenotype is often different and less severe than if the mutant is an amorph, possessing none. We have seen how in *C. elegans* one may determine, using nonsense suppressor backgrounds, whether certain mutations are completely deficient in activity and hence null mutations. (There is a slight distinction between the terms "amorph" and "null mutant" in that the former refers to mutants without a morphologically detectable phenotypic effect of residual gene activity while the latter refers to mutations for which a biochemical test indicates zero residual gene activity.)

In *Drosophila*, however, a much older and simpler method than the use of suppressor mutants is available. The method involves constructing lines with different dosages of the mutant and wild-type alleles (Muller, 1932). Such strain construction is comparatively easy to do in *Drosophila* because of the existence of a large number of small deficiencies and duplications that collectively cover almost the entire genome. Furthermore, the positions of these deficiencies and duplications can be placed very precisely by examination of their polytene chromosomes. By constructing strains with the appropriate variations in dose of wild-type and mutant genes and comparing the phenotypes of these different strains, it is possible to establish unambiguously the character of the mutant. This set of comparisons and their possible outcomes and interpretations are summarized in Table 4.3.

The crucial comparison for *bic* is between the hemizygous (*bic*/Df) and homozygous (*bic*/*bic*) mothers. When the test is performed, it is found that the penetrance and expressivity of the mutant is much higher in the hemizygote (Nüsslein-Volhard, 1977). The *bic* mutant allele evidently possesses some residual activity and must therefore be a hypomorph. Furthermore, the more extreme the reduction in maternal wild-type gene activity is, the more extreme the resultant mutant embryonic phenotype. In fact, a single dose of bic^+ activity is insufficient to give full wild-type development: bic^+/Df mothers produce a small percentage of *bic*-type embryos. In other words, despite the classification of the mutant allele as a recessive, it exhibits some degree of haplo-insufficiency.

The semidominant mutants *BicC* and *BicD* similarly produce whole-scale alterations along the a-p axis and with the same range of phenotypes seen with *bic* mothers (Mohler and Wieschaus, 1986). By the Müllerian tests described, *BicC* is a dominant

---

Table 4.3. Genetic Tests to Determine the Type of Mutant Expression[a]

If $m/m = m$/Df, then the mutant is an amorph
If m/Df $> m/m$, then the mutant is a hypomorph
If $+/m = +/+/m$, then the mutant is a neomorph
If $m/m > +/m/m$ (or m/Df), then the mutant is an antimorph
If $+/m =$ mutant phenotype but $+/+/m =$ wild-type, then the mutant is a dominant haplo-insufficient

[a]m, mutant gene; $+$, wild-type allele; Df, deficiency (deletion of $+$ allele); $>$ signifies more severe mutant expression.

of the haplo-insufficient type while the two *BicD* mutant alleles described by Mohler and Wieschaus are "antimorphs," that is, neomorphic mutant forms that can compete with the wild-type product. Thus, homozygous *BicD* females are fertile but produce a much higher percentage of affected embryos than heterozygotes. The *BicD* gene has been cloned and found to encode a cytoskeletal component related to myosin; in wild-type oocytes this protein is uniformly distributed, but in oocytes from *BicD* mothers, it is concentrated at the anterior pole, anchoring a posterior "determinant" there (Wharton and Struhl, 1989; Lehmann and Nüsslein-Volhard, 1991). (The nature of the anchored molecule will be discussed later in this chapter, in connection with the posterior gene group.)

The mutant with an effect opposite to the *bic/Bic*-type mutants, *Dicephalic* (*Dic*), creates the opposite effect, as its name implies, that of "two-headed" embryos. The *Dic* syndrome seems to result from the relative dispositions of oocytes and nurse cells within the egg chambers. In the mutant egg chambers, the oocyte does not occupy its typical posterior postion but tends to be between nurse cell clusters (Lohs-Schardin, 1982). Although few larvae develop from the oocytes of such egg chambers, those that do exhibit reversed anterior halves. The *Dic* syndrome confirms that one important factor in setting a-p polarity is the relative positioning of oocyte and nurse cells within the egg chamber.

Despite the dramatic transformation produced in embryos by the *bic* class of maternal effect mutations, and the molecular characterization of *BicD*, these mutants have proved less informative about the morphogenetic system than might have been hoped. In contrast, the analysis of the *bicoid* (*bcd*) gene, discovered in the mid-1980s, has provided one of the triumphs of developmental genetics in the last decade. *bcd* has been found to encode a molecule that forms an "instructive" gradient, which runs from anterior to posterior and plays a crucial role in specifying head and thoracic structures.

The first clues to the action of the *bcd* gene product were provided by the mutant phenotype of embryos derived from homozygous *bcd* mothers for strong (amorphic or near amorphic alleles). These lack both head and thoracic segments and show a replacement of head structures by tail structures such as telson, anal plate, tuft, and spiracles (Frohnhöfer and Nüsslein-Volhard, 1986). Females with hypomorphic alleles yield embryos with less severe transformations, but even the strongest reductions of head and thoracic structures, seen with amorphic maternal alleles, are not accompanied by the appearance of abdominal segments in the anterior half as seen in embryos from *bic* mothers. Instead, there is a slight shift of abdominal segments forward, and for embryos produced by females with strong (amorphic) alleles, abdominal segmentation defects appear, the particular defects being characteristic of the allele and not always showing a strict a-to-p sequence in their production (Frohnhöfer and Nüsslein-Volhard, 1987). An embryo exhibiting an extreme *bcd* phenotype is shown in Figure 4.17B.

The primary alteration in embryos produced from *bcd* mothers is thus the elimination of head and thoracic structures (with secondary defects produced in the abdominal segments). The primary defects thus resemble those produced by simple removal of anterior cytoplasm, described earlier (Fig. 4.15). These segmental defects thus reflect a primary change in the fate map, a change that is signaled as early as the first gastrulation movements, occurring immediately after the blastoderm stage and well before visible segment formation. The position of one of these infoldings, the cephalic fold (CF), an invagination running transversely in an anterior direction from the dorsal side, in fact provides a measure of the extent of maternal *bcd* activity (Frohnhöfer and Nüsslein-Volhard, 1986).

In embryos from wild-type mothers containing the normal diploid dosage of *bcd* product, the CF appears at the dorsal edge of the embryo at 65% EL, two-thirds of the distance from the posterior pole. However, in hemizygous mothers (containing only one *bcd*$^+$ allele), the CF is shifted anteriorly, to 71% EL, whereas with three maternal doses, the CF is moved posteriorly to 62% EL. (Surprisingly, the latter embryos subsequently regain normal proportioning and give rise to normally proportioned adults, showing that there are compensatory mechanisms that operate after gastrulation.) The results demonstrate that there are distinct quantitative relationships between the amount of maternally supplied *bcd* activity and the positioning of anlagen for anterior structures within the fate map.

An even more direct kind of proof of the relationship between quantity of *bcd* product and long-range effects on structure along the a-p axis comes from cytoplasmic rescue experiments. This approach involves transplanting various amounts of cytoplasm from embryos made from *bcd*$^+$ mothers to embryos from *bcd*$^-$ mothers and assaying for rescue, namely elimination of the *bcd* phenotype (Frohnhöfer and Nüsslein-Volhard, 1986). The results show that rescue is produced only when cytoplasm from the anteriormost (100–85% EL) part of the embryo is used and that the degree of development of head and thoracic structures is a function of the amount of cytoplasm injected. No rescue is obtained with cytoplasm from homozygous *bcd*$^-$ mothers, proving that rescue is indeed a function of the wild-type gene product. In addition, the degree of rescue of anterior structures in embryos is proportional to the maternal gene dosage for *bcd*$^+$, for a given amount of cytoplasm, as predicted from the correlations between CF postion and maternal gene dose. Finally, rescuing activity is highest in unfertilized eggs, declining with the age of the embryo and disappearing near the end of cleavage; recipient embryos are similarly only responsive during the preblastoderm period. The results show, altogether, that 1) the active (rescuing) *bcd* gene product is localized at the anterior end; 2) it acts sometime during the period preceding the cellular blastoderm period (by which the first instructional steps of determination have occurred); and 3) though localized anteriorly, it has long-range effects on pattern ranging from the extreme anterior end to the anterior abdominal region, the pattern being responsive to the amount of maternal *bcd* product deposited in the egg.

Intriguingly, while the *bcd* gene product has the character of an anterior determinant of some kind, the kind and pattern of the structures induced is as much a function of *where* along the a-p axis the material is injected as of the quantity of the injected cytoplasm. Thus, cytoplasm from wild-type mothers which, when injected at the anterior end, rescues head and thoracic structure development in the normal a-p sequence (though the head structures are less well developed than the thoracic) will, when injected at the middle of the embryo (50% EL), induce reversed, mirror-image thoracic structures on either side of head structures developing at the site of injection. Significantly, injection at the posterior end does not induce either head or thoracic structures, leaving abdominal development essentially intact but with some compression of the abdominal segments (Frohnhöfer and Nüsslein-Volhard, 1986). These results indicate that 1) head structure development is induced by the highest concentrations of *bcd* gene product and thoracic structures are induced by lower concentrations and 2) there is an influence of posterior pole material that is antagonistic to the action of the *bcd* product. This last deduction is consistent with the finding, cited earlier, that injection of posterior pole material into anterior regions of wild-type embryos antagonizes the development of anterior structures.

The results suggest, but do not prove, that the *bcd* gene product exists and functions as an a-p gradient within the early wild-type embryo. The proof has come from molecular methods involving the cloning of the gene, which occurred as a fortuitous event, based on the sequence-relatedness of *bcd* to the zygotically expressed gene *paired* (*prd*). As will be discussed in chapter 7, *prd* is a member of one of the main classes of zygotically expressed genes necessary for development of the segment pattern, the so-called "pair-rule" genes. Each pair-rule gene is expressed in a spatial array of seven stripes of transcript at blastoderm, the spacing of the stripes being at roughly alternate segmental intervals, while the precise spatial array is characteristic for each pair-rule gene in wild-type embryos.

bcd was, in fact, isolated in two independent searches. The first was by Scott et al. (1983), as part of the cloning of the region around the *Antp* gene, but *bcd* was not further characterized at that time. The second time, however, proved informative. In a search for genes with sequence-relatedness to *prd,* Frigerio et al. (1986) isolated a gene in the chromosomal region in which *bcd* mutants were known to map. When the mRNA of this gene was localized by *in situ* hybridization in early embryos, it was found to be localized in the 15% most anterior region of early cleavage embryos and to disappear after the blastoderm stage. These characteristics are essentially those expected of a gene product with the early anterior patterning properties of *bcd*.

That this cloned gene was, indeed, *bcd,* which was beginning to be genetically characterized at the same time, was indicated by the finding that two "point mutants" (mutations which show no obvious cytological defects in polytene chromosomes) are actually small deletions within the gene identified by the clone. A second, and even stronger proof, was the demonstration that the 8.7 kb fragment can rescue the *bcd* maternal defect (Berleth et al., 1988). This kind of experiment involves placing the gene to be tested in a derivative of a P element, one of *Drosophila*'s transposable elements, and producing *in vivo* transformants; the chromosomes containing the insert are then placed into the mutant genotype and the resulting animals are scored for rescue of the mutant defect. The procedure begins with making the hybrid DNA construct, which contains the gene of interest, the P sequences needed for transposition, and usually a marker gene for screening or selecting for transformants. Copies of the purified construct are then injected into the posterior end of a cleavage-stage embryo before pole cell formation, with resultant transformation of some of the pole cells and, hence, of the germ line. (Insertion of the construct can take place in any of many possible sites within the chromosomes of the pole cells.) The survivors (not all injected embryos survive) are then bred to obtain transformant *Drosophila* lines, each containing the construct at a particular novel site (Rubin and Spradling, 1982). Different transformants are then crossed to lines carrying the genetic defect of interest. The strategy for making germ line transformants by this technique is diagramed in Figure 4.19 (and a review of the technique and its uses can be found in Spradling [1986]).

In the case of *bcd,* the transformants were crossed to lines that produce homozygous *bcd* females. Females were obtained that were homozygous for *bcd* but which were found to carry the presumptive *bcd* construct. Three transformant

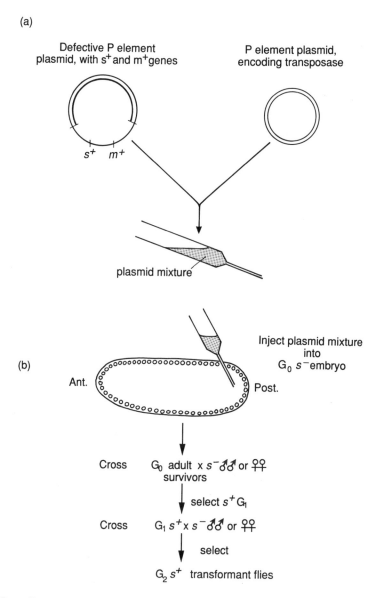

Fig. 4.19. Germ line transformation. m^+, the gene of interest; s^+, the gene for screening or selection of transformants. See text for description.

strains were recovered, in each of which the maternal *bcd* defect was suppressed, showing that the construct contains functional *bcd* (bcd^+) activity (Berleth et al., 1988).

The *bcd* gene contained within the 8.7 kb cloned fragment encodes two size classes of mRNA: A major transcript of 2.6 kb, which is present only during the critical period in which *bcd* is required, and a minor, 1.6 kb, species which is present throughout development; it is undoubtedly the former which is the biologically significant one for early embryogenesis. However, cDNA clones of the major species reveal a small element of differential splicing within it, yielding two different proteins of expected coding potential of 489 and 494 amino acids; both contain four exons, but the second form contains an additional five amino acids at the 5' end of the third exon, through the use of an alternative splice site (Berleth et al., 1988), and apparently is the more abundant form.

Isolation of the *bcd* gene permitted direct analysis of the distribution of the *bcd* protein(s) during early development. High titer antibodies were made to the *bcd* portion of a fusion protein, synthesized from the cloned gene, and then used to detect *bcd* protein in early *Drosophila* embryos (Driever and Nüsslein-Volhard, 1988a). The antibodies to the *bcd* portion of the DNA construct detect two specific proteins, of apparent MWs of 55 and 57 kDa, in good agreement with the expectation from the DNA sequence data of MW 53.9 kDa. (The second, higher MW variant possibly is a posttranslationally modified version of the protein.) In agreement with the developmental and genetic characterization of *bcd* action, these proteins are at their highest abundance 2–4 hr after egg laying, similarly in agreement with the predictions from the developmental studies. Given that the mRNA is present in unfertilized eggs, there must be some initial block to its translation in the oocyte. Egg laying evidently relieves this block and translation of the mRNA leads to a buildup of *bcd* protein during the nuclear cleavage divisions. That the detected protein species are the relevant ones for *bcd* activity is shown by their absence or reduction in the embryos produced by mothers homozygous for 9 of 11 *bcd* mutants tested (Driever and Nüsslein-Volhard, 1988a).

A critical finding, with respect to *bcd* action, is that the *bcd* protein first appears at the extreme anterior end, shortly after egg laying, increasing in amount during early cleavage and then spreading in a decreasing gradient from the anterior tip posteriorly to about 30% EL, where it can no longer be detected by immunostaining (Driever and Nüsslein-Volhard, 1988b) (Fig. 4.20). Staining remains strong until the end of the cleavage divisions, and then decreases, vanishing at about the time of gastrulation; the evident degradation of the protein is triggered by the developmental program, rather than by a simple temporal clock, as unfertilized eggs retain their *bcd* protein significantly longer than developing embryos (Driever and Nüsslein–Volherd, 1988a).

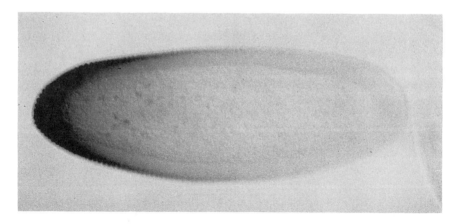

Fig. 4.20. Gradient of Bcd protein. (Photograph courtesy of Dr. C. Nüsslein-Volhard.)

The evidence that this protein gradient is directly essential for development comes from several strong correlations showing clear relationships between the *particular concentration* of *bcd* protein and the development of particular structures. In the first place, there is a strong, though not absolute, three-way correlation between maternal *bcd*+ dosage, amount of *bcd* protein, and position of the cephalic fold (Driever and Nüsslein-Volhard, 1988b). Secondly, there are distinct correlations between the amount of residual *bcd* protein in the anterior regions of embryos produced by *exuperantia* (*exu*), *swallow* (*swa*), and *staufen* (*stau*) mothers and the extent of development of head structures in these embryos (Driever and Nüsslein-Volhard, 1988b).

The reduction in Bcd protein in embryos from both *exu* and *swa* females, which exhibit *bcd*-like defects in the early embryo, seems to reflect the reduced concentrations of *bcd* mRNA at the anterior end of these embryos; the primary defect in these embryos appears to be a reduced localization of the *bcd* mRNA at the anterior end, perhaps through disruption of one or more key cytoskeletal components in the early embryo (Berleth et al., 1988). For both *swa* and *stau*, the defects in *bcd* localization are probably reflections of more general defects. *swa* is highly pleiotropic, and its first mutant was picked up in a general screen of female-steriles. *stau* was initially classified as a posterior gene group mutant (Nüsslein-Volhard et al., 1987), but its embryos clearly show anterior head defects as well and the *stau* protein has been shown to localize to the anterior pole of the embryo, where it probably serves to bind *bcd* mRNA (St. Johnston et al., 1991). (The localization of the *bcd* mRNA to the anterior pole of the oocyte involves, of course, not only the correct cytoskeletal configuration but also a specific sequence on the mRNA itself; experiments involving deletions of the *bcd* gene have identified this localization signal to a 625 bp region in the 3' untranslated region of the gene [MacDonald and Struhl, 1988].)

If it is *bcd* protein concentrations that are crucial, then one might expect to see corresponding alterations in the fate map of embryos of *bcd* mothers and of females homozygous for other mutations of the anterior group. Such alterations in the fate map have indeed been shown for the various mutants that alter the distribution of Bcd protein (Frohnhöfer and Nüsslein-Volhard, 1987). These alterations in the fate map have been found by a method that does not rely on morphological markers, which, of course, could not provide an independent test. The method utilizes a sensitive assay of certain early molecular markers of fate and reveals that there are, in fact, global shifts in fate map pattern in embryos from *bcd*, *exu*, and *swa* mothers, not simply local anterior changes (Frohnhöfer and Nüsslein-Volhard, 1987; Lehmann, 1988). The method makes use of the transcript patterns of the zygotically expressed pair-rule genes (chapter 7). As mentioned earlier, each pair-rule gene produces a characteristic striped array of transcripts at the blastoderm stage in the wild-type embryo. In effect, the stripe pattern of each pair-rule gene prefigures the future segmental pattern and serves as an early "snapshot" of the fate map; an example, that of *fushi tarazu* (*ftz*) (whose name is said to mean "too few segments" in Japanese) is shown in Figure 7.8.

The kinds of shift shown in these transcript arrays, using the *ftz* pattern, are diagramed and illustrated for embryos derived from wild-type and mothers representative of the three a-p maternal systems (*bcd, nos*, and *tor* mothers) (Fig. 4.21). In the case of *bcd*, it can be seen that there is a distinct expansion of the anteriormost stripes and a slight anterior displacement of all the rest. For *exu* and *swa* (not shown), there

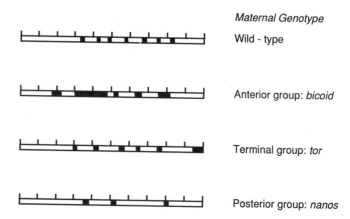

Fig. 4.21. Schematic of fate map shifts in progeny embryos of different maternal a-p patterning systems. These are idealized pictures of the redistribution of *ftz* stripes in progeny embryos of mothers deficient in activities of the anterior (*bcd*), posterior (*nos*), and terminal (*tor*) systems. For the anterior system, see Frohnhöfer and Nüsslein-Volhard (1987); for shifts associated with posterior system mutants, see Lehmann (1988); for the fate map shift associated with maternal terminal system deficiency, see Casanova and Struhl (1989).

is a slight posteriorward expansion of the more anterior stripes, reflecting the expanded, shallow *bcd* gradients in these mutants (Frohnhöfer and Nüsslein-Volhard, 1987).

Indeed, global alterations in stripe pattern are characteristic for all the maternal effect genes affecting a-p pattern that have been tested (Driever and Nüsslein-Volhard, 1988a; Lehmann, 1988). This discovery, in turn, has altered some of the previous terminology. In the earlier literature, there was a distinction made between "coordinate mutants" (Sander, 1976), like *bic*, which visibly alter the entire pattern of the embryo in an unmistakable fashion, and "maternal gap genes" (Lehmann and Nüsslein-Volhard, 1986a), such as those of the anterior and posterior groups, which were believed to alter only localized regions of the embryo. Since *all* of the maternal mutants of the anterior and posterior classes are now known to produce global shifts in the fate map, as assayed by pair-rule gene stripe patterns, this distinction has lost its validity and fallen out of use.

The elucidation of the *bcd* gradient has settled one major question about *Drosophila* embryogenesis: "Instructive" gradients *are* involved in early pattern formation and the more traditional "mosaic egg" hypotheses about *Drosophila* can be discarded. The immediate question raised in turn, however, concerns the nature of *bcd* product "instruction": How does the *bcd* protein gradient determine the pattern of head and thoracic segments? The answer, in brief, is that the Bcd protein is a transcriptional regulator and one that controls the expression of several key zygotically expressed segmentation genes, whose activities (ultimately) lead to the characteristic head and thoracic segmentation pattern. The first clue was provided by a part of the sequence of the gene: *bcd* encodes, within the third exon, a DNA binding domain of 60 amino acids termed a "homeobox." The significance of such homeobox sequences, named for the fact that they were first discovered in certain *Drosophila* homeotic genes (McGinnis et al., 1984a), is that the encoded peptide motif, termed a "homeodomain," is strikingly similar to the DNA binding and DNA sequence recognizing motifs of some well-characterized prokaryotic regulator genes transcriptional regulatory genes (Laughan and Scott, 1984; Ptashne, 1986). The essential point, with respect to *bcd*, is that its identification as a homeobox-containing gene was the first indication that it acts as a transcriptional regulator. Its particular activities in regulating specific segmentation genes will be described in chapter 7.

Maternal Setting of the Anteroposterior Pattern: The Terminal Genes and Their "Pathway"

The second major class of maternal genes that affect embryonic pattern along the a-p axis comprises the so-called "terminal genes." The discovery of these genes, a group of five autosomal genes and one X-linked exclusive maternal effect gene (Table 4.4),was a complete surprise. The mutant phenotype of the embryos produced from homozygous mothers (who themselves show no morphological defect) is that of embryos lacking the most anterior structures (collectively the acron) and the most posterior (the telson) (Fig. 4.17D). In contrast to *exu* and *swa*, however, the anterior defects in these mutants do not stem from the absence of mislocalization of *bcd* mRNA, which is normal in embryos derived from these mutants (Driever and Nüsslein-Volhard, 1988b) but must reflect deficiencies of other molecules. Secondly, the existence of these mutants shows that the distinctive morphological differences between acron and telson mask some fundamentally shared elements of their specification. That the terminal gene activities are superimposed on those of the anterior genes themselves, at the anterior tip, is shown by the existence of a telson at the extreme anterior end of embryos from *bcd* mothers.

   torso (*tor*) and *trunk* (*trk*) are representative of the terminal gene class in their phenotypic effects. Embryos from *tor* or *trk* mothers lack certain of the most anterior structures, such as the labrum and the anterior part of the brain, and the reduction of others (the pharyngeal arms of the internal mouth skeleton), while more posterior head and gnathal structures remain intact; at the posterior end, all the structures of

Table 4.4. Maternal Terminal Group Genes of *Drosophila melanogaster*

Gene	Map position (cytogenetic)	Phenotype	References
torso (*tor*)	2-57	Acron and telson defects; germ band nearly normal	Nüsslein-Volhard et al. (1982); Schüpbach and Wieschaus (1986b)
trunk (*trk*)	2-36	As for *tor*	As above
torsolike (*tsl*)	(93F)	As for *tor*	
fs(1)polehole fs(1)ph)	(5CD)	As for *tor*	Perrimon et al. (1986)
fs(1)nasrat (fs(1)N)	1-0.0	As for *tor*	Degelmann et al. (1986)
lethal(1)polehole (l(1)ph)	1-(?)	As for *tor*	Perrimon et al. (1985); Ambrosio et al. (1989)

the telson are absent (anal pads, tuft, terminal spiracles, and the structures known as the filzkorpers) (Schüpbach and Wieschaus, 1986b).

Furthermore, the fate map of blastoderm embryos from *tor* and *trk* mothers is subtly shifted, with an expansion of more centrally located anlagen toward the poles. Thus, for example, the CF, normally located at about 65% EL, is at 69% EL in *tor* and *trk* embryos, while the posterior end of the ventral furrow, which normally is located at 12–15% EL, extends to 0–5% EL (Schüpbach and Wieschaus, 1986b). These fate map shifts can also be detected by the alteration of the *ftz* transcript stripe patterns (Strecker et al., 1988). Thus, as in the case of the anterior group genes, the absence of key maternal components for specifying particular embryonic regions leads to compensatory global shifts in the fate map.

The central question, of course, is how the terminal genes participate in the specification of the embryonic termini. Again, as in the case of the anterior genes, clues have been provided by the cloning of the genes. In contrast to the *bcd*-based system, however, the mechanism seems to involve an elaborate signal transduction pathway, which begins *exterior to the oocyte* (at the termini) and which, ultimately, results in the terminal activation of particular transcriptional activators.

The first indication that a signal transduction pathway is operative came from the molecular characterization of the *tor* gene. In an initially surprising discovery, *in situ* hybridization experiments with the cloned gene showed that the *tor* transcript is not restricted to the termini but uniformly distributed throughout the embryo in early cleavage stages and becomes distributed uniformly in the periplasm, underneath the nuclei, during syncytial blastoderm (Sprenger et al., 1989). Furthermore, the protein itself is present throughout the cell membrane (Casanova and Struhl, 1989). Consistent with a localization in the membrane, the deduced amino acid sequence of *tor* indicates that it is a transmembrane protein possessing a small region rich in hydrophobic residues, characteristic of such proteins, and, in addition, a set of domains within the cytoplasmic region that are signature sequences of tyrosine kinase proteins (Sprenger et al., 1989).

One possible way to reconcile the uniform distribution of *tor* with the terminal-specific defects seen in the embryos of mutant mothers is to posit that the gene product is normally only activated, and required, at the termini. Activation, for a tyrosine kinase, would involve specific phosphorylation at tyrosine residues in one or more target proteins, whose activity, in turn, would be essential for normal development of the termini. If this hypothesis is correct, then other members of the terminal gene group, such as *torsolike* (*tsl*), *fs(1)Nasrat* (*fs(1)N*), and *fs(1)polehole* (*fs(1)ph*), may be involved either "upstream" or "downstream" of *torso* in a chain of signal transmission and transduction that ultimately leads to development of the termini. This hypothesis, in effect, has two parts: 1) that the *tor* gene product is only activated at the termini in wild-type embryos and 2) that it is part of a chain of signals, in which the other terminal genes participate. Genetic evidence supports both propositions.

The principal evidence that *tor* is only active at the termini comes from certain dominant *tor* mutants, whose maternal expression produces embryos lacking normal thoracic and abdominal segmentation (Klingler et al., 1988; Strecker et al., 1989). This abolition of normal segmentation is preceded, and foreshadowed, by the elimination of the early *ftz* stripe pattern in this region (Klingler et al., 1989; Strecker et al., 1989). These dominant mutations are evidently hypermorphs, because an additional maternal dose of *tor*+ exaggerates the effects while maternal hemizygosity for the dominant reduces them. The simplest explanation is that, in the hypermorph, the overexpression of the gene product causes suppression of segmentation. The results also imply that in the wild-type, *tor* suppresses segmentation at the termini as well as, presumably, actively promoting terminal structures.

How does *tor* action relate to that of the other terminal genes? Every developmental process requires not only a particular set of gene products but their action *in a particular temporal sequence.* Such a sequence of requisite genetic and cellular events is termed the "genetic pathway" for that process. Clearly, *tor* activity is part of the developmental pathway that leads to normal development of the acron and telson, but one would like to know both the sequence of action of the other members of the terminal gene group and the biochemical functions of these gene products. Since the circumstantial evidence reviewed so far suggests that the *tor* protein is part of a signal transduction pathway, it is important to determine which members of the terminal gene group encode gene products that act "upstream" of (before) *tor* and which act "downstream" of (after) it.

There is a simple strategem for determining this sequence and it makes use of the genetic property of "epistasis," namely the overriding of one mutant phenotype by another in double mutants. By comparing the epistatic relationships among pairs of mutations in different genes in the pathway, one can order the steps in the sequence of wild-type gene actions. The basic requirement for constructing such a pathway is that the pairs of mutant phenotypes differ in at least some characteristic fashion. However, to interpret the results, which, of course, is the point of the exercise, one needs two further kinds of information. The first is knowledge about the character of each mutation, whether it involves a loss of function or a gain of function. The second concerns the nature of the steps in the pathway, whether they form an *obligatory* sequence or a *conditional* one, in which the steps can be uncoupled in certain circumstances.

This distinction, between obligatory and conditional sequences, is a crucial one and therefore merits some further explanation. Any linear pathway can be diagramed as follows:

$$A \rightarrow B \rightarrow C \rightarrow D \rightarrow E$$

Now, if one is dealing with a sequence of direct conversions of state in which the occurrence of B depends on the prior occurrence of A, or that of D on the prior occurrence of C, then the loss of any early state will automatically predominate over, or be epistatic to, the loss of any later state. Thus, for instance, if one had a double mutant in which both steps B and D were blocked, and where both phenotypes were distinguishable, the double mutant would display the B phenotype.

Biochemical pathways have this property and, indeed, the first use of epistasis in genetics was to solve the relative order of two steps in the synthesis of the brown eye pigments, the ommochromes, of *Drosophila*. Beadle and Ephrussi (1937) found that *cinnabar* (*cn*) host larvae could supply a diffusible metabolite to transplanted developing eye discs from *vermillion* (*v*) larvae, but that the reciprocal rescue could not occur; *v* hosts cannot supply a metabolite to allow *cn* eyes to develop wild-type eye color (although wild-type hosts can do so). These results show that *v* is epistatic to *cn*, and the pathway can be rationalized within the following pathway scheme:

v mutant cn mutant

precursor ⫫→ v^+ metabolite ⫫→ cn^+ metabolite ⟶ brown pigments

Thus, for a sequence of obligatory conversions, early (mutant) steps are epistatic to late ones. (A small historical note is that this work on genetic blocks in eye pigment metabolism was highly influential in Beadle's thinking about the relationships between genes and enzymes and led to his later work with Edward Tatum, which resulted in the one gene–one enzyme hypothesis.)

However, many kinds of developmental pathways do not have this character, but involve a series of signals or regulatory switches, which, in principle, can be genetically dissociated from one another in certain mutant conditions. In a pathway of this character, A would normally *trigger* B, which would normally trigger C, and so forth. However, if a mutation allows B to take place without the prior signal from A, then the occurrence or nonoccurrence of the early step will be of no consequence. Similarly, if B is making a constitutive signal but C is deficient, the hypermorphic character of B will make no difference. In both instances and, indeed, in general for signaling or regulatory pathways, mutations in the *later* step are epistatic to mutations in earlier ones.

For the maternal terminal gene pathway, the signaling model is clearly more appropriate as a description than the substrate conversion pathway and, with that premise, the pathway can be deduced from epistasis results. Double mutant combinations of maternal genes were constructed, in which each had a *tor* dominant hypermorph and a loss-of-function mutation in one of the other terminal genes, and the embryos were scored for the presence of the alternative phenotypes (embryos lacking central segmentation, showing the dominant *tor* phenotype, and embryos showing central segmentation but aberrant termini). By these tests, *trk, tsl, fs(1)N*, and *fs(1)ph* appear to act upstream of (before) *tor* because the embryos of the double mutant mothers show the *tor* dominant phenotype (Casanova and Struhl, 1989). In contrast, *l(1)ph*, which is epistatic to the maternal *tor* dominant phenotype, acts downstream of (after) *tor* (Ambrosio et al., 1989). Thus, the pathway can be drawn with respect to *tor* as:

$$\left.\begin{array}{l} trk \\ tsl \\ fs(1)N \\ fs(1)ph \end{array}\right\} \rightarrow tor \rightarrow l(1)ph$$

While *tor* itself is a tyrosine kinase, it is of interest that *l(1)ph* is also a kinase, but of the serine/threonine type, and is homologous to the vertebrate oncogene D-*raf* (Mark et al., 1987; Nishida et al., 1988). The simplest interpretation is that the signal transmission involves the phosphorylation of *tor* protein, which then phosphorylates *l(1)ph* protein, the latter then acting on other gene products, further downstream, whose activities, in turn, promote development of the termini.

The pathway, as shown above, provides, of course, only a schematic outline: Among the four genes on the left, it is not possible to determine the sequence of interactions, since the epistasis test relies on distinguishable phenotypes. Nor should the results be taken to imply that the *only* functions of these genes are in the pathway of terminal development; mutants in *fs(1)N, fs(1)ph,* and *l(1)ph,* for instance, all show pleiotropic defects in oogenesis.

In addition to the complete sequence of gene actions, one would also like to know the cellular sites of action of these genes. In principle, any maternal effect in oogenesis that creates defective embryos can result from defectiveness in either of two cellular sites: the germ line cells (the nurse cells and the oocyte itself) or the somatic cells (the follicle cells) of the egg chamber. There is a special technique for making this determination, one that involves pole cell transplantation from mutant embryos to generate germ line chimeras.

The method is diagramed in Figure 4.22 for the particular case in which one is testing an exclusive maternal effect mutation (*m*). Homozygous mutant mothers are crossed to mutant fathers (both sexes will be phenotypically normal, for an exclusive maternal) to produce *m/m* embryos and, using a fine-bore pipette, pole cells are removed from pregastrulation embryos and inserted into the posterior pole of genetically marked host embryos of the same stage. Surviving embryos are allowed to complete development and the adult females among them are crossed to appropriately marked males; the embryos resulting from these crosses are then scored for the presence or absence of the phenotype associated with embryos from *m/m* mothers. If mutant embryos are seen, then the mutation must act in the germ line (since all oocytes develop surrounded by follicle cells of the host). On the other hand, if normal embryos bearing markers from the original donor strain appear, then the mutation must act only in the soma (which will be *m⁺* in the chimeras).

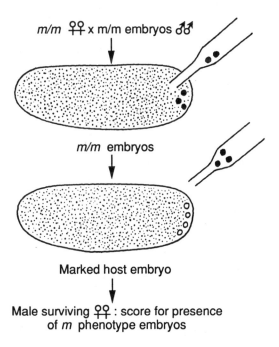

Fig. 4.22. Pole cell transplantation experiment to test for tissue site of maternal mutant defect. See text for description.

By such tests, *tor, trk, fs(1)N,* and *fs(1)ph* are all germ line dependent—when pole cells homozygous for these mutations are transplanted into nonmutant hosts, they give rise to embryos showing the terminal mutant phenotype (Perrimon and Gans, 1983; Schüpbach and Wieschaus, 1986a). In contrast, *tsl* germ line cells in a wild-type soma give rise to wild-type embryos and, therefore, the site of action of *tsl* is identified as the follicle cells (Stevens et al., 1990).

Thus, of the four genes shown to act prior to *tor*, only *tsl* acts in the follicle cells. This begins to provide a cellular context for the first part of the genetic pathway, outlined above. The simplest interpretation is that the initial signal in the terminal gene pathway is elaborated in the follicle cells, mediated in some way by the *tsl* gene product, and is then transmitted to the oocyte, being transduced by the *tor* gene product (perhaps via the other three gene products that act prior to *tor*). Given that *tor* protein is distributed throughout the oocyte membrane, one must further postulate that the initial signal is released specifically from the follicle cells at the poles of the oocyte. Since the pole cell transplant experiments establish only that *tsl* acts in the follicle cells, as opposed to the germ line cells, one needs a further method to determine whether *tsl* activity is supplied specifically by these terminal follicle cell groups. Ideally, one would like to make different mosaics, in which the terminal follicle cells, but not others, were mutant; one could then determine whether these particular mosaics, but not others, showed the mutant phenotype.

The genetic technique for creating such mosaics makes use of the phenomenon of mitotic recombination. As its name implies, mitotic recombination is the exchange of chromosome segments between paired homologous chromosomes in mitotically dividing cells. In contrast to standard meiotic recombination, which accompanies the production of haploid gametes, mitotic recombination produces diploid recombinant cells; it can take place in either somatic cells or premeiotic diploid germ line cells, and with suitable markers, its occurrence can be readily detectable in *Drosophila*. Although it occurs spontaneously, it can also be stimulated by X-ray treatment, which, by creating chromosome breaks, facilitates exchange of segments between the paired chromosomes. Like meiotic recombination, mitotic exchange takes place between replicated chromosomes when each chromosome consists of two chromatids tied together at the centromere.

In homozygous wild-type cells, the occurrence of mitotic recombination will be without genetic consequence. However, in cells that are heterozygous for recessive mutations, mitotic recombination creates homozygous diploid cells that develop into marked clones. When the recombining cell is heterozygous for just one recessive genetic marker and its wild-type allele, recombination gives rise to a clone homozygous for the mutation and a second clone homozygous for the wild-type allele. However, if the cells are heterozygous for genetic markers in *trans* on the two homologs, recombination can produce two marked daughter cells that grow into two marked clones; such double clones are termed "twin spots." Mitotic recombination is diagramed in Figure 4.23 and explained in more detail in the legend.

To determine whether *tsl* is specifically required in the terminal follicle cells, Stevens et al. (1990) induced homozygous *tsl* clones by X-ray treatment at various stages of development in females and then allowed the X-rayed animals to mate and produce embryos. Since the frequency of induced clones in any particular location is relatively low, even with X-rays, more than 10,000 embryos were examined and scored

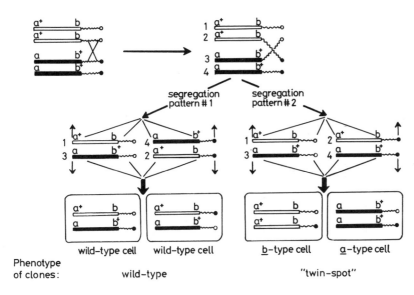

Fig. 4.23. Mitotic recombination in a *trans* double heterozygote. Exchange always occurs in the postreplicative (four-strand) stage, and usually in the centromeric heterochromatin (as shown). Following exchange, the chromosomes can segregate in either of two equally likely patterns. In segregation pattern 1, both nonrecombinant chromosomes are segregated to one daughter cell and both recombinants to the other; the consequence is the production of two genetically wild-type daughter cells. In segregation pattern 2, each daughter cell receives one recombinant and one nonrecombinant chromosome; the two daughters therefore become homozygous for the two opposing genetic markers, generating a twin spot. For a recombination between *a* and *b*, all the daughters from segregation 2 would be heterozygous (and wild-type) for *b*, and half would be homozygous for *a;* the latter would generate single *a* clones.

for the terminal phenotype; to correlate the occurrence of the phenotype with the location of the homozygous mutant follicle cells, a follicle cell marker linked to *tsl,* which produces a thin and distinctive chorion (termed *fragile chorion [fch]*), was employed. Of the seven embryos found exhibiting a mutant terminal phenotype at the posterior pole, six were clearly associated with an overlying *fch* patch at that end and the estimated clonal sizes ranged from 6 to 30 terminal follicle cells. The results demonstrate a strong correlation between inability to produce *tsl*[+] gene product specifically in the terminal follicle cells and the failure of normal posterior terminal development. They do not, however, establish whether *tsl* directly produces the putative ligand molecule that interacts with *tor* or is only indirectly required for that synthesis.

The analysis of the maternal terminal genes, reviewed here, provides an impressive example of the combined use of mutant hunting, genetic inference, and molecular characterization to elucidate a developmental pathway. In this instance, the pathway constitutes a signal transduction chain, which begins in the terminal follicle cells of the egg chamber and proceeds within the termini of the oocyte. The maternally specified apparatus in this transmission, however, only provides the first part of the story. As we shall see in chapter 7, the maternal signal chain activates the localized synthesis by the zygotic genome of at least two transcriptional regulator molecules at the termini. It is the activities of these molecules that promote the normal development of the acron and the telson.

Maternal Setting of the Anteroposterior Pattern: The Posterior Genes

The third system of maternal genes affecting the a-p pattern are the so-called "posterior group genes." Their unifying characteristic is that the embryos from homozygous mutant mothers display abdominal defects. These range from complete deficiencies for the abdominal region to minor pattern abnormalities in one or a few of the central segments; the extent of the abnormalities reflects the degree of residual expression rather than the particular locus involved (Lehmann and Nüsslein-Volhard, 1986; Schüpbach and Wieschaus, 1986b).

Altogether, there are nine genes currently placed in this group; they are listed in Table 4.5. The mutants of all except two, *oskar* (*osk*) and *tudor* (*tud*), are associated with various pleiotropic defects, and, hence, these genes are not devoted exclusively to abdominal development (Lehmann and Nüsslein-Volhard, 1991). In particular, mutants of three of these genes have effects in the "domains" of two of the other maternal patterning systems. Embryos from *stau* mothers, as we have seen, exhibit morphological defects in the head or thorax region and are defective in "anchoring" *bcd* mRNA at the anterior pole (St. Johnston et al., 1991). Embryos from amorphic or near amorphic *cappuccino* (*capu*) and *spire* (*spir*) mothers are strongly dorsalized (Manseau and Schüpbach, 1989). In the latter aspect, *capu* and *spir* could be classified as part of the d-v maternal system, discussed later in this

Table 4.5. Maternal Posterior Group Genes of *Drosophila melanogaster*

Gene	Map position	Embryo phenotype[a]	References
oskar (*osk*)	2-48.5	Abdominal and polar plasm defects	Lehmann and Nüsslein-Volhard (1986, 1991)
staufen (*stau*)	2-83	As for *osk*	Schüpbach and Wieschaus (1986b); Frohnhöfer and Nüsslein-Volhard (1986b)
tudor (*tud*)	2-90	As for *osk*	Nüsslein-Volhard et al. (1982); Boswell and Mahowald (1985)
valois (*vls*)	2-53	As for *osk*	Schüpbach and Wieschaus (1986b); Lehmann and Nüsslein-Volhard (1991)
vasa (*vas*)	2-51	As for *osk*	Schüpbach and Wieschaus (1986b); Lasko and Ashburner (1990)
cappucchino (*capu*)	2-8	Abdominal, polar plasm, and d-v defects	Manseau and Schüpbach (1989)
spire (*spir*)	2-54	As for *capu*	Manseau and Schüpback (1989)
nanos (*nos*)	(91F)	Abdominal defects; pole cells present	Nüsslein-Volhard et al. (1987); Lehmann and Nüsslein-Volhard (1991)
pumilio (*pum*)	3-48	As for *nos*	Lehmann and Nüsslein-Volhard (1987b, 1991)

[a]Refers to phenotype that places gene in posterior group. Severe loss-of-function mutants of these genes show other phenotypes as well. For descriptions of the various phenotypes associated with mutants of these genes, see Lehmann and Nüsslein-Volhard (1991).

chapter, though the dorsalizing effects are undoubtedly secondary ones (Manseau and Schüpbach, 1989).

Within the posterior gene group as a whole, there is a further phenotypic divide. For seven of these genes, the progeny display an abnormal posterior polar plasm, which lacks polar granules and which fails to give rise to pole cells. The two genes in the group whose embryos develop normal pole cells, although exhibiting the abdominal defects characteristic of the group, are *nanos* (*nos*) and *pumilio* (*pum*). The link between the abdominal segmentation defects and the pole cell defects in seven of the mutants suggests that there is a shared early process in the formation of abdomen and pole cells. However, the fact that mutations in two genes can create the abdominal defects without the pole cell deficiency suggests that the latter mutants are affected after the common step(s). In reviewing the posterior gene group, it is convenient to separate the somatic and germ line defects. We will first look at the basis of the abdominal defects in the posterior gene group mutants, and then discuss the nature of the polar plasm defect, in these and other mutants, in the next section.

The abdominal phenotype of the affected embryos is a deficiency for segments A1 to A8, the exact extent of the deficiency being variable and, primarily, a function of the degree of residual activity of the gene. However, the range of abdominal phenotypes is remarkably similar for mutants of different members of the group (Lehmann and Nüsslein-Volhard, 1986; Schüpbach and Wieschaus, 1986b). For hypomorphic alleles, the segment most likely to be missing is A4, with the central region in general (A3–A6) being most sensitive to deficiencies of activity. For somewhat more severe reductions in activity, segments A2 and A7 are the next most likely to be deleted, with A1 and A8 deleted only in the amorphic condition. The strong posterior gene phenotype is shown for an embryo from an *osk* mother (Fig. 4.17C). (*osk* was the first well-characterized gene of the group and was named for the diminutive hero of Gunter Grass's novel *The Tin Drum*.)

Given the phenotypic series of abdominal defects, one might have assumed that the wild-type activity is normally localized in the presumptive abdominal region of early embryos. Surprisingly, this is not the case. When cytoplasm from wild-type embryos is injected into embryos from *osk* mothers, the strongest rescuing activity is found in the posterior polar plasm (Lehmann and Nüsslein-Volhard, 1986), although some activity can be detected in the cytoplasm from the abdominal region (Lehmann and Nüsslein-Volhard, 1987a). This difference in activity between posterior pole and abdominal region itself suggests that there is a posterior gradient of some kind, emanating from the posterior pole, which is involved in abdominal specification.

If there is a "posterior morphogen" of some sort, the question immediately arises as to which member of the posterior gene group, if any, encodes it. The identification of this gene has been approached by means of the polar plasm rescue test, utilizing as recipients embryos derived from *osk, nos, pum, stau, vas*, and *vls* mothers, which can all be rescued by wild-type polar plasm (Lehmann and Nüsslein-Volhard, 1987a, 1991). However, surprisingly, polar plasm from *pum* embryos was also found to have strong rescuing activity, whether the injection was into the abdominal regions of *pum* or *osk* embryos (Lehmann and Nüsslein-Volhard, 1987a). In contrast, there is no rescuing activity from the polar plasm of embryos from *nos, vls, stau, vas*, or *osk* (Lehmann and Nüsslein-Volhard, 1991). The positive result with *pum* embryo polar plasm shows that the *pum* defect cannot be in the manufacture of the posterior morphogen but must be, rather, in the transmission of the signal from the posterior pole. Indeed, for embryos from *pum* mothers, it appears that the intervening pretelson region acts as a block to signal transmission; *pum* embryos that lack this region, produced by making the mothers homozygous simultaneously for *pum* and the terminal group gene *tsl* (which eliminates the telson anlagen in the progeny), show a partial suppression of the abdominal defect (Lehmann and Nüsslein-Volhard, 1987a).

A further surprising finding is that when cytoplasm from an earlier stage of oocyte development, that of stage 10 egg-chamber nurse cells, is used to test for rescue of abdominal development, using embryos from *osk* or *nos* mothers as recipients, rescuing activity *is* found not only for *pum* but for *osk, stau, vas, vls,* and *tud* (Lehmann and Nüsslein-Volhard, 1991). Of the tested genes (*capu* and *spir* were not included), only *nos* egg chambers fail to give rescuing activity (Lehmann and Nüsslein-Volhard, 1991).

These results strongly implicate *nos* as necessary specifically for manufacture of the rescuing activity. Indeed, *nos* mRNA synthesized *in vitro* and injected into eggs rescues the abdominal defects in a concentration-dependent manner (Wang and Lehmann, 1991). In contrast, those posterior maternal genotypes that can supply the rescuing activity from early oocytes, but not from the posterior polar plasm of embryos, must be defective in localizing the substance(s) at the posterior pole. This failure of localization, in the case of *osk, vas, stau, vls,* and *tud,* is presumably a secondary consequence of the failure of normal polar plasm development in these embryos. Putting all the facts together, one can frame a relatively simple scheme for the posterior gene group pathway (Lasko and Ashburner, 1990). In this hypothesis, *osk, stau, vas, vls,* and *tud* are needed (along with *capu* and *spir*) to form a functional posterior polar plasm, perhaps the polar granules themselves; this posterior polar plasm then "anchors" the *nos*-dependent morphogen; finally, maternally specified *pum* activity is required for transport of the morphogen into the abdominal region. One prediction of this hypothesis is that *nos* mRNA should be found at the posterior pole of the embryo. Using the cloned *nos* gene as a probe, Wang and Lehmann (1991) have found that the *nos* mRNA is, indeed, tightly localized at the posterior pole in wild-type embryos and incorporated into the pole cells (Wang and Lehmann, 1991). Similarly, *osk* mRNA and *vas* protein are also localized at the posterior pole, and their localization is essential for subsequent *nos* mRNA localization there (Lasko and Ashburner, 1990; Ephrussi et al., 1991).

With the identification of *nos* as the posterior morphogen gene, the next question concerns the mode of action of this substance. By analogy with *bcd,* one might predict that the *nos* mRNA generates a gradient from the posterior pole, which, in turn, creates a differential regulation of one or more zygotic segmentation genes. As we shall see in chapter 7, this is precisely how the *bcd* gradient works. In this view, the a-p (longitudinal) axis of the embryo would be organized (except for the termini) by two opposing gradients running from opposite poles, each carrying out a differential regulation of one or more region-specific segmentation genes.

Several experiments indicate, however, that this picture, while pleasing, is not a valid one. The findings indicate, rather, that the *Drosophila* posterior gradient serves solely to repress the activity, in the posterior part of the embryo, of a second maternal a-p gradient, that of the product of the *hunchback* (*hb*) gene (Hülskamp et al., 1989; Irish et al., 1989; Struhl, 1989). *hb* is one of the so-called "gap genes," which are expressed from the zygotic genome in broad domains and which are crucial for the development of the normal segment pattern (chapter 7), but unlike the other major gap genes, *hb* is also expressed maternally, forming an mRNA and protein gradient in the late cleavage divisions from anterior to posterior (Tautz, 1988).

In contrast to the *bcd* gradient, which is essential for normal embryonic development, the maternal *hb* gradient proves to be dispensable; expression from a single zygotic copy of *hb* is sufficient to give normal development in the absence of prior maternal *hb* product. This was shown by Lehmann and Nüsslein-Volhard (1986b), who performed pole cell transplants of homozygous *hb* pole cells (from generated *hb/hb* embryos) into embryos carrying a dominant female-sterile mutation (in the gene *ovo*^{D}). Since in all females surviving the transplantation the only germ line cells capable of giving rise to embryos will be those derived from the donor, all fertile host females must carry donor germ line cells. Such females were mated to *hb*$^+$ males and found to produce viable, normal offspring, showing that the maternal *hb* gradient is not required for embryonic development. (In such pole cell transplant experiments, the donor embryos are usually a mixture of homozygous *hb* embryos and *hb/+* heterozygotes. The usual procedure, for the cross that produces the embryos, is to employ a balancer chromosome, bearing distinctive dominant markers, for the + allele chromosome. Pole cell chimeras derived from the heterozygous donor embryos can then be distinguished by the presence of the balancer in some fraction of the offspring.)

Though the maternal *hb* gradient is evidently dispensable, it is not without activity. The essential clue to its role in early development was the observation that in embryos from *pum* or *nos* mothers, and hence showing the posterior group defect, there is a uniform level of *hb* protein throughout the embryo (Tautz, 1988). The fact that the a-p gradient of maternal *hb* RNA is seen in these embryos suggests that the posterior group genes somehow repress the translation of this RNA (Tautz, 1988).

If repression of translation of maternal *hb* RNA is the significant developmental function of the posterior gradients, then embryos lacking *both* the maternal *hb* product and the posterior gene activity should show essentially normal development. This is exactly what is observed. When maternal germ lines are made homozygous for both *hb* and *nos* and pole cell chimeras are constructed, fully normal, viable embryos are formed, and these can give rise to viable adults (Hülskamp et al., 1989; Irish et al., 1989) (Fig. 4.24).

Furthermore, the genetic inference that *nos* acts to inhibit *hb* translation has been confirmed directly by molecular means. While control early embryos (pole bud stage) from *nos* mothers show uniform levels of Hb protein, Hb protein expression is abolished in the abdominal region when *in vitro* synthesized *nos* mRNA is injected there (Wang and Lehmann, 1991).

The implication of all the evidence is that, in the absence of maternal *hb* and the posterior *nos* activity, the *bcd* gradient is sufficient to initiate a normal segmental pattern. The *nos* morphogen does not "instruct" the details of abdominal development but is an inhibitor of an activity (Hb protein) that would, otherwise, prevent normal abdominal development. As will be described in chapter 7, this effect of maternal *hb* activity is an inhibition of one or more zygotic genes in the posterior region, whose expression is required for abdominal development.

It would appear that *Drosophila* has evolved something rather complicated for itself, namely the elaboration of a posterior-based activity designed to antagonize the action of a single maternally expressed gene. The precise answer to why this should be so may be lost in the mists of evolutionary time, but one possibility is that the system

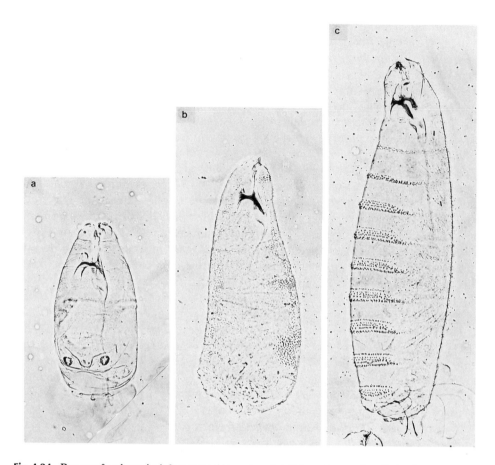

Fig. 4.24. Rescue of embryonic defect, caused by maternal deficiency for *nos*, by elimination of maternal *hb* activity. **a**: Embryo derived from *nos− hb+* mother. **b**: Embryo derived from mother homozygous for *nos−*, but hemizygous for *hb*; there is a partial rescue of the abdominal defects caused by maternal *nos* deficiency; **c**: Embryo from *hb−* germ line clones of *nos* mother; rescue of defects is complete. (Reprinted with permission from *Nature*, vol. 338, pp. 646–648 © 1989, Macmillan Magazines Ltd.)

reflects the rather odd developmental pattern that the Diptera use, namely the "long germ band" system in which segments are laid down simultaneously throughout the presegmental region (the germ band). In contrast, the early ancestors of the Diptera probably used a different mode, the "short germ band" arrangement in which segments are laid down sequentially in an a-p direction and which is found in other, more primitive insect systems today.

It may be that the maternal *hb* gradient is ancient in evolutionary terms and, in more primitive insects, inhibits what would be the "premature" development of abdominal segments, thus facilitating their appearance in the short germ band manner. If such were the case, then *nos* (and, perhaps, *pum*) evolved in the Diptera to suppress posterior *hb* activity, allowing simultaneous abdominal segment development and, hence, the long germ band mode. Given the pleiotropy of *nos* and *pum* mutants (Lehmann and Nüsslein-Volhard, 1991), one or both genes might well have been present in the early insects but serving other functions.

This evolutionary scenario is not the only one that can be imagined, but it can help to rationalize what appears to be, in strictly functional terms, an unnecessarily complicated system. As homologs of the various genes are isolated from short germ band insects and their expression patterns studied, the probable evolutionary origins of the *Drosophila* posterior system should become clearer.

Germ Line Determination: Is It a Special Case?

The precise nature of the relationship between somatic cell and germ line determination in the early *Drosophila* embryo has long been a contentious issue but seems to be nearing resolution. That there are links between posterior somatic development and polar plasm formation is clear from the phenotypes of seven of the nine posterior group mutants, where both processes are aberrant, and from the examination of the so-called *grandchildless* (*gs*) mutants (discussed below). On the other hand, it is clear that the linkage is not obligatory. Thus, the *bic* class of maternal effect mutants can completely transform the soma of the anterior blastoderm in progeny embryos into abdominal segments without the accompanying formation of anterior polar plasm or pole cell formation (Fig. 4.25) (Nüsslein-Volhard, 1979a; Manseau and Schüpbach, 1989; Lasko and Ashburner, 1990). Indeed, *nos* abdomen-rescuing activity has been shown to be redistributed to the anterior pole in the absence of polar plasm development in *BicD* mutants (Wharton and Struhl, 1989; Lehmann and Nüsslein-Volhard, 1991). Conversely, polar plasm and pole cell formation appear fully normal in *nos* and *pum* embryos, although these embryos show the typical abdominal defects of the posterior gene mutants.

At present, the simplest explanation of the connection between posterior polar plasm formation and abdominal development is that something in the polar plasm, quite possibly the polar granules themselves, acts as an anchor for *nos* activity. If polar plasm assembly is disrupted, the anchor is lost and *nos* activity will not be sufficiently concentrated at the posterior pole (Lasko and Ashburner, 1990). Under this interpretation, the dominant bicaudal mutants, *BicC* and *BicD*, provoke the development of anterior abdominal structures by distributing sufficient *nos* activity to the anterior pole (while leaving sufficient amounts at the posterior pole for posterior abdominal development) (Wharton and Struhl, 1989; Lehmann and Nüsslein-Volhard, 1991). In this connection, it is of interest that in *BicD* mutants there is also a bipolar distribution of *osk* RNA, though the anterior *osk* RNA is not as tightly localized to the anterior pole as that at the posterior pole (Ephrussi et al., 1991). Perhaps the polar granules are particularly good anchors for *nos* mRNA but the anterior pole has some intrinsic anchoring capability.

Apart from its role in abdominal development, however, posterior polar plasm is crucially important in its own right for germ line development. Intriguingly, it appears to act as a qualitatively distinct "determinant" for pole cell, and hence germ line, formation. The evidence also comes from polar plasm transplantation experiments. When posterior polar plasm was transplanted to anterior or ventral locations in early cleavage-stage embryos, pole cells were produced at these locations (Illmensee and Mahowald, 1974; Mahowald et al., 1979). When these induced

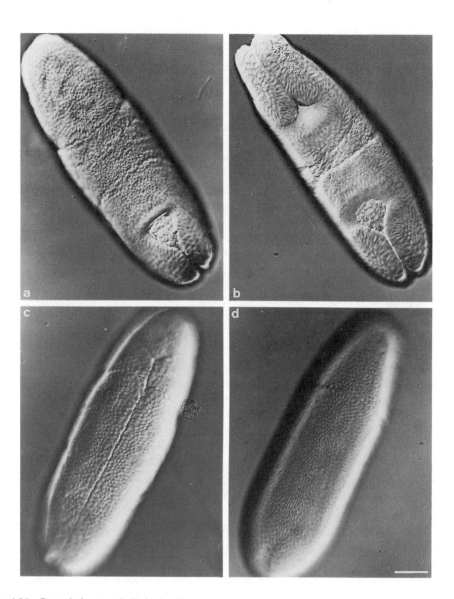

Fig. 4.25. Gastrulation morphologies in wild-type and embryos from *bic* and *dll* + mothers. **a**: Embryo from wild-type mother. **b**: Embryo from *bic* mother. Both embryos are seen from the dorsal side, with the anterior end in the upper left corner. Pole cells are carried forward by posterior midgut invagination. In the *bic* embryo, the comparable invagination occurs at the anterior end, but there is no cluster of pole cells. **c**: Wild-type embryo seen from the ventral side. **d**: *dl* dominant embryo seen from the ventral side. Note the absence of midventral invagination. (Reproduced from Nüsslein-Volhard, 1979a, with permission of the publisher.)

pole cells were transplanted into the posterior polar regions of genotypically distinct host recipient embryos, they gave rise to substantial portions of the germ line of the hosts, as shown by progeny tests. The experiment is diagramed in Figure 4.26. The results show unequivocally that posterior polar plasm has some special quality that specifically induces germ line formation. Whatever the nature of this property, it is present in stage 13 oocytes, as shown by polar plasm transplantation (Illmensee et al., 1976).

(It should be noted that these experiments were done before the posterior gene group was discovered and prior to the various transplantation experiments, described above, that show the capacity of posterior pole plasm to rescue somatic abdominal developmental defects in embryos of posterior group mothers. These later results, however, raise the retrospective question of why, in the Illmensee-Mahowald work, induction of abdominal cells was not discovered. It may be that in the earlier work the amounts of cytoplasm were insufficient to show somatic pattern alterations or, alternatively, that such cells were formed but, not being sought, were not observed.)

Other and earlier genetic work also reinforced the notion of the distinctiveness of the posterior polar plasm as a germ line determinant and its independence from somatic development. These observations concerned the group of mutants collectively designated *grandchildless* or *gs* mutants. These are recessive maternal effect mutants that produce fully viable but sterile progeny; the progeny possess gonads but agametic ones, there being no functional germ line in these animals. In all cases, the failure of gamete formation stems from a failure of pole cell formation or function. The first *Drosophila* mutant of this type to be isolated was the *gs* mutant of *D. subobscura* (Spurway, 1948). The characterization of this mutant subsequently prompted a search for comparable *gs* mutants in *D. melanogaster*, of which a number have been isolated (Table 4.6).

Although all of these mutants were isolated on the basis of the canonical *gs* phenotype—the production of morphologically normal progeny that are sterile—

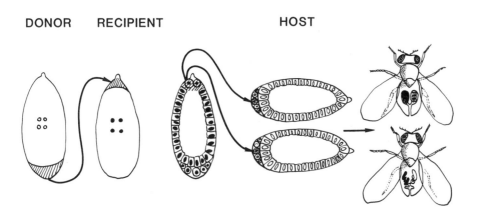

Fig. 4.26. Procedure for inducing pole cells at nonposterior locations and using induced pole cells to create germ line chimeras. The genotypes of the donor, recipient, and host are all different to permit unambiguous delineation of the source of the induced pole cells. (Reproduced from Mahowald et al., 1979, with permission of the publisher.)

Table 4.6. *Grandchildless* Mutants of *Drosophila melanogaster*

Locus	Chromosomal position	References
fs(1)nas A	1-0.0	Counce and Ede (1957)
gs[87]	1-20.0	Thierry-Mieg (1976)
gs(1)N26	1-33.8	Niki and Okada (1981)
gs(1)N441	1-39.6	Niki and Okada (1981)
gs(2)M	Autosomal	Mariol (1981)

nearly all of the mutants have been found to have some somatic defects. Furthermore, some temperature-sensitive *fs* mutants show a *gs* phenotype when they produce eggs at permissive temperature (Gans et al., 1975). These observations, coupled, of course, with the discovery of the posterior group genes, suggest that the germ line defect in these mutants may be a secondary effect.

For the *gs*-type mutants, the link between germ line and somatic defects may be in the cleavage behavior of the mutants, since many, including the original mutant in *D. subobscura*, show delayed nuclear migration into the polar plasm (Mahowald et al., 1979; Niki and Okada, 1981). Such delayed nuclei might therefore suffer the same fate as their blastoderm-destined sibs. Furthermore, these abnormalities of nuclear migration, which are not always restricted to the posterior pole region, could be responsible for the somatic structure defects seen in some of the *gs* mutants. Interestingly, two allelic temperature-sensitive *fs* mutants that show a *gs* phenotype at the permissive temperature produce embryos at the restrictive temperature with exaggerated mitotic waves and other cleavage abnormalities (Gans et al., 1975); this observation suggests that the permissive temperature (*gs*) phenotype reflects a less severe cleavage defect. Indeed, the correlation between cellularization defects and particular alterations of somatic and/or germ line development has been made for a number of other maternal effect genes. These include *BicC* (Lasko and Ashburner, 1990), *capu* and *spir* (Manseau and Schüpbach, 1989), and three (pleiotropic) terminal group genes, *fs(1)N* (Degelmann et al., 1985), *fs(1)ph* (Perrimon et al., 1986), and *l(1)ph* (Perrimon et al., 1985).

One way to interpret such results is that nuclei must enter the polar plasm within a certain time during cleavage or they experience something that prevents them from producing functional pole cells. From this perspective, the function of the posterior polar plasm may not be to *instruct* pole cells to become germ line cells but to *protect* pole cell nuclei from events of determinative restriction; as a result of this shielding by the polar plasm, the pole cells would retain their totipotency and hence the capacity to be germ line precursor cells (Kalthoff, 1979). In the fruit fly, at least, this characteristic may be established early by removal of the germ line precursor cells from the somatic determination system. Once set aside as pole cells, however, the germ line cells of *Drosophila* are no longer susceptible to a change in developmental capacity when transplanted to new environments in the embryo (Technau and Campos-Ortega, 1986b).

Further analysis of the posterior group genes, and their activities, should help to elucidate both the pathway of polar plasm construction and the nature of the germ line determinant. Some information is already available, from studies of *vasa* (*vas*), whose activity is needed not only for pole cell differentiation but for early steps of oocyte differentiation. *vas* has been cloned and its sequence shown to be related to certain known RNA binding proteins, in particular, eukaryotic initiation factor 4A, which serves as an initiation factor through the RNA helicase activity it possesses (Lasko and Ashburner, 1988). Presumably, the role of the *vas* gene product in germ line determination and elaboration of the posterior morphogen involves an RNA binding activity and may involve the sequestration of particular RNA molecules in pole cells.

However, while proper *vas* protein localization is essential for pole cell development, it is far from sufficient; embryos from *tudor* (*tud*) and *valois* (*vls*) show normal *vas* protein localization in the pole cells (Lasko and Ashburner, 1990). In contrast, embryos from *osk, stau, cap*, and *spir* mothers show altered *vas* protein localization and their activities are clearly essential for polar plasm formation. A provisional pathway of polar granule formation, with respect to posterior gene activities, based upon these and various other observations, is shown in Figure 4.27.

Further characterization of the other posterior group genes, in terms of their roles in pole cell formation, should help to clarify the nature of germ line determination. To date, it has been impossible to dissociate pole cell formation from polar granule formation, suggesting that the substance(s) necessary for germ line determination are the granules themselves or one of their components. This issue, unresolved for decades, may yield a solution soon, given the recent progress in analyzing the pathway of polar plasm formation.

A final thought, perhaps worth considering, is the intriguing coincidence that in both *C. elegans* and *Drosophila* there is a cytoplasm at the posterior pole that is associated with the formation of germ line precursor cells. Furthermore, in many other animals, such as the mouse (chapter 8), these cells also arise at the posterior tip of the

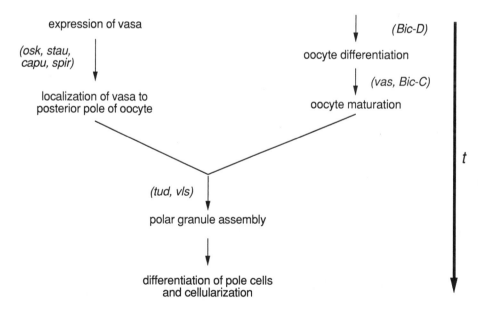

Fig. 4.27. A pathway scheme for the formation of polar granules in developing pole cells, involving some of the known gene activities. For discussion, see text and Lasko and Ashburner, 1990.

embryo proper. Is there a hint, in all of this, of some connection between "posterior-ness" and capacity for totipotency and, if so, what is its precise basis?

MATERNAL COMPONENTS THAT ORGANIZE THE DORSOVENTRAL AXIS OF THE EMBRYO

A notable feature of maternal effect mutants that disrupt embryonic patterning along the a-p axis is that, for most, the progeny embryos are substantially unchanged along the d-v axis, and hence in their germ layer assignments (Fig. 4.10). From such observations, and from the discovery of mutants in the *dorsal* (*dl*) gene, whose maternal effects disrupt patterning along the d-v axis while leaving segmentation unimpaired, came the early inference that the two axes are genetically specified by separate maternal systems (Nüsslein-Volhard, 1979a).

Indeed, subsequent work has largely confirmed this inference. A group of 18 maternal effect genes has been found, whose members act primarily or exclusively in setting d-v pattern without altering the basic segmental pattern. In terms of their gross mutant phenotypic effects, these d-v patterning genes can be divided into two catego-ries (Table 4.7). Class A, the smaller group, includes those maternal genes required for setting both egg chamber and embryonic d-v polarity, while class B genes affect only

Table 4.7. Maternal Effect Genes Affecting Dorsoventral Polarity in *Drosophila melanogaster*

Gene	Map position	Loss-of-function phenotype	Molecular characteristics
Maternal effect genes required for eggshell and embryonic polarity			
fs(1)K10 (*K10*)	1-0.5	Dorsalizing	Transcription factor
cappuccino (*capu*)	2-8	Dorsalizing	
spire (*spir*)	2-54	Dorsalizing	
torpedo (*top*)	2-100	Ventralizing	EGF receptor homolog
gurken (*grk*)	2-30	Ventralizing	
cornichon (*cni*)		Ventralizing	
Maternal effect genes required for embryonic polarity only			
nudel (*ndl*)	3-17	Dorsalizing	
pipe (*pip*)	3-47	Dorsalizing	
windbeutel (*wind*)		Dorsalizing	
gastrulation defective (*gd*)	1-37	Dorsalizing	Serine protease
snake (*snk*)	3-52	Dorsalizing	Serine protease
easter (*ea*)	3-57	Dorsalizing	Serine protease
spatzle (*spz*)	3-92	Dorsalizing	
Toll (*Tl*)	3-91	Dorsalizing	Transmembrane receptor
pelle (*pll*)	3-92	Dorsalizing	
tube (*tub*)	3-47	Dorsalizing	
dorsal (*dl*)	2-52.9	Dorsalizing	Ventral morphogen, *rel* family
cactus (*cact*)	2-51	Ventralizing	Cytoplasmic anchor for *dl* morphogen

Source. Adapted from Govind and Steward (1991).

embryonic d-v polarity. To introduce these genes, two of the genes first identified as affecting d-v pattern will be described, *fs(1)K10* (class A) and *dl* (class B).

K10 was first identified as a loss-of-function female-sterile mutant, characterized by the production of abnormal eggs (Wieschaus et al., 1978). Eggs produced by *K10* and wild-type mothers are shown in Figure 4.28. In normal eggs, the principal landmark of the dorsal surface is a pair of long dorsal appendages that are secreted during the final stages of oogenesis. In many eggs from homozygous *K10* mothers, these appendages are greatly extended around the embryo both laterally and ventrally, being distally fused in the ventral midline; together they form an incomplete ring, with only the topmost part of the dorsal surface left unencircled (Fig. 4.28). The extension of these structures effectively produces a "dorsalization" of the chorionic pattern. This change in d-v polarity is also shown by the follicle cell "footprints," the traces left by the follicle cells when they are sloughed off at the end of oogenesis. In normal eggs, these imprints are considerably elongated on the dorsal side, relative to the ventral and lateral sides. In the anterior portion of *K10* eggs, the ventral and lateral footprints display the elongated character typical of wild-type dorsal footprints. (Whether there is comparable dorsalization in the posterior end is difficult to judge because the differences in the wild-type pattern at the posterior end are less pronounced). *K10* is variably expressive, with some eggs showing a lesser degree of dorsalization than that shown in the figure.

A surprising aspect of *K10* action is its site of expression. Because the chorion is secreted by the follicle cells, it might be expected that *K10* is expressed in these cells. Pole cell transplantations from *K10* mothers show that *K10*, whose phenotype can be scored directly by inspection of the eggs, is expressed in the germ line; eggs derived from *K10* pole cells, produced within a wild-type soma, produce *K10* eggs (Wieschaus et al., 1978). Indeed, *K10* is not only expressed in the germ line but, as shown by molecular studies, is one of the few known genes to be expressed by the oocyte nucleus during early oogenesis (Haenlin et al., 1987). Given the pronounced effect of *K10* on the follicle cells, there must be some form of cellular communication between the developing germ line cells of the egg chamber and the follicle cells.

K10's effects, however, are not limited to those on the eggshell, but involve a dorsalization of the embryo as well. This was demonstrated by culturing imaginal cells derived from such embryos. The embryos are broken up, cultured as fragments in larvae, and then allowed to metamorphose along with the host larvae; the imaginal implants are then scored. While implants derived from embryos of wild-type mothers yield predominantly (about 80%) structures derived from midlateral discs (eye, antenna, wing, and leg), with only a few showing dorsal disc (labial and humerus) or genital disc provenance, implants from embryos produced by *K10* mothers are predominantly genital disc material, with a secondary contribution of dorsal disc structures and none from the midlateral discs (cited in Wieschaus, 1979).

Of the six maternal effect genes that affect d-v patterning of both egg and embryo, loss-of-function mutations in three (*K10, capu,* and *spir*) produce dorsalization. (*capu* and *spir* embryos, as noted earlier, also exhibit defects along the a-p axis.) In the other three genes of class A (*torpedo* [*top*], *gurken* [*grk*], and *cornichon* [*cni*]), loss-of-function mutations in the maternal genome produce ventralization of the embryo. These embryos show much more mesoderm than do normal embryos, and the region that gives rise, in wild-type, to amnioserosal membranes appears to be lateral ectoderm. Like *K10*, all except one of these genes act in the germ line (Schüpbach, 1987;

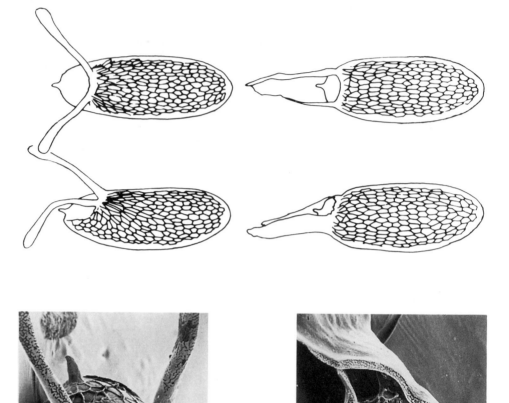

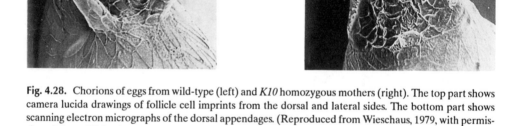

Fig. 4.28. Chorions of eggs from wild-type (left) and *K10* homozygous mothers (right). The top part shows camera lucida drawings of follicle cell imprints from the dorsal and lateral sides. The bottom part shows scanning electron micrographs of the dorsal appendages. (Reproduced from Wieschaus, 1979, with permission of the publisher.)

Manseau and Schüpbach, 1989; Govind and Steward, 1991). The exception is *top*, which acts in the soma (the follicle cells) (Schüpbach, 1987). The somatic expression of *top* shows that the signaling between germ line and soma, first indicated by the findings on *K10*, involves active somatic cell participation. The role of *top* may be, indeed, to transmit the germ line-derived signal(s), ultimately created by the other genes of class A, from the cellular interface between soma and germ line. That *top* may be acting as some form of receptor is indicated by its sequence. The *top* gene encodes a protein homologous to the receptor for the vertebrate growth factor EGF (epidermal growth factor) (Price et al., 1989; Schejter and Shilo, 1989).

Of the two classes, A and B, it seems probable that the members of class A, affecting both egg and embryo, act prior to class B, those affecting just the embryo.

This supposition has been tested by determining epistatic relationships between mutants of the two classes that differ in phenotype and has been confirmed. When females are made doubly homozygous for either *grk* or *top* (class A), which ventralize, and for *dl* (class B), the embryos are found to be dorsalized, showing that *dl* is epistatic to both *grk* and *top* and acts downstream of them (Schüpbach, 1987).

Of the genes in class B, *dl* was the first to be identified (Nüsslein-Volhard et al., 1980). *dl* and the other 10 genes whose maternal loss-of-function phenotype creates dorsalized embryos are referred to as the "dorsal group"; the remaining member of category B is *cactus* (*cact*), whose maternal loss-of-function phenotype is ventralization of the embryo.

The phenotype of embryos from homozygous *dl* mothers is typical of dorsal group mutants. The first sign of abnormality in embryos is that they do not show the midventral invagination at the beginning of mesoderm development (see chapter 7), and thus never develop mesoderm. The dorsalizing effect can be seen at blastoderm itself. In normal embryos, cellularization of the ventral cells is completed 5 min before that of the dorsalmost cells, while in *dl* embryos, cellularization is simultaneous at the dorsal and ventral midlines (Anderson and Nüsslein-Volhard, 1983). Later-stage *dl* embryos lack the characteristic ventral denticle bands of the wild-type embryos, and consist of segmented tubes of hypoderm (which, as in *C. elegans*, secretes the larval cuticle) (Fig. 4.29). They are covered by the fine hairs that are characteristic of the dorsal side of normal embryos and enclose a yolk cylinder, lacking internal tissues (since the endoderm and mesoderm are derived from blastodermal cells at lateral and ventral positions).

dl shows an interesting dosage effect and this has been used to explore the gene's mode of action. At 29°C, the original *dl* allele (which is an amorph) shows a partial dominant phenotype; *dl/+* females grown at this temperature can produce embryos that are partially dorsalized (lacking the midventral mesodermal invagination) (Fig. 4.25 c,d). That this phenotype is a true change in specification to more dorsal values is revealed in two ways. First, the embryos show a narrowing of the ventrally located denticle bands. Second, the fate map of the ventral and lateral hypoderm on the blastoderm embryo can be shown to have contracted ventrally. In normal embryos, the ventral hypoderm of the larva is derived from blastodermal cells located within the two strips of the so-called "neurogenic ectoderm," which lies on either side of the band of cells that develop into the mesoderm (Fig. 4.10). The neurogenic ectodermal regions generate both the epidermal cells that give rise to the hypoderm and precursor cells of the neural system, the neuroblasts. In wild-type blastoderm embryos, localized laser beam irradiation to the midventral line never results in defects in the larval hypodermis (Nüsslein-Volhard et al., 1980). However, when blastoderm embryos from *dl* dominant mothers are irradiated in the ventral midline, a high percentage develop ventral hypodermal defects, indicating that some dorsalization of the ventralmost cells has taken place.

The *dl* dominant phenotype and the effects of hypomorphic alleles of other members of the dorsal group show that there can be degrees of dorsalization, that it is not an all-or-nothing process. A simple hypothesis to explain this is to posit a bilateral gradient of ventralizing morphogen, with highest values at the midventral line of blastoderm embryos and a decreasing concentration in morphogen, on both sides, as one moves dorsally (Nüsslein-Volhard, 1979a). Loss-of-function mutants of

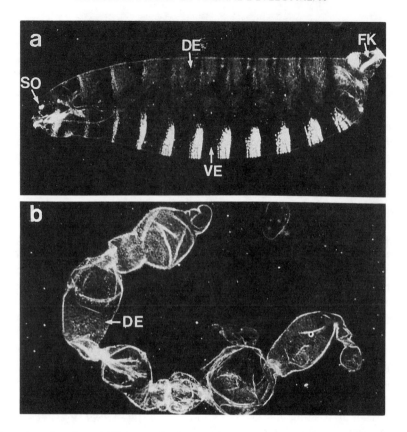

Fig. 4.29. a: An embryo from a wild-type mother. **b:** An embryo derived from a dl mother; all ventral pattern elements are missing, the embryo consisting of a tube of dorsalized hypoderm surrounding a yolk center. DE, dorsal ectoderm; VE, ventral ectoderm; FK, filykoryser. (Figure courtesy of Dr. C. Nüsslein-Volhard.)

the dorsal group would either abolish the gradient entirely or diminish it, as in the *dl* dominant phenotype.

In this interpretation, all the genes of class B are viewed as either required for localizing or synthesizing the ventralizing morphogen. Since the proposed substance promotes ventralization, the *cact* gene, the sole member of class B whose loss-of-function phenotype is ventralization, cannot be the morphogen gene itself but must be part of the localization apparatus. In contrast, the gene encoding the morphogen must be one of the 11 genes of the dorsal group. Which one is it and what, precisely, are the other genes of the dorsal group doing?

Again, epistatic relationships can be informative. If the dorsal group genes are viewed as creating a pathway that leads to a morphogen gradient, the synthesis of the morphogen should be at or near the end of the pathway and loss-of-function mutations in the morphogen encoding gene (assuming that one gene in the group has this role) should be epistatic to most or all of the other dorsal group gene mutants. To perform such tests, one needs mutants of distinguishable phenotype; in this case, the existence of dominant ventralizing, gain-of-function mutations in the *Toll* (*Tl*) gene provide a contrasting condition to that of the dorsal group phenotype (Anderson and

Nüsslein-Volhard, 1984). By such tests, *snake* (*snk*), *easter* (*ea*), *spatzle* (*spz*), and *gastrulation defective* (*gd*) are found to be upstream of *Tl* (that is, *Tl* ventralizing mutations are epistatic to the dorsalizing mutations in those genes), while *pelle* (*pll*), *tube* (*tub*), and *dl* itself are epistatic to *Tl* dominants and hence act after *Tl* (Anderson et al., 1985; Anderson, cited in Govind and Steward, 1991).

Of the three genes whose products act after *Tl*, it is *dl* that is central to the establishment of embryonic d-v polarity. The cloning of the gene and subsequent molecular studies have revealed that it is the *dl* gene product that comprises the gradient, while molecular studies of the other gene products have allowed, in combination with the genetics, the construction of a model of the d-v pathway.

Molecular Studies

The first indication that *dl* activity may itself comprise the putative d-v morphogen was obtained from cytoplasmic rescue experiments, using cytoplasm derived from wild-type embryos and injecting it into embryos from *dl* mothers. These experiments revealed a degree of localization of *dl*+ rescuing activity between early cleavage and syncytial blastoderm toward the ventral side of the embryo (Santamaria and Nüsslein-Volhard, 1983). These experiments did not reveal whether the rescuing species was RNA or protein. Subsequent experiments, however, failed to show significant rescue with poly A+ mRNA fractions for *dl*, though the presence of *dl* mRNA could be demonstrated, by *in situ* hybridization, in embryos. In contrast, rescue with RNA fractions could be obtained for *snk, ea, tub, pll* and *Tl* (Anderson and Nüsslein-Volhard, 1984). By inference, the *dl* rescuing species reported by Santamaria and Nüsslein-Volhard was probably protein.

Indeed, while the maternally synthesized *dl* RNA is found to be uniformly distributed in the embryo at cellular blastoderm, *dl* protein, traced by specific antibody staining, is found to be distributed in a ventral-to-dorsal gradient by late syncytial blastoderm. Most significantly, this gradient is a nuclear gradient, in which ventral nuclei are heavily stained for protein, with little or no staining in either dorsal nuclei or cytoplasm by cellular blastoderm (Steward et al., 1988). The *dl* nuclear gradient is shown in Figure 4.30B.

The formation of the nuclear gradient takes place during late embryogenesis. Prior to cleavage stage 10, the protein is uniformly distributed in the cytoplasm along the d-v axis. Nuclear localization begins around cleavage stage 11, with ventral nuclear localization becoming pronounced by stage 12 and complete by stage 14 (Rushlow et al., 1989; Steward, 1989). At cellular blastoderm, there is a strip 14–15 nuclei wide of high, and possibly uniform, ventral nuclear *dl* staining, with a sharp decrease on both sides of this strip, over a region about 5 nuclei wide (Rushlow et al., 1989).

These observations, in themselves, implicate nuclear localization of *dl* in ventral nuclei as a key event in establishing d-v polarity by the *dl* gradient. However, genetic observations establish it beyond doubt. Mutations in maternal genes upstream of *dl* that create dorsalization all weaken or eliminate the nuclear gradient, causing *dl* protein to remain cytoplasmic. Conversely, ventralizing conditions, such as occur in embryos from *cact* mothers, are associated with more extensive, or uniform, nuclear localization of *dl* along the d-v axis (Roth et al., 1989; Steward, 1989) If Dl protein, and specifically that part of it localized to the nucleus is the effective species, then *dl*

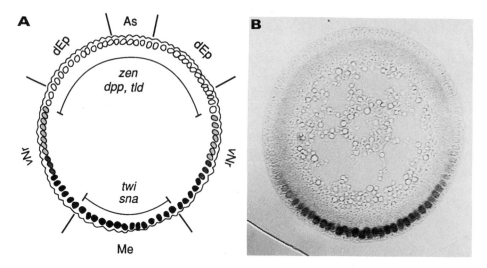

Fig. 4.30. The *dl* protein nuclear gradient is shown in **B**. The gene activities shown in **A** will be discussed in Chapter 7. (Reproduced from Govind and Steward, 1991, with permission of the publisher.)

should be epistatic to *cact*, and such is the case: Embryos derived from *cact, dl* mothers are fully dorsalized (Roth et al., 1989). The results confirm the role of *dl* as the active species and suggest that the role of the *cact⁺* product is to restrict *dl* nuclear localization to the ventral side.

These findings raise further questions, of course. The first concerns how the Dl protein works. The answer, almost certainly, is that *dl* is a transcription factor and, hence, its differential nuclear localization should promote differential gene transcription. The first indication that *dl* is a transcriptional regulator was the discovery that it is homologous to the avian oncogene v-*rel* (Steward, 1987), the *rel* gene itself being a member of a family of transcriptional regulator genes (Govind and Steward, 1991). Subsequently, *dl* was shown to be a DNA binding protein that recognizes specific promoters (Ip et al., 1991). In chapter 7 we will take a further look at transcriptional control along the d-v axis, in which *dl* plays a part.

The second set of questions concern the roles of the other dorsal group genes. Three of the upstream genes, *snk, gd,* and *ea*, are known to encode serine proteases (see Table 4.7), and it has been proposed that this part of the pathway involves a proteolytic cascade, such as is found in the blood-clotting process (De Lotto and Spierer, 1986). Indeed, *snk* bears some sequence resemblance to the mammalian blood clotting factor, factor IX. Furthermore, these genes are upstream of *Tl* and it is possible that the cascade involves a proteolytic modification of *Tl*, which, perhaps coincidentally, is also homologous to one of the factors in the blood clotting cascade, namely the α chain of platelet glycoprotein 1b (Hashimoto et al., 1988).

The *Tl* gene product is itself a large transmembrane protein, with a long extracellular domain, and appears to be uniformly distributed around the cell membrane of the embryo (Hashimoto et al., 1988). If Tl is a substrate for the proposed proteolytic cascade, then the participation of Tl in establishing d-v polarity would involve a differential ventral activation by partial proteolysis. (The situation would be somewhat analogous to the differential terminal activation of *tor* in the terminal gene pathway, though in the case of Tor, the differential activation does not involve proteolysis).

Another scenario for the localized activation of *Tl* can also be envisaged; it involves localized release of a factor near the ventral side of the embryo that produces a localized activation of the Tl protein. It has been found, using pole cell transplantation, that three members of the dorsal group, namely *nudel (ndl), pipe (pip)*, and *windbeutel (wind)*, do not act in the germ line but in the follicle cells (in contrast to the other tested members of the dorsal group) (Stein et al., 1991). Evidently, these three genes contribute to elaborating a somatic signal required for correct ventralization. Given the further discovery that injected perivitelline fluid (the fluid between the egg cell membrane and the vitelline membrane) can partially restore ventral development to embryos from *ndl, pip*, and *wind* mothers near the site of injection (Stein et al., 1991), it appears that the signal produced by the follicle cells is secreted into the perivitelline fluid. In light of these observations, one can envisage that the ventrally localized activation of Tl protein involves a prior localized, temporally regulated, secretion of a ligand, determined ultimately by maternal *ndl, pip*, and *wind* activity. It is possible that the *ea* and *snk* gene products perform a partial proteolysis of the released ligand, rather than acting on Tl itself.

The activation of *Tl*, by whatever mechanism is involved, would in turn promote, possibly through the activities of *pll* and *tub* (two downstream genes), the differential nuclear localization of the D1 protein. The mechanism that produces this localization is unknown but it may also involve a proteolytic event. Removing even as few as six of the amino acid residues from the C-terminal end of the *dl* protein promotes its nuclear localization (Rushlow et al., 1989). The action of *cact* in the pathway seems to be to act as a cytoplasmic anchor for *dl* protein; in the absence of Tl activation, *cact* protein would continue to hold *dl* protein in the cytoplasm (Govind and Steward, 1991).

A model that encompasses the various observations and purported roles of the various gene products is shown in Figure 4.31. Though it is provisional, it provides a satisfactory picture of the findings to date. Questions yet to be answered concern the nature of the ligand in the perivitelline fluid and the nature of its action, the precise roles of the three genes that act in the follicle cells to produce it, and the factors that cause the ventral follicle cells, specifically, to produce it.

SOME CONCLUSIONS ABOUT "DETERMINATION" AND GRADIENTS IN EARLY *DROSOPHILA* DEVELOPMENT

The last five years have witnessed a remarkable vindication of the ideas that morphogenetic gradients, rather than mosaic determinant systems, govern early *Drosophila* determination. These ideas were based, in part, on both early genetic studies (Nüsslein-Volhard, 1977, 1979a) and surgical manipulations of primitive insect systems (Sander, 1975, 1976) and of *Drosophila* itself (Schubiger and Newman, 1981). However, a major additional stimulus was a theoretical contribution, the models of gradient formation proposed by Meinhardt (1977, 1982).

The evidence for fate-assigning gradients along the a-p axis is now incontrovertible. The decisive findings involve the identification of the a-p gradient of the protein

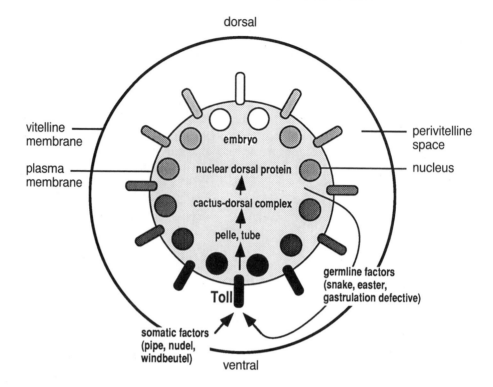

Fig. 4.31. A model of the sequence of maternal gene-mediated effects in the development of dorso-ventral polarity of the *Drosophila* embryo. A somatic signal generated on the ventral side, and transmitted through the perivitelline fluid, activates the *Toll* transmembrane protein. The *snake, easter* and *gastrulation defective* gene products, secreted by the embryo, may act directly on the ligand or on the Toll-ligand complex. The activated *Toll* product, acting through the *pelle* and *tube* gene products, causes the localized release of *dorsal* protein from a cytoplasmic complex with the *cactus* gene product, to produce a nuclear localization of *dorsal* protein. (The figure indicates a graded series of nuclear *dorsal* concentrations but the gradient may not extend as far as is indicated.) See text for further discussion. (Reproduced from Govind and Steward, 1991, with permission of the publisher.)

encoded by *bcd*. The specification of the abdominal segmental region does not appear to involve a comparable gradient, but rather the suppression in the posterior region of the maternal *hb* transcript by the *nos* gene product. As we will see, in chapter 7, the actual specification of posterior pattern seems to involve a complex set of interactions of *zygotically* expressed segmentation genes, whose initial spatial frame is set at its anterior edge by gene expression events triggered by *bcd*.

   The terminal gene system, like the anterior *bcd*-based system, may also involve molecular gradients, though ones of shorter range; these seem to involve graded levels of *tor* activity that may differentially activate particular zygotic gene activities (Casanova, 1990). Finally, there is the case of the *dl* gradient, whose differential nuclear localization underlies the subsequent differential zygotic gene transcription along the d-v axis. Both the terminal and d-v systems involve signal transduction processes that are initiated by localized follicle cell groups outside the embryo but which result in differential gene activations within the embryo. The zygotic gene activities triggered by the *tor* and *dl* gradient systems will be reviewed in chapter 7.

Leaving aside the special aspects of the posterior region, one may ask: Does the existence of the *bcd* protein gradient and its known properties essentially solve the problem of "determination" along the a-p axis? The evidence that the *bcd* gene product(s) differentially regulates two or more zygotically expressed segmentation genes at the syncytial blastoderm stage has been referred to and we will examine the evidence in chapter 7. Such events, however, may not comprise the entire phenomenon of determination. Determination, as noted earlier (see chapter 2), involves at least two seemingly distinct elements: the differential selection of a particular array of developmental fates, events that may be broadly termed ones of "instruction," and the imposition of stable somatic inheritance of those assignments, a process of "commitment" (Sander, 1976). While the activation or repression of gap genes by maternal factors may well constitute a set of primary instructional events, they would not seem to suffice to explain the stability of blastodermal developmental cells states.

That blastodermal cells are, indeed, committed to particular developmental fates, both with respect to eventual segmental assignment along the a-p axis and with respect to histotype (the assignments along the d-v axis), has been discussed in this chapter. These are *cellular* states of stability as assayed with intact, transplanted cells. The significant question from the point of view of mechanism is what these states mean at the *nuclear* level. One possibility is that particular genes, once activated, continue to activate their own expression by means of their own gene products. There are precedents for this in both prokaryotic biology (Ptashne, 1986) and for some homeotic genes in *Drosophila* (chapter 7). In effect, this is one way of creating a stable developmental state by means of dynamic gene expression. An alternative possibility which also involves dynamic gene expression is that a particular cascade of gene expressions is set in train in intact cells, once particular genes are activated, and that these cascades are irreversible. There are, indeed, transcriptional cascades of zygotic gene activities in early *Drosophila* development and they may well contribute to the stability of blastodermal cell states.

There is, however, a third possible mechanism for producing commitment at the nuclear level, namely some form of stable chromatin structure, which creates a permanent repression of those genes not selected for activation and which reinforces the original selected state(s) (reviewed in Weintraub, 1985). This view has, to some extent, gone out of fashion but there is some evidence to support its relevance. These states may be fairly local or involve broad chromosomal regions. With respect to relatively localized chromatin inactivations, a large number of genes whose activities are required for stabilization of segmental identity states has been found, the so-called *Polycomb* (*Pc*) group, and the properties of a few indicate that they are chromatin proteins of some kind (chapter 7).

The existence of larger-scale, regional inactivations is indicated by the appearance of many regions of chromosome condensation within the euchromatic chromosomal arms during the final syncytial blastoderm stages (Foe and Alberts, 1985); such heterochromatic regions could be the chromosomal correlate of repression at the gene level. One advantage of sequestering large sections of the genome in permanently repressed structures is that the "search" time that activator proteins must make for the remaining "open" sequences is shortened and, hence, facilitated.

If a chromatin-based repression mechanism is part of the process of determination, it would nevertheless have to be rapid, given the speed of events in late syncytial blastoderm, operating globally or over broad regions at least, and be sequence and region specific. One possibility is that it involves mitotic waves (Kauffman, 1973) or waves of DNA initiation, followed by conversion of replicative states to transcription potentials (Wilkins, 1976). The possibility of converting replicative states to transcription potential states has been discussed in other systems by von Ehrenstein et al. (1979) and Brown (1984).

Indeed, it seems probable that all three kinds of gene regulation noted above are involved in the creation of blastodermal cell states: initially transient gene activations which are enforced by positive feedback, progressive cascades of gene activation, and, perhaps, one or more modes of chromatin state fixation, leading to quasi-permanent or, at least, stable repression of certain gene activities. As studies on the transcriptional control events, initiated by the various maternally specified molecular gradients, progress, the relationship(s) between the as yet poorly characterized nuclear and chromatin states, on the one hand, and the induced transcriptional changes, on the other, may become clearer. The elucidation of what "determination" signifies at the chromatin level continues to be a question of interest for developmental biologists.

FIVE

MUS MUSCULUS

From Oocyte to Implanting Blastocyst

It seems to us that we will achieve a more coherent understanding of development by identifying cellular mechanisms, rather than by searching for elusive and possibly illusory determinants.

M.H. Johnson et al. (1986)

Fertilization is a developmental node. It may be regarded as the point in time and space from which a new individual develops; but the egg and spermatozoon that meet at that point are themselves each the culmination of a developmental process, the product of its own cell lineage, with an initial genetic endowment developing in interaction with the environment of the parent organism.

A. McLaren (1979)

INTRODUCTION

The sources of contemporary mammalian developmental genetics can be found in the study of mouse mutants in the first part of this century and, before that, in the "mouse fancy" of the late 19th century. In contrast to *Drosophila*, where developmental studies were so long slighted in favor of transmission genetics, developmental hypotheses about and phenotypic characterization of mouse mutants have been an intrinsic part of murine genetics from its early 20th-century beginnings.

Despite their different experimental histories, the mouse and the fruit fly have had an unusual reciprocal relationship as research subjects. Fruit fly work began in 1908, as the result of a suggestion to T.H. Morgan by William Castle, a mouse geneticist, that Morgan take up *Drosophila* for genetic experiments. The features that recommended *Drosophila* were its wealth of surface characteristics, short life cycle, and easy maintenance. There were, however, no mutants of *Drosophila* at the time, and as a genetic model system it started, in effect, as the poor cousin of the mouse. Yet today, one of the most remarkable features of developmental genetics has been the effect of recent *Drosophila* work on ideas and approaches taken in the mouse. These departures stem from the discovery of strongly conserved sequences between certain homeotic genes in the fruit fly and particular genes in the mouse (chapter 8).

Undoubtedly, when Castle made his suggestion, he felt that progress in the genetics of the simpler organism might aid his own research, but the connections being established today could hardly have been envisaged even 10 years ago, let alone 80. The beneficial effects also run the other way: The discovery of related genes and aspects of their genetic organization between the mouse and fruit fly has helped to validate the use of *Drosophila* as a model system for animal development in general.

As in every mammalian embryo, the developmental sequence of the mouse divides naturally into two phases: preimplantation development, during which the fertilized egg undergoes cleavage and forms the so-called blastocyst, and postimplantation development, in which, following implantation of the blastocyst into the uterine wall, the major events of embryonic and fetal development unfold. The two phases differ dramatically in complexity and outcome. Preimplantation development, lasting about 4.5 days, generates the 120-cell blastocyst, which consists of four distinct cell types, of which only one, the embryonic ectoderm, will contribute directly to the fetus or embryo proper. Postimplantation development, although only three to four times as long as the preimplantation phase, produces the fully developed animal, in all its intricacy and complexity of organization of cells, tissues, and organs.

The two stages also differ greatly in terms of their accessibility to genetic analysis. The most informative mutant analyses that have been carried out on the mouse pertain predominantly to postimplantation development. Only a handful of point mutants (as opposed to more complex chromosomal changes) are known to affect preimplantation development (reviewed in Magnuson, 1986), and their contribution to understanding the events of this first phase has thus far been less than that of direct experimental manipulation. (As mutant hunting becomes more efficient in the mouse, through the new mutagenesis methods, described in chapter 8, this situation is very likely to change.)

Furthermore, there are even fewer conventional maternal mutant effects of the kinds detailed in the previous two chapters. This scarcity reflects, at least in part, the activation of the zygotic genome in mouse embryos at an early point in development, the two-cell stage, with the expression of zygotic genes being essential for development beyond the two-cell stage. In consequence, the reliance on maternally supplied products is diminished in the early mouse embryo, relative to that of animals such as the nematode and fruit fly.

Where genetic analysis has played a major part in the analysis of early development is in the discovery of a special aspect of mammalian development, which has precedent only in a number of nonvertebrate systems. It is now known that during gametogenesis in both sexes in the mouse, and in other mammals, certain expressional properties are "imprinted" onto the haploid paternal and maternal genomes. These imprinting events involve differential restrictions of genetic expression, which can be perpetuated throughout development. As a result, both the maternal and paternal genomes are required to complement each other's expressional deficiencies and both are essential for complete embryonic development (in contrast to species that are capable of parthenogenetic development).

In this chapter we look at three aspects of preimplantation development: 1) its cellular basis, in particular the roles of cell lineage and cell interactions in establishing the four major cell types of the preimplantation embryo, 2) the evidence for early zygotic genome activation and its essentiality, and 3) the discovery of chromosome imprinting, and its subsequent analysis, in early development. This material will be introduced by a short section on gametogenesis and early development. An overview of mouse genetics will be deferred until chapter 8, where postimplantation development and its investigation by genetic and molecular techniques will be described.

THE LIFE CYCLE AND EARLY DEVELOPMENT: AN OVERVIEW

Embryonic and fetal growth in the mouse takes approximately 20 days from fertilization to birth, of which the first 4.5 days, on average, are devoted to preimplantation development. In contrast to most animals, the pace of the first mammalian embryo cleavage divisions is slow. In the mouse, the time between fertilization and first cleavage is 22–26 hr, depending upon the strain; the second cleavage cycle is 18–22 hr; and succeeding cleavage cycles during preimplantation development are approximately 10 hr each (reviewed in Pedersen, 1986). However, the rates of cell division and developmental change increase following implantation and accelerate further with the establishment of the placental connections to the mother's circulatory system and nutrient supply (McLaren, 1976b). During the first stages of postimplantation development, the three germ layers (ectoderm, mesoderm, and endoderm) form within that portion of the embryonic structure that gives rise to the fetus, and by days 8–9, organ formation has begun. By 7.5 days, only 3 days after the 120-cell stage, the embryo proper consists of approximately 10^5 cells (McLaren, 1976b). Cell proliferation and tissue and organ development continue, of course, during the period of fetal development, which begins with organ formation. Birth occurs at 20 days after fertilization. At 3 months, sexual maturity has been attained.

Because the embryo develops inside and attached to its mother's body, a special question arises in mammalian development: Does the physical attachment of the embryo via the placenta to the uterine wall play an "instructive" role in the embryo's development, or is the maternal role purely "permissive" for a process that is intrinsic to the embryo itself? This question has been investigated by culturing embryos in various ectopic (nonuterine) sites, particularly in males. Although such embryos show some abnormalities, the results indicate that the embryonic developmental sequence is organized by the embryo itself; the principal function of the maternal environment after fertilization is to provide adequate nutrition to the developing individual (reviewed in Graham, 1973; Johnson, 1979).

Nevertheless, there is now evidence for the involvement of maternally supplied growth factors, both of the insulin gene family (reviewed in Heyner et al., 1989) and of various growth factors that were first associated with the lymphohematopoietic system (reviewed in Pampfer et al., 1991). Such influences may account for some of the earlier reported effects of maternal uterine environment on embryo growth, detected by transferring early embryos to surrogate mothers of a different strain (McLaren, 1979).

In general, preimplantation development is most readily analyzed *in vitro*. To obtain sufficient numbers of embryos for study, females are induced to "superovulate" by injection with gonadotropic hormones (horse chorionic gonadotropic hormones, hGC) prior to mating. Normally, a female will release up to 10–20 eggs per ovulatory cycle; superovulated females can release several times this number of mature ova. Following injection, females are placed with males and mating takes place; in the standard protocol, the time of fertilization is generally taken as occurring

13.5 hr after the hormone injection. The fertilized eggs produced can be readily removed from the female reproductive tract for further experimental analysis or embryos can be examined after particular intervals of *in vivo* preimplantation development.

Although preimplantation development to the blastocyst stage can be studied *in vitro*, such experiments are feasible only with a small number of inbred strains, and, even for these, the best *in vitro* growth rates are less than the comparable *in vivo* developmental rates (Harlow and Quinn, 1982). For the majority of strains, preimplantation development *in vitro* stops at the two-cell stage (Goddard and Pratt, 1983). The nature of this "two-cell block" is poorly understood but may involve, in part, strain-specific differences in requirements for maternally supplied growth factors (Pampfer et al., 1991).

GAMETE FORMATION, FERTILIZATION, AND PREIMPLANTATION DEVELOPMENT

Egg and Sperm Formation

As in all animal systems, it is the female gamete, the oocyte, that contributes the bulk of the nonnuclear constituents to the early embryo. The development of the mouse oocyte takes place in two discrete temporal phases (reviewed in Schultz, 1986). The first phase extends from mid-embryogenesis until 5 days after the birth of the female animal and extends from the formation of the oogonial stem cells by the primordial germ cells to the formation of immature oocytes, which are held in late diplotene of meiosis I. The second phase begins with the attainment of sexual maturity and involves oocyte growth and maturation. This takes place in small batches of oocytes within each estrous cycle.

In its general aspect, the initial development of the female germ line in the mouse bears a certain resemblance to that of *Drosophila*. The forerunners of the germ line are the primordial germ cells (PGCs), which are distinguishable from the other cells by their round shape and diffuse chromatin, characteristics similar to the pole cells of *Drosophila*. However, they are most readily detected and tracked by their high alkaline phosphatase content. As in the fruit fly embryo, the PGCs migrate from their posterior site of origin to the gonad rudiment. This migration, between 8 days postconceptus (pc) and 13.5 days pc, begins at the caudal end of the embryo proper (the primitive streak) and continues through the hindgut to the genital ridges (the somatic rudiment of the gonad). In female embryos, the arrival of the PGCs stimulates the proliferation of the epithelial cells of the genital ridges; the PGCs also commence multiple rounds of cell division to form oogonia.

The oogonia subsequently undergo two waves of inward migration. Most of those that participate in the first wave degenerate, but the cells of the second wave come to rest in the peripheral (cortical) region of the developing ovary. These cortical oogonia continue to proliferate, although they remain connected by many intercellular bridges. As we have seen, this syncytial arrangement is common in early germ line cells, and presumably serves to facilitate intercellular molecular transfer and synchrony of nuclear divisions.

During the migration to the ovarian cortex, groups of oogonia become surrounded by prefollicle cells of mesodermal origin. Eventually, each pro-oocyte is covered by a monolayer of these somatic cells to form a "unilaminar follicle" (Fig. 5.1a). Pro-oocytes enter the first meiotic division but arrest at late diplotene until just prior to ovulation. Development of the unilaminar follicle is accompanied by the loss of syncytial connections between pro-oocytes. The further division of follicle cells and the secretion of extracellular material serve to additionally isolate the follicles from one another.

The great majority of oocytes remain quiescent until sexual maturation, when they are hormonally stimulated to commence growth and development in small groups. Although the estrous cycle takes 5 days, each preovulatory growth period lasts for 20 days, with most growth taking place between 20 and 6 days prior to ovulation. During

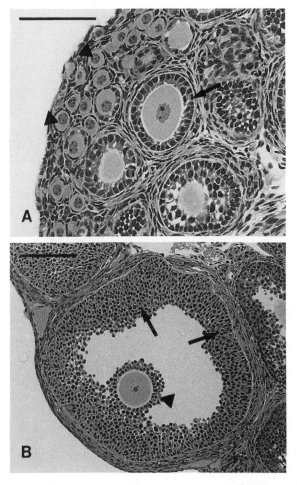

Fig. 5.1. Two stages of oocyte development within the mouse ovarian follicle. **A:** Primordial follicles (arrowheads) and a later-stage (preantral) follicle. **B:** Preovulatory antral follicle, showing the cumulus-enclosed oocyte (arrowhead) and the granulosa cells lining the follicle (arrows). Scale bars: 100 μm. (Reproduced from Eppig, 1991, with permission of the publisher. © ICSU Press.)

this period, each stimulated oocyte increases in diameter from 15 to 80 μm. This growth is markedly different from somatic cell growth in that cell size dramatically increases without triggering cell division; throughout this phase, the chromatin remains in a diffuse, transcriptionally active state.

While the oocyte itself remains in the dictyate stage of meiosis I, the surrounding follicle cells continue to divide, reaching approximately 50,000 per follicle, and the oocyte comes to occupy a fluid-filled chamber, the "antrum," within the follicle (Fig. 5.1b). This developmental progression of the ovarian follicle involves a series of interactions between the growing oocyte and the surrounding follicle, or granulosa cells. In particular, the granulosa cells act as feeder cells for the oocytes; 85% or more of the metabolites required by the oocyte are passed to it through the follicle cells (Heller et al., 1981). The granulosa cells also supply factors that produce the meiotic arrest of the oocyte. The oocyte, in turn, influences the behavior of its immediately surrounding granulosa cells, the cumulus cells, and thereby influences the organization of the developing follicle as a whole. These two-way interactions between follicle cells and oocytes are reviewed in Eppig (1991) and illustrated schematically in Figure 5.2.

The growth phase of the oocyte is accompanied by its progressive differentiation. These include characteristic sequences of differentiative changes in the major cytoplasmic organelles (ribosomes, endoplasmic reticulum, nucleoli, mitochondria) and the synthesis and accumulation of histone proteins (Wassarman and Mrozak, 1981) and of tubulins (Schultz et al., 1979), these proteins being employed in the early cleavage divisions of the embryo. The active growth of the oocyte is also accompanied by active transcription of both ribosomal and informational RNAs (Bachvarova and De Leon, 1980). The synthesis of poly A$^+$ mRNA is part of this transcriptional activity, coming to comprise about 20% of the total RNA, as measured by [^{3}H]uridine incorporation (De Leon et al., 1983). Indeed, the fraction of stored RNA that is in the form of mRNA is substantially higher in the mouse oocyte and in the mature egg than in the sea urchin or in the frog *Xenopus*, where the proportion is closer to 1–2% of the total (Davidson, 1986).

The period of oocyte growth and differentiation ends about 6 days prior to

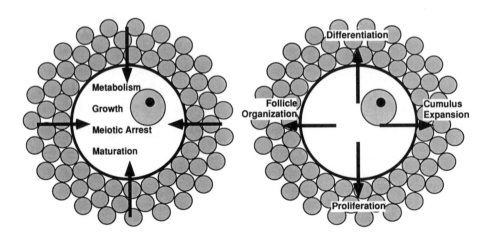

Fig. 5.2. Two-way interactions between the developing oocyte and the surrounding granulosa cells. See text for discussion. (Reproduced from Eppig, 1991, with permission of the publisher. © ICSU Press.)

ovulation. In the last 9–10 hr before ovulation, the oocyte reenters meiosis and completes the first meiotic division, giving rise to the first polar body. The oocyte nucleus then proceeds directly to meiosis II without re-formation of the nuclear membrane, but halts at metaphase until fertilization. In contrast to the period of oocyte growth, when there is a net accumulation of the various RNA species, the period of maturation, lasting about 12 hr, is characterized by both transcriptional quiescence and a partial degradation of both rRNA and mRNA. Total RNA declines by about 19% during this period, and poly A$^+$ mRNA exhibits an even greater relative depletion, coming to represent only 10% of the total RNA, down from 20% in the premature oocyte (Bachvarova et al., 1985).

Ovulation involves release of the ovum from the follicle into the oviduct. It is surrounded by a polysaccharide-rich envelope, termed the "zona pellucida," which plays a critical role in fertilization, and by the cumulus cells, the "cumulus oophorus," which are shed during the passage down the oviduct, prior to fertilization.

Although the sequence of events in sperm formation, spermatogenesis, produces a very different kind of cell—one that has lost most of its cytoplasm, consisting of little more than a packaged nucleus and a tail—it, like oogenesis, produces haploid gamete cells and begins with PGCs. As in female fetuses, the PGCs migrate from a caudal position in the primitive embryonic ectoderm to the genital ridge, where they proliferate to form numerous stem cells of the germ line, primitive spermatogonia. In contrast to oogenesis, however, the immediate precursor cells of the male gametes require a further sequence of differentiative cell divisions, which continue throughout the remainder of fetal life and into the first days of postnatal life. The final precursor cell type, which gives rise to the spermatids, the so-called B-type spermatogonia, appear by 9 days after birth (reviewed in Hecht, 1986).

The mature spermatogonia then commence meiosis. In comparison to oogenesis, therefore, the premeiotic history of the male germ is a comparatively long sequence of mitotic and differentiative events. The meiotic cells are termed spermatocytes and the complete process is prolonged, taking 11 to 12 days (and longer in several other mammalian species). In contrast to oocyte development, there is no "holding" of cells at particular stages of meiosis; the process of meiotic events is continuous, resulting in the production of haploid cells, the spermatids, which then commence the final process of sperm differentiation, spermiogenesis. Spermiogenesis takes 14 days and involves the differentiation of the initially round, haploid spermatid into the final cellular form, the mature haploid spermatozoan. As in many animals, the spermatids produced from a single spermatogonium remain connected cytoplasmically, forming a small syncytium.

As in any multicellular differentiation, the entire process of spermatogenesis is accompanied, and driven by, a sequence of differential gene expression events. Its particularly novel feature (relative to oogenesis) is the occurrence of a prolonged haploid differentiative phase, that of spermiogenesis, including both *de novo* transcriptional events and translational control of mRNAs (reviewed in Erickson, 1990). The final phase of sperm differentiation involves the progressive condensation of the chromosomes and the replacement of histones by protamines in the condensed chromosomes.

Underlying the numerous cytological differences between oogenesis and spermatogenesis are significant differences in molecular biology. The existence of a haploid gene expression phase in spermatogenesis, noted above, is one. A second molecular difference of note, because of its possible bearing on the different roles of the maternal and paternal genomes in early embryonic development, is the occurrence of different DNA methylation patterns in oocytes and sperm, involving the cytosine residues of CG doublets. There is a substantially elevated level of total genome methylation in sperm relative to oocytes (Monk et al., 1987) and an increased methylation in sperm of a number of multigene families (Sanford et al., 1987), though both sperm and oocyte DNA sequences are somewhat undermethylated globally, compared to somatic tissue (Monk et al., 1987). These differences in DNA methylation are detectable as early as prophase of meiosis I in both gamete types.

DNA methylation is potentially significant because, within promoter regions, it is correlated with transcriptional inactivation (Riggs, 1975), although the precise molecular basis of such inhibition is unknown. Furthermore, methylation of cytosines in CG doublets prompts a ubiquitous enzyme, the so-called maintenance methylase, to methylate the opposite strand on the diagonally placed cytosine, a possibility predicted by Riggs (1975) and Holliday and Pugh (1975). This simple enzymatic fact, when coupled with the effect of methylation on transcription, permits, in principle, the perpetuation through successive cell divisions of particular states of gene expression and, hence, stable and gamete type-specific parental genome contributions to the embryo. Nevertheless, demethylation can occur, in characteristic ways, during development, and the process is therefore subject to occasional "programmed" reversals (Monk and Grant, 1990).

From Fertilization to Blastocyst

Fertilization, the union of mature oocyte and sperm, following mating begins the process of embryonic development. Although fertilization looks like a simple and continuous process, involving penetration of the sperm first through the cumulus oophorus (which is shed during passage down the oviduct) and then the zona pellucida (ZP), it involves numerous molecular interactions of high specificity. In these events, the ZP, the thick but transparent outer envelope of the egg, plays a key role. First appearing in immature small oocytes, the ZP undergoes its maximum growth in concert with the oocyte during the latter's principal growth phase, reaching a final thickness of about 7 μm and consisting of about 3 ng of glycoproteins. The three principal glycoprotein species, designated ZP1, ZP2, and ZP3, play distinct roles in fertilization. ZP3, in particular, is especially important, acting as both the key sperm receptor and the activator of the acrosome reaction (the acrosome being the crescent-shaped cap of the sperm whose contents help mediate passage of the sperm through the ZP) (reviewed in Wassarman, 1990).

Following entry, the sperm head is converted to the male pronucleus and its chromosomes then lose their complement of protamines, which are replaced by histones and acidic proteins from the maternal cytoplasm. Fertilization triggers resumption of meiosis in the oocyte nucleus, and completion of meiosis is quickly followed by extrusion of the second polar body. The two pronuclei then move slowly into apposition, over a period of 12–18 hr. (DNA replication takes place in the pronuclei and is completed during this lengthy migratory phase.) As the pronuclei come together, their chromosomes become mingled on the metaphase plate and first cleavage quickly ensues, to give rise to the two-cell embryo. The duration of the two-cell stage is, as noted earlier, about 18 hr, and successive cleavage divisions take place at approximately 10 hr intervals.

The eight-cell stage is a developmentally critical juncture. It is at this point that the heretofore loosely knit blastomeres undergo the process of compaction. During compaction, the cells transform from spherical to columnar and begin to develop a variety of intercellular junctions (Ducibella, 1977; reviewed in Fleming and Johnson, 1988). The net result is the drawing together of the ball of cells. The eight-cell stage, before and after compaction, is shown in Figure 5.3. Subsequent cleavages take place in the compacted embryo.

It is during the fourth cleavage division, as the 8-cell embryo becomes a 16-cell embryo, that the first important cell divergence takes place, with the formation of distinct inner and outer cell populations. By the 16- to 32-cell stage, fluid has begun to be secreted into the developing central cavity, the blastocoele. By the 60-cell stage, approximately 3.5 days after fertilization, the embryo is at the early blastocyst stage and consists of two visibly differentiated cell populations. The outer layer of approximately 45 cells constitutes the trophectodermal (TE) layer, and the internal 15 cells comprise the inner cell mass (ICM). Formation of the early blastocyst is followed by reexpansion of the embryo to fill the ZP (Fig. 5.4).

The final phase of preimplantation development involves the production of two further cell groups. The ICM gives rise to a group of endodermal cells that eventually cover the entire inside of the blastocoele, while the TE diverges into a cap of "polar" TE cells, which remain diploid and are situated at the top of the ICM, and the "mural" TE cells, which comprise the sides of the blastocyst. By the late blastocyst stage, some of these mural TE cells have begun to become polyploid; eventually, these large polyploid cells transform into "primary giant cells" when the blastocyst implants into the uterine wall. The sequence of cellular events during preimplantation development is illustrated in Figure 5.5, and the time course of events is summarized in Figure 5.6.

EXPRESSION OF THE ZYGOTIC GENOME DURING PREIMPLANTATION DEVELOPMENT AND CHROMOSOME IMPRINTING

Activation of Zygotic Genome Expression

Despite the existence in the newly fertilized egg of large stores of maternal mRNAs, the poly A$^+$ mRNAs amounting to 10% of the total stores, there is a shift from primary reliance on these messages to zygotic genome transcription early in the development of the embryo, the two-cell stage. The time course of changeover from maternal to zygotic genome control in the early mouse embryo has been tracked by several means. The first approach involved prelabeling the stores of maternal mRNA, with tritiated uridine or adenine, and then following the label during preimplantation development. This labeled maternal RNA pool is found to decline in parallel with the total stores of prelabeled maternal RNA during early postfertilization development. The results showed an initial depletion of prelabeled mRNA to 60% of the prefertilization content during the first 24 hr of development (to the early two-cell stage), followed by a more gradual decline between the four-cell and early blastocyst stages to a final level of 30% (Bachvarova and De Leon, 1980).

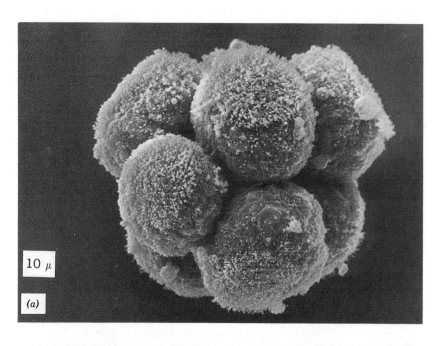

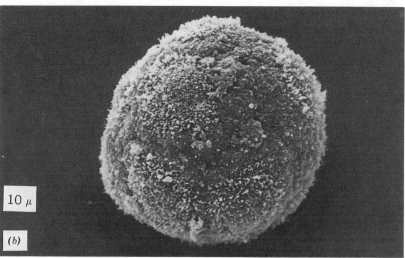

Fig. 5.3. Compaction in the mouse embryo. **a**: Uncompacted eight-cell embryo. **b**: Compacted eight-cell embryo. (Photographs courtesy of Dr. M.H. Johnson.)

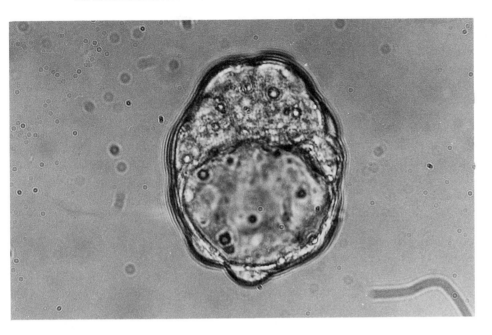

Fig. 5.4. An expanded 3.5 day blastocyst. Diameter: 70–80 μm. (Photograph courtesy of Dr. M.H. Johnson.)

The rate of loss of *functional* maternal mRNA species is, however, considerably greater than indicated by measurements of labeled molecules. The first indication came from experiments that showed that new protein synthesis becomes susceptible to inhibition by exposure to α-amanitin, starting in the two-cell stage (Flach et al., 1982; Pratt et al., 1983). The findings indicated that a rather rapid inactivation or depletion of maternal mRNA stores takes place by the mid–two-cell stage. One form of functional inactivation apparently entails the clipping off or shortening of poly A tails from certain mRNA species, a process that may, in some manner, interfere with their translational initiation (Bachvarova et al., 1985). Since α-amanitin treatment from the two-cell stage onward stops further cell division, further cleavage divisions must be dependent on new transcription.

This early shift to zygotic genome dependence seems, at first, to be in marked contrast to the early development of sea urchin and amphibian embryos, where actinomycin D treatment during early cleavage has little immediate effect on either protein synthesis or cell division (the effects showing up at the gastrula stage). In these embryos, of course, the maternal mRNA stores persist to a considerably later developmental stage, a fact that accounts for their seeming resistance in early stages to RNA synthesis inhibitors. However, when one considers the absolute time periods from fertilization to dependence on the zygotic genome, there is little difference between the mammalian system, on the one hand, and the sea urchin/frog systems, on the other (Davidson, 1986); by the end of 24 hr, the mouse embryo is still in the two-cell stage, while at the same time point the *Xenopus* embryo may be in the late neurula stage and fully dependent on zygotic transcription.

Besides inhibitor studies, other evidence identifies the two-cell stage as the point

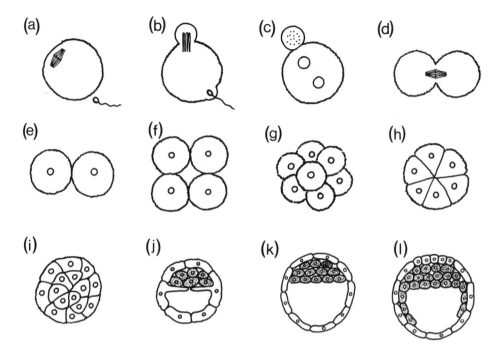

Fig. 5.5. Diagrammatic representation of the stages of preimplantation development. **a:** The meiosis II-arrested oocyte is fertilized. **b:** The egg nucleus spindle rotates, meiosis resumes, and the second polar body is extruded. **c:** Pronuclei are formed, 4–6 hr after fertilization, and migrate to the center during the next 8–14 hr. **d:** Pronuclear membranes break down and chromosomes mingle on the first cleavage division spindle. **e:** The 2-cell stage, activation of the embryonic genome. **f:** The 4-cell stage. **g, h:** The 8-cell stage and the occurrence of compaction. **i:** The 16-cell morula stage, first separation into inner and outer cell populations. **j:** The 32-cell stage, with blastocoele. **k:** Later blastocyst, with inner endodermal layer. **l:** Migration of endodermal cells on inside of blastocoelic cavity. (Adapted from Fleming and Johnson, 1988.)

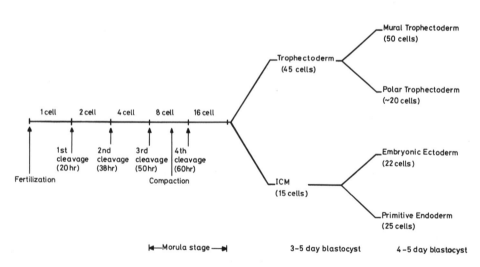

Fig. 5.6. Time course and stages of preimplantation development. During postimplantation development, the mural and polar TE and the primitive endoderm will give rise to the extraembryonic tissues (EET), while the embryonic ectoderm will give rise to the fetus itself and part of the allantois; see Figure 8.8. (Adapted from Gardner, 1978.)

at which zygotic genome transcription begins. One kind of experiment that demonstrates this involves a genetic approach, the tracking of enzymatic activities specified by a paternal allele during early development. Since only the zygotic genome can be the source of a paternally encoded enzymatic difference, the detection of such differences in the late two-cell stage or just after the second cleavage division indicates this period as the approximate starting point of zygotic genome transcription (see, for instance, Chapman et al., 1976; Esworthy and Chapman, 1981).

More sensitive determinations of the start of zygotic genome expression involve the labeling of new RNA synthesis by radioisotope incorporation. Using [^{3}H]uridine incorporation, Knowland and Graham (1972) placed the beginning of transcription in the embryo at the middle of the two-cell stage. With the same label, Levey et al. (1978) similarly located the beginning of nonribosomal nuclear RNA and of poly A$^+$ mRNA at this point. The earliest detected protein products expressed by the zygotic genome are several high-molecular-weight proteins (68 and 70 kDa) of the heat shock protein class, these species appearing early in the two-cell stage (Flach et al., 1982; Bensaude et al., 1983). These proteins are not present in the oocyte and there is no evidence for preexisting maternal mRNAs for them; hence their appearance at the early two-cell stage signifies the onset of transcription of their genes from the zygotic genome. The heat shock proteins were first discovered, as their name indicates, as a group of proteins that appear when eukaryotic cells are suddenly exposed to higher-than-normal temperatures (or under a variety of other traumatizing conditions); they probably mediate one or more protective responses. The appearance of these proteins in the early mouse embryo could be a response to the *in vitro* conditions present during labeling.

In terms of transcript utilization and synthesis, the mouse embryo's pattern is fairly clear: A dependence on maternal mRNAs up to the early two-cell stage, followed by a depletion of maternal mRNA stores and a shift to complete dependence on zygotic genome transcription prior to the second cleavage. Nevertheless, maternally encoded protein products and oocyte organelles (mitochondria) persist until late into development (Pratt et al., 1983), and undoubtedly some, at least, of these maternal contributions continue to be needed for the embryo's development.

Concomitantly with this changeover from maternal to zygotic gene transcripts, there are large changes in the pattern of proteins synthesized, as detected by high-resolution two-dimensional gels (Latham et al., 1991). The periods of greatest change during early preimplantation development are the late one-cell and the mid–two-cell stages, with 60–85% of the detected proteins showing at least a twofold level of change and many significantly greater ones (Latham et al., 1991). The switch from maternal to zygotic transcripts affects the rate of synthesis of virtually every protein detected. Earlier work has indicated that much posttranslational modification of proteins also takes place in the preimplantation embryo (van Blerkom, 1981), and such modifications undoubtedly contribute to the changes seen with these two-dimensional separations.

From the patterns observed and then compared between fertilized eggs and nonfertilized, aging oocytes, it appears that there are two "programs" at work in the early embryo: an endogenous, maternally based set of changes that occurs in oocytes independently of fertilization and a second set, dependent on fertilization, that is superimposed on the maternal "program" (Howlett and Bolton, 1985). Such a superimposition of fertilization-based changes on a pattern of oocyte-based protein changes is, however, by no means unique to the mouse. To take one of many possible examples, the *bcd* mRNA of *Drosophila* begins to be translated upon oviposition and the protein gradient forms independently of whether of not fertilization has taken place (although degradation of the protein is evidently keyed to events triggered by fertilization [Driever and Nüsslein-Volhard, 1988a]).

Between the 8-cell compaction stage and the early blastocyst, a number of significant changes in gene expression occur. The first is a key regulatory event in the late morula stage, as the embryo goes from a 16-cell structure to one of 32 cells, which is necessary for the secretion of fluid that takes place during blastocyst formation. This fluid initially accumulates between cells and then collects in the developing blastocoele, the entire process being referred to as "cavitation." The onset of cavitation is dependent on the prior occurrence of five replication cycles (Smith and McLaren, 1977) and on transcription (Braude et al., 1979), though the specific genetic and molecular requirements have not yet been defined.

The second set of events concerns the divergence of the first two distinct cell populations, the ICM and the TE cells. Van Blerkom et al. (1976) were the first to detect differences in patterns of polypeptide labeling between these cell groups, using separated ICM and TE cell fractions labeled *in vitro* with [^{35}S]methionine. Although the majority of spots on two-dimensional separations were shared between ICM and TE cells, as expected from the general constancy of labeled polypeptides in whole embryos between the 8- and 64-cell stages, a few distinctive ICM and TE spots were detected. Handyside and Johnson (1978) subsequently showed that the "ICM-specific" and "TE-specific" cells are first detectable between the early morula (12- to 25-cell) and late morula (25- to 30-cell) stages. When the inner cells from embryos of these stages were isolated by immunosurgery (involving antibody- and complement-mediated lysis of the TE cells), these inside cells were found to have just the shared and ICM-specific polypeptides and not the TE-specific ones. (Some of these differences may reflect posttranslational modification rather than new gene expression *per se*.) As discussed below, the cells that will go on to form ICM and TE have already begun to display certain cytological differences.

The sequence of changes in gene expression and macromolecule composition that take place in the preimplantation embryo are summarized in Figure 5.7. The entire sequence shows much complexity, but the salient points are the shift from dependence on maternally encoded gene products to reliance on zygotic genome products during the late two-cell stage and the beginnings of detectable molecular diversification between presumptive ICM and presumptive TE by the early morula stage.

DEVELOPMENTAL ASSIGNMENTS IN THE PREIMPLANTATION EMBRYO

Nature of the ICM-TE Divergence

The first set of cellular differences to arise in the early mouse embryo are between those cells that will form the ICM and those that will go on to form the TE of the early blastocyst. The first published hypotheses about the source of this cellular divergence were as different as such ideas can be. The initial speculation was that of Dalcq (1957) and was clearly influenced by ideas derived from invertebrate models. He proposed that there are separate "determinants" for the ICM and the TE in the mouse egg and

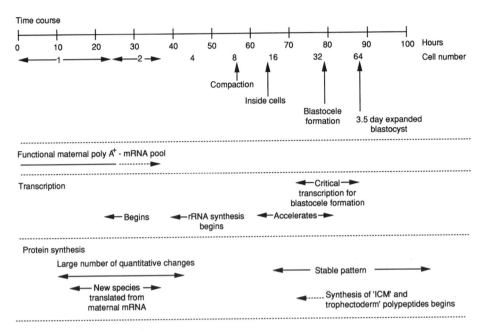

Fig. 5.7. Summary of the time course of gene expression and gene product changes during preimplantation development. See text for discussion.

that a physical segregation of these determinants takes place by the eight-cell stage. Dalcq formulated his hypothesis on the basis of his examination of fixed preparations of mature mouse oocytes and reported that these specimens exhibited differentiated "dorsal" and "ventral" cytoplasm, which were, apparently, differentially segregated to blastomeres by the eight-cell stage. In Dalcq's hypothesis, the dorsal cytoplasm eventually gave rise to the ICM and the ventral cytoplasm to the TE. Subsequent workers were unable to confirm this putative cytoplasmic segregation and, in fact, it appears that the regional differentiation of fixed eggs reported by Dalcq was an artifact of the fixation process (Wilson and Stern, 1975).

In the 1960s, a new hypothesis came to the fore and was, indeed, the antithesis of Dalcq's proposal, being a form of environmental instructionism. It was based on the obvious difference between the ICM and TE cells in their relative physical positions, with the ICM cells being on the inside of the blastocyst and the developing TE cells on the outside and directly exposed to the external environment. As first emphasized by Mintz (1964), this difference in physical position directly establishes a difference in "microenvironment"—an "inside–outside" difference—that might shape the initial divergence in fate and developmental capacity between the two cell types.

The first experimental findings to support the inside–outside hypothesis of determination were presented by Tarkowski and Wroblewska (1967). They examined the products of *in vitro* development of isolated blastomeres from both four-cell and eight-cell embryos and concluded that all cells at these stages had the capacity to give rise to vesicular (blastocyst-like) forms. They concluded therefore that there was no evidence for the segregation of specific determinants for the ICM to any of the blastomeres at these two stages. Significantly, however, the isolated one- and two-cell products of eight-cell embryos were found to form a higher percentage of purely trophoblastic vesicles (lacking ICM) than did single blastomeres from four-cell embryos. This difference is explicable if fluid secretion from the outside cells, and hence blastocyst formation, tends to occur only after a particular number of cleavage divisions, for instance, after the fifth cleavage division. Under these circumstances, cell clusters derived from single blastomeres of eight-cell embryos would necessarily be smaller at the time of fluid secretion than mini-embryos derived from single cells of four-cell embryos and would be less likely to have inside (proto-ICM) cells. In this view, the capacity to form ICM-type cells is a direct consequence of the progenitor cells being inside at the time fluid secretion begins.

Chimera Experiments

One form of experimental support for the inside–outside hypothesis came from mouse embryo chimera experiments. In chimera experiments, single or multiple early blastomeres are physically brought into contact with complete early embryos, either by injection or aggregation, in which the donor cells are genetically or biochemically distinguishable from the recipient cells. The composite, or "chimeric," morulae or blastocysts are then transferred to the uterus of a pseudopregnant foster mother, where implantation and subsequent development take place; chimera formation by injection is illustrated in Figure 5.8. (The foster mothers are usually prepared for receipt of the embryos by prior mating to vasectomized mice.) Analysis of the composition of the embryos or newborn chimeras permits one to determine, within the limits of accuracy associated with the marker, the degree of contribution by the donor cells to particular tissues or regions. Markers that can be scored in individual cells, and which autonomously reflect the genotype of those cells, are said to be "direct" markers. Biochemical markers that cannot be distinguished *in situ* but only estimated after dissection and relative quantitative measurement are said to be "indirect" markers. A typical indirect marker would be an allelic enzyme difference, which creates a difference in mobility during electrophoresis. For mouse studies, one of the most commonly used is that between two GPI-1 isozymes, the "a" and "b" forms (Fig. 5.9).

In the first test of the inside–outside hypothesis using chimeras, cells from four- or eight-cell embryos were dissociated from one another and placed singly or in groups, either internally or externally, on carrier embryos. The donor cells were marked either by prior labeling of their nuclear DNA with [^{3}H]thymidine (a direct cell marker) or with the GPI-1 isozyme difference and the chimeras were scored, after they had been allowed to develop, for contribution to the ICM or TE at the blastocyst stage (Hillman et al., 1972). The results, with both kinds of cell marking, showed a clear biasing of cell fate toward either TE or ICM development as a function of donor cell positioning on the recipient embryos.

In another kind of chimera experiment, cells were tested for their developmental capacity as a function of developmental stage. One set of such experiments demonstrated that all blastomeres at the eight-cell stage are totipotent, but that by the time well-defined ICM and TE cell populations have formed (the early blastocyst stage), cells are restricted to one fate of the other, while at intermediate times there is a degree of developmental lability. The experiments thus provide no support for a rigid and early segregation of determinants, as predicted by Dalcq's hypothesis. Instead, there appears to be a progressive restriction of fate, in accordance with relative position.

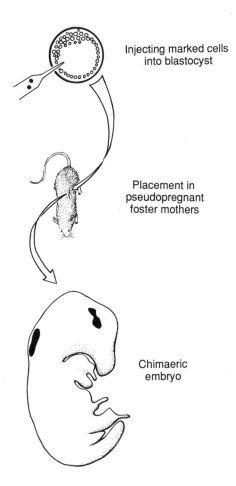

Fig. 5.8. Chimera construction and placement of chimeras in pseudopregnant foster mothers.

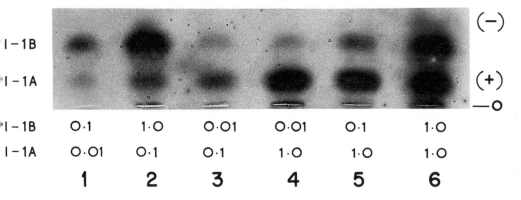

Fig. 5.9. Electrophoresis of mixtures of GPI electrophoretic variants. Kidney homogenates from strains homozygous for the two allelic enzyme forms were prepared, and mixtures were made in the proportions indicated; the samples were electrophoresed and stained for GPI activity. (Figure courtesy of Dr. V.M. Chapman.)

The work of Kelly (1975) established that cells of the eight-cell embryo are totipotent; this is shown in Figure 5.10. She dissociated four-cell embryos *in vitro* and then allowed each blastomere to divide once, to give four "octet pairs." The members of each pair were then separately aggregated with individual eight-cell carrier morulae and placed in pseudopregnant foster mothers. Donor cells carried the *Gpi-1ᵃ* allelic form, while host morulae carried *Gpi-1ᵇ*. In addition, the donor strain was albino, so that chimeras that developed to term could be distinguished by coat color. Fourteen chimeras, derived from seven octet pairs, had donor cell contributions in the fetuses. In five of the pairs, individual donor cells had participated in the formation of both

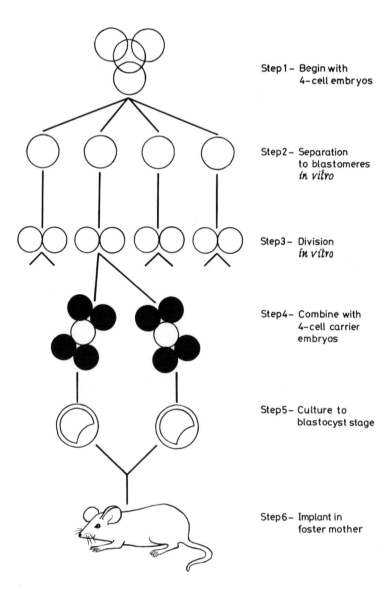

Step 1 – Begin with 4-cell embryos

Step 2 – Separation to blastomeres *in vitro*

Step 3 – Division *in vitro*

Step 4 – Combine with 4-cell carrier embryos

Step 5 – Culture to blastocyst stage

Step 6 – Implant in foster mother

Fig. 5.10. Test for totipotency of blastomeres at the eight-cell stage in the mouse embryo. See text for discussion. (Adapted from Kelly, 1975.)

ICM- and TE-derived fractions. A substantial donor cell contribution was also seen in all chimeras allowed to develop to term. Evidently, the blastomeres of the eight-cell stage do not show segregation of developmental potential for either ICM or TE. This stage, it will be recalled, marks the last point prior to the emergence of distinct inner and outer cells, which arise in the fourth cleavage division.

However, by the 3.5 day expanded blastocyst stage, cells are committed definitively to either ICM or TE fates. At this stage, the blastocyst consists of 60–64 cells and the ICM and TE differ in several distinctive cytological features. These differences include the kinds of cellular junctions each displays, the presence of microvilli on TE cells and their absence from ICM, and visible differences in mitochondrial morphology. That cells are determined by this point was first shown in an experiment by Rossant (1975). In her experiment, donor ICMs were first aggregated with recipient eight-cell morulae, transplanted to the uterine horns of pseudopregnant foster mothers, and then allowed to develop for periods ranging from 8.5 to 18.5 days. The genetic marker used for tracing the ICM-derived cells was GPI-1. Following termination of development, the fetuses and associated embryonic structures were isolated and dissected into the ICM-derived embryo plus membranes fraction and the TE-derived fraction. The chimeras showed a substantial and highly preferential contribution of donor GPI enzyme to the ICM-derived tissues, with little or no contribution to the TE-derived fraction. Of the 15 embryos showing a donor cell contribution, 12 showed an exclusive contribution to the embryo plus membranes fraction. The three apparent exceptions showed only a minor contribution, probably reflecting contamination with ICM-derived tissues.

Comparably, TE cells from 3.5 day blastocysts are also restricted in their developmental potential. When TE vesicles are emptied of their ICM cells before fostering, they give rise to trophoblast cells and implant. However, they cannot give rise to ICM or to those structures normally derived from the ICM. Furthermore, injections of TE cells from mature blastocysts into 3.5 day blastocysts never yields embryos with a donor cell type contribution (Gardner, 1978). Altogether, the results indicate that by the expanded blastocyst, 60- to 64-cell stage, the cells of the ICM and TE are committed to distinct and restricted pathways of development.

If cells are totipotent at the 8-cell stage, before the creation of inner and outer cells at the fourth cleavage, but completely restricted to the ICM or TE pathway by the expanded blastocyst stage, are these restrictions absolute by the 16-cell stage, when the first inner and outer cell populations have arisen? Both chimera and cell labeling experiments have addressed this question and the answer is unequivocal: Inner cells can give rise to outer cells that contribute to the TE, and outer cells can contribute to the ICM, until at least the early blastocyst stage (32 cells) (Handyside, 1978; Hogan and Tilly, 1978; Johnson and Ziomek, 1983; Fleming, 1987). It is also clear, however, that by the early blastocyst stage, there is relatively little cross over between the ICM and TE precursor pools (Dyce et al., 1987). Hogan and Tilly (1978) have argued that the developmental restriction is a progressive, gradual process with some capacity for reversal until, or even slightly beyond, the time of expanded blastocyst formation.

Direct Examination of the Cells

The chimera experiments thus help to define the period in which the restriction event(s) takes place, but the most informative work about the nature of these changes, in terms of cellular phenotype, have come from detailed, direct observations of the cells themselves, from the time of compaction, at the eight-cell stage, through the two successive cleavage divisions.

A key finding is that the process of compaction, the drawing together or self-compression of the eight-cell embryo, is crucial for the subsequent formation of the ICM. The importance of compaction in the separation of ICM and TE has been demonstrated experimentally using treatments that reversibly inhibit compaction during development. With treatments that inhibit compaction without affecting cleavage, a delay in compaction beyond a certain time produces blastocysts that lack characteristic ICM-type cells, although the cells can synthesize the full spectrum of abundant proteins characteristic of normal blastocysts, including those diagnostic of the ICM (reviewed in Johnson et al., 1981). Under these conditions, the gene expression pattern characteristic of normal development is apparently activated, but in the absence of cell compaction, none of the cells proceed to develop as ICM-type cells.

The molecular basis of compaction lies in the action of a special transmembrane molecule, one of a family of protein molecules that function specifically to promote cell adhesion. This molecule, a 120 kDa protein, was originally termed "uvomorulin" (Hyafil et al., 1981), but has been referred to by other names, such as E-cadherin and CAM 120/80 (reflecting its detection on other cell types). It is present at a low level on the surface of unfertilized eggs but is first synthesized at the two-cell stage. Initially distributed essentially uniformly prior to the eight-cell stage, it has, by the eight-cell stage, come to be localized preferentially at the basolateral surfaces where cells are in contact. The mutual or "homophilic" binding of uvomorulin molecules by their extracellular domains literally pulls the cells together, causing compaction. Furthermore, once localized at the basolateral surfaces, it can act as an anchor and, hence, a localizing device for other membrane proteins, creating polarized distributions of these molecules (McNeill et al., 1990).

Indeed, the immediate consequence of compaction on the properties of the cells of the eight-cell embryo is the creation of a general *polarized phenotype* within these cells, followed by division with the appropriate cleavage orientation of these polarized cells into two morphologically distinct cell types. The first detectable signs of this axial polarization are a columnarization of the cells, from a previously more spherical shape, and the appearance of a distinctive cap of microvilli at the apical (externally facing) end. In addition to this morphological differentiation of the apical and basal ends, there is an external biochemical differentiation, in that the microvilli can bind a fluorescent ligand (fluorescein isothiocyanate–concanavalin A) and the polarized phenotype is maintained *in vitro* upon disaggregation of compacted eight-cell embryos. Besides the external signs of polarization, there is an internal polarization of elements of the cytoskeleton along the apical-basal axis as well; these cytoplasmic events can be experimentally dissociated from external polarization by various treatments, showing that they are not obligately linked (reviewed in Fleming and Johnson, 1988).

The potential relevance of the polarized phenotype at the compacted eight-cell stage is that when these cells divide they can give rise to two different cell types: an outer polarized cell (possessing the ligand binding site, like the mother cell) and an inner apolar daughter cell. The polar:apolar phenotypes can be seen in 2/16 cell couplets (pairs of cells at the 16-cell stage) (Fig. 5.11). Although not all of the dividing cells of the eight-cell compacted embryo perform such "differentiative" divisions *in*

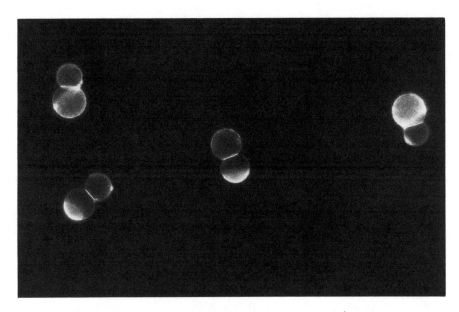

Fig. 5.11. Polarity in 2/16 couplets (cell pairs derived from division of individual blastomeres at the eight-cell stage). The outer polarized cell shows binding of fluorescent ligands; the inner cell is non-polarized. (Figure courtesy of Dr. M.H. Johnson.)

vivo, a significant number do, while *in vitro*, the majority of the isolated polarized cells divide to give a polar and an apolar cell (Johnson and Ziomek, 1981). Subsequently, these two types of cells tend to form two discrete lineages, polar cells usually dividing to give pairs of polar cells, and apolar cells, in general, dividing to give apolar cells (Ziomek et al., 1982). Since the outer polar cells give rise, predominantly, to the cells of the TE and the inner apolar cells to those of the ICM, this division behavior contains the possible origins of two distinct cell lineages.

These lineages are only presumptive, however, not definitive. If the fourth cell division, critical as it appears to be, provided the definitive foundation for two cellular lineages, then one might predict that from the 16-cell stage onward, polar (presumptive TE) cells would generate only polar cells and apolar (presumptive ICM) would give rise only to apolar cells. In fact, experiments that trace the descendants of cells labeled with readily traceable biochemical markers, both in isolated cells and in the intact embryo, show that from the 16-cell to the early 32-cell stage (preblastocoele formation), polar cells will sometimes divide to give an apolar cell and vice versa (Johnson and Ziomek, 1983; Pedersen et al., 1986; Fleming, 1987). Indeed, differentiative divisions of outside (polar) cells (such divisions giving apolar cells) at the 16-cell stage can be as frequent, or nearly so, as among polarized cells of the 8-cell embryo (Pedersen et al., 1986). It is only at the expanded blastocyst stage that labeled outside cells give rise only to outside cells (Pedersen et al., 1986).

These results provide visual confirmation of the developmental plasticity of cells, indicated by the chimera experiments, between the 16-cell and early blastocyst (32- to 40-cell) stages. Furthermore, the tracking of patterns of differentiative divisions during this period strongly suggests that direct cell interactions between and among polar and apolar cells strongly influence the frequency of such divisions during this period (Johnson and Ziomek, 1983; Fleming, 1987). A cell of either type that is not in contact with a cell of the other type is more likely to undergo a differentiative division than otherwise. The net result is that, despite some large initial inequalities in presumptive ICM cells generated by the 8-cell embryo (a range of 2–7 such cells in 16-cell embryos), successive divisions tend to equalize the number of ICM cells (10–14) between embryos at the 32-cell stage (Fleming, 1987).

The cell lineage experiments do not, however, definitively establish the state of commitment of cells of similar cytological appearance (polar or apolar) generated at different times. The data are most simply interpreted as indicating that cells of a given type generated at later times have a greater propensity or degree of commitment to develop along the lines of "their" lineage than earlier cells of the same type, but this is not known with certainty.

The central conclusion, however, is an important one: While the first cellular divergence in the preimplantation embryo is correlated with a segregation of two distinctive cellular types, the developmental fates of these two cell types are, initially, propensities rather than hard-and-fast restrictions of potential.

Late Developmental Events in the Preimplantation Embryo: Formation of the Endoderm and the Polar and Mural TE

Although there remain some large questions about the mechanisms that generate the ICM-TE cell divergence, still less is known about the subsequent development of cell types within the ICM and TE cell populations in the final 24 hr of preimplantation development. One of the few established facts is that the delamination of the internal layer of endoderm from the ICM also involves a determinative restriction. The endoderm cells of the late blastocyst differ from the (ectodermal) cells of the remaining ICM in three respects: They show more histochemical staining; they possess a more extensive endoplasmic reticulum; and their surface membranes have a "rough" character that contrasts with the "smoother" surface of the ectodermal ICM on dissociation *in vitro* under the appropriate conditions.

Gardner and Rossant (1979) purified rough (endodermal) and smooth (ectodermal) inner cells and injected them singly or in small groups into 3.5 day blastocysts, the latter marked with a different GPI-1 electromorph, then dissected tissues from the fetus and surrounding extraembryonic membranes and scored for donor cell contribution. The results are summarized in Figure 5.12. Early endodermal cells contribute solely to the endodermal layer of the yolk sac, while ectodermal cells contribute solely to the mesodermal layer of the yolk sac and the fetus proper, but never to the endoderm of the yolk sac. (The structure of the early postimplantation embryo will be described in chapter 8.) Evidently, by the time the two ICM-derived cell populations of the late blastocyst have separated, they have diverged both in fate and in their respective developmental capabilities.

The case of the two TE cell populations may be slightly different. A principal difference between polar and mural TE cells is in their nuclear character. Polar TE cells remain diploid and retain their division capacity, while mural TE cells cease dividing and become polyploid, endoreduplicating their genomes up to 500-fold, becoming so-called giant cells. The nature of this polyploid state was, for many years, a subject of controversy, but some experiments employing *in situ* hybridization

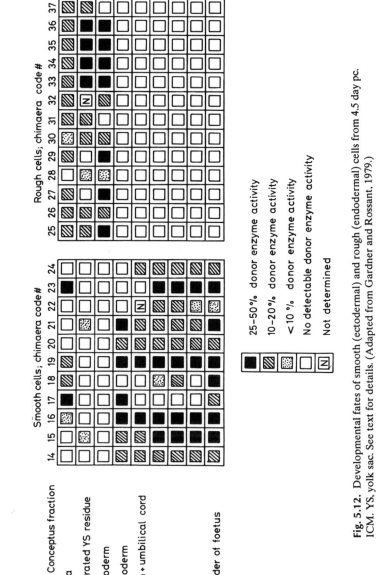

Fig. 5.12. Developmental fates of smooth (ectodermal) and rough (endodermal) cells from 4.5 day pc. ICM. YS, yolk sac. See text for details. (Adapted from Gardner and Rossant, 1979.)

suggest that the chromosomes of the trophoblastic giant cells are polytene, although not as tightly packed as *Drosophila* polytene chromosomes (Varmuza et al., 1988). The experiments show single hybridization spots per nucleus for a hemizygote of a single-site marker (present in multiple tandem copies at that site) and two spots per nucleus for the homozygote. This kind of pattern is the one expected for polytene chromosomes but not for conventional polyploids, which should show numerous sites of hybridization for such single-copy markers (as, indeed, liver cells, which are polyploid, do).

The retention of division capacity by polar TE cells seems to be a consequence of their association with the ICM (Gardner, 1978); in some manner, this contact serves to maintain the polar TE as diploid cells. However, while mural TE cells of the expanded blastocyst apparently cannot become polar TE cells, the reverse transformation can and does occur, as polar cells are pushed to the side and assume positions away from the ICM. Determination in the TE cell populations is thus a one-way street, not the mutually exclusive pair of possibilities exhibited by the ICM-TE and ICM-endoderm divergences.

GENOMIC IMPRINTING

Preimplantation development generates four cell types possessing two different general functions. As shall be discussed in more detail in chapter 8, the cells of the trophectoderm and the primitive endoderm give rise to the extraembryonic tissues (EET), while the internal embryonic ectoderm gives rise to the embryo proper. Beneath their surface differences, however, these tissues also possess a set of intrinsic distinctive genomic specializations, which limit and determine their abilities to perform their respective developmental roles. These genomic states derive, ultimately, from differences in the original maternal and paternal chromosome sets inherited by the zygote and involve gene-specific restrictions on expression within each chromosome set. The restrictions arise during gametogenesis and confer a requirement for the paternal genome in the development of the EET and, on the maternal genome, a crucial role in the development of the embryo proper.

As noted in the introduction, such gamete-of-origin marking of genomes is termed chromosomal or genomic "imprinting" (Crouse, 1960). Although the general phenomenon of genomic imprinting was first described more than three decades ago in certain insect groups, its existence in mammalian development was unsuspected until the late 1970s and only definitively established during the 1980s. Its discovery was the product of an unusual conjunction of two independent lines of research, each directed at solving a particular genetic puzzle.

Indications From Genetics: T^{hp} and Disomy Studies

The initial evidence for different roles of the two input parental genomes involved the discovery that for certain genes and for certain chromosomal regions, it matters profoundly whether the embryo receives functional copies solely from the mother or solely from the father. The first case of this kind to be explored involved a particular mutant of the short-tail of *Brachury* (*T*) locus, the "hairpin-tail" mutant (T^{hp}).

The critical initial observation was that the two reciprocal crosses, involving one heterozygous parent for the dominant mutant allele and one homozygous wild-type parent, give different results (Johnson, 1974). For the cross T^{hp}/+ males × +/+ females, one obtains progeny with the hairpin-tail phenotype and wild-type mice in nearly the expected 1:1 Mendelian ratio (there being a slight deficit of the former attributed to some postimplantation mortality). However, for the reciprocal cross, +/+ males × T^{hp}/+ females, the progeny show a vast excess of wild-type to hairpin-tail. (In the results described by Johnson [1974], the ratio was 144 wild-type to 12 hairpin-tail progeny.) Furthermore, the phenotypes of the hairpin-tail progeny are different in the two crosses: In the former, one obtains viable, fertile mice which are essentially normal except for the defining hairpin-tail defect, while in the latter, the hairpin-tail progeny show general swelling, a tendency toward polydactyly on the rear limbs, and die shortly after birth. These survivors are the rare "escapers"; the majority of the heterozygotes in this cross die during postimplantation development (Johnson, 1974). Evidently, transmission of the mutant chromosome through the female germ line is associated with exaggerated mutant effects not seen when it is transmitted from the father.

This difference between the two crosses involves neither meiotic abnormalities in the mother (Johnson, 1974, 1975) nor a conventional maternal effect involving deficiency for a cytoplasmic product, with late-onset, long-term effects on development. One observation that eliminates a cytoplasmically based maternal effect is that when heterozygous males and females are intercrossed, T^{hp}/+ males × T^{hp}/+ females, one obtains *both* kinds of hairpin-tail progeny, with a vast excess of the milder (viable) phenotype class, as expected from the differential viabilities observed in the earlier crosses (Johnson, 1974). Were the mutation-producing effects solely a standard maternal effect, only the severely affected class would be expected.

The definitive proof, however, that a conventional maternal effect is not involved was obtained by reciprocal nuclear transplantation experiments between one-cell embryos from T^{hp}/ + and +/+ mothers, followed by transfer to pseudopregnant females (McGrath and Solter, 1984b). Transfers of +/+ nuclei into the cytoplasm of T^{hp}/+ eggs showed a high level of development to term, while the reciprocal transfer showed only a low frequency of T^{hp} progeny and the one survivor showed the severe phenotype. The results show that the T^{hp} defect is associated with a maternal *nuclear* state, not a cytoplasmic one (McGrath and Solter, 1984b).

An important question is whether the lethal maternal inheritance pattern reflects a dominant neomorphic property of the mutant allele or rather the absence of a gene that must be inherited from the maternal genome in order to be expressed in the embryonic one. Genetic analysis of the T^{hp} defect has answered this question: The mutant "allele" is, in reality, a deletion (Bennett et al., 1975). Furthermore, smaller deletions show that the relevant gene is not *T* itself, but a separate locus, *Tme* (for *T*-associated *m*aternal *e*ffect locus), situated distally to *Brachyury* on chromosome 17, between *quaking (qk)* and *tufted (tf)* (Winking and Silver, 1984). (The probable identity of *Tme* will be discussed below.)

As pointed out by McLaren (1979), in an early speculation on the nature of the T^{hp} phenomenon, the results are explicable if the paternal allele of the essential locus is normally "inactivated" in some manner, with the gene being expressed solely from the maternally derived allele during wild-type development. Under this hypothesis, the deletion would remove this wild-type maternal allele, and the paternal allele, being unexpressed, cannot supply the needed gene product. In contemporary terms, *Tme* is imprinted during spermatogenesis, to prevent its subsequent expression in the embryo, while the maternal allele retains its expression potential.

Evidence for the existence of other imprinted genes was obtained in the 1970s and early 1980s, using translocations that allowed the production of offspring with either two paternally derived or two maternally derived chromosomal regions. These experiments have served to identify more than half a dozen other chromosomal regions in the mouse genome that appear to be differentially imprinted in the maternal and paternal genomes.

The procedure has employed a special kind of translocation, Robertsonian fusions. In the mouse, all chromosomes are normally "acrocentric," meaning that their centromeres are located very near one end. Rare recombination events in the germ line, however, will sometimes "fuse" two nonhomologous chromosomes near their centromeres, within the genetically silent heterochromatin, to produce a metacentric chromosome. Such translocations, possessing a centrally located centromere and two chromosome arms derived from nonhomologous chromosomes, are termed Robertsonian fusions.

Individuals that possess such a Robertsonian fusion and two normal acrocentrics, for the two constitutive arms, are said to be translocation heterozygotes and, like all such heterozygotes, experience pairing and segregation difficulties during meiosis, which lead to an elevated frequency of nondisjunction events. The consequence is the production of gametes that possess either two or zero copies for one of the constitutive chromosomes. When such a gamete combines with a normal gamete the result is either trisomy or monosomy, respectively, for the particular chromosome, both being lethal combinations. If either unbalanced gamete, however, is fertilized by the reciprocal type, in crosses between two such translocation heterozygotes, the result will be a zygote that is balanced in terms of chromosome composition, and hence viable, but which has received two copies of the chromosome from one parent and none from the other. To detect those individuals that have inherited two maternal chromosomes, one makes one parent homozygous for a recessive viable marker and the other homozygous for the wild-type allele; those progeny (sometimes a few percent) who show the homozygous recessive phenotype are disomics. The reciprocal disomy, involving homozygosity for the wild-type, can be distinguished from normal (heterozygous) sibs by testing for transmission of the marker (and, in some cases, is detected by a different developmental syndrome). The genetic scheme for producing such disomies is illustrated for the case of chromosome 11 in Figure 5.13. For chromosome 11, there are essentially opposite effects for the two different parental disomies: Maternal disomy gives reduced growth, and paternal disomy gives enhanced growth, relative to normal sibs.

When all the results are compared, a notable outcome is the uneven distribution of chromosomes associated with strong paternal or maternal disomic effects, with many showing no evidence of such effects. Thus, monoparental inheritance from either parent for chromosomes 1, 3, 4, 5, 9, 13, 14, 15, 16, and 19 (of the 20 chromosome pairs in the mouse genome) is fully compatible with embryonic survival and development to the adult stage (Lyon et al., 1976; Gropp et al., 1981; Searle, 1985; Cattanach and Beechey, 1990). Hence, there are no genes on any of these chromosomes whose differential imprinting markedly affects normal development. (Comparatively subtle differences, involving imprinting of less essential genes, might not have been detected.)

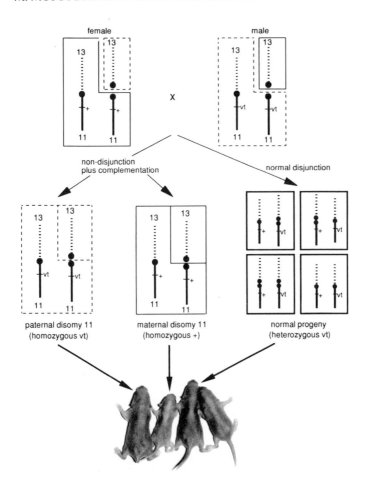

Fig. 5.13. The production of maternal or paternal disomy, using Robertsonian (Rb) translocations. The particular translocation illustrated involves chromosomes 11 and 13. Rb(11.13)4Bnr and the results of nondisjunction leading to disomy for chromosome 11 are shown. Maternal disomy is associated with reduced size (about 70% of normal) and paternal disomy with increased size (about 130% of normal). (Reproduced from Cattanach and Beechey, 1990, with permission of the Company of Biologists Ltd.)

In contrast, strong differential effects have been seen for monoparental inheritance of chromosomes 2, 6, 7, 11, and 17 (Cattanach and Kirk, 1985; reviewed in Cattanach and Beechey, 1990). Furthermore, the particular region(s) of each chromosome involved can be delineated where suitable reciprocal translocations between nonhomologous chromosomes exist; such translocations can be used to generate animals that have received both proximal or distal regions of a given chromosome, with respect to the translocation breakpoint, from one parent (Searle and Beechey, 1978). With such additional information, the nonrandom distribution of imprinted regions becomes even more striking: six (or possibly seven) of the eight (or nine) differentially imprinted regions are located on just three chromosomes, numbers 2, 7, and 17. (The uncertainty in the precise number stems from the uncertainty as to whether one or two regions of the distal part of chromosome 7 are differentially imprinted.)

In the analyses of both kinds of translocations (Robertsonians, reciprocals), the absence of maternal duplication/paternal deficiency classes (or defects associated with such classes) indicates a requirement for expression from the paternally derived genome during development (and their imprinting in the maternal genome). The reverse situation, the absence of paternal duplication/maternal deficiency, indicates the existence of a requirement for genes whose expression is required from the maternal genome (and that are imprinted in the paternal genome). For instance, the finding that paternal disomy for chromosome 6 permits viability but that maternal disomics for chromosome 6 are not found indicates the existence of one or more genes on this chromosome whose expression is required from the paternal genome and which does not take place from the maternally donated chromosome (Tease and Cattanach, 1986). A map of the autosomal regions implicated as differentially imprinted, with some of the associated effects, is shown in Figure 5.14.

The X chromosome is not included in the map, but it too is differentially imprinted. In female mouse embryos, the paternal X is preferentially inactivated in the EET of the embryo (Takagi and Sasaki, 1975; West et al., 1977), while there is a slight bias for inactivation of the maternally derived X in the cells of the fetus (Cattanach and Beechey, 1990). The function of such X chromosome inactivation is "dosage compensation" of the X and equalizes expression for this chromosome between males (XY) and females (XX) (Lyon, 1961). In contrast to the autosomes that experience imprint-

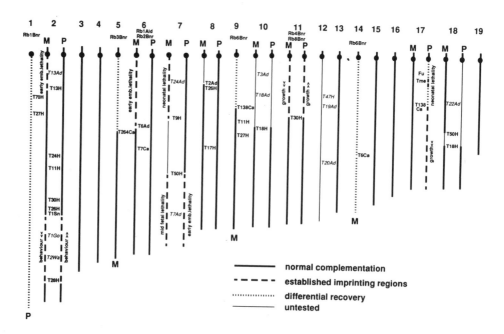

Fig. 5.14. Imprinting map of the autosomes of the mouse. Robertsonian fusions that have been used are listed at the top, reciprocal translocations that have been used are given in boldface (at the positions of their breakpoints), and reciprocal translocations that might be used in the future are given in italics. Where particular defects are associated with imprinting for those regions, they are given alongside the indicated region. (Reproduced from Cattanach and Beechey, 1990, with permission of the Company of Biologists Ltd.)

ing, the inactivated X becomes fully heterochromatic, becoming the darkly staining entity in the nucleus of XX cells termed the Barr body. There is an initial site of imprinting, the so-called X inactivation center (*Xce*) and the inactivation event spreads from that center (Cattanach, 1975). Whether for autosomes there is a comparable imprinting initially at specific sites, followed by spreading effects, is not known, but it is a possibility (Cattanach and Beechey, 1990). For none of the identified autosomal imprinted regions is it known how large the region is, or how many genes are affected. If imprinting involves a spreading phenomenon, it will be important to establish what processes set limits to the size of imprinted autosomal regions, in contrast to the X.

Nuclear Transplantation Experiments

The second line of investigation that revealed the existence of imprinting involved nuclear transplantation experiments. The motivation for this work was to discover why mammalian embryos cannot undergo parthenogenetic development. While such embryos, derived from activated but not fertilized eggs, can begin development and sometimes reach intermediate stages of fetal development, they often fail in early development and never complete the sequence of fetal development. The failure of mammalian parthenogenotes is surprising in light of the fact that various other animals, including a number of insects and all other vertebrate groups, can exhibit full development as parthenogenotes.

Two reasons were commonly advanced for the failure of mammalian parthenogenesis to yield viable offspring (Graham, 1974). The first was that parthenogenotes, whether haploid or diploid, when derived from the same postreductional meiotic nucleus, would be effectively homozygous at most or all loci; their failure to develop might reflect the "exposure" of one or more deleterious recessive alleles in this situation. The second hypothesis was that the sperm might make an essential but nonchromosomal ("extragenetic") contribution, for instance, a cytoskeletal component, and that parthenogenotes, beginning development without benefit of such a component, would necessarily be defective for the part(s) of the developmental sequence in which this component was required. Both hypotheses are subject to test by means of nuclear transplantations between fertilized eggs.

The fundamental technique for transplanting nuclei between mouse embryos was developed by McGrath and Solter (1983). The first steps involve placing the donor and recipient embryos in a solution of cytochalasin and Colcemid (to relax the cytoskeleton and thereby facilitate transfer of nuclei without damage to the embryos), placing a pipette near the nucleus to be transferred, and aspirating it into a solution of Sendai virus (a membrane binding agent that promotes membrane fusion). The donor nucleus is then placed into the perivitelline space, the region under the zona pellucida, by means of the pipette tip, and fusion of the karyoplast and recipient embryo takes place within 1 hr; when done carefully, survival rates of the recipient embryos are high, on the order of 90% (McGrath and Solter, 1983).

To test the possibility that the death of parthenogenotes reflects the absence of an essential "extragenetic" component, normally supplied by the sperm, Mann and Lovell-Badge (1984) transferred nuclei between fertilized eggs and parthenogenotes. The experiment involves the exchange of pronuclei between newly fertilized eggs and activated diploid parthenogenotes, followed by transfer of the operated embryos to pseudopregnant foster mothers; the experiment is diagramed in Figure 5.15. In 22 of 34 parthenogenetic cytoplasmic environments that had received zygotic nuclei, development to term of the operated embryos occurred, whereas of 34 fertilized eggs which had their nuclei substituted by parthenogenetic nuclei, none developed to term (though 21 implanted). Since the fertilized eggs would have acquired any "extragenetic" component that might be brought in by the fertilizing sperm, yet still fail to develop when their nucleus is of parthenogenetic origin, the developmental failure must involve a nuclear defect, not the absence of a sperm-supplied cytoplasmic component.

The other hypothesis for the failure of parthenogenetic development, that such failure reflects the deleterious effects of recessive lethals "exposed" in haploid or diploid parthenogenetic nuclei, has been tested by carrying out nuclear transplantations of male and female haploid pronuclei between embryos from different inbred lines; such embryos, derived from two different genetic backgrounds, will necessarily be heterozygous at nearly all loci. Where the two haploid component genomes are both paternal in origin, such embryos are termed diploid androgenetic embryos, and where both are maternal in origin, they are termed diploid gynogenetic embryos. Control embryos consist of nuclear transplantations where the transferred nucleus is of opposite parental origin to the remaining pronucleus (as in normal fertilized embryos). The experiment is shown in Figure 5.16.

While all interembryo pronuclear transfers between embryos from different strains will produce highly heterozygous zygotes, virtually eliminating the possibility of lethality through exposure of recessive lethals, the striking result is that the only

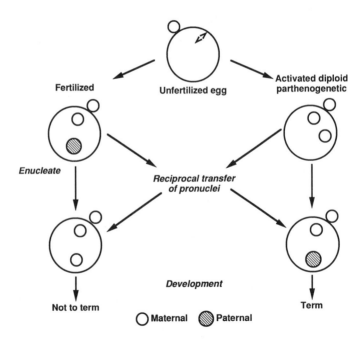

Fig. 5.15. Failure of cytoplasm from a fertilized egg to rescue development of embryos with parthenogenetic nuclei. (Adapted from Surani, 1986.)

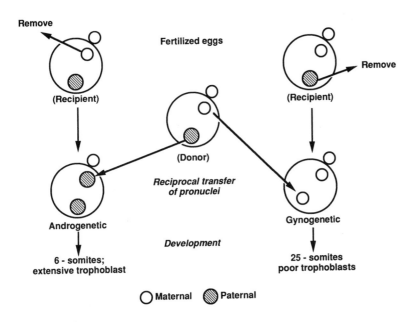

Fig. 5.16. Different developmental consequences of biparental androgenetic and gynogenetic eggs, formed by nuclear transplantation. (Adapted from Surani, 1986.)

nuclear combinations to produce embryos that develop to term (and yield viable adults) are those involving a male and female nucleus. In contrast, both diploid androgenetic and diploid gynogenetic embryos fail to yield any viable progeny (McGrath and Solter, 1984a; Surani et al., 1984, 1986a; Solter et al., 1985). The results show, unambiguously, that successful embryonic development requires both a male and a female pronucleus. Two pronuclei from the same germ line, whether both are male or both female, cannot "complement" to yield normal development.

Despite the incapacity of both diploid androgenetic and gynogenetic embryos to undergo full development, the specific causes of failure in the two cases are very different and show an interesting complementary relationship. In the diploid gynogenones, the embryo proper (the fetus) can develop essentially normally until fairly late stages, up to the 25-somite stage (postembryonic stagings are described in chapter 8), but its extraembryonic membranes, derived from the TE and primitive ectoderm, are highly underdeveloped. The immediate cause of developmental failure is undoubtedly, in part, a failure of nutrition produced by the underdevelopment of the support system. This cannot be the sole cause, however, of their failure to develop fully; ICMs from parthenogenetic embryos placed in blastocysts derived from fertilized eggs develop better but, still, not to term (Barton et al., 1985).

In contrast to the gynogenetic embryos, diploid androgenetic embryos show very poor development of the embryo itself, but extensive, relatively normal development of the extraembryonic membranes, particularly the trophoblastic (implanting) membranes (Surani et al., 1986b). The differences between the two developmental syndromes are illustrated in Figure 5.17. The results, in sum, indicate that there are strong preferential requirements for the maternal and paternal genomes, respectively, in

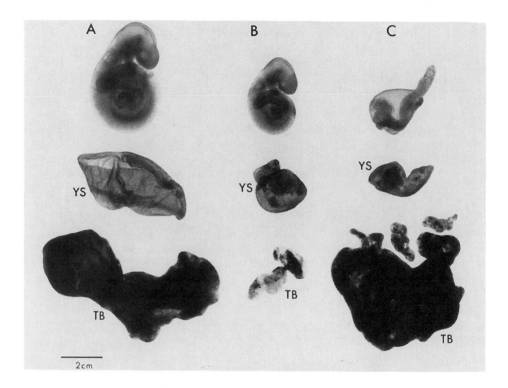

Fig. 5.17. Illustration of different developmental syndromes in androgenetic and gynogenetic embryos, at 10 days pc. **A**: Control embryo. **B**: Gynogenetic embryo. It is smaller than the control but has developed to the 26-somite stage; it has poorly developed yolk sac (YS) and trophoblast (TB). **C**: Androgenetic embryo. It is poorly developed, shows 6–8 somites, and has a poorly developed YS but comparatively well-developed TB. (Reproduced from Surani et al., 1986b, with permission of the publisher. © Cell Press.)

fetal and EET development but that both genomes are required for full, normal development of both sets of tissues.

It is not clear how many genes or what percentage of the genome may be differentially imprinted and required in this "division of labor" between the maternal and paternal genes for embryonic development as a whole. From the genetic translocation studies, it is clear that the majority of imprinted regions identified to date affect postimplantation stages (rather than preimplantation stages), even up to late fetal or neonatal stages (Cattanach and Beechey, 1990). From both the disomy syndromes (Cattanach and Beechey, 1990) and studies of chimeras between gynogenetic or androgenetic cells and cells from normal embryos, it appears that imprinting affects genes particularly involved with growth and size regulation (Cattanach and Beechey, 1990; Surani et al., 1990). The chimera studies indicate that imprinted states are not erased over many cell generations and that many of the effects are cell autonomous. (The chimera studies will be discussed in more detail in chapter 8.)

Although much investigation has been directed to the precise molecular basis of imprinting, its mechanism is still unknown. Methylation, with subsequent inhibition of transcription, is certainly part of the answer but it cannot be the whole story, given several findings that methylation occurs after the initial inactivation event (Lock et al., 1987; Monk and Grant, 1990). An apparent (and rare) case of imprinting has been reported in *Drosophila* (Kuhn and Packert, 1988), yet *Drosophila* is known to lack DNA methylation. In effect, methylation in mammals may serve to "confirm" or "lock in" an initial chromosomal imprinting event and thereby help to maintain it. This initial event may be common to many regional chromosomal inactivations in many species, while the reinforcing mechanism(s) may differ among species, with methylation being involved in some but not others. The initiating processes in mammalian imprinting, however, remain to be discovered.

Transgenic mice, those that have received DNA constructs as one-cell embryos and integrated them, will almost certainly help to resolve the matter. A number of transgenes have been found to undergo imprinting, which is reversible during gametogenesis in the opposite sex, and the imprinting events correlate with methylation (Surani et al., 1989).

Imprinting: What Is It For? What Are the Genes?

A central question about imprinting is why it exists. The answer is, inevitably, connected to its evolutionary origins, which remain a matter of speculation. One line of argument is that its function is to prevent parthenogenesis and the resultant loss of genetic variation that parthenogenetic systems experience. Although the prevention of parthenogenesis is undoubtedly a consequence of the existence of imprinting, it is a moot point whether some feature or process can evolve in order to benefit the species some time in the future. Such arguments fall into the "group selectionist" category and evolutionists are, in general, skeptical of such hypotheses.

Another, and more cogent, kind of explanation relates imprinting to the unique relationship of the embryo to the mother, involving formation of the placenta and nutritional supply through the mother (Hall, 1990; Moore and Haig, 1991). The link is suggested by the fact that among vertebrates, only mammals are incapable of parthenogenesis and utilize placentation. Moore and Haig (1991) have proposed that the existence of imprinting reflects a parental "tug-of-war": It is in the (evolutionary) interests of the male to promote the growth of his offspring in the female, while it is in the interests of the female to limit the growth of any one litter, so as to be fit to mate and produce offspring sired by other males. In this hypothesis, genes supplied by the male will have been selected to promote placentation (accounting for the ability of androgenetic embryos to produce nearly normal EET development), while selection will have favored imprinting processes in the female that reduce EET development for any embryo (which accords with the observation that gynogenetic embryos show poor development of the EET).

The identification of specific genes that are differentially imprinted, and clarification of their roles, should prove helpful in evaluating this idea. Three endogenous genes of the mouse genome (as opposed to transgenes) have been identified as undergoing imprinting. One of these, designated simply *H19*, encodes an abundant mRNA and is imprinted in the father but is active when transmitted from the mother (Bartolomei et al., 1991). Though the function of the *H19* gene product is unknown, the states of imprinting are maintained through neonatal stages in at least two tissues, liver and muscle (Bartolomei et al., 1991). It maps to the distal part of chromosome 7, one of the known imprinted regions (Fig. 5.14).

The two other identified endogenous genes are that for a growth factor, insulin-like growth factor 2, (*Igf-2*) and the gene for the receptor of this growth factor (*Igf2r*); they are imprinted in opposite fashions in male and female gametogenesis and may have a bearing on the hypothesis of Moore and Haig. The growth factor gene is on distal chromosome 7 in a known region of imprinting and is imprinted in the mother but not in the father (De Chiara et al., 1991). It maps near *H19* but is imprinted in the opposite direction (which indicates that imprinting is probably a gene-by-gene affair rather than involving broad regions). Since this growth factor is believed to be a general stimulator of growth of undifferentiated cells (Heyner et al., 1989), if the mother represses expression of the maternally derived allele, there would be less growth factor to stimulate placental growth of the embryo. In contrast, the receptor gene, which maps to distal chromosome 17, is expressed in the mother but imprinted in the father. The receptor is known to transport certain molecules to be degraded in the lysosomes (Kornfeld and Mehlman, 1989), and Moore and Haig (1991) speculate that maternal expression of this gene could lead to the mopping up of surplus Igf-2 protein, thereby limiting growth of the embryo. *Igf2r* maps to the location of *Tme* and shows its expression pattern; it could be *Tme* or could be one of several imprinted genes that give the *Tme* effect.

These results are only a beginning but they are provocative. It will be interesting to see if other growth factors and their receptors are imprinted in comparable fashions. As noted earlier, many of the genetically identified instances of imprinting seem to involve genetic controls on growth.

EARLY EMBRYOGENESIS IN THE MOUSE: CLASSIC THEMES, NOVEL VARIATIONS

The aspects of early embryogenesis that have always captured the interest of embryologists are 1) the relationship between oocyte substructure and blastomere fates, 2) the respective importance of cell inheritance and cellular interactions in the acquisition or modification of blastomere assignments, and 3) the nature of the determined states that appear, sooner or later, in the cells of the embryo. All of these themes are relevant to the phenomenology of the early mouse embryo, but this embryo exhibits—and, by extension, mammalian embryos in general—some novel variations on these themes.

With respect to the relationship between oocyte structure and blastomere fate, it appears probable that there is *some* degree of preexisting organization within the egg that contributes to the first cell divergence, that between polar and apolar cells (presumptive TE and ICM, respectively). This might involve some preferential inheritance of oocyte membrane by the polar cells (Pratt, 1989) or it may be a cytoplasmic property, if only one that acts as a precondition for compaction. One particular hypothesis about preexisting structure was the "polarization hypothesis" of Johnson et al. (1981). The idea posits that the inside–outside differences between cells that appear during the division of the eight-cell stage stem from a radial asymmetry within the embryo in which individual cells are polarized with respect to molecular and cellular properties at their (outer) apices and (inner) bases. In this view, cleavage in the compacted eight-cell embryo, occurring perpendicularly to the radial axis, separates cells with "inner" (apolar cell) and "outer" (polar cell) values of this radial gradient.

This hypothesis has not been disproved, but it is now clear that the whole process of generating apolar and polar cells is considerably more dynamic than would be suggested by the model, given that, as discussed above, apolar cells can generate polar cells and vice versa, until at least one cell division later than the one in which polar and apolar cells first appear. Discrete ICM and TE lineages *do* arise in the embryo but they are not set aside at the first division that gives rise to presumptive ICM- and TE-type cells; true cell lineages, each restricted in developmental capacity, have only arisen by the expanded blastocyst stage. Although determined states exist in the mouse embryo, the first ones come into existence comparatively gradually, over an interval of several cell division periods and through a series of cellular interactions that first create and then reinforce these restrictions. As noted by Johnson et al. (1986), and quoted at the beginning of this chapter, an understanding of the first cell fate divergences in the mouse embryo is probably most advisedly to be sought in the analysis of the cell interactions that take place rather than in the inheritance of determinants from the oocyte.

Although in this respect there is a strong contrast to *Drosophila*, where maternally specified and prelocalized molecules are known to play a key part and where, with the appearance of the first cells at the cellular blastoderm stage, there is an accompanying array of different determined states throughout the embryo (chapter 4), there may be some similarity to the nematode embryo in these cell interactive and regulative features. Although the nematode embryo has seemed, for many decades, to be the archetypal "mosaic" system, the increasing evidence for the importance of ongoing cell interactions in fate setting in this embryo (see chapter 3) has substantially modified this picture. The phenomenon of cell polarization, as a prerequisite to divisions that create differing sister cells, is, at least on the surface, an important commonality between the embryo of the nematode and that of the mouse.

In the phenomenon of genomic imprinting, however, the mouse, and, in general, the mammalian embryo, is clearly different from those of the nematode and fruit fly. If imprinting arose in mammals in connection with placentation, then its functional relevance to other animal systems is questionable. However, as a chromosomal phenomenon, it extends beyond mammals to a number of insects, plants, and even fission yeast (Monk and Surani, 1990). Furthermore, a form of chromosomal imprinting is seen in the phenomenon of variegated position effect in fruit flies (Baker, 1968; Locke et al., 1988). Although the special need for imprinting in the mouse embryo may be related to the peculiarities of mammalian development, its molecular basis may have broad relevance to the long-term maintenance of chromosomal states and, hence, to the general phenomenon of determination.

An important element in the further characterization of early development in the mouse must be more intensive genetic analysis. Indeed, for the geneticist, the striking feature of the study of preimplantation development is the relative scarcity of mutant studies, in comparison to the analysis of early development in *Caenorhabditis elegans* and *Drosophila*. There are, to be sure, a small number of mutants known to affect preimplantation development, and they have received some attention. One of these is the autosomal dominant mutation *Oligosyndactyly* (*Os*), which, when heterozygous, causes various limb defects. When homozygous, it causes death of preimplantation embryos, following a cell-autonomous mitotic block, which becomes manifest around the 32-cell stage (Magnuson and Epstein, 1984). A second is the classic allele, *yellow* (A^y) (Papaioannou and Gardner, 1979), which when homozygous, creates defects in the preimplantation embryo. Nevertheless, the study of these and the few other extant mutants that affect preimplantation development has not yet contributed substantially to understanding the mechanisms of the ICM-TE divergence or the subsequent bifurcations of cell type and developmental capacity within these lineages. The principal hope for a breakthrough lies in the new and intensive mutant hunts that can be carried out in the mouse; these have already served to identify new lethal complementation groups, including some affecting preimplantation development (chapter 8).

As new preimplantation lethal point mutants are obtained, they can be tested for their respective roles in the ICM and TE by means of reciprocally reconstituted blastocysts with wild-type (Papaioannou and Gardner, 1979) and also tested for their effect on these components in cell culture (Papaioannou, 1988). Given the apparent involvement of fundamental cellular processes in the first allocations of fate, one may anticipate that many of the preimplantation lethals will be in genes required for general cellular functions. Some, however, may be required specifically for polarization and the first divergences between polar and apolar cells.

In addition, further genetic and molecular analysis of genomic imprinting should provide more clues to its molecular basis as well as its function in mammalian development. Some of the *Drosophila* genes now being characterized that participate in stabilizing chromosomal states may have homologs in the mouse that perform a comparable role. The identification of homologs for some of the key pattern formation genes of *Drosophila* in the mouse (see chapter 8) makes this, at least, a possibility.

In sum, while the contribution of genetics to the analysis of developmental mechanisms in mammalian preimplantation development has, thus far, been comparatively slight, one may anticipate a considerably larger genetic component in future studies of the early mouse embryo.

SIX

CAENORHABDITIS ELEGANS
Early Embryo to Adult

...both the cellular anatomy and the pattern of cell divisions from the single-celled zygote to the adult are essentially invariant among individuals....Mutations affecting this rigidly determined cell lineage offer one way of answering a variety of questions about the partitioning of the genetic program for *C. elegans* development. For example, is it possible to isolate mutants altered in a specific cell division or set of divisions? If so, how many genes are involved in each cell division? In how many cell divisions is each such gene involved? What other features are common to that set of cell divisions affected by a given gene? Can mutants be isolated in which cell fate is transformed so that a cell follows a lineage that is normally that of another cell?

H.R. Horvitz and J.E. Sulston (1980)

INTRODUCTION: AN OVERVIEW OF EMBRYOGENESIS AND POSTEMBRYONIC DEVELOPMENT

As in the prepupal development of *Drosophila,* the development of *Caenorhabditis elegans* can be divided into a period of embryogenesis, which takes place within the eggshell, and an extended period of postembryonic larval development, which occurs in several stages. Embryogenesis involves the period of most rapid and extensive change in the nematode, in which the basic internal anatomy and form of the animal come into being, and consists of several overlapping processes: the sequence of cleavage divisions, which produce the 558 cells present in the newly emerged hermaphrodite larva, and the events of morphogenesis and differentiation, which begin during cleavage and extend beyond its completion. The second phase, postembryonic development, follows the emergence of the newly formed first-stage larva, the L1, from the eggshell and consists of four larval stages. It is characterized by growth, development of the gonads and the secondary sexual characteristics, and the elaboration and reorganization of a few other somatic structures not directly connected with sexual maturation. Each larval stage displays characteristic cell division and developmental events. The animal that emerges from the fourth larval cuticle is the nearly fully mature adult, which, within a few hours, can commence reproduction. ·

Although embryogenesis is the period of most profound developmental change, it has also been less accessible to investigation than postembryonic development; it takes place within the eggshell and early embryos experimentally liberated from the eggshell for further study undergo only partial development. Nevertheless, the principal facts and many of the details are now clear.

During the first half of embryogenesis, cleavage takes place and is accompanied by many morphogenetic cell movements and by the development of the principal somatic tissues and structures of the larva; during this period, all the major organ systems except the gonads are formed. In the second half of embryogenesis, the animal, rotating within the eggshell, experiences a threefold elongation, acquiring the shape of the first-stage or L1 larva, and completes the functional maturation of the cells required for larval development. The embryo is completely formed about 2 hr before hatching and the newly hatched larva has fully functional digestive and juvenile motoneuron systems. Though all of the organ systems in the animal continue to develop and mature, the L1 has much of the behavioral repertoire of the adult. The L1 hermaphrodite larva is shown in Figure 6.1.

Hatching of the embryo marks the beginning of postembryonic growth and the start of autonomous growth. This phase of development consists of four distinct larval stages, which are separated by molts of the larval cuticle and which altogether produce a fivefold increase in length and the sexual maturation of the animal. Each larval stage is distinguished by a particular pattern of further cell divisions and developmental changes, in both somatic and germ line tissues; the net result of postembryonic development is the emergence, from the final larval cuticle, of the sexually mature animal. The adult hermaphrodite and male forms are shown in Figure 6.2.

The distinctive structures of the adult, in particular the gonads and the secondary sexual characteristics appropriate to each sex (such as the vulva in the hermaphrodite and the rayed tail of the male), ultimately arise from a special set of precursor cells, the postembryonic blast cells. These cells, analogous in function to the six embryonic blast cells that give rise to the major cell lineages in the embryo, are produced in the AB and MS embryonic lineages, which also furnish the greatest variety of cell types in the embryo itself. The postembryonic cells comprise about 10% of the cells of the newly emerged L1 animal and are found at specific locations in the L1 (Fig. 6.3). By convention, they are designated by capital letters with or without numerals, depend-

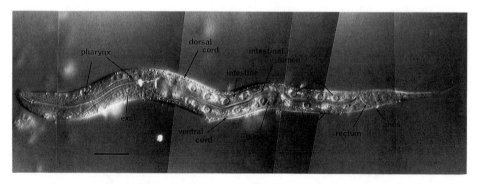

Fig. 6.1. The L1 hermaphrodite of *C. elegans.* (Reproduced from Sulston and Horvitz, 1977, with permission of the publisher.)

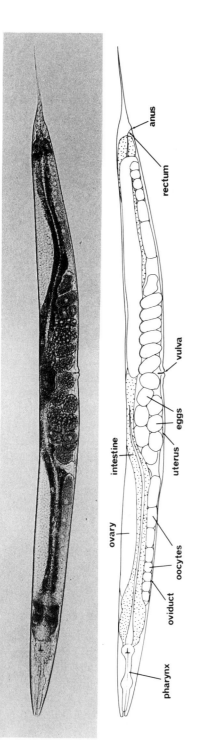

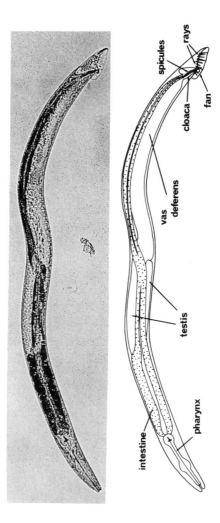

Fig. 6.2. Lateral views of the adult hermaphrodite (top) and adult male (bottom) of *C. elegans*. Scale bar: 20 μm. (Reproduced from Sulston and Horvitz, 1977, with permission of the publisher.)

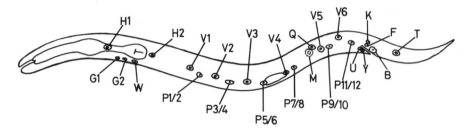

Fig. 6.3. Positions of postembryonic blast cells in the young L1 larva. Blast cells G1 and G2 and the male-specific B, U, Y, and F, which are found in both sexes but divide only in males, are located medially; M is located on the right side and K on the left; all others exist as pairs located on both sides (but with only the left representative shown). (Adapted from Chalfie et al., 1981.)

ing upon whether they are unique or have homologs (e.g., W, Q, P1–P12, etc.) and generate lineages, characterized by individually stereotyped cell division and differentiation patterns during the four larval stages.

In this chapter we examine both the cellular events that comprise the development of *C. elegans* from mid-embryogenesis through postembryonic development, and the principal genetic analyses that have been carried out on these developmental events. Because so much of the development of the animal consists of the summed properties of the individual cell lineages, a major emphasis will be on the genetic analysis of the formation and properties of cell lineage in *C. elegans*. In addition, since the development of the gonad in both sexes involves some novel features not seen in the other tissues and organ systems, the principal part of this chapter will be divided into separate reviews of nongonadal development and gonadal development. We will then look at the genetic analysis of sex determination in this organism, a system of genetic controls which involves both somatic and germ line tissues, and then, briefly, at an alternative developmental stage that can take place during postembryonic development, dauer larva formation. The chapter concludes with a discussion of some of the principal conclusions from this body of work and some questions about the nature of genetic control of nematode development.

EMBRYOGENESIS

Embryonic Cell Lineages

As discussed briefly in chapter 3, the cellular foundations for the composition of the embryo are laid down with the formation of the six embryonic blast cells. These cells are the precursors, collectively, of the major cellular lineages of the embryo (Fig. 3.11), and, as noted above, two (AB and MS) are also precursors of the postembryonic blast cells.

The contributions of the embryonic blast cells to the cellular composition of the embryo can be briefly summarized. The AB lineage gives rise to most of the external tissue covering the animal, the hypodermis, the major part of the neuronal system, and part of the pharynx. The MS lineage gives rise to part of the musculature of the pharynx, some secretory and some scavenging cells, and, together with the D lineage, the major part of the body muscle system of the larva. The C lineage contributes the part of the hypodermis not formed from AB-derived cells and the remainder of the body muscle system of the larva. The two remaining blast cells, the E and P_4 blast cells, are the precursor cells of the intestine and germ line, respectively; these tissues are the only ones to originate from single cells (clonally) in the nematode.

Morphogenesis and Histogenesis

In contrast to many embryonic systems, morphogenetic movements begin in the *C. elegans* embryo long before cleavage itself is completed. Cleavage continues during these first events of morphogenesis, continuing to expand the cellular lineages. The first movements of morphogenesis consist of a series of ventral invaginations and external cellular migrations, and as described earlier in connection with the *emb-5* mutants, the first event of morphogenesis is gastrulation, the ventral invagination of the daughters of the E cell. These two cells subsequently divide and the four resultant E lineage cells give rise eventually to the juvenile intestine, consisting of 20 cells. Following the first invagination, the germ line precursor P_4 and the cells of the MS lineage enter through the ventral invagination. P_4, after completing its migration, divides to form two postembryonic blast cells, Z2 and Z3, which remain quiescent throughout the remainder of embryogenesis, but which will give rise subsequently to all the gamete cells, in both hermaphrodites and males.

From the eight MS cells originally present on the ventral surface of the embryo, all internal MS lineages increase in cell number during the first half of embryogenesis. As noted above, some contribute to the formation of the pharyngeal muscles in the future head region, and the remainder form body muscle cells, coelomocytes (which may function as phagocytic cells), a number of neurons, and the ring gland of the head region.

The ventral invagination zone progressively widens and lengthens, after entry of the MS cells, first in a posterior direction, engulfing the C and D blast cell progeny, and then anteriorly, as AB blast cell-derived pharynx muscle cell precursors enter. The complete sequence of cell invaginations takes place between 100 and 290 min from the first cleavage; the zone of invagination then closes. Throughout the entire period, the surface AB-derived cells that give rise to the juvenile hypodermis and ventral nervous system spread both posteriorly and ventrally. Concurrently, the C blast progeny destined to give rise to hypodermis spread anteriorly and dorsally, covering the remainder of the external surface. (The embryonic hypodermis is composed of 11 syncytia and a number of discrete cells; the largest syncytium, hyp 7, comes to encircle most of the body and consists of nuclei contributed by both AB and C blast cell descendants.)

These external movements are paralleled by the formation of the central cylinder of the animal, consisting of pharynx cells anteriorly and intestinal cells medially. The body myoblasts, or muscle precursor cells, which are derived from the MS, C, D, and AB lineages, move into position between this central cylinder and the outer cells, the surface AB myoblasts entering last. The progeny of these blast cells form the four longitudinal muscle strips of the juvenile body (seen in cross section in Fig. 6.4A).

The neural system of the juvenile is produced from AB-derived external neuroblasts. The ventral side of the animal, after the first cell invaginations, is largely

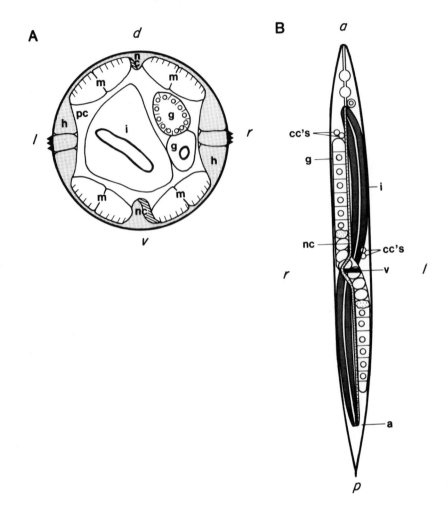

Fig. 6.4. Some morphological left–right asymmetries in the adult hermaphrodite of *C. elegans*. **A**: Cross section through anterior half, between the uterus and pharynx, viewed toward the anterior tip. **B**: Ventral view, longitudinal section. Anterior (*a*), posterior (*p*, left (*l*), right (*r*) as shown. There are pronounced left-right asymmetries for both the intestine (i) and the arms of the gonad (g). The ventral cord is located on the midline except for its rightward bulge at the vulva (v). a, anus; cc's, coelomocytes; h, hypodermis; m, body muscles; nc, nerve cord; pc, pseudocoelomic cavity. (Reprinted by permission from *Nature*, vol. 349, pp. 536-538. © 1991 Macmillan Magazines Ltd.)

comprised of these cells; they divide and sink inward, being replaced externally by the sheet of hypodermis. The ventral neuroblasts form the juvenile ventral nerve cord. Dorsal and lateral cells, located in the hypodermis, similarly sink in from their external positions and are covered by hypodermis.

In addition to these movements, a number of cells undergo extensive longitudinal migrations. These include the MS-derived precursor cells of the somatic portion of the gonad, Z1 and Z4. These cells move from an anterior position to the posterior half of the embryo, where they assume positions proximate to the germ line precursors, Z2 and Z3. In addition to such movements along the anteroposterior (a-p) axis, a few circumferential cell movements also occur, as do extensive crisscrossing transverse nuclear movements within the syncytial juvenile hypoderm.

The final major event of morphogenesis takes place in the second half of morphogenesis and consists of the lengthening of the embryo within the eggshell to produce the distinctive elongated form of the animal. This lengthening is apparently produced by circumferentially oriented bundles of actin fibers in the cells of the dorsal and ventral hypodermis; shortening of these filaments, presumably produced by contraction, is correlated with and appears to cause a lengthening of both the hypodermal cells and the embryo as a whole (Priess and Hirsh, 1986). Either elimination of the external hypodermis by laser ablation of its cell precursors or interference with actin filament contraction with cytochalasin D prevents embryo elongation (Sulston et al., 1983; Priess and Hirsh, 1986).

Programmed Cell Death

In addition to the fixed events of cell division and morphogenesis, another class of invariant event is that of "programmed" cell death. In the hermaphrodite embryo, there are precisely 113 cell deaths, and the usual pattern involves the death, after a specific cell division, of one specific sister cell. Each such death is presaged by the development of blebs on the nucleus, which then turns refractile and then dissolves (Sulston and Horvitz, 1977). In the great majority of these cases, the deaths are autonomous, not involving the actions of neighboring cells; for these autonomous cell deaths, their fates are written in their lineages (reviewed in Horvitz et al., 1982; Sulston et al., 1983). Not surprisingly, there are specific genetic requirements for cell death, with about a dozen *cell death* (*ced*) genes known to be required for the execution of autonomous cell deaths (reviewed in Driscoll and Chalfie, 1992). Loss-of-function mutations in any of these genes bring a halt to nearly all programmed cell deaths. Two of the genes, *ced-3* and *ced-4*, are necessary for the initial stages, whereas seven others (*ced-1, ced-2, ced-5, ced-6, ced-7, ced-8,* and *ced-10*) are responsible for the second stage in which nuclei become refractile. The gene *nuc-1* is required for the last stage, the degradation of the nuclei of dying cells (Ellis and Horvitz, 1986). Cells that have been "saved" by an early acting *ced* mutation will often differentiate and resemble their normally surviving sister cells.

The function of cell death is illustrated by one of the few examples of sexual dimorphism seen in the embryo, the existence of six sex-specific neurons of the AB lineage. Four of these, the cephalic companion neurons (CEM cells) are produced in the head and constitute part of the male adult sensory apparatus. They may serve to mediate chemotaxis of the male toward the hermaphrodite. The other two neurons are hermaphrodite-specific neurons (HSN cells) and are formed posteriorly and migrate anteriorly toward the midpoint of the embryo; they innervate the vulva of the adult hermaphrodite for the function of egg laying. All six neurons are formed in both sexes through identical lineage division patterns. However, only the neurons specific to the sex of the individual survive, the others undergoing programmed cell death. In a developmental sequence featuring great precision and economy of cell use, it is evidently less costly to retain general patterns of cell division and to expunge unneeded cells than to modify the basic division programs individually so as to save and utilize every cell.

Nevertheless, the existence of programmed cell death in this animal is not essential for the development or survival of the individual animal. Loss-of-function mutants of the *ced-3* and *ced-4* genes have essentially unimpaired viability and normal behavior, despite the survival of those cells that would ordinarily die (Ellis and Horvitz, 1986). Furthermore, in animals in which sisters of the normally dying cells have been eliminated by mutation or physical ablation and the cells programmed to die have been allowed to survive by means of a *ced* mutation, the spared cells can differentiate and function in place of their sisters (Ellis and Horvitz, 1986).

The existence of a number of genes required specifically for cell death, combined with the seeming dispensability of the function of these genes and of embryonic cell death itself, seems to pose an evolutionary paradox: If loss of the trait does not visibly affect the individual animal, why might the trait have been selected during the course of evolution? Like most paradoxes, the problem is, undoubtedly, a spurious one: The elimination of unneeded cells probably confers a slight selective advantage per generation, one that is not readily detectable by gross inspection of individuals. As is well known from evolutionary studies, even small selection coefficients can, over time, produce a highly uniform trait or capacity in a species. In principle, the existence of such slight selective advantage, associated with the elimination of unneeded cells, should be experimentally obtainable by the equivalent of "population cage" comparisons of *ced* mutants versus wild-type. Some evidence suggests that a degree of selective advantages will, indeed, be found for programmed cell death. *ced-3* and *ced-4* animals are less proficient in chemosensory responses than wild-type; they also mature at a slightly slower rate (cited in Driscoll and Chalfie, 1992).

The "Rules" of Embryogenesis: Some General Conclusions and Considerations

Although the formation of the juvenile worm involves many distinct events, related in complex fashions, three general features can be discerned. The first is that of invariance. From the first cleavage divisions through the final steps of functional maturation, the individual cells show highly reproducible behavior in the wild-type strain. Cells undergo fixed division patterns and fixed movements (with a few exceptions where cells of comparable lineage have interchangeable roles). The highly structured set of cellular events that comprise embryogenesis is thus based upon highly structured cell lineage patterns. From the 50-cell stage onward, individual cells can be selectively killed by means of a highly focused laser beam. For the great majority of these embryonic ablations, there is no replacement by other cells or compensatory regulation by novel routes (Sulston et al., 1983). Yet such experiments do not necessarily signify that cells cannot replace each other under other circumstances. In chapter 3, the interchangeability of ABa and ABp (Priess and Thomson, 1987), which normally possess very distinct fates, was discussed, and an experiment, to be described below, shows that this is not a unique situation.

The second point concerns the cellular origins of particular tissues. Although two major lineages, the E (intestine) and P_4 (germ line), are clonal in origin, most tissues and defined structures arise from groups of specially set-aside cells, or "polyclones," to use a term first employed in *Drosophila* (Crick and Lawrence, 1975). In this respect, the embryonic and postembryonic programs show some differences. Sensilla, for instance, arise in both stages of development, each sensillum consisting of neuronal and supporting (glial) cells. In postembryonic development, however, a number of sensilla arise clonally from single mother cells, while in embryogenesis, each sensillum consists of two or more clones.

A consequence of polyclonality is that most cell lineages give rise to multiple cell types and structures. This is especially true for the AB and MS lineages. Within the AB ectodermal lineages, for example, there arise a few muscle cells in terminal embryonic cell divisions that also produce sister neuronal cells (Sulston et al., 1983). Binucleate muscle cells, which result from the fusion of single AB-derived and MS-derived cells, also sometimes develop. It is apparent that "ectodermal" and "mesodermal" characteristics can be simultaneously present in certain lineages, even at very late stages. Indeed, the derivation of mesodermal-type cells from ectodermal precursors is not unique to the nematode but also occurs in vertebrates, as instanced by the formation of migrating neural crest cells from the ectodermal neural tube that give rise to various nonectodermal cell types as well as neural cells.

The final point concerns the development of symmetry in the embryo. The L1 larva is largely bilaterally symmetrical and much of this symmetry is derived from comparable, though often slightly different (see below), division programs of lineally equivalent cells on the left and right sides. Nevertheless, beyond the bilaterally symmetric features, there are also elements of threefold and sixfold rotational symmetry. The origin of some of these symmetries is intricate and not reducible to simple rules. The cell division and utilization patterns that give rise to the triangular pharynx are a particularly complex case (Sulston et al., 1983). Clearly, different strategies of cell formation and recruitment are employed for the development of these different symmetries. Evidently, specific *ad hoc* evolutionary modifications of basic cell deployment patterns may have been incorporated into initially simpler ancestral cell lineage patterns.

A central issue, however, can be discerned through this complexity, and that is the question of how much of this complexity is produced by the inherent properties of cells within lineages and how much by the close-range cellular interactions that each cell experiences. Though invariance of cell lineage is usually taken to imply the existence of some form of autonomous somatic inheritance within each lineage, it is equally possible that each step within each lineage is governed by the set of cell contacts made, and that the complete sequence reflects the constraints of particular cell contacts that arise following each division.

This matter has been illuminated by some experiments of Wood (1991), who explored the relationship between certain cell asymmetries in the early embryo and the resulting alterations in lineage and subsequent development as a consequence of reversing one left–right asymmetry. In the adult hermaphrodite, it should be noted, there are two organ systems, the intestine and the gonad, that show pronounced left-right asymmetry (Fig. 6.4). Thus, in the anterior half, the gonad arms are on the right side, while in the posterior half, they are on the posterior side; the intestine shows the reverse left–right asymmetry. These differences ultimately stem from different cell lineage patterns on the left and right sides of the early embryo, which produce contralateral analogs by different routes. Many of these differences in turn reflect an initial difference in position between the daughter cells of AB.a and AB.p (the two daughter cells of the AB embryonic blast cell). At the six-cell stage, AB.al and AB.pl are slightly anterior to their right homologs, AB.ar and AB.pr, in the normal embryo.

Wood (1991) reversed their relative positions by micromanipulation and found that the lineage patterns in the early embryos showed reversed left–right patterns *and* that these embryos developed into fully left–right reversed, but otherwise normal, hermaphrodite adults.

Because a number of the seemingly symmetrical left–right differences in structure arise from non-identical cell lineage programs on the left and right sides (Sulston et al., 1983), and given that these cell lineage patterns are presumably reversed as well, the results indicate that much of the invariance of cell lineage patterns must be due to the sequence of cell contacts rather than to inherent properties passed down through particular cell lineages. They dramatically confirm and extend the results discussed in chapter 3 that show how important cell interactions are in development of the early nematode embryo. Evidently, much of the invariance of the cell lineage is produced by stereotypical cell interactions, each one set in motion by the previous ones, rather than by autonomous cell inheritance.

POSTEMBRYONIC DEVELOPMENT

Emergence of the L1 larva from the eggshell marks the beginning of postembryonic development. Like the adult, the immature worm is encased by a cuticle, secreted by the underlying syncytial hypodermis. The body musculature consists of four longitudinal strips of muscle. Two of these strips are located on either side of the dorsal midline and two are in corresponding positions on either side of the ventral midline (Fig. 6.4A); three consist of 20 cells, one of 21 cells. The head of the L1 larva includes the pharynx and is fully formed. The intestine is functional but contains only 20 nuclei (14 fewer than the adult intestine, which is partly syncytial).

Postembryonic development of the animal takes place during a series of four larval growth periods, separated by intervening molts. Its most obvious features are the large increase in body size and the development of the gonads and secondary sexual characteristics. The size increase results mainly from increased cell size and secondarily through an increase in cell numbers. In the gonads, however, a major increase in cell and nuclear number takes place, particularly in the germ line.

The net result of postembryonic development is the building of an adult superstructure on top of the larval scaffolding. Part of this developmental sequence involves a remodeling of the larval foundation, with some major changes in preexisting neural and muscle connections for the innervation and movement of new somatic sexual structures. However, most of the events that serve to convert the larva to an adult entail principally an addition of complexity rather than a dramatic reworking of the basic material. (In contrast, the emergence of the adult that takes place in *Drosophila* involves a profound metamorphosis and consists largely of a substitution of the adult structures for the larval ones.)

In the somatic cell lineages of the wild-type nematode, the pattern of postembryonic cell divisions is, in general, as stereotyped as it is in embryogenesis. It involves oriented cell divisions, precise relative timing of cell division, oriented cell movements, and programmed cell death. This fixity leads to precisely constant numbers of somatic cell nuclei in the adult. In the hermaphrodite, for instance, the L1 larva contains 558 cell nuclei (not counting those eliminated by cell death); the adult hermaphrodite contains exactly 959 somatic cell nuclei, of which about 200 are in various-size syncytia (in the hypodermis, certain muscles, and the gut). The numbers are somewhat different in the male, but the same general invariance of the somatic division program is observed; in a few instances, pairs of cells have reciprocal "choices" but the net outcome is still a fixed pattern of cells in specific places. (However, as will be discussed later, the germ line presents a different picture, with the patterns of nuclear and cellular division being essentially proliferative and "filling" the gonad.)

These highly specific division programs originate from the postembryonic blast cells (Fig. 6.3), each of which undergoes an invariant number of rounds of division, ranging from one to eight. For some lineages and for some structures, the increases are correspondingly modest or great. For the head, which contains the majority of cells at hatching (305–310), there is virtually no increase in content of its constituent cells (pharynx, sensilla, connecting neurons, etc.) during postembryonic development. Other structures, such as the secondary sexual structures (e.g., the vulva in the hermaphrodite, the elaborate tail structures of the male), only come into existence during postembryonic development.

As with the embryonic cell lineages, the central questions concern the mechanisms by which these invariant cell division programs are produced. The development of the vulva, the ventrally located transverse slit in the hermaphrodite through which eggs are laid, provides an illustration of the roles of cell lineage and cell interactions in nematode development. As a case study in the cellular events of postembryonic development, we will next examine those events in development of the vulva, the opening for the eggs in the hermaphrodite. This developmental process has proved to be a particularly informative one, lending itself not only to a thorough description of its cellular basis, but, recently, of the genetic and molecular events that underlie the cellular changes.

Vulval Development: The Roles of Cell Lineage, Cell Equivalence Groups, and Induction in a Postembryonic Event

The vulva is a ventral structure derived from the central P blast nuclei, which are situated initially in the lateral hypodermis of the embryo. The hypodermis, a collection of surface syncytia and cells, consists of four longitudinal ridges in the L1—one dorsal, one ventral, a lateral right, and a lateral left. The four groups are connected by thin sheets of cytoplasm interposed between the internal musculature and the external cuticle.

The six pairs of lateral P nuclei (Fig. 6.3) migrate ventrally through cytoplasmic connections to the ventral ridge. This migration takes place about the middle of L1, the anterior nuclei moving first, followed in sequence by the others. The 12 migrant P nuclei insert themselves between the 15 juvenile motor neurons already present in the ventral cord.

The particular member of a left–right pair of P nuclei that assumes the relative anterior position, within each pair, is variable, but having reached the ventral cord, they maintain their positions and produce a line of cells which then divide. In the hermaphrodite, each of the 12 P cells gives five neurons and one hypodermal cell. As in many of the lineages of the larva, the division planes are perpendicular to the axis, producing an anterior and a posterior daughter. The convention for designating cells formed during larval development is similar to that for embryonic cells: The division history of a particular cell is indicated by the sequence of letters, each signifying the resulting position from a given division, following the name of the founder cell, e.g., P2.app is the posterior daughter of the posterior daughter of the anterior division product of the P2 blast cell.

The cell lineages for P1–P12 in the hermaphrodite are diagramed in Figure 6.5. It can be seen that the anterior daughters of the first P cell divisions give rise to several progeny neurons, all of which contribute to the ventral nerve cord. These divisions of Pn.a daughters are complete by the L1–L2 transition. Several of the posterior daughters, the Pn.p cells, however, have a different fate and contribute cells to the vulva in the hermaphrodite (and to the preanal ganglion of the male). In the hermaphrodite, the posterior daughters of P1, P2, P9, P10, and P11 all form ventral hypoderm (vh) cells which fuse with the syncytial hypoderm cord, but the middle Pn.p cells (P3.p– P8.p) divide once more. This additional division is followed by further divisions of the P5.p, P6.p, and P7.p lineages. This last set of divisions, taking place during the L3 stage, produce the 22 cells that comprise the vulva. The cell divisions of the vulva are complete by the early L4 stage and all 22 cells have arrived in their final positions by late L4; the sequence of events is shown in Figure 6.6. The formation of these 22 cells is followed by a number of cell fusions, reducing their number to 12. The vulval opening is at the center of a stack of six toroidal-shaped cells in this group, and the opening and closing of the vulva is mediated by a separately derived group of eight innervated muscle cells.

As with any cellular developmental sequence under study, one would like to know something about the states of developmental capacities of the cells that form the vulva. To explore this matter, individual cells and groups of cells were ablated by the laser microbeam technique and the consequences for vulval development were examined (Kimble et al., 1979; Sulston and White, 1980; Sternberg and Horvitz, 1986). The results show a well-defined but larger set of cells with the capacity to contribute to the vulva, more cells than actually do so. Specifically, if P5.p, P6.p, or P7.p is eliminated, cells are recruited from three Pn.p lineages which normally do not contribute to the vulva—namely P3.p, P4.p, and P5.p.

For instance, if the central vulval precursor cell P6.p is destroyed, then P4.p, instead of forming just two cells, forms seven, all of which contribute to the vulva, and P5.p., normally a source of seven cells, undergoes an extra division to give eight; the net result is a 22-nucleus vulva. If both P7.p and P5.p are destroyed, then both the P4.p and P8.p lineages are recruited and give a 22-nucleus vulva. If, however, all three standard precursor cells (P5.p, P6.p, and P7.p) are destroyed, then both P3.p and P4.p are recruited but only manage to construct a 16-nucleus vulva. In contrast, Pn.p cells outside the P3.p–P8.p never contribute to the vulva, showing that there are strict limits to the set of cells that can so contribute. A group of cells whose members can substitute for each other in development has been termed a cell equivalence group (Kimble et al., 1979).

Cell lineage provides a distinct clue to membership in the vulval cell lineage group and also delineates differences between members of the group. The within-group identifying key is that the Pn.p cells that belong to the vulval equivalence

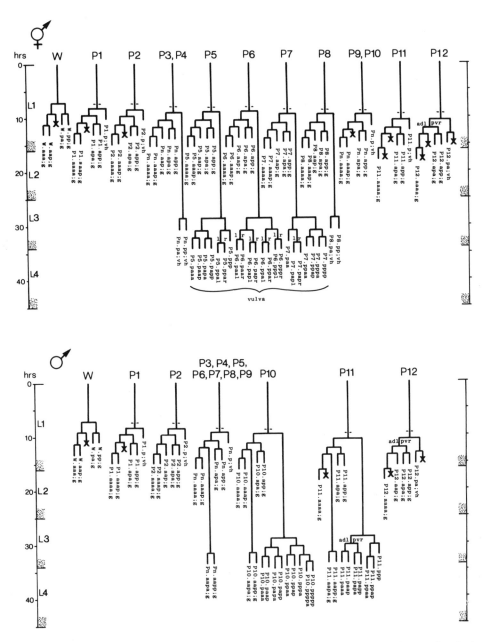

Fig. 6.5. Postembryonic division patterns in the P blast cell lineages. Nomenclatural rules are given in the text. Anterior and left daughter cells are placed to the left, posterior and right daughters to the right. g, neuronal or glial cell; vh, ventral hypodermal cell; X, programmed cell death. (Reproduced from Sulston and Horvitz, 1977, with permission of the publisher.)

209

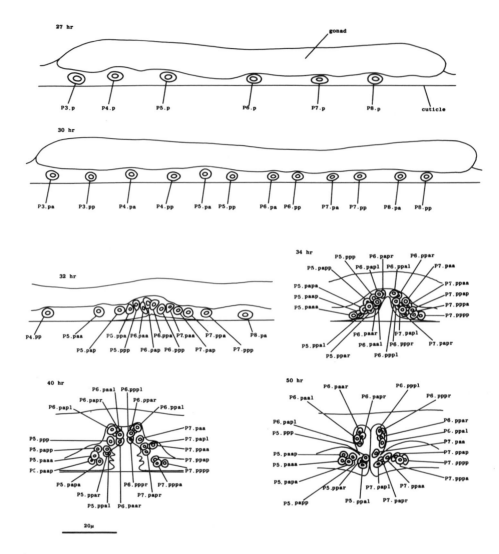

Fig. 6.6. Formation of the hermaphrodite vulva. Times shown are in hours from hatching. (Reproduced from Sulston and Horvitz, 1977, with permission of the publisher.)

group undergo one or more cell divisions, in contrast to the Pn.p cells (P1.p., P2.p, P9.p) outside the group (Fig. 6.5). For the recruitable but normally nonparticipatory cells (P3.p, P4.p, and P8.p), this consists of one extra division. Within the subgroup of cells that do form the vulva in wild-type development, P5.p, P6.p, and P7.p are distinguishable in terms of the number of cells each contributes to the vulva (P6.p contributing eight, P5.p and P7.p each six) and in the pattern of cell division planes (longitudinal vs. transverse vs. nondividing) during the final round of cell divisions.

These distinctive cell division patterns within the equivalence group are also associated with a distinctive pattern of preferential replacements following cell ablation, reflecting a hierarchy of fates (Kimble et al., 1979; Sulston and White,

1980). Thus, cells further from the central member (P6.p) can replace more central cells within the group but the reverse replacements happen less readily (Sternberg and Horvitz, 1986). These patterns permit a simple designation of the cell fates associated with each member of the equivalence group. The P6.p division pattern is termed 1°, the lineages of P5.p and P7.p as 2°, and the normally nonvulval lineages of P3.p, P4.p, and P8.p as 3°. Since all the cells in the P3.p–P8.p group have the potential to contribute to the vulva, they are referred to collectively as vulval precursor cells or VPCs.

The vulval equivalence group is just one of more than a dozen in *C. elegans* development, the majority being in postembryonic cell groupings (Sulston, 1988). The preanal ganglion of the male provides a second specific example. In the male, the cells that correspond to the vulval equivalence group (P3.p–p8.p) plus P9.p all contribute to the ventral hypoderm (Fig. 6.5). However, P10.p and P11.p undergo a specific division sequence to form the preanal ganglion, one of the set of ganglia required for innervating the male tail (used in clasping the hermaphrodite during mating). Nevertheless, P9.p (but not P8.p) is part of the preanal ganglion equivalence group, as shown by laser ablation experiments (Sulston and White, 1980). These results thus reveal a subtle but real boundary of developmental capacity in the male, that between P8.p and P9.p, a boundary that is an overt one in the cell lineage of the hermaphrodite (Fig. 6.5). Indeed, despite many overt differences in somatic cell lineages between the hermaphrodite and the male, there are many underlying connections, with the male-specific features often based or superimposed upon the hermaphrodite construction plan (Sulston et al., 1980). Several of the larger equivalence groups are depicted in Figure 6.7.

However, while equivalence groups point up the importance of cell lineage relationships in setting developmental capacities, other factors besides lineage are important in postembryonic development. These additional factors are in the general category of cell-cell interactions. Again, the development of the vulva is illustrative.

For the vulva, these interactions are of two sorts. The first involves interactions within the group, described by the hierarchy of fates, and is probably mediated by direct cell contacts. The second kind of cellular interaction involves the influence of an inductive, diffusible signal from the anchor cell. The anchor cell is one of the somatic cells of the gonad and is close to, but probably does not touch, P6.p during the start of cell division within the vulval precursors. If the anchor cell is destroyed by microbeam irradiation prior to these divisions, the vulva does not form; instead, all members of the vulval cell equivalence group divide once and the daughter cells then join the ventral hypodermis (Kimble, 1981a).

Is the anchor cell's influence limited to the cell which acquires the 1° fate (with 2° and 3° fates being determined secondarily, perhaps by direct interactions) or does it extend to other members of the vulval cell equivalence group? To answer this question, Sternberg and Horvitz (1986) studied the relationship between anchor cell position and division pattern in isolated Pn.p cells within the ventral hypodermis. They obtained such isolated cells by growing a temperature-sensitive postembryonic mutant (*unc-84*), defective in P nuclear migration (see Table 6.1), at intermediate temperatures and analyzed the consequences for vulval development. Under these conditions, variable nuclear migration takes place and individual Pn.p cells can be obtained. Two findings of note were made. First, it was found that cells of 2° fate can form in the absence of cells with a 1° fate; this shows that the fates of P5.p

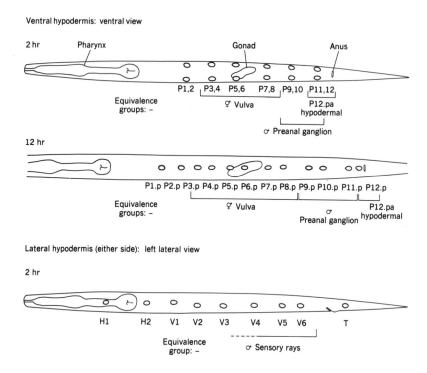

Fig. 6.7. Some of the cell equivalence groups in the ectodermal lineages of the developing L1 larva. The times are given from hatching. Note the changes in equivalence cell groupings of the P lineages between the early (2 hr) and late (12 hr) L1 form. (Reproduced from Sulston and White, 1980, with permission of the publisher.)

and P7.p need not be set by interaction with P6.p. Second, distance from the anchor cell was tightly correlated with fate; Pn.p cells closest to the anchor cell acquired a 1° fate, those further away a 2° fate. The simplest interpretation of the results is that a diffusible signal emanates from the anchor cell and is critical in determining both 1° and 2° fates, with closer proximity and, hence, strength of signal predisposing toward the 1° fate. (A more complex explanation is that two signals are emitted and that the second signal spreads more broadly, with cells of the equivalence group responding to it, in the absence of the first signal, by adopting the 2° fate.)

The picture of vulval development that emerges is of a complex mix of factors that influence developmental processes and cell assignments. The results show the combined importance in postembryonic development of both cell lineage constraints and cell–cell interactions of various kinds. The cellular phenomena also help to frame the kinds of genetic questions one wants to ask: Which gene activities are crucial for specific features of postembryonic development, such as that of the vulva, and how do their gene products act? To answer such questions, and indeed to probe the bases of the various cellular events in postembryonic development, mutants affected in these developmental processes must be obtained. We will first look at some of the general features of the genetic analysis of postembryonic development and some of the particular genes and processes of special interest, and then return to the specific case of vulval development and the elucidation of the genetic pathway that underlies the visible cellular events.

GENETIC ANALYSIS OF POSTEMBRYONIC DEVELOPMENT: NONGONADAL STRUCTURES

Though the genetic analysis of events in embryogenesis is possible, as we have seen (chapter 3), and is proceeding in several laboratories, it is postembryonic development that has received the most intensive genetic investigation. The larval stages are, quite simply, easier to inspect and hence to analyze. However, while the emphasis in what follows is on mutants identified on the basis of postembryonic developmental abnormalities, it should be kept in mind that a large number of these are also affected in embryonic development. Several that exhibit dramatic aberrations of postembryonic development and initially seemed to be normal as embryos have been found, on closer inspection, to have slight aberrancies in embryogenesis. Furthermore, other genes known to be important in postembryonic development but whose mutants show no visible effects in the embryo may in fact be required in the embryo; the homozygotes may show no zygotic effects because of maternal supplies of the gene products. As mentioned earlier, some results support this idea: Within a group of temperature-sensitive mutants isolated on the basis of their larval defects, a substantial percentage show maternal rescue for their defects (Wood et al., 1980). In effect, a general caution in interpreting gene function is in order: Genes may be initially identified on the basis of one phenotype, but their gene products may be required in several or, even numerous, cell types and stages, such additional requirements only being detected subsequently upon closer inspection or by alternative methods of analysis.

A second general observation is that while most of the postembryonic mutants that have been isolated possess a distinctive phenotype, many, including those shown to be null mutants by the amber suppression test, show highly variable penetrance and expressivity; thus, not all of the mutant individuals show the mutant phenotype and not all of the affected individuals show the extreme phenotype manifested by others. This fact suggests that, for many of these genes, other genes can supplement their functions to various extents, the phenomenon of functional redundancy.

The first large mutant hunt for postembryonic defective mutants was carried out by Horvitz and Sulston (1980). The focus of the hunt, given the importance of cell lineage in setting developmental assignments, was on obtaining mutants affected in specific lineages; the problem at the outset was in not knowing whether specific gene activities are set aside for specific lineages and, if so, what mutant phenotypes should be sought. The general strategy was therefore to look for a variety of nonlethal postembryonic mutant phenotypes that might be caused by specific lineage defects and to use a variety of screening methods that might pick up mutants altered in cell number or type (Horvitz and Sulston, 1990).

The first step was to mutagenize hermaphrodites with EMS and screen isolated F_1 and F_2 individuals, each in a separate petri dish, for defects. (Screening of F_1 will detect dominant mutants; screening of F_2, recessive mutants.) This search was limited to mutants that create visible abnormalities but which permit survival and at least partial fertility; in consequence, the mutants obtained can be perpetuated without great difficulty. Many such defects would be the consequence of specific postembryonic cell lineage defects in secondary sexual structures or involve postembryonic neural development. For instance, a distinctive part of the postembryonic somatic development of the hermaphrodite concerns the vulva and vulva-associated structures. Any early defects in the P blast lineages should produce vulva defects, and mutants affected in these lineages can be detected either under the dissecting microscope as lacking a vulva or will be revealed as egg-laying defectives. Such animals will not be sterile, however, because internal fertilization will occur, as in the wild-type, and the resulting embryos hatch inside the mother, turning her into a "bag of worms," which bursts and releases the progeny.

Another class of postembryonic defect is Unc defects (though some of these will reflect defects developed during embryogenesis in the neuromusculature). The initial examination procedures included inspection of living animals by Nomarski optics (to determine if postembryonic lineages had abnormal cell numbers or arrangements), fluorescence microscopy after formaldehyde treatment of fixed animals (to detect all dopamine-producing neurons generated during larval development), and Feulgen or Hoechst staining of fixed animals, procedures which can be used to detect either alterations in individual nuclear DNA content of postembryonic cells or the total number of nuclei and which are especially useful for identifying defects in the vulva and ventral nerve cord.

This first search for postembryonic mutants yielded 24 mutants, defining 14 complementation groups in all (Horvitz and Sulston, 1980). All showed defects in cell lineages, either as direct or indirect effects; those with seemingly specific lineage defects were termed *lin* mutants. In general, the mutants isolated can be divided into simple defectives—those displaying a deficiency for one or more cell types or processes—and those showing a transformation in developmental fate in one or more cell types, a category that can be broadly described as "homeotic." The major categories of phenotypic defect which have been identified in the various searches for postembryonic defectives are listed, along with some specific examples of identified genes, both from the first mutant hunt and from later ones, in Table 6.1. Some of the findings of interest are discussed below.

Mutants Defective in General Processes

Mutants in this first category include those that are defective in a fundamental nuclear or cell division property or behavior; for most of the mutants in this category, the defects occur in homozygous worms of both sexes. Although, in general, simple defectives are not highly informative about developmental mechanisms, they can be of use in determining whether the particular process or event they abolish is essential for other developmental events or processes.

One class of postembryonic defectives is the nuclear migration defectives. Nuclear migration occurs in both the embryonic and postembryonic hypodermis. *unc-83* and *unc-84* are defective in both sets of migrations; their postembryonic defect is in the migration of P blast nuclei, and, as would be expected, aberrant migration of these nuclei produces defective vulval and ventral cord development. (One use of *unc-84* to analyze the nature of vulval cell commitments has been described above.) Mutant

Table 6.1. Some Postembryonic (pe) Defective Phenotypes in *Caenorhabditis elegans* and Some Representative Genes

Category	Locus	Mutant defects	References
GENERAL CELL PROCESS DEFECTIVES			
Nuclear division	*lin-5 (II)*	No pe nuclear divisions; polyploidyin pe lineages	Sulston and Horvitz (1981); Albertson et al. (1978)
	lin-6(I)	No pe DNA replication in somatic cells; small somatic cells; nuclei	Sulston and Horvitz (1981); White et al. (1978)
Nuclear migration	*unc-83(V), unc-84(X)*	Embryonic and pe P lineage nuclear migration delayed/blocked.	Sulston and Horvitz (1981)
Cell division	*unc-59(I), unc-85 (II)*	Defective cytokinesis and polyploidy	Sulston and Horvitz (1981)
HOMEOTICS			
Reiterative stem cell pattern	*unc-86 (III)*	*Reiterative divisions in Q, T, and V5 cell lineages*	*Sulston and Horvitz (1981);* Chalfie et al. (1981)
Regional transformation	*mab-5(III)*	Transformation of posterior region lineages in body to anterior region lineages	Hodgkin (1983a); Kenyon (1986); Costa et al. (1988)
Binary choice	*lin-12 (III)*	Transforms cell fates in various alternate fate choices	Greenwald et al. (1983); Greenwald (1985); Yochem et al. (1988)
Heterochronic	*lin-14 (X)*	Skipping or reiteration of L1 or L2 lineage programs	Ambros and Horvitz (1984); Ruvkun and Giusto (1989)
Vulval		See Figure 6.15	
Sex		See Table 6.3	

males are less visibly defective, but show abnormalities in the preanal ganglion, derived from two hypodermal cell lineages. Clearly, the failure of nuclear migration in the hypodermis has severe phenotypic consequences.

Two other postembryonic defectives have been studied to explore the relationships between cell division and DNA replication, on the one hand, and states of cell commitment on the other. As is the case in embryogenesis, the geometry of cell division in postembryonic development is correlated with the fates of the daughter cells. In symmetrical cell divisions, both daughters are similar in size and immediate developmental behavior. In asymmetrical divisions, the daughter cells differ in size and shape and in their immediate developmental fates.

Such relationships raise the question of whether cell division *per se* is necessary for either the expression or the segregation of developmental potential in postembryonic development. We have seen that in embryonic development, neither cell division nor nuclear division appear to be required for the expression of such potentials. An analysis of the mutant *lin-5* indicates that, similarly, neither cell nor nuclear division is required for the expression of differentiated postembryonic characteristics but may be required for their segregation in sublineages.

In the *lin-5* mutant, the ventral cord is populated by 12 P blast nuclei, but because nuclear division is blocked in this mutant (Table 6.1), the nuclei do not divide but become polyploid. Nevertheless, these blocked precursors show signs of the normal differentiation program (Albertson et al., 1978). For example, those Pn precursor cells which in the wild-type exhibit a cell death in one of their descendants (W, P1, P2, P9–P12) (Fig. 6.5) show partial or complete nuclear degeneration in the mutant. In contrast, the P3–P8 lineages, which do not exhibit programmed cell deaths in the wild-type, do not show nuclear degeneration.

The second sign of normal differentiation in the blocked cells concerns neuronal character. In the wild-type strain, the ventral nerve cord consists of five different classes of motoneuron (a motoneuron being one that innervates a muscle), and each is distinguished by its axon directions, the position of muscles innervated (dorsal or ventral), and the kinds of synaptic connections it makes (whether chemical or gap junctions, the types of interneurons with which it connects, etc.). Furthermore, each neuron type is derived via a particular sequence of divisions from the Pn precursor (Sulston, 1976). In the *lin-5* ventral nerve cord, the division-blocked P blast precursor cells each develops eight to 16 neuronal processes (each wild-type neuron has only two), and these form neuromuscular junctions (NMJs) is approximately the right numbers and with the correct spacing. In addition, each Pn precursor develops characteristics similar to those that define the five normal classes of ventral cord motoneuron; each division-blocked Pn cell has partially intermediate neuronal character but only one or two types predominate for a given cell.

Thus, each kind of differentiation potential normally associated with a given member of a Pn ventral cord lineage can be expressed in division-blocked polyploid cells. It thus appears that the expression of differentiation potential is independent of the occurrence of actual cell division sequences, with their well-defined a-p polarities and times of occurrence; the result is comparable to that seen in certain embryonic cell lineages when cell division is blocked (chapter 3).

However, the *segregation* of different potentials within a lineage seems to require cell division, as suggested by the occurrence of intermediate neuronal characteristics in the blocked cells. One can imagine a direct role for cell division asymmetry in this process. For instance, it is possible that some "regulatory" molecular influence is distributed unequally in each asymmetric cell division and that this quantitative absolute difference is amplified by subsequent events to give a strong differential bias in cell fate. In this situation, the change of an asymmetric division to a symmetric one should produce like daughters. Furthermore, reversal of the size difference between daughters should be accompanied by a reversal of daughter cell fates. Such a reversal of cell fate following reversed placement of the division plane has occasionally been observed in cell ablation experiments (Sulston and White, 1980; Kimble, 1981a). However, the precise cause–effect relationship between cell division geometry and cell fate specification is unknown, as it is in embryogenesis. The placement of division plane could be a consequence rather than a cause of a primary fate determination.

Since neuronal differentiation in the blocked *lin-5* mutants apparently takes place on schedule, it would appear to be governed by some kind of intracellular "clock," running independently of cell or nuclear division *per se.* One obvious possibility is that the ticking of the "clock" is the number of rounds of DNA replication. However, at least some events in postembryonic development are known to occur independently of DNA replication. For instance, in normal development, certain of the juvenile moto-neurons—those formed during embryogenesis and present in the ventral nerve cord at the beginning of L1—change their connectivity pattern from the innervation of ventral muscles to that of dorsal muscles and form new connections with postembryonic P lineage neurons. In the *lin-6* mutant, which is blocked in DNA replication (Table 6.1), yet whose cells continue to divide, similar changes in juvenile neuron connectivity occur, although these neurons fail to receive new connections from the late-developing ventral neurons whose formation is blocked in the mutant (White et al., 1978). Thus, certain differentiations can proceed independently of the formation or DNA synthesis of other cells of the wild-type program with which they are normally associated.

Homeotic Mutants

The second general category of postembryonic defectives may be classed as homeotics. The members of this group differ markedly from one another with respect to the kinds and numbers of cell types affected and the nature of the developmental substitutions observed. In all members of this group, however, particular cell types are replaced by recognizable other cell types and all are zygotic, rather than maternal, in action. Although they differ in the apparent magnitude of their phenotypic effects, some of the larger effects are produced by initial changes in just one or a few cells, whose consequences involve gross alterations in development and morphology.

Four examples of mutant homeotic changes are described below, along with the molecular characterizations of the genes. The genes discussed have been chosen for their intrinsic interest, because they illustrate some of the important strategies for genetic analysis in this organism and because they serve to illustrate two broadly different molecular bases of homeotic change in *C. elegans.*

lin-12

One of the best characterized sets of homeotic transformations are those produced by mutations in the *lin-12* gene. Initially classified as a *lin* gene in terms of its observed effects on the cell lineages that give rise to the vulva, further analysis showed that *lin-12* transformations involve a number of postembryonic cell lineages, both ecto-dermal and mesodermal, in males as well as in hermaphrodites. In addition, it became apparent that the gene was unusual in terms of its genetic properties, being capable of mutating to two different states of activity with opposing effects on these cell fate specifications. *lin-12* is evidently an important "binary switch" gene.

Unlike many of the other Lin mutants, whose gene activities were characterized on the basis of their recessive phenotypes, the first *lin-12* mutants were semi-dominants, detected initially by their vulval abnormalities. In these strains, which were designated *lin-12(d)*, the vulval defects (some mutants showing a single nonfunctional vulva, others showing multiple small "pseudovulvae") result in an inability to lay eggs, the Egl phenotype. The vulval defects stem from a conversion of the members of the vulval equivalence group (P3.p–P8.p) to the 2° fate; though such cells are "vulval," they cannot construct a functional vulva.

In addition, other cell transformations affecting particular pairs of cells were found to take place in the *lin-12(d)* mutants (Greenwald et al., 1983). Each of these pairs constitutes an equivalence group, in which each member has a distinctive fate in the wild-type but can take on the other's, if the latter is destroyed, or in which either member can initially take on either fate with the other then assuming the alternative. An example of the latter type is the two-cell equivalence group, Z1.ppp and Z4.aaa, derived from the postembryonic blast cell precursors of the somatic portion of the gonad (Z1 and Z4, respectively). In normal development, Z1.ppp and Z4.aaa are in close contact and either can become the anchor cell (ac), the other cell becoming a ventral uterine (vu) cell (Kimble and Hirsh, 1979). In *lin-12(d)* mutants, both become vu cells; thus, such animals lack the ac signal for vulval induction, contributing further to the creation of the Egl phenotype in these animals.

Subsequent to the isolation of the original mutants, recessive loss-of-function mutants, designated *lin-12(o)*, were isolated as "revertants" of the semidominants. These double mutants effectively lack *lin-12* activity and, as heterozygotes with the wild-type allele, are recessive, giving a wild-type phenotype. Homozygotes of these null mutants exhibit an overly prominent single vulva structure and were found to give opposite cell transformations to those of the semidominants, within each of the *lin-12*-affected equivalence groups (Greenwald et al., 1983). In *lin-12(o)* mutants, both Z1.ppp and Z4.aaa develop as acs and P5.p–P7.p develop the 1° vulval fate (Greenwald et al., 1983). Both types of cell transformation would conspire to produce a large, protuberant vulval structure.

The *lin-12* results emphasize the significance of cell equivalence groups as devel-opmental units and confirm the importance, suggested by the laser ablation experi-ments, of cell–cell interactions in creating or maintaining the particular developmen-tal propensities shared by the cells in each such group. To interpret the meaning of the genetic findings, however, one needs to know the character of the *lin-12(d)* mutants. In principle, dominant or semidominant mutations can reflect either excess hypermorphic activity or a novel neomorphic activity. The two possibilities can be distinguished by the Müllerian test of altering wild-type gene dosage in strains containing the mutation. In a series of comparisons of gene dosage effects, it was found that strains possessing two + alleles in addition to a single *lin-12(d)* allele had increased penetrance for the Egl defect, while *lin-12(d)/lin-12(o)* strains had lower penetrance than *lin-12(d)/+* (Greenwald et al., 1983). The results show that the semidominants have excess *lin-12* activity and identify these mutants as hypermorphs. High *lin-12* activity thus tends to promote one type of fate within each affected equivalence group, and low or zero activity the alternative fate.

Because most of the reciprocal fate conversions involve neighboring cell pairs, a reasonable interpretation is that in most or all of the cells affected by *lin-12,* close-range cellular interactions determine cell fates in the wild-type by regulating the respective levels of *lin-12* activity in the two cells. High *lin-12* activity in one cell produces one fate, while low activity in the other produces the opposite cell fate. In the mutants, in contrast, the mutations result in both cells being "stuck" at one level of activity or the other, with both being assigned the cell fate appropriate to that *lin-12* activity level and their respective cell lineage. If the *lin-12* gene product mediates cellular interactions, the simplest explanation is that the gene product is some kind of signal-receiving device at the level of the cell membrane. The molecular evidence, which comes from the cloning and sequencing of the gene, has borne out this interpretation.

To clone *lin-12,* Greenwald (1985) made use of the existence of a class of transposable elements in *C. elegans,* the Tc1 elements, and the fact that these elements were suspected to possess mutator activity in the Bergerac strain of *C. elegans* (Moerman and Waterston, 1984). In Bergerac, there are approximately 300 copies of Tc1 and the element undergoes frequent transposition; in contrast, in the standard laboratory strain, Bristol, in which most of the genetics has been done, there are only about 30 copies of Tc1 and these neither undergo transposition nor exhibit mutator activity.

The first step in the cloning of the *lin-12* gene involved a semidominant non-egg-laying (Egl) *lin-12(d)* mutant isolated in the Bristol strain. This mutation was transferred to the Bergerac strain and rare egg-laying heterozygous revertants were isolated; such mutants were presumptive *lin-12(o)* (null) mutants produced by insertion of a Tc1 element into the *lin-12(d)* allele. These were then crossed back into Bristol, and recombinants in the region of *lin-12* were selected and put through a series of backcrosses to Bristol, selecting for retention of associated markers, in order to progressively eliminate all Tc1 elements from Bergerac which were not linked to *lin-12* while retaining the *lin-12(o)* mutation itself. In effect, a *lin-12* Tc1-defective mutant was isolated in the mutator Bergerac background, then put back into the Bristol genome, which is relatively Tc1 deficient, with its putative associated Tc1 element, with much of the extraneous Bergerac background (*Ber) crossed out, to produce a novel Tc1-associated *lin-12* mutation. The procedure is outlined in Figure 6.8. One such Tc1-generated *lin-12* mutant was then tested for linkage by a molecular test, screening for a new Tc1-associated band on a Southern blot (diagramed in Fig. 6.8). A novel 3.0 kb *Hind*III fragment was detected and a *Hind*III digest of the DNA from this strain was size-fractionated, the appropriate-size fragments then being cloned and screened for hybridzation to Tc1. The DNA region that flanks the Tc1 element, and which must be part of *lin-12,* was thereby identified and subsequently sequenced.

Conceptual translation of the *lin-12* sequence revealed a most unexpected property, namely the presence of multiple repeats of a sequence found in a mammalian growth factor, epidermal growth factor (EGF) (Greenwald, 1985). This first clone did not contain the whole gene and the results were consistent with either of two possibilities, that *lin-12* either encodes a diffusible EGF-like molecule or that it encodes a transmembrane protein that acts as a receptor via its EGF-like sequences.

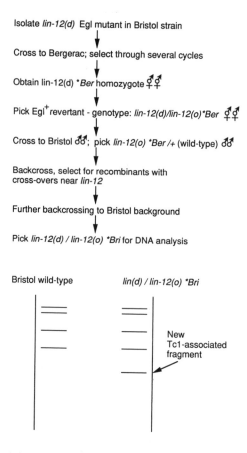

Isolate *lin-12(d)* Egl mutant in Bristol strain

↓

Cross to Bergerac; select through several cycles

↓

Obtain lin-12(d) *Ber homozygote ♂♂♀♀

↓

Pick Egl⁺ revertant - genotype: lin-12(d)/lin-12(o)*Ber ♀♀♂♂

↓

Cross to Bristol ♂♂; pick lin-12(o) *Ber /+ (wild-type) ♂♂

↓

Backcross, select for recombinants with cross-overs near *lin-12*

↓

Further backcrossing to Bristol background

↓

Pick *lin-12(d) / lin-12(o)* *Bri for DNA analysis

Bristol wild-type   lin(d) / lin-12(o) *Bri

New
Tc1-associated
fragment

Fig. 6.8. Scheme for cloning *lin-12*; the steps are described in the text.

Subsequent cloning of the complete gene has settled the issue; the sequence unambiguously indicates that *lin-12* encodes a transmembrane protein, in which the EGF repeats are extracellular and comprise part of the external reception machinery (Yochem et al., 1988).

One can now begin to interpret the results of the genetic analysis of *lin-12* with respect to molecular properties: The *lin-12* gene product evidently acts as a cell surface receptor molecule for signals which emanate from other cells. Reception of the signal(s), either diffusible or cell surface borne, is transduced into a message that affects cell fate of the recipient cell. If insufficient signal is received, as in *lin-12(o)* mutants, cells in the interacting cell pairs assume one of the alternative fates; if excess receptor activity is present, as in the *lin-12(d)* mutants, the other fate is assumed. Presumably, in wild-type, reception of the signal by one of the cells within a *lin-12*-responsive equivalence group produces a second interaction with the neighboring cell(s) of the group, reducing its *lin-12* activity and causing it to assume the other fate.

Further analysis of the *lin-12(d)* mutants has provided additional clues to the nature of the *lin-12* signal reception mechanism. Sequencing of five such mutants has shown that all are located in the extracellular region of the protein just beyond the transmembrane region but before the EGF repeats. Indeed, three are located within four codons of one another (Greenwald and Seydoux, 1990).

The most interesting aspect, however, was revealed by complementation tests. It suggests that *lin-12* activity is triggered by association of *lin-12* molecules at the surface, an association that is prompted by an external signal molecule, and that *lin-12(d)* mutants permit *lin-12* association in the absence of the signal(s) (Greenwald and Seydoux, 1990). The conclusion follows from the finding that heterozygotes of certain of the semidominants and a particular double mutant (a "revertant" of a dominant) have higher *lin-12* activity (detected as a more penetrant Egl defect, reflecting higher-frequency ac-to-vu conversion by *lin-12* activity) than heterozygotes between those semidominants and the wild-type allele (Greenwald and Seydoux, 1990). The simplest interpretation of how a loss-of-function form can promote higher activity than a wild-type, in combination with a hypermorphic form, is that *lin-12* molecules are activated by association to form dimers or higher-order oligomers and that the loss-of-function monomer cannot activate itself but can activate other forms. In wild-type, this activation would be in response to an external signal while the "spontaneous" higher activity associated with the semidominants would reflect a lack of dependence on this signal. If this model is correct, then one would predict that semidominant loss-of-function mutants might also be found. In a comparable cell surface signaling system in mouse cells, which also may require association of the receptor molecules for activity, certain semidominant mutants are believed to produce such inhibitory interactions (see chapter 8).

The sequence of molecular interactions that take place at the cell surface involving the *lin-12* gene product must be, of course, only the first events in a cascade that leads, ultimately, to the subsequent fate "decisions" of the responding cells. These decisions presumably involve alterations in specific gene activity to create the final cellular phenotypes. Although the full sequence in this molecular cascade is unknown, the results that have been described above clarify the role of the *lin-12* product in its initiation and help to explain the nature of the cell homeotic changes produced by mutations in this particular binary switch gene.

unc-86

The second example of a homeotic gene, *unc-86,* was first identified on the basis of the defective movement of its first mutants, which was subsequently shown to have specific cell lineage defects. Loss-of-function mutations in *unc-86* produce a characteristic repeated reenactment of a particular cell division sequence in three postembryonic cell lineages (Chalfie et al., 1981) and certain embryonic cell images (Finney and Ruvkun, 1990). The precursor cells for the affected postembryonic cell lineages, all situated in the posterior region of the animal, on both left and right sides—the Q, V5.paa, and T.pp cells—participate in the production of specific neurons in a sequence of two divisions (Fig. 6.9). In each of these lineages, the posterior daughter of each of these cells gives rise to an anterior cell, which either becomes a neuron or gives rise in further divisions to neurons, and a posterior cell that dies.

In *unc-86* mutants, the affected cell, instead of producing a posterior daughter that dies, undergoes the division pattern of its parent cell one or more times, thereby generating more neurons (Chalfie et al., 1981; Finney et al., 1988). In effect, the posterior daughter reiterates its parental cell's division program; this division pattern

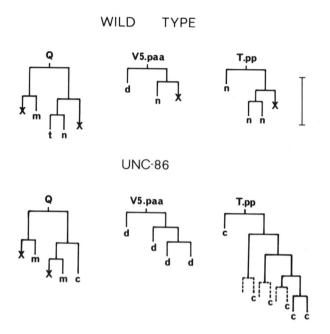

Fig. 6.9. Reiterative cell divisions in the *unc-86* mutant. See text for description. c, compact nucleus of a neuronal structural cell; d, dopaminergic neuron; m, migrating neuron; n, neuron; t, neuronal structural cells; X, programmed cell death. (Reproduced from Chalfie et al., 1981, with permission of the publisher. © Cell Press)

is that of a "stem cell," a cell which at each division produces one daughter that goes on to differentiate and a second daughter that possesses the mother cell's division pattern. In the case of the V5.paa lineage, the supernumerary neurons can be detected histochemically by fluorescence after formaldehyde treatment, which stains dopamine. In the wild-type hermaphrodite, there is only one dopaminergic neuron, the anterior daughter produced by V5.paa. In *unc-86* hermaphrodites, multiple dopaminergic neurons are produced, and each of these sends processes into the ventral nerve cord, just as the single cell in the wild-type does.

Clues to the nature of action of the *unc-86* gene product came from the cloning and sequencing of the gene (Finney et al., 1988). In contrast to the cloning of *lin-12*, where it was possible to screen for a Tc1-linked mutation in the gene based on a phenotypic consequence of the insertion, a less direct procedure had to be employed. The method, however, also utilized the multiple polymorphisms for Tc1 elements between the Bergerac and Bristol strains, and the cloning procedure involved three stages. In the first, the wild-type *unc-86* gene in the Bergerac strain was mapped with respect to a linked group of DNA markers, the series of Tc1 transposons which surround the wild-type gene in the Bergerac strain of *C. elegans*, and the most closely linked, and flanking, Tc1 elements were identified. The procedure involves correlating inheritance of particular genetic alleles with particular Tc1-associated bands and is described in detail in another paper, that of Ruvkun et al. (1989). In the second stage, the two Tc1 elements were cloned and their unique flanking sequences were used to identify cosmid and phage lambda clone banks, which were then used for "chromosome walking" to isolate the sequences between the two Tc1 elements. Finally, a cosmid clone in the region showing an altered restriction fragment hybridization pattern in a translocation strain that has a break in an adjoining gene (*unc-36*) was identified, and subsequently neighboring cosmid clones were screened in various *unc-86* mutants for alterations of their restriction fragment patterns. This last step identified a cosmid that contains the *unc-86* gene.

The gene itself consists of five or six exons (the number depending upon whether a particular splice site is used) and encodes a protein of 467 or 429 amino acids. The striking feature, as discussed by Finney et al. (1988), is the existence of a conserved region of 158 amino acids in the 3′ terminal end, consisting of three domains, which is homologous to several well-characterized transcription factors identified from mammalian cells and includes a homeodomain (Herr et al., 1988). The entire region has been termed the "POU domain" (an acronym for the Pit, Oct, and Unc genes that share it). The evidence strongly suggests, therefore, that *unc-86* itself encodes a transcription factor. The reasonable extrapolation is that one or more of the genes whose transcription is regulated by *unc-86* is essential for ensuring the correct division pattern and/or fate assignments in those lineages affected by *unc-86* mutations. The larger implication is that homeotic transformations in *C. elegans* can, as in *Drosophila,* be produced by mutational alteration of specific transcription factors.

Further molecular studies have extended these findings (Finney and Ruvkun, 1990). Immunochemical tracking of the *unc-86* protein shows that it is localized to the nucleus, as expected for a transcription factor. Furthermore, consistent with its mutant phenotypes, it appears only in neuronal lineages, including both neuroblasts and differentiating neurons, but appears in many more cells, both embryonic and postembryonic, than are obviously affected in the mutants. This last fact indicates that the *unc-86* product must act combinatorially, rather than unilaterally, within the context of particular other transcription factors. Finally, in the asymmetric neuroblast divisions of those lineages in which one daughter cell is affected but the other is not, it appears in the susceptible daughter but not the other (Finney and Ruvkun, 1990), showing that its expression is determined by the cell division program in those lineages.

mab-5, lin-22, and pal-1

The *mab-5* gene was initially identified by Hodgkin (1983a) in a search for mutants affected specifically in male development (*mab* stands for male abnormal). The *mab-5* mutant was identified during a screen of mating ability in the male progeny of F_2 hermaphrodites derived from mutagenized *him* hermaphrodites (*him* standing for *h*igh *i*ncidence of *m*ales, the phenotype resulting from high levels of X chromosome nondisjunction in the hermaphrodites). When normal males produced from *him* mothers are mixed with homozygous *dpy* or *unc* hermaphrodites, a significant proportion of the resulting progeny are wild-type. In contrast, a clone of male mating defectives will be revealed, in such a test, by a low proportion of wild-type progeny and a predominance of the mutant self-type progeny of the hermaphrodite.

Like the other *mab* mutants, the *mab-5* mutant was observed to be grossly defective in the distinctive tail and genital structures. In particular, *mab-5* males have a reduced tail, lacking some of the sensory rays associated with the fan-shaped tail; a picture of the wild-type tail structure is shown in Figure 6.10. The sensory and morphological tail defects of *mab-5* males lead to their ability to copulate.

Inspection of the cell lineages that generate the 18 rays of the tail revealed *mab-5* to be defective in some of these lineages. In the wild-type, the rays are generated by cells derived from the postembryonic blast cells V5, V6, and T (Fig. 6.11). However, in the original *mab* mutant (and other loss-of-function *mab-5* mutants subsequently isolated by Kenyon [1986]), those rays generated from the V lineages but not from T are missing. Instead, these lineages produce seam cells, which secrete the lateral hypodermal ridges known as alae (see Fig. 9.1), as do the seam cells generated by the V1–V4 lineages.

Clearly, *mab-5* is a *lin* mutant, but several other aspects of the mutant phenotype bear no obvious relationship to this particular lineage defect. These include several cell migrations in early postembryonic development, which are altered in both sexes, and, further, abnormalities in the coelomocytes in *mab-5* hermaphrodites (Hodgkin, 1983a; Kenyon, 1986). Thus, while the reduction in the male tail can be seen as a partial hermaphroditization, and hence a sexual transformation, a variety of other changes cannot be so classified.

The unifying feature of the defects is that they all originate from cells located in the posterior body of the nematode. Furthermore, in the case of the various lineage defects, involving the P1–P12, V1–V6, Q1, and M blast cell lineages, all the defects can be categorized as involving transformation to corresponding anterior fates (Ken-

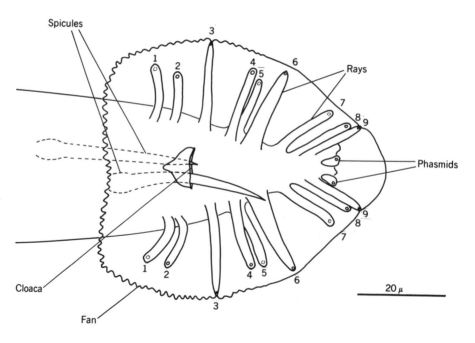

Fig. 6.10. Tail structure in the adult male worm. The tips of each ray emerge on either the dorsal surface (dotted circle) or the ventral surface (solid circle). Phasmids are male-specific sensory elements; the copulatory spicules are also shown. The fan is an acellular cuticular product secreted by the hypodermis. (Adapted from Sulston and Horvitz, 1977.)

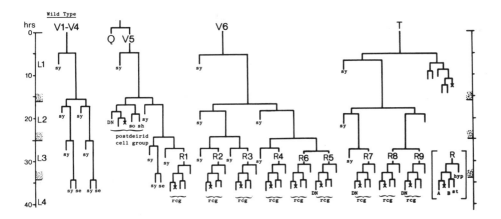

Fig. 6.11. Postembryonic cell lineages in the rays of the male tail. DN, dopaminergic neuron; rcg, ray cell group; Rn, ray precursor cell; se, hypodermal seam cell; sy, nucleus of a large hypodermal cell; X, programmed cell death. (Reproduced from Sulston and White, 1980, with permission of the publisher.)

yon, 1986). This pertains even to the cell migrations. Thus, for example, while the QL neuroblast normally migrates posteriorly in both sexes, this cell initially migrates posteriorly for a short distance and then migrates anteriorly, like its sister cell, the QR neuroblast, in *mab-5* individuals (in both males and hermaphrodites).

In effect, therefore, the *mab-5* defects involve a defective *regional* behavior, specifically of the posterior body region; the diverse cell lineage defects observed in the *mab-5* mutants are a consequence of this regional defect. In principle, the defect could be either of two kinds. It could be in the form of an altered general property, external to the cells, which signifies "posteriorness" or it could involve an incapacity in the various affected posterior cells to respond to such a signal. To understand the *mab-5* defect, it is important to distinguish between these two alternatives.

There is a genetic test which allows one to do this; it involves a form of mosaic analysis, in which some of the cells normally affected by *mab-5* are mutant for the gene while others of this set are wild-type for the gene. If the *mab-5* gene creates some kind of general signal extrinsic to cells in the posterior region, then in some of those mosaics in which the signal-generating cells are wild-type, there will be genetically *mab-5*-deficient cells which nevertheless display wild-type behavior. Conversely, in other mosaics, those in which the signal-generating cells are mutant, certain of those cells dependent on *mab-5* activity and containing the wild-type allele will, nevertheless, be phenotypically mutant. If, on the other hand, the *mab-5* gene product is required within the *mab-5*-dependent cells to enable them to respond to "posterior-ness" (whatever this means in molecular terms), then the response will be strictly "cell autonomous." In other words, if *mab-5* is part of the reception apparatus, the pheno-type of the cells within the mosaic will be a direct reflection of the genotype of those cells.

To create such mosaics, use was made of the technique devised by Herman (1984). In *C. elegans,* it is possible to create so-called "free duplications" of virtually any region of the chromosome by X-irradiation. Such duplications can be passed on through mitotic and meiotic divisions because all chromosomal regions of the *C.*

elegans genome can attach themselves to cell division spindles, a condition known as "holocentric," and thus be inherited through somatic or germ line divisions (Albertson and Thomson, 1982). Nevertheless, such duplications are lost with a certain slight but significant probability during cell division. If the duplication (Dp) carries a wild-type gene which its chromosomal homolog is deficient for, a mosaic animal will be detected as an individual showing partial or complete expression for the trait (the degree to which the mutation is "uncovered" depending upon which cell lineages lack the Dp and on which and how many cells are required to express the wild-type gene in order to produce a wild-type phenotype).

Kenyon (1986) performed such a test on *mab-5,* identifying potential *mab-5* mosaics as those progeny of the Dp-containing mothers that showed either a partial or strong Mab-5 phenotype or those which showed partial or complete expression for either of the linked markers. The principle of the experiment is illustrated in Figure 6.12. Thirty-four mosaics were identified out of 2500 animals examined and seven different mosaic patterns associated with *mab-5* defects were identified. The critical finding is that for all except one animal, the pattern of *mab-5* defects showed strict cell autonomy with respect to the known lineage patterns. In other words, in each of 33 out of the 34 animals, the cells showing a Mab-5 defect were all lineally related, in terms of the known cell lineage patterns. The exceptional animal (class VII) was most probably produced as the result of two independent but relatively late and, hence, not readily scorable losses of the Dp (in one V6 lineage and in the M-derived lineage).

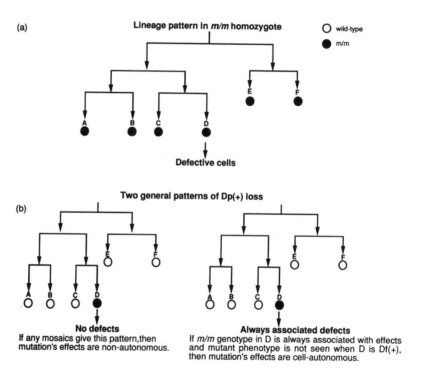

Fig. 6.12. Mosaic analysis in *C. elegans,* using loss of a duplication (Dp) to "uncover" a recessive defect. See text for description.

The results indicate that *mab-5* is not involved in generating a posterior-specific signal but rather in mediating cellular responses, in the indicated cells, to such a positional property.

The subsequent cloning of *mab-5* supports this conclusion. The gene is, like *unc-86*, a homeobox-containing gene, and hence, in all probability, a transcriptional control factor (Costa et al., 1988). Furthermore, studies with a *mab-5*-promoter-*lacZ* coding reporter gene shows that *mab-5* is expressed more widely than just the lineages affected by *mab-5* genetic deficiency, whereas heat shock–induced ectopic expression of an *hsp70-mab-5* fusion gene produces different cell migratory behaviours in Q lineage cells (Salser and Kenyon, 1992). These results indicate that, as in the case of *unc-86*, the gene creates its effect within a particular gene expression context (that which exists in the responding cells) and does not, in itself, "control" or "determine" posteriorness. One molecular difference between *unc-86* and *mab-5* is that, while both are homeobox-containing genes, *mab-5* is not a member of the POU family but rather more similar in its homeodomain to certain *Drosophila* homeotic genes (Costa et al., 1988). Homeobox genes, it should be noted, are not uncommon in *C. elegans.* Searches by hybridization at low stringeny have uncovered approximately three dozen, and it is estimated that there may be as many as 60 in the *C. elegans* genome (Burglin et al., 1989).

If *mab-5* acts autonomously as part of a response mechanism to posteriorness, then other elements of a cell signaling system must exist. Two further genes, *lin-22* and *pal-1,* have been identified as part of this system governing the development of posterior body pattern. *lin-22* mutants exhibit the opposite phenotype to *mab-5* males, namely an anterior extension of ray-producing lineages, involving V blast cells in the male that normally do not produce them (V1–V4). That the effects of *mab-5* are linked in a developmental pathway involving *lin-22* is indicated by the fact that mutations in either tend to suppress the mutant phenotype of the other and that raising *mab-5* dosage enhances the anterior extension of ray development seen in *lin-22* (Kenyon, 1986). Furthermore, in strains lacking both gene activities, *mab-5; lin-22* double mutants, the boundary between alae and rays becomes highly variable, suggesting that neither gene activity is necessary *per se* for one activity or the other but that some form of competition between them sets the alae-ray boundary (Kenyon, 1986).

In gross appearance, *pal-1* (for posterior *alae*) mutant males resemble *mab-5* males in showing diminished ray structure and a further extension of the alae into the tail region than seen in wild-type males. The alteration reflects a transformation of the V6 lineage, which normally produces five rays on each side, into a V1–V4 lineage, to produce seam cells and hence alae (Waring and Kenyon, 1990). Superficially, this resembles a sexual transformation (as does the *mab-5* reduction of the male tail), since hermaphrodites have similar V6 and V1–V4 lineages. However, the double mutant combination *lin-22; pal-1* reveals the transformation to be one of lineage rather than sexual type: *lin-22* alone converts V1–V4 but not V6 into a V5-type lineage (producing postdeirid cell groups; Fig. 6.11) and the double mutant shows the majority of V6s similarly producing postdeirids, indicating a conversion of V6 first to V1–V4 (by *pal-1*), then a conversion to V5-type lineage (by *lin-22*). The results also show that *lin-22* is epistatic to and, hence, acts after or "downstream" of *pal-1*.

In fact, *pal-1* seems to be involved specifically in overriding a signal from T to V6 to produce alae. If T is ablated in *pal-1* males, then V6 produces a normal complement of rays (Waring and Kenyon, 1990). Ablation of V5, just anterior to V6, in *pal-1* males has no such effect. Apparently, in the absence of T, V6 exhibits its normal propensity to produce rays; the *pal-1* wild-type activity is not needed to permit ray production in V6 if T has been removed. Thus, V6 and T, which are in contact, undergo a cellular interaction that helps to organize where the alae-ray border will be (although in wild-type males, the border does not appear there but within the V5 lineage).

Indeed, the respective positioning of alae and rays can be seen as involving a competition between *lin-22* and *mab-5*, in which there are regional "weightings" as to which wins, while the final pattern is "refined" by gene products such as *pal-1* (Waring and Kenyon, 1990). Thus, in *lin-22; mab-5* double mutants, there is still a tendency for alae to form anteriorly and rays posteriorly, while in wild-type, the competition and the refining mechanisms always set the boundary precisely at the boundary between the V4 and V5 domains. The genetic pathway and the postulated cellular interactions are shown in Figure 6.13.

lin-14, a Heterochronic Gene

In addition to the homoeotics in which cells take on the behavior and fates of cells in other places within the same stage of development, there is a group of homeotics in which cells take on the lineage characteristics of earlier or later stages within their

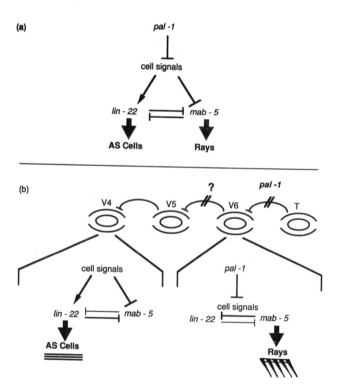

Fig. 6.13. The relationships between *pal-1*, *mab-5*, and *lin-22* in the generation of adult seam (AS) cells and rays of the male tail. **a:** The regulatory relationships between the three genes. **b:** Hypothesized interactions that produce stereotypical V cell patterns. See text for discussion. (Reproduced from Waring and Kenyon, 1991, with permission of the publisher. © Cell Press.)

lineages. Such mutants have been termed "heterochronics" (Ambros and Horvitz, 1984), a reference to the evolutionary phenomenon of heterochrony, in which somatic development becomes either accelerated or retarded relative to sexual maturation (Gould, 1977). Heterochronic mutants may be thought of as exhibiting "temporal homeosis," in which cells, tissues, or structures appear out of their normal time sequence.

The first heterochronic to be isolated was a mutant of the *lin-4* gene, whose phenotype consists of a reiteration of the L1 cell division patterns; because of this division pattern, *lin-4* mutants do not produce a normal sequence of vulval cell divisions and do not make a normal vulva (the characteristic that led to the mutant's isolation). This repetition of L1 cell lineages is reflected in the cuticular composition of mutant individuals, who maintain a larval cuticular structure rather than the more complex cuticular pattern of the adult (see Fig. 9.1). These *lin-4* defects represent amorphic or hypomorphic states and hence reflect effects produced by insufficient *lin-4+* activity.

Following the isolation and characterization of the *lin-4* mutant (Chalfie et al., 1981; Sulston and Horvitz, 1981), three other heterochronic genes (*lin-14, lin-28,* and *lin-29*) were identified (Ambros and Horvitz, 1984). Mutants of each are characterized by a particular pattern of precocious appearance or repetition/retardation of many cell lineages. Of these, *lin-14* has been best characterized. Like *lin-4, lin-14* mutant effects are widespread; tissues affected by *lin-14* include mesodermal tissues, the intestine, the lateral hypodermis, and the male posterior ectoderm. In contrast to *lin-4,* however, but similarly to *lin-12, lin-14* can mutate to either of two states, which produce opposite effects. Semidominant mutants mimic the *lin-4* defect, causing a reiterated L1 development with each molt. Recessive mutants, on the other hand, retain a normal L1 stage, but cause one or more of the following larval stage division programs (that of L2, L3, or L4) to be skipped in whole or in part. In effect, the recessives cause a premature expression of later stages and a loss of earlier ones, with each mutant falling into one of several different classes depending upon which stages are reiterated or skipped (Ambros and Horvitz, 1987).

Genetic tests show that the semidominant mutations are hypermorphs and the recessives are amorphic or nearly so, comparably to the *lin-12* classes (Ambros and Horvitz, 1987). Evidently, high *lin-14* activity causes the reiteration of early stages, while reduced *lin-14* activity causes early stages to be skipped. One may extrapolate from this genetic characterization to form the following hypothesis about *lin-14* activity in wild-type development: that *lin-14+* activity is high during early development in the wild-type, this high activity being essential to ensuring the completion of early stages, and decreases during postembryonic development, this diminished activity permitting later stages to take place (Fig. 6.14). This model may, however, be something of an oversimplification: By both phenotypic criteria and intragenic complementation patterns, *lin-14* may in fact consist of two different activities, which can be independently mutated. These distinguishable activities may represent two distinct molecular species, perhaps produced through differential splicing or differential posttranslational modification, or, alternatively, they may represent different thresholds of a single activity, where the amount of that activity is differentially altered in the different mutant classes (Ambros and Horvitz, 1987).

Molecular data supports and extends the picture of *lin-14* action derived from the genetics. The gene was mapped with respect to multiple Tc1 elements, as described

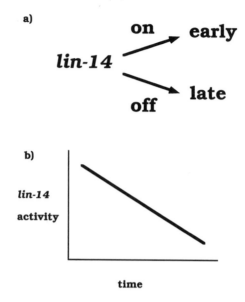

Fig. 6.14. Hypothesized *lin-14* expression pattern (**a**) and activity (**b**) with time. (Reproduced from Horvitz, 1988, with permission of the publisher.)

above for *unc-86,* and then identified in a set of contiguous DNA clones by molecular analysis of an intragenic recombinant, revealed phenotypically by the unmasking of a semidominant mutation from a double mutant (Ruvkun et al., 1989). The gene product was then synthesized in quantity, following induction of a *lacZ-lin-14* fusion gene in *Escherichia coli,* and antibodies to *lin-14* protein were prepared. Immunochemical staining of animals at different stages reveals that the *lin-14* gene product is a nuclear protein and is present in the nuclei of the nuclei of all the affected somatic lineages (plus some terminally differentiated cell types) (Ruvkun and Giusto, 1989).

Most significantly, the staining patterns confirm the genetic model: The protein first appears in late embryos and is present in highest concentration in the hypodermal and intestinal lineages in early L1 and then diminishes in those lineages (peak concentrations in neuronal and Pn.p [including the vulval] lineages appear somewhat later but then also diminish). Furthermore, while a null mutant lacked nuclear staining, two semidominants were found to give high staining in the L2 through L4 stages, well past the times at which staining is seen in the wild-type (Ruvkun and Giusto, 1989). The basis of the hypermorphic phenotype for at least two such mutants, however, is not persistent *lin-14* transcription because the mutational defect in these strains is a deletion in the 3' terminal exon (Ruvkun et al., 1989). In some manner, perhaps through reduced degradation of the mRNA, these deletions provoke elevated and continuing presence of the protein.

Though a nuclear protein, the nature of action of *lin-14* still awaits elucidation. The gene sequence does not show obvious homologies to known transcriptional factor families (Ruvkun et al., 1991), but the evidence collectively suggests that it is a gene regulatory protein, acting presumably in transcription or mRNA processing. The construction of double mutants between *lin-14* mutants and those of other heterochronic genes, to determine epistatic and hence "pathway" relationships, as well as further molecular characterization should help to elucidate the basis of temporal homeosis in *C. elegans.*

The Vulval Development Pathway

The analysis of genes first identified in general mutant searches, genes such as *unc-86, lin-12, mab-5,* and *lin-14,* has provided invaluable information about aspects of postembryonic development. However, attempting to understand the genetic basis of postembryonic development from mutants isolated in this fashion is analogous to trying to visualize an entire landscape from a small number of narrow-aperture photographs. For the thorough understanding of a postembryonic developmental process—whether it be development of the male tail, the hermaphrodite vulva, or the gonad in either sex—one needs to know all the genes whose activities are crucial, their times of action, and the roles that their gene products play in the observable events of that developmental process. For such an approach, one must first isolate mutants of all the genes whose products play key or rate-limiting roles in the developmental process of interest and then proceed to analyze the aberrations in cell behavior in each mutant relative to the wild-type; the aggregate picture allows one to construct a picture of the gene-based events that generate the developmental process.

The pathway of vulval development is ideal for such an analysis and has been studied in this manner. Four conditions recommend vulval development for study. First, the visible cellular events are easy to track and are well documented, as we have seen (Fig. 6.6). Second, the cellular "rules" and phenomena, involving the cell replacement rules within the cell equivalence group and the role of the anchor cell in inducing the vulva, are known. Third, the critical divisions from the immediate precursor cells take place in a brief interval, 5 hr, facilitating analysis. Fourth, it is comparatively easy to isolate and identify mutants affected in vulval development by inspection. The vulva is not an essential structure and, therefore, complete loss-of-function mutations in genes required solely or principally for its development would not compromise survival. Genes that are required for other developmental processes as well might reveal their roles in vulval development through hypomorphs.

The first extensive mutant hunts, following the initial work of Horvitz and Sulston (1980), were carried out by Ferguson and Horvitz (1985); a search for various Egl mutants by Trent et al. (1983) also contributed some of the genetic source material for the analysis. Following EMS mutagenesis, hermaphrodite F_2's were examined and those displaying Egl, Vul, or Muv defects were isolated. To date, hundreds of mutations, defining more than 40 complementation groups, have been identified (reviewed in Ferguson et al., 1987; Horvitz and Sternberg, 1991). As might be expected, the genes are scattered over the genome, with all six chromosomes represented.

The genetic characterization of these mutants reveal several points of interest (Ferguson and Horvitz, 1985). The first is the fact of variability for each mutant: Most, including some of those classified as amorphs, give less than 100% penetrance or expressivity, a property which we have seen to be a general one among postembryonic mutants. Furthermore, while each mutant is classified as Vul or Muv, many with incomplete penetrance present some fraction with the opposite phenotype as well as some with the wild-type phenotype. Evidently, there are not only partial compensatory mechanisms for many mutant defects, but each genetic defect has a certain indeterminacy in its translation into phenotype.

Second, while the mutants were isolated on the basis of their vulval defects in the hermaphrodite, relatively few are devoted to vulval development exclusively. This conclusion follows from several observations. For instance, the majority of the mutations produce defects in the male, either in terms of obvious morphological abnormality or in some degree of loss of mating ability. Most of the mutants, in fact, display defects in specific cell lineages in both sexes and are, like so many postembryonics, defective in particular cell lineages (Lin mutants). Furthermore, mutants of a number of the genes represented by single alleles are hypomorphs, as judged by the Müllerian test; more deficient alleles of several of these produce sterility or lethality in homozygotes. Indeed, the class of genes most likely to be those devoted exclusively to vulval function are those represented by null alleles that give Muv or Vul phenotypes with no further hermaphrodite defects or male defects. Of the 22 genes described by Ferguson and Horvitz (1985), only four fall into this category. Given the numbers of repeat isolates of mutants from the same genes, the exclusively vulval-function class has probably been saturated. One may provisionally conclude that the pathway of vulval development relies, like the majority of developmental pathways, on a large battery of genes, relatively few of which are employed exclusively in that pathway.

In the first systematic analysis of the vulval pathway, mutants affecting vulval development in 28 genes were characterized with respect to their cellular defects, and 23 genes in this group were ordered within this genetic pathway (Ferguson et al., 1987). In assigning a presumptive role to each gene, two critical questions were posed: 1) At what step in the cellular lineage do the first divergences from wild-type development become apparent in the mutants? and 2) Does the gene act in the hypodermis itself, some of whose cells directly form the vulva (the vulva precursor cells or VPCs), or in the anchor cell of the gonad, which produces an inductive signal for vulval development in the hypodermis?

Direct observations of the cellular deviations from wild-type behavior in the mutants, and their time of onset, allow one to categorize the genes as essential for one of three processes: 1) precursor cell formation, 2) determination of vulval cell fate (with respect to the 1°, 2°, and 3° lineages), and 3) (correct) expression of a particular vulval cell fate (Fig. 6.15). Thus, for instance, mutants of unc-83 and unc-84, which prevent the P blast nuclei from migrating to their standard ventral positions, clearly prevent formation of the VPCs. Another gene whose product is essential for precursor cell formation is lin-26. The Vul lin-26 mutant, a hypomorph of an essential gene (Ferguson and Horvitz, 1985), switches the Pn.p cells into the production of neural-like cells. Clearly, the lin-26 gene wild-type product is required for the generation of VPCs.

The largest subset of the identified genes permits the formation of vulval precursor cells in the appropriate positions but alters their cell fate with respect to those positions. Altogether, more than 30 genes have so far been placed in this category. For instance, the entire vulval equivalence group P3.p–P8.p in the amorphic or hypomorphic Vul mutants of lin-2, lin-3, lin-7, lin-10, and let-23 all show only the 3° pattern of cell divisions. One may conclude that these gene activities are necessary for the acquisition of 1° and 2° fates. In contrast, lin-12 activity, as we have seen, is necessary specifically for the 2° fate and also for development of the ac, required for vulval induction.

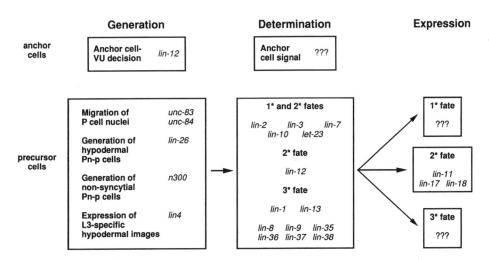

Fig. 6.15. Genetic pathway of vulval development. See text for description. (Adapted from Ferguson et al., 1987.)

A critical question about those mutations that alter cell fate is whether they act in the VPCs directly, or in the ac (in the latter case, altering the intensity or range of its signal), or in the surrounding hypodermis (which might, in principle, affect the behavior of the VPCs.) To take one example, two simple alternative explanations for the mutants that produce a Muv phenotype (*lin-1, lin-12(d), lin-13, lin-15*, or the double mutant *lin-8; lin-9*) are that the mutations make the VPCs independent of the ac signal or that they exert their effect through the ac, perhaps amplifying its signal. To decide this issue, one can eliminate the ac, by laser microbeam ablation of the somatic gonad precursor cells, and then score the phenotypes of these mutants to see if they are maintained. It is found that the Muv phenotype still develops, showing that the developmental defect for each makes them independent of the ac signal and must be in the hypodermal cell lineages, presumably, though not necessarily, in the cells from which the vulval cell equivalence group arises (Ferguson et al., 1987).

If the cell fate-altering Vul mutants also act in the hypodermis, then one might expect such mutants to be epistatic to the signal-independent Muv mutants, since the Vul defect in such strains presumably eliminates the capacity of the hypodermis to produce VPCs. The appropriate double mutant combinations were constructed and the phenotypes examined. For mutants in *let-23, lin-2, lin-7, lin-10*, and *lin-3*, an abolition or diminution of the Muv phenotype of several signal-independent Muv mutations was found. The results suggest that the defect in these Vul mutants is, indeed, outside the ac and presumably therefore in the hypodermis.

Finally, mutants would be predicted that allow VPCs to carry out their normal fates, with respect to 1°, 2°, and 3° character, but in which the particular character of the cell division pattern for that fate was found to be aberrant. Genes required specifically for the execution of 1° and 3° fates have not yet been identified, but three genes have been found to be essential for expression of the normal 2° division pattern (*lin-11, lin-17*, and *lin-18*). In *lin-11* mutants, for instance, all four granddaughters of each of P5.p and P7.p adhere to the ventral cuticle following a longitudinal cell division, while in the wild-type only two sister cells (within each group of four descendant cells) have this property. The *lin-11* gene is now known to be a homeobox gene and therefore a presumptive transcriptional control gene (Freyd et al., 1990). Its transcriptional regulatory activity must be necessary to produce a specific difference in the two daughters of both P5.p and P7.p.

The placement of 23 genes in the vulval developmental pathway is diagrammed in Figure 6.15. The analysis shows that it is possible to assign fairly specific cellular roles to gene products for a number of genes in the creation of this particular postembryonic structure. Thus, from an initial description of a mutant as either Vul or Muv, one can, using genetic techniques, identify with reasonable precision what part of the overall process the wild-type gene product participates in.

On the other hand, it is important to remember that this description pertains only to a limited set of genes, those that can mutate to give a vulval defect, either as amorphs or hypomorphs. Any gene whose partial loss of activity is not rate-limiting for vulval development, but for an earlier stage of development, would not have been picked up and would therefore not be included in the scheme, although its activity might be essential for development of the vulva. The absence, so far, of identified genes for particular processes, such as ac signal generation (as opposed to ac formation) or 1° fate expression, presumably signifies only that few or no genes function exclusively for those processes. Clearly, it would be desirable to find genetic means for identifying such functions. One procedure that may help to detect other genes involved in vulval development entails looking for mutants that suppress or enhance the effects of known mutants in the pathway.

One exploration of this kind was based on the finding that an early Muv isolate owed its phenotype to mutations in two separate genes. The *lin-8; lin-9* strain is Muv but homozygosity for either mutation alone gives a wild-type phenotype (Horvitz and Sulston, 1980). The existence of one such "synthetic" phenotype suggests that others might exist and that finding them might be as simple as looking for new mutations that give the Muv phenotype by mutating *lin-8* or *lin-9* strains independently.

The results of such a search, by Ferguson and Horvitz (1989), produced 18 new mutations, identifying seven other genes that could interact with either *lin-8* or *lin-9* to produce the Muv phenotype. Four genes (*lin-35, lin-36, lin-37,* and *lin-38*) had not been picked out before as having roles in vulval development. Most, though not all, of the mutations were "silent," that is, not capable of producing a Muv phenotype when homozygous on its own, and each could be placed in either of two classes, A or B, depending upon its interactions. Double-A or double-B mutants do not produce a Muv phenotype, while any A mutation combined with any B mutation does do so. The results argue for the existence of functionally redundant pathways in vulval development, or at least for that part of the total pathway governing the establishment of 3° fates (Ferguson and Horvitz, 1989). One of the genes in this group, and the only one that can mutate to either the A or the B class, *lin-15*, functions in producing a signal from *outside* the VPCs that inhibits 1° and 2° fates (see below). The functional redundancy that exists for the pathway might exist primarily to restrict vulval development among the VPCs to only those that normally experience it (P5.p–P7.p).

It is, of course, molecular characterization, in combination with the genetics, that should definitively characterize the nature of the events in the pathway and, in particular, the various cellular interactions critical for vulval development. Altogether, it appears that at least three different signals, or types of signal, are involved. The first is, of course, the ac-derived inductive signal(s). Though its precise character is still unknown, there is some circumstantial evidence concerning the cell surface molecule that may act as its receptor. While null mutations in the *let-23* gene produce a larval lethal phenotype, certain hypomorphic mutations produce a Vul phenotype and certain hypermorphs a Muv phenotype. The gene has now been cloned and shown to be a member of the EGF-receptor tyrosine kinase family (Aroian et al., 1990), like *top* in *Drosophila.* Furthermore, the product of a second gene, *let-60*, which gives the same spectrum of mutant phenotypes, may act as part of the system that transduces the *let-23* signal. *let-60* shows homology to the human *ras* protein, one of the so-called G proteins, involved in signal transduction from the cell surface (Han and Sternberg, 1990). If *let-60* helps to transduce a signal received by the *let-23* gene product on the surface of the VPCs, then *let-60*, by definition, must act downstream of *let-23*. Consistent with this notion, transformation by injection of wild-type *let-60* into the syncytial gonad of *let-23* eggs (such DNA injection producing multicopy extrachro-mosomal arrays) rescues the *let-23* defect (Han and Sternberg, 1990).

The second kind of signal or cellular control comes from *outside* both the vulval equivalence group and the ac and is dependent, at least in part, on the activity of the *lin-15* gene. *lin-15* loss-of-function mutants show a Muv phenotype, with an alteration of 1° and 2° lineage characteristics in the vulval equivalence group; as with other Muv mutants, this development can take place in the absence of the ac. Herman and Hedgecock (1990) have shown by mosaic analysis, employing a Dp containing *lin-15* and two independent markers, that the *lin-15* phenotype is expressed even when members of the P3.p–P8.p group contain the wild-type allele. The result shows that wild-type *lin-15* activity, in cells other than the VPCs, inhibits vulval development. One possibility is that this inhibitory activity is expressed in the principal hypodermal syncytium, hyp 7; daughters of the VPCs that do not contribute to the wild-type vulva fuse with hyp 7. Thus, it is possible that the primary action of the ac signal(s) is to inhibit such fusion of the daughters of P5.p–P7.p (Herman and Hedgecock, 1990). In the latter cells, continued division to generate the vulval cells would then be a "default" response, following failure to fuse.

The third signal, or class of signal, governs interactions among the Pn.p cells that contribute to the vulva, hence those that adopt 1° or 2° fates. Several pieces of evidence indicate that these interactions fall into the category of "lateral inhibition," in which election by one cell of a particular pathway causes it to inhibit its immediate neighbors from following the same route. (The *lin-12*-mediated interactions are all examples of lateral inhibition, and further instances will be given in *Drosophila* development, in chapters 7 and 9.) Within the vulval equivalence group, lateral inhibition probably follows assignment of a cell to the 1° fate (in wild-type, the P6.p cell), which then inhibits its immediate cellular neighbors from taking on this fate, these assuming the 2° fate instead.

The best evidence for lateral inhibition among the VPCs comes from observations and experiments on several Muv mutants, where the ac signal can be eliminated as a complicating factor. Thus, in many of these, one sees the alternation of 1° and 2° fates, noted above for *lin-15*. When both the ac and varying numbers of the VPCs are laser ablated, in a *lin-15* mutant background, it is seen that a VPC that never takes on the 1° fate can do so, if it is isolated from others, and two neighboring VPCs will only both develop the 1° state if there is physical distance between them; when in contact, the pattern always seen is for one to be 1°, the other 2° (Sternberg, 1988). Given the involvement of *lin-12* as a cell surface molecule required for the 2° fate, it seems highly probable that it is involved in this system of lateral signaling, being activated in the responding cell.

The final steps of organogenesis, in which the neuronal and muscular systems necessary for egg laying are integrated with the development of the vulval hypodermal cells, also involve cell and tissue interactions. These steps have been investigated by using either cell ablation or mutants to remove or shift cells, and then determining which events are obligatorily linked and which are independent of one another (Li and Chalfie, 1990). The results indicate that both the pattern of neuronal branchings from the motor neurons of the VNC and the positioning of the vulval muscles respond primarily to cues from the VPCs rather than from the gonad or neuronal-muscle signalling. Furthermore, comparisons between different *lin* mutants that differentially reiterate 1° or 2° lineages show that 1° lineages are more effective in inducing branching of the motor neurons from the VNC (Li and Chalfie, 1990).

The vulval developmental pathway, first characterized thoroughly at the cellular level, is rapidly becoming one of the better-characterized developmental sequences at the genetic and molecular levels. The particularly exciting feature is the increasingly precise specification of the molecular events that lie at the basis of the known cellular interactions. The pathway also provides, as we have seen, an example of the importance of genetic functional redundancy in a developmental process.

GONADAL AND GERM LINE DEVELOPMENT

The gonad in both sexes is the major organ system in the animal; in the hermaphrodite, for instance, the syncytial, germ line nuclei of the gonad comprise more than two-thirds of the total nuclei in the adult animal. Although the gonads develop from what appear to be identical sets of four precursor cells, the respective patterns of gonad development show some major differences (Fig. 6.16). The obvious structural difference between the hermaphrodite and male gonads is one of shape. The hermaphrodite gonad is bilaterally symmetrical, with twofold rotational symmetry and two gamete-generating arms that meet in the center, while the male gonad is asymmetric, consisting of one reflexed gamete-generating arm that runs posteriorly. This difference is initiated early in the L1 stage, with an asymmetric cell migration in the male gonad primordium that does not occur in the hermaphrodite. The difference becomes especially obvious during L3, when the hermaphrodite gonad, which has been growing both anteriorly and posteriorly, undergoes a symmetric reflexion in both arms; the male gonad, growing only anteriorly, makes a single bend and then continues growing posteriorly, eventually joining the cloaca.

Despite the pronounced morphological and growth differences, the gonads of hermaphrodites and males share two important organizational features. First, the gametes in both sexes mature from the most distal (outward) tips of the gonads toward the proximal (central or midline) end(s). Second, the major somatic structures in both gonads are located proximally.

The developmental processes that underlie this similarity of organizational polarity are revealed by lineage analysis of the gonadal somatic cells (Kimble and Hirsh, 1979). Both types of gonad originate from just four precursor blast cells, designated Z1, Z2, Z3, and Z4, which are located midventrally. In both sexes, the Z2 and Z3 cells (daughter cells of the germ line embryonic blast cell P_4) give rise to the gametes, while

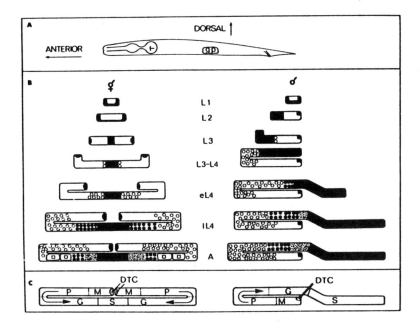

Fig. 6.16. Gonadogenesis in the *C. elegans* hermaphrodite (left) and male (right). **A:** In the newly hatched L1 of both sexes, the initial position of the gonadal primordium (gp) is midventral and the same. **B:** Gonadal morphogenesis during the four larval stages (eL4 is the early larval stage; lL4 is the late larval stage). See text for discussion. **C:** The adult gonad in both sexes. DTC, distal tip cell; G, gamete-forming regions; M, mitotic nuclei; P, pachytene nuclei; S, somatic tissue. (Reproduced from Kimble and White, 1981, with permission of the publisher.)

Z1 and Z4 (derived from the MS blast cell) give rise to the somatic cells of the gonad. The major differences in gonadal growth pattern between the sexes involve the activities of certain key somatic cells, derived from Z1 and Z4, that are generated in the first divisions of the gonad primordia during L1.

Two are referred to as "distal tip cells" (dtcs) in both sexes, because of their location at the distal (outermost) point in each gonad, where the germ line precursor cells undergo premeiotic divisions. In the developing hermaphrodite gonad, each dtc leads one of the two arms of the developing gonad toward its final distalmost point. In the male, the two dtcs position themselves at the posterior (and future distalmost point) of the gonad primordium during late L1 and remain there, marking the stationary point of male gonad growth. Despite this difference in behavior, the dtcs are formed early in L1 in both sexes and occupy similar positions in the Z1 and Z4 lineages, being the anteriormost Z1 and the posteriormost Z4 descendants in both gonad primordia (Z1.aa and Z4.pp in the hermaphrodite and Z1.a and Z4.p in the male primordium). An additional similarity is a functional one: In both sexes, the dtcs promote mitosis, suppressing meiosis (and hence gamete maturation) in the diploid germ line precursor cells at the distal end of each gonad. (This aspect will be discussed further below.)

The third key regulatory cell in the male gonad is the "linker cell" (lc). The lc of the male provides the growing point of the male gonad: It leads gonadal growth throughout development and its death at the end of its migration provides a junction with the cloaca (through which sperm will pass). The lc thus carries out the leader function that is performed in the hermaphrodite gonad by the dtcs. Interestingly, though very different in role from the ac of the hermaphrodite, these two cells are morphologically similar (Kimble and Ward, 1988) and related in lineage, arising from either of two cells: Z1.ppp or Z4.aaa in the hermaphrodite and Z1.paa or Z4.aaa in the male (Kimble and Hirsh, 1979). The "choice" in both cases seems to be mediated by a cell interaction between the two cells, which is, as we have seen, influenced by the state of *lin-12* activity. (There are many other resemblances in the details of lineage program between the two gonadal soma and these have been described by Kimble and Hirsh [1979].)

Like the somatic cell division programs in nongonadal tissues, those in the gonad are highly precise. In hermaphrodites, the gonadal soma consists of 143 cells, and in males, 56 cells. In contrast, the germ cells, the descendants of Z2 and Z3, do not show this rigidity of division pattern. These cells begin multiplying without fixed orientations or precise schedules in L1 and continue to do so throughout development and even into adulthood in the distal arm of each gonad. In the hermaphrodite, meiosis begins during the L3/L4 intermolt in the proximal arm of each gonad, and the first gametes to differentiate are sperm, arising from about 40 primary spermatocytes in each gonad arm. Gametogenesis then switches to oogenesis and all the remaining gametes formed are oocytes. In the male, meiosis begins somewhat earlier than in the hermaphrodite, during L3, also within the proximal (and posterior) region of the gonad, and during the lifetime of the male, several thousand sperm can be produced if matings are frequent.

Despite the absence of fixed cell or nuclear lineage patterns in the germ line, germ cells are under a form of division control. If the dtcs are ablated in either sex, the germ cells nearest the distal tips cease mitosis and enter meiosis (Kimble and White, 1981). Evidently, the dtcs exert some sort of meiotic inhibitory action on their germ line cell neighbors. One possible inference is that the whole polarity of gamete maturation *away* from the distal tips toward the proximal somatic structures is a function of this influence. This hypothesis receives support from the results of certain cell ablation experiments. When sister cells of dtcs are ablated, the dtcs sometimes assume abnormal positions. The resultant gonads grow in aberrant directions but gamete maturation is always polarized away from the dtcs, as occurs in normal gonads (Kimble and White, 1981).

Linear polarity in gamete maturation away from precursor mitotic stem cells is also seen in *Drosophila* oogenesis. The distal tip of the ovary is populated by mitotic stem cells that grade into meiotic stem cells down the length of the ovariole. Such linear polarity of gamete maturation is not universal but it may be common (see discussion by Kimble and White, 1981).

Although the absence of strict division control of germ line cells differs from the somatic cell division programs, it makes sense in terms of the requirements of germ line development. In the nematode, somatic structures are constructed accurately and with economy, employing precise sequences of somatic cell division. Such accuracy is not required in the germ line cells, as each functions independently. Furthermore, it is in the interest of reproductive efficacy to produce many gametes in each sex. Thus, germ line cells divide rapidly at their distal tips, to ensure the production of numerous gametes, and the gonad is organized in a polar array such that only mature gametes are discharged through the appropriate somatic structures. Beyond this, no accurate control of lineage is required. The developmental organization of the germ cells in the nematode fulfills these requirements.

There is still relatively little genetic analysis of the mechanisms underlying the pattern of germ line development in *C. elegans*, but one crucial gene has been identified, *glp-1* (for germ line proliferation control) on chromosome III, mapping close to *lin-12*. In a search for sterile hermaphrodites, Austin and Kimble (1987) identified a zygotic mutant in which few germ line nuclei are produced, these entering meiosis and maturing into sperm precociously. Subsequently isolated steriles mapping to the same region of chromosome III were tested for complementation with the original mutant, and a single complementation group, *glp-1*, was identified. To examine the nature of the defect further, temperature shifts were done on a ts *glp-1* allele. It was found that animals shifted from the permissive to the restrictive temperature from mid-L2 onward produced increasing numbers of gametes, including oocytes. The defect is thus not one of oogenesis *per se* but rather on the shift from mitosis to meiosis; the initially observed *glp-1* defect (gonads with a small number of sperm) reflects the fact that in normal hermaphrodite development, spermatocytes are produced first, gamete production then switching to oogenesis.

The defect in adult *glp-1* homozygotes looks intriguingly similar to that produced by physical ablation of the dtcs. Yet, in such animals, the dtcs are present and appear morphologically normal (Austin and Kimble, 1987). Nevertheless, the deficiency might be in the dtc-generated signal. To explore this, a mosaic analysis was performed, in which a Dp was used that carries *glp-1*, a nucleolar size marker, and an *unc* gene (whose expression is required in the AB lineage); loss of the Dp during embryonic development "uncovered" mutant alleles for all three genes.

The results showed that the *glp-1* gonadal defect can appear in mosaic animals that are *glp-1*+ in the MS lineage (from which arise the precursors of the somatic gonad, Z1 and Z4, and hence of the dtcs), and, conversely, that animals lacking the Dp (and *glp-1*+) in the somatic gonad (and hence in the dtcs) can have normal gonads (Austin and Kimble, 1987). It follows that the *glp-1* function is not expressed in the dtcs, and therefore does not involve the generation of the dtc signal, but in another tissue, and presumably the germ line itself. (Because the markers used cannot be scored directly in the germ line, the conclusion that *glp-1* acts in the germ line remains an inference but a strong one.) If the *glp-1* product acts in the germ line, its most probable function is that as a receptor for the dtc signal or as some part of the signal-processing cellular machinery. Subsequent molecular evidence supports this; like the adjacent *lin-12*, the *glp-1* gene encodes a transmembrane protein possessing multiple EGF repeats in the cytoplasmic domain (Yochem and Greenwald, 1989).

There is, however, one further point about *glp-1* which should be stressed: Its function is not limited to the mitosis/meiosis decision or to expression in the germ line. In a search for maternal effect mutations which cause early lethality, Priess et al. (1987) isolated several *glp-1* mutants in this category, showing that *glp-1* is expressed not only in the zygotic genome but the maternal germ line as well. The point of arrest in the embryos from homozygous mothers varies with the mutant allele under study, but a characteristic defect is the absence of that part of the pharynx derived from ABa (the remainder, derived from the MS lineage, is present). In addition, there is disorganization of the hypodermis associated with the presence of extra hypodermal cells. Taken together with its presumptive identity as a membrane signal receptor, the results suggest that maternal germ line-derived *glp-1* functions in cell membrane-mediated interactions in several embryonic tissues. The precise nature of the signal(s) and the transduction machinery, for processing the signal(s), remain to be elucidated.

A further point of interest about *glp-1* concerns its evolutionary and functional relationship to *lin-12*. Both genes are highly similar in overall structure and amino acid sequence (greater than 50%), though *glp-1* is slightly shorter, containing 10 EGF-type repeats to *lin-12*'s 13 (Yochem and Greenwald, 1989) (Fig. 6.17). Their close proximity on chromosome III and evident sequence relationship suggest that they derive from an ancient gene duplication event.

Despite their structural similarities, the phenotypic effects associated with mutants in either are very different from those of the other; *lin-12* mutants show various somatic defects and *glp-1* various germ line or germ line-derived effects. Nevertheless, there is genetic evidence of similar intrinsic functional capacities of these gene products (Maine and Kimble, 1990). A gain-of-function *glp-1* mutant, *glp-1(q35)*, was found to mimic one *lin-12* gain-of-function defect, the Muv phenotype (Austin and Kimble, 1987), indicating that *glp* can be mutationally activated in the soma, triggering VPC proliferation. Indeed, *glp-1(q35)* can mimic the Muv phenotype of dominant *lin-12* mutants in a *lin-12*-deficient background (Mango et al., 1991). Analysis of the mutation shows that it is a stop codon which causes the production of a truncated *glp-1* protein, missing 122 amino acids from the C-terminal end (Mango et al., 1991). The results indicate that the C-terminal end of the wild-type protein inhibits *glp-1* activity in somatic cells. That *glp-1* is expressed in the soma is shown by the presence of *glp-1* mRNA, in low levels, in somatic tissues (Austin and Kimble, 1989).

Evidence for such functional redundancy has been reported by Lambie and Kimble (1991). Double mutant homozygotes die as early L1 larvae, in contrast to either single mutant, and show a range of defects which are, individually, not seen or only rarely seen with the single mutants, the so-called Lag phenotype (for *l*in-12 *a*nd glp-1 phenotype). The latter includes a twisted nose, absence of the excretory cell and/or anus, and a few specific posterior muscle defects. These defects are also seen in single mutant homozygotes for two separate genes, termed *lag-1* and *lag-2*. Though the precise relationships between *lin-12*, *glp-1*, *lag-1*, and *lag-2* are unknown, one attractive possibility is that the *lag* genes are involved in transmitting the signals received by the Lin-12 and Glp-1 transmembrane receptors. Other possibilities, however, have not been excluded (Lambie and Kimble, 1991).

The larger significance of the observations is that they provide yet a further example of functional redundancy. The gene products are functionally interchangeable, at least to a degree, and the differences in mutant phenotype reflect their

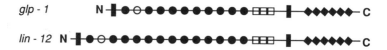

Fig. 6.17. Comparison of *lin-12* and *glp-1* sequences. The EGF-like repeats are shown as filled circles; another repeated sequence motif, found in the cytoplasmic domain, is indicated by the diamond symbol; the transmembrane domain is indicated by the shaded bars; the open square represents a third repeated motif, the LNG motif. (Reproduced from Maine and Kimble, 1990, with permission of the publisher. © ICSU Press.)

incorporation into different regulatory networks and expression patterns. Yet, even so, they provide a degree of backup for one another.

SEX DETERMINATION AND DOSAGE COMPENSATION IN CAENORHABDITIS ELEGANS

Beyond the question of how gonadal growth and germ line development are organized in both sexes, there is a more fundamental developmental question, namely how the "decision" between the two different kinds of gonads (hermaphrodite vs. male) and their associated sets of secondary sexual characteristics is made in the first place, the fact of sex determination itself: What are the genetic factors that play a key role in the determination of sex and how do they work? A second question concerns how this initial decision is translated into either of two specific patterns of sexual differentiation.

The problem of sex determination is the oldest in developmental genetics; it was identified as a problem and investigated well before there was a recognizable discipline of developmental genetics. T.H. Morgan and E.B. Wilson and their colleagues first turned to it in the early years of this century with the discovery that the sexes in several insect species differ in a particular chromosome or chromosome pair, these sex-distinguishing chromosomes being termed the sex chromosomes. This finding simultaneously prepared the way for the modern chromosomal theory of inheritance, following the finding by Morgan that the inheritance of the *white* gene of *Drosophila* and several other genes is coincident with that of the major sex chromosome (the phenomenon of sex linkage). With respect to sex determination in general, we know now that for most animal species, a difference in sex chromosome composition is the critical determiner of sex (though in some fish and reptiles, the primary determinant involves a threshold in an environmental signal).

The extent of sexual dimorphism between adult hermaphrodites and males is large, as can be seen in Figures 6.2 and 6.16. The cells and tissues affected include the germ line (in both sexes a substantial fraction of the total cellular mass), the somatic cells surrounding the germ line, the accessory gonadal structures, and a variety of secondary sexual structures important for gamete delivery, such as the vulva in the hermaphrodite and the copulatory apparatus in the male, and perhaps for sexual recognition. In the somatic tissues alone, the fraction of cells whose phenotype is affected by the sex of the animal is approximately 30% in hermaphrodites and 40% in males (Hodgkin, 1988). Clearly, the assignment of sexual identity has major developmental consequences.

In *C. elegans*, this assignment depends on the X chromosome composition of the individual: The hermaphrodite possesses two X chromosomes (the XX state) and the male only a single X (the XO state). However, the sex determination decision in *C. elegans* is based, as it is in *Drosophila* (Bridges, 1916, 1925), not on the absolute number of Xs but on the ratio of Xs to autosome sets. This was demonstrated by Nigon (1951) in a study of tetraploid animals carrying different numbers of X chromosomes. He found that tetraploids possessing two X chromosomes per cell (a genotype denoted as 2X;4A) are male and not hermaphrodite, as they would be if the number of Xs were crucial, while 4X;4A and 3X;4A animals are hermaphrodite.

Nigon's findings were confirmed and extended by Madl and Herman (1979). Their results are summarized in Table 6.2 and show that an X:A ratio of 1 to 0.75 produces the hermaphrodite state while a ratio of between 0.67 (2X;3A) to 0.5 (1X;1A) yields males. Furthermore, by adding partial X chromosome duplications to a 2X;3A genotype, one can shift the sexual phenotype away from maleness to femaleness, producing intersexual animals, those showing a mixture of male and female structures.

How is this ratio measured or sensed? In principle, the process could take place in either of two ways (Fig. 6.18). In the first, the X chromosomes produce one or more substances in a dose-dependent manner which react(s) with the autosomes: Low relative doses of X substance(s) would turn on male differentiation or turn off female differentiation while higher doses of the X chromosome gene product(s) would do the opposite. In the second mechanism, it is the autosomes that produce something which is titrated by the X chromosomes; in this case, a low X:A ratio would produce a relative saturation of X chromosomes by these autosomal products and provoke male differentiation while a high ratio would titrate out these substances, triggering female differentiation.

One line of evidence somewhat favors the second alternative, the titration of autosomal products by X chromosome sites. By adding duplications of parts of the X to genotypes with a borderline X:A ratio (0.67 or slightly greater), one can shift the phenotype toward femaleness (Madl and Herman, 1979). From a comparison of the extent to which different duplications, containing different portions of the X, can shift sexual phenotype (from no shift to some to conversion to a normal hermaphrodite condition) in a 2X;3A genotypic background, it may be calculated that there are at least 20 critical loci or sites on the X chromosome involved in sensing or measuring the X:A ratio (Madl and Herman, 1979; Hodgkin, 1988).

Comparably, McCoubrey et al. (1988) obtained feminizing effects by injecting different gene clones derived from the X chromosome into the gonads of tetraploid hermaphrodites at a critical chromosomal threshold (males of 2X;3A and carrying the duplication mnDp8). The effects were found for several independent X chromosomal clones, and not for a number of autosomal DNA clones tested. Furthermore, the active region in one X chromosome clone was found to lie within a fragment consisting of 130 bp of intronic sequence plus 1 kb of exon. This finding suggests that a gene site rather than a product is crucial, the implication being that the Xs may be titrating one or more autosomal products. The results are noteworthy but caution is advisable in extrapolating to the normal sex determination mechanism; the signifi-

Table 6.2. X:A Ratio and Sex Determination in *Caenorhabditis elegans*

Sexual phenotype	X chromosomes	Autosomes	X:A ratio
Hermaphrodite	2X	2A	1
Hermaphrodite	3X	3A	1
Hermaphrodite	4X	4A	1
Hermaphrodite	3X	4A	0.75
Male	2X	3A	0.67
Male	2X	4A	0.5
Male	1X	2A	0.5

Source. Data summarized from Madl and Herman (1979).

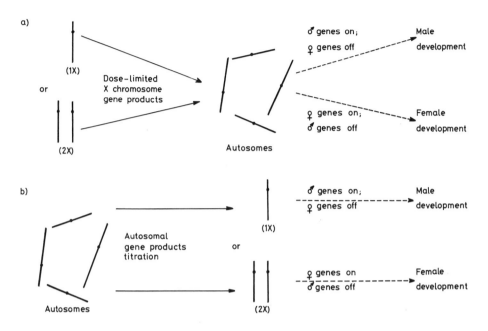

Fig. 6.18. Two modes of measuring the cellular X:A ratio. See text for discussion.

cance of results obtained with threshold X:A ratio genotypes in *Drosophila* have proved difficult to interpret (Laugé, 1980).

By whatever means the X chromosome participates in setting the X:A ratio, the identity of the A "denominator" elements has remained elusive. In principle, the relevant autosomal region(s) should also be detectable by a dosage effect. In XO animals, hemizygotes for such a locus should be shifted toward feminization, while XX animals carrying a duplication for the locus (and hence three doses) should be partially masculinized. Though much of the genome is now "covered" by duplications and deficiencies, no region has yet been found to have this effect (Hodgkin, 1987a).

The X:A ratio must also be involved in the setting of one other critically important process, that of dosage compensation, the equalization of X chromosome activities between the two sexes. In general, organisms are sensitive to large shifts in chromosome composition because such changes in gene dosage change the ratios of gene products to one another; if the "genetic imbalance" in chromosomal material is great enough, lethality is usually the result. The difference in X chromosome composition between the sexes might be expected to pose just such a problem in genetic imbalance. *Caenorhabditis elegans* has solved this problem through "dosage compensation," a mechanism that equalizes total X chromosome gene products between the two sexes by differential regulation of the expression of the X. mRNA levels for three genes located on the X were found to be the same in both sexes, suggesting that the equalization process is, as might be expected, achieved through transcriptional control (Meyer and Casson, 1986). Interestingly, however, not all X chromosome genes are dosage-compensated between the two sexes; Meyer and Casson (1986) found one gene that was not, showing twofold higher expression in XX than XO animals, and Hodgkin (1985) has reported that two sex-linked suppressor tRNA genes are not equalized between the two sexes. Thus, dosage compensation of X chromosome genes is probably the norm but is not required for every gene on the X. If zygotic X gene products play a role in setting the X:A ratio, then these genes must not be dosage-compensated. We will return to the phenomenon of dosage compensation after first looking at the genetics of sex determination.

Sex Determination: Genes and Mechanisms

The key to the analysis of sex determination has been the identification of genes which, when mutated, produce the "wrong" sexual phenotype for a given X chromosome constitution. (Historically, as we shall see in chapter 7, *Drosophila* work provided a precedent for this approach in the nematode.) Sex-transforming mutations in *C. elegans* fall into three categories: transformer (*tra*) mutations, which convert XX individuals into male or masculinized animals; hermaphroditization (*her*) mutants, which do the reverse, converting XO individuals into hermaphrodites; and feminization (*fem*) mutants, which cause both males and hermaphrodites to develop into full females (individuals which produce only eggs and no sperm).

Since the sexual transformations produced by many of these mutants produce individuals with a completely or nearly normal outward phenotype, some explanation about their isolation is required. The *tra* mutants were found as part of a search for individual hermaphrodites that produce a large number of male progeny. (Typically, as mentioned earlier, hermaphrodites produce about one male per 500 progeny, as the result of rare nondisjunction events.) This screening for overproducers of males yielded two kinds of mutants: *him* mutants, which give large numbers of normal (XO) male progeny through increased nondisjunction of the X chromosomes, and *tra* mutants, which produce masculinized XX individuals. The two kinds can be distinguished genetically: Hermaphrodites that are heterozygous for *tra* mutations (*tra*/+) produce 25% male (*tra*/*tra*) progeny through normal Mendelian segregation (Hodgkin and Brenner, 1977), while *him* heterozygotes produce the normal low numbers of males (*him* being recessive). The *him* and *tra* mutants can be distinguished further by tests that are based on the presence of two X chromosomes in the converted male progeny of the latter.

To isolate *her* mutants, which transform XO males to XO hermaphrodites, the expression properties of the *dpy-21* gene were employed. *dpy-21* is involved in dosage compensation (see below) and is one of four genes whose wild-type activity reduces X chromosome expression (Meneely and Wood, 1987). Mutants of *dpy-21* over-express X chromosome gene products with the consequence that homozygous hermaphrodites, possessing two Xs, are Dpy; homozygous males, however, carrying only a single X, are normal.

A *dpy-21* strain that contains the *him-5* mutation (which increases the production of XO individuals) normally segregates just Dpy (XX) hermaphrodites and wild-type (XO) males. Any mutation that produces hermaphroditization of XO animals should lead to the production of wild-type XO hermaphrodite progeny (readily distinguishable from wild-type males, under the dissecting microscope, by their very different tails). Using this screening procedure, Hodgkin (1980) isolated eight allelic, recessive *her-1* mutants.

Other methods have relied on obtaining mutants that reverse (suppress) the effects of previously isolated mutants in the set. The various mutants of the three *fem* genes were for the most part isolated in this fashion, as mutants that restore feminization to *tra* gene transformed XX individuals (Nelson et al., 1978; Doniach and Hodgkin, 1984; Hodgkin, 1986). For all of the approaches mentioned here, the majority of the mutants are hypomorphs or amorphs but, in addition, a handful of dominant hypermorphic mutants have been obtained; their sexual transformation effects are, in general, opposite to those of the corresponding null mutants.

The seven principal sex determination genes identified by these various approaches are listed in Table 6.3, along with the phenotypes of their loss-of-function mutants in both XX and XO animals. (Hypomorphic alleles for all of these genes produce less complete transformations.) All act zygotically, but for four genes, *tra-3* and the three *fem* genes, there is maternal expression which can compensate (to differing extents) for zygotic genome deficiencies of these genes. An interesting feature is that there are differing degrees of transformation of the soma and the germ line for null mutants of several of the genes. This partial autonomy of soma and germ line indicates that neither is dependent upon the other for specification of its own state and confirms the idea that sexual phenotype in the nematode does not involve dominating hormonal or other systemic influences. It also provides a hint that some aspects of sex determination of the germ line may differ from that of the soma, an idea that has been borne out by subsequent analysis.

Among the *tra* genes, whose mutants transform XX animals into phenotypic males, the most complete somatic transformations are obtained with *tra-1* null mutants (though the *tra-1* germ line transformation is only partial). In contrast, null alleles of *tra-2* and *tra-3* produce incomplete somatic transformation, yielding males with only partially transformed tail structures and no mating behavior. Furthermore, the most deficient *tra-3* alleles are the weakest of all, producing effects comparable to hypomorphic *tra-2* alleles. Several lines of evidence indicate that the *tra-3* product is required in only small amounts, supplied either maternally or zygotically, and it seems likely that the function of the *tra-3* gene product is to act as some kind of cofactor for the *tra-2* gene product (Hodgkin, 1980).

Table 6.3. Sex Determination Genes of *Caenorhabditis elegans* and Their Loss-of-Function Phenotypes

Genotype	Linkage group	XX phenotype	XY phenotype
Wild-type		Hermaphrodite	Male
isx-1	IV	Female (25°C)	Incomplete male (25°C)
tra-1	III	Male (with gonadal defects)	Male
tra-2	I	Incomplete male	Male
tra-3[a]	IV	Incomplete male	Male
her-1	V	Hermaphrodite	Hermaphrodite
her-2	III	Female	Female
fem-1	IV	Female	Female
fem-2	III	Female	Female
fem-3	IV	Female	Female

[a]*tra-3* also possesses a maternal effect.

The *her-1* gene, whose null mutants transform XO animals into fully functional hermaphrodites, appears to be required only for male development. *her-1* XX animals appear to be fully wild-type hermaphrodites in all respects. One dominant, apparently hypermorphic allele has been isolated and has the opposite effect to the null mutants: It masculinizes XX animals, with the effect being stronger, though not complete, in homozygotes (Trent et al., 1983).

The first *fem* mutant was a mutant in *fem-1* and was isolated fortuitously as a temperature-sensitive sterile; at the restrictive temperature, it produces spermless XX hermaphrodites (females, effectively) and converts XO individuals into partially feminized animals (Nelson et al., 1978). Other *fem-1* mutants, and subsequently isolated *fem-2* and *fem-2* mutants, have broadly similar effects, though these three genes differ in the extent to which maternal expression can rescue mutant defects (Doniach and Hodgkin, 1984; Hodgkin, 1986).

The development of a sexual phenotype is the result of two sequential processes, the initial sex determination decision and the execution of this decision in the differentiation of sex-specific structures. There are two grounds for classifying the mutants described above as sex determination mutants rather than sex differentiation mutants. The first is that for temperature-sensitive alleles of *tra-2* (Klass et al., 1976), *fem-1* (initially called *intersex-1* (isx-1) (Nelson et al., 1978), *her-1* (Hodgkin, 1984a), and *fem-2* (Kimble et al., 1984), the temperature-sensitive periods (TSPs) occur well in advance of visible sexual differentiation. The temporal separation between TSP and differentiation is evident, for instance, for the *tra-2* mutant, with the TSP ending approximately 12 hr after hatching. This is long before the male and hermaphrodite gonads can be histologically distinguished but concurrently with the first divisions of the gonadal somatic primordium and the establishment of the decisive difference in progenitor cell position between the two gonads. The second indication that the genes affect the sex determination decision is that the wild-type alleles of the *tra* genes and of *her-1* appear largely, if not entirely, dispensable in wild-type males and hermaphrodites, respectively.

A crucial question about this set of seven genes is whether they act independently of one another—which would indicate that there are many routes to changing the sex determination decision—or, alternatively, whether they act in some kind of sequence or pathway. As in other situations where this question arises, phenotypic characterization of double mutants can resolve this issue. If the gene products act independently of one another, then combining two mutations of opposite phenotypic effect in the same genome should either show highly deranged sexual development, in the pattern of neither mutant, or perhaps exhibit a canceling out of both mutant defects, in effect, a reversion to wild-type. On the other hand, if the gene products of the seven genes are part of a pathway, then there should be a clear set of epistatic relationships between different mutant pairs. The results show that the latter is the case (summarized in Hodgkin, 1980, 1988). There are distinct epistatic relationships, and these can, in fact, be ordered into two similar, though not identical, pathways of sex determination, for the soma and germ line, respectively.

Thus, for example, *tra-1; her-1* double mutants are phenotypic males in XX (as well as XX animals) instead of hermaphrodites; clearly, *tra-1* is epistatic to *her-1*, eliminating the latter's phenotypic effect. Similarly, animals that are doubly mutant for *tra-1* and either *tra-2* or *tra-3* show the complete transformation typical of *tra-1* XX animals, rather than the partial ones of *tra-2/tra-3*, showing that *tra-1* is epistatic to these other masculinizing mutants. When the somatic transformations are taken as a whole, the sequence can be diagramed as:

$$her\text{-}1 \longrightarrow \left.\begin{array}{l} tra\text{-}2 \\ tra\text{-}3 \end{array}\right\} \longrightarrow \left.\begin{array}{l} fem\text{-}1 \\ fem\text{-}2 \\ fem\text{-}3 \end{array}\right\} \longrightarrow tra\text{-}1$$

In this depiction, *her-1* is at the beginning of the sequence and *tra-1* is at the end (Hodgkin, 1980, 1987a) (those genes which apparently act at the same point being grouped vertically).

For the germ line tissues, nearly the same sequence is obtained, but there is one difference: The three *fem* genes are epistatic to *tra-1*. This sequence can be depicted as:

$$her\text{-}1 \longrightarrow \left.\begin{array}{l} tra\text{-}2 \\ tra\text{-}3 \end{array}\right\} \longrightarrow tra\text{-}1 \longrightarrow \left\{\begin{array}{l} fem\text{-}1 \\ fem\text{-}2 \\ fem\text{-}3 \end{array}\right.$$

Leaving aside for a moment the difference in *fem* genes/*tra-1* sequence between soma and germ line, the critical question concerns what the pathways mean in molecular and cellular terms. One conceivable interpretation is that they reflect a sequence of "substrate" conversions to a final sex-determining substance, required in one sex and whose absence promotes development of the other sex. Under this interpretation, earlier-acting genes would be epistatic to the later-acting ones and, if such were the case, the sequences should be written in the reverse direction. However, this hypothesis seems most improbable as it would seem to demand that successive intermediate compounds have opposite effects on sex determination and that each gene in the pathway should be essential in the sex whose development its mutants promote in the "wrong" chromosomal genotype. In fact, as we have seen, the *tra* genes and *her-1* are largely dispensable in the sex whose conversion they promote in the opposite sex. Furthermore, *tra-1* gain-of-function mutants are epistatic to *tra-2* loss-of-function mutants (Hodgkin, 1980), and this result is impossible to reconcile with the substrate conversion hypothesis.

The simplest interpretation of the two related patterns of epistasis is that they reflect sequences of regulatory switches, culminating, in each sex, in the determination of sexual phenotype of each affected cell (Hodgkin, 1980). From this viewpoint, an alteration in a later switch will always dominate—be epistatic to—an earlier step and the direction of the pathways is that shown in the above diagrams. Thus, from an initial pair of opposite settings of the first gene in such a regulatory cascade, in this case *her-1*, the switches are thrown into alternating opposite states in each sex, culminating in the final gene-controlled step being different in hermaphrodites and males.

Although this model is purely a formal one and does not specify the molecular nature of the switches, one can deduce, from the mutant phenotypes, which regulatory state corresponds to high activity and which to low, or zero activity, in which sex. Thus, from the fact that *her-1* null mutations are without effect in hermaphrodites, one may reasonably surmise that *her-1*—essential for male development—is "on" in males and "off," or in low activity, in hermaphrodites. Since *her-1* is at the beginning of the hierarchy and hence presumably at the beginning of the sequence, it seems likely that

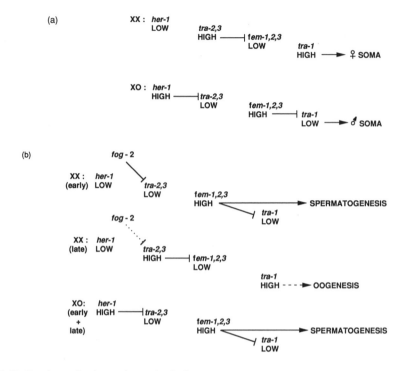

Fig. 6.19. Sex determination pathways in *C. elegans.* **a:** Somatic pathway. **b:** Germ line pathway. See text for discussion. (Adapted from Hodgkin, 1990.)

the X:A ratio sets *her-1* activity high and low, respectively, in males and females. In turn, high *her-1* activity in males represses *tra-2* and *tra-3* (which are not needed in males), while the absence of high *her-1* activity in females allows expression of these genes in hermaphodites (where their activities, as shown by the mutant phenotypes, are required).

At the last step in the pathway, in somatic tissues, *tra-1* is turned on in hermaphrodites and is off or in low activity in males. It is ultimately the state of *tra-1* activity that determines sex, according to this model (Hodgkin, 1980). In accordance with this view, a dominant hypermorphic *tra-1* allele (initially designated *her-2*), epistatic to *tra-2* and *tra-3* alleles in XX animals, has the reverse effect of *tra-1* amorphs, causing the feminization of XO animals (Hodgkin, 1980, 1987b). Furthermore, as might be expected for a key gene activity of this sort, its segregation alone can act as a determiner of sex. Thus, segregation from heterozygous *tra-1(o)/tra-1(d)* females, either XX or XO (where *tra-1(o)* is a null and *tra-1(d) 2(d)* the hypermorph), when these animals are mated to *tra-1* null homozygous males (either XX or XO) produces a self-perpetuating two-sex stock, consisting of males and females, in which X chromosome composition is irrelevant (Hodgkin, 1983b).

The model for sex determination in the soma is diagramed in Figure 6.19a, along with some additional features of the system that have been deduced from additional findings. In this scheme, the X:A ratio initiates the sequence, acting via one or more gene products (of genes to be described below), to produce a differential setting of *her-1* activity in hermaphrodites and males, which in turn trigger opposite states of activity in the cascade in the two sexes, culminating in two different states of *tra-1* activity (high in hermaphrodites, low in males).

The development of the germ line, however, has a special requirement: A transient capacity to make sperm demands a particular regulatory device in the germ line (Fig. 6.19b). Because hypermorphic *tra-1* activity can convert XX hermaphrodites to full females (Hodgkin, 1980, 1987b), it follows that *tra-1* needs to be temporarily turned off in the germ line to allow sperm formation. Such transient repression is, judging from the epistasis results, produced by transiently high levels of the *fem* gene activities, which would produce low *tra-1* activity in the germ line, favoring spermatogenesis. These elevated *fem* gene activities appear, in turn, to be produced by the temporary repression of *tra-2* activity, as inferred from the observation that dominant hypermorphic *tra-2* mutants eliminate spermatogenesis in XX animals and act, evidently, by preventing *fem* gene activity at the critical time for spermatogenesis (Doniach, 1986).

If *tra-2* is modulated in wild-type hermaphrodite development to allow *fem* gene activity, *its* activity must be modulated by one or more genes. One candidate gene, required only for spermatogenesis in the hermaphrodite germ line but not in males, *fog-2* (for feminization of the germ line), and whose mutants produce only oocytes in hermaphrodite bodies, has been described (Schedl and Kimble, 1988a). (In contrast to *fog-2*, mutants of the *fem* genes produce only oocytes in the germ line of both sexes.) Other *fog* genes have also been found and the complete regulatory network allowing transient sperm production in the hermaphrodite has yet to be worked out. Ultimately, much of the complexity of the regulatory cascade is probably involved with the need to coordinate somatic and germ line sexual development, involving a sequence of cell–cell interactions, in both sexes, but particularly with respect to the special needs of the hermaphrodite.

The final genes in the regulatory hierarchy, *tra-1* and the *fem* genes, must, in turn, regulate specific gene activities that are, ultimately, the source of the visible sex-specific differences. Though less is known about such downstream gene functions, a few facts are clear. For instance, some of these genes are specific to the production of one gamete type or the other. Thus, of the six yolk (vitellogenin) genes that have been identified, five are on the X chromosome and one is autosomal, with the expression of all six restricted to the intestine of hermaphrodites (Heine and Blumenthal, 1986). With respect to sperm production, the major sperm proteins (MSPs) are encoded by a large gene family in a total of six clusters on chromosomes II and IV (Ward et al., 1988). It is a reasonable to suppose that the *fem* and *fog* genes regulate, directly or indirectly, the expression of the MSP genes. Beyond such gamete-specific genes, however, there must be differential regulation of various *lin* genes to produce the sex-specific lineage patterns from postembryonic blast cells possessed by both sexes.

Dosage Compensation and Its Relationship to Sex Determination

In addition to the obvious phenotypic differences between males, on the one hand, and hermaphrodites or females, on the other, there is the subtler, but nevertheless important difference in the way the two sexes regulate X chromosome expression. Its net result is the equalization of expression of most X chromosome genes between males and hermaphrodites, despite the twofold higher dosage of these genes in the latter. This is the phenomenon of dosage compensation alluded to earlier. The fact that such equalization takes place signifies that it must have functional importance. However, there is a stronger argument that dosage compensation has adaptive significance: Mutants exist that show altered expression of the X chromosome specifically, in one or both sexes, and show consequent developmental abnormalities.

The first of such mutants to be described were those in the *dpy-21 V* gene. This gene was designated as *dpy* because homozygous hermaphrodites display a Dpy phenotype. Intriguingly, however, male homozygotes are wild-type in phenotype. This difference, however, is not a function of sexual phenotype *per se* but of the difference in X chromosome constitution between the two sexes, since *tra-1*-transformed *dpy-21* XX animals, which are outwardly male, still display the Dpy phenotype (Hodgkin, 1980, 1983b). A further clue to the basis of the *dpy-21* defect is that it mimics one of the effects of X chromosome hyperploidy, the acquisition of a Dpy phenotype in 3X;2A animals of otherwise wild-type genotype; this defect evidently arises from the increase in X chromosome gene products relative to autosomal ones. That the *dpy-21* defects arise from an enhanced expression of X chromosome genes has been shown directly: When transcripts from four X chromosome genes were measured, elevated levels were found, relative to autosomal genes, for three genes of this set that are normally dosage-compensated, but not for a gene which is not (Meyer and Casson, 1986). However, this elevated expression was found in both XX and XO animals, though the effect is greater in the former (Meyer and Casson, 1986). These results imply that *dpy-21* is required for dosage compensation although its absence affects X chromosome gene expression in both sexes.

dpy-21 is not unique in its effects on X chromosome gene expression: A number of genes have subsequently been identified whose mutants cause a Dpy phenotype in XX animals (independently of phenotypic sex) and overexpression of X chromosome genes, while homozygous XO animals are relatively unaffected. Three of these genes are *dpy-26, dpy-27,* and *dpy-28* and are collectively designated the X-dependent *dpys* (DeLong et al., 1987; Meneely and Wood, 1987; Plenefisch et al., 1989). In contrast to *dpy-21* mutants, however, there is a strong effect on viability associated with the mutants: XX mutant homozygotes for these four genes exhibit very poor viability when they are derived from homozygous mothers, the rare escapers showing Dpy and Egl phenotypes (DeLong et al., 1987; Plenefisch et al., 1989). A further difference between these genes and *dpy-21* is that the latter is strictly zygotic in its action and requirements, while the other four genes show a strong maternal rescue effect on viability for animals derived from heterozygous hermaphrodite mothers. (These survivors, nevertheless, show the defining Dpy trait.)

The effects of these genes on X chromosome gene expression have been assayed in two ways. The first method is the more direct one: measuring transcript levels of dosage-compensated X chromosome genes in the homozygous XX and XO animals (normalizing to a reference autosomal gene transcript) (Meyer and Casson, 1986). The second method involves a genetic test, assaying for the suppression of hypomorphic X gene defects. The rationale for the latter is quite simple: If X chromosome genes are expressed at relatively greater levels in these mutants, then hypomorphs should show elevated expression as well and approach closer to the wild-type phenotype. A particularly sensitive test was carried out with *lin-14* hypomorphs (DeLong et al., 1987). This gene, it will be recalled, is on the X chromosome, and loss-of-function mutations in it cause lateral hypodermal seam cells to make alae precociously. Enhanced expression of *lin-14* can be scored in terms of suppression of this mutant defect, such suppression reducing the number of seam cells making alae precociously (DeLong et al., 1987). By these tests, X chromosome expression is elevated specifically, relative to wild-type, in the X-dependent *dpy*s in XX animals, but not in XO animals (DeLong et al., 1987; Plenefisch et al., 1989).

The simplest interpretation of the function of the wild-type gene products of the X-dependent *dpy*s is that they preferentially reduce X chromosome gene expression in XX animals specifically, reducing it to the level of that of XO animals, where presumably these genes are "off" or expressed at low levels. Nevertheless, they may also have slight effects in 1X (male) animals despite the absence of the Dpy phenotype (DeLong et al., 1987; Meneely and Wood, 1987).

Although these genes have generally similar phenotypic effects and expression characteristics, they also differ among themselves in various ways (reviewed in Meneely, 1990). A genetic test of this distinctness is that extra doses of wild-type alleles of one cannot substitute for missing activities of any of the others (Plenefisch et al., 1989). *dpy-21*, in particular, seems different on several counts, with its lack of dramatic effects on viability and the fact that it may be required in 1X animals to repress certain X chromosome genes (Meyer and Casson, 1986) and activate others (DeLong et al., 1987).

Dosage compensation in *C. elegans* would seem, therefore, to involve largely a reduction of total X chromosome activity in XX animals to that of XO animals. It is possible, however, that it also involves an element of selective activation of X chromosome gene expression in both sexes. To date, no gene has been identified whose function is to boost expression of X chromosome genes solely in XO animals. (The latter mechanism is the primary one of X chromosome equalization in *Drosophila*, as discussed in the next chapter.)

A general characteristic of the X-dependent *dpy* genes is that their mutants affect dosage compensation but not sexual phenotype in diploids. In contrast, the sex determination genes described earlier affect phenotypic sex in diploids without changing the dosage compensation level appropriate for the chromosomal sex (XX vs. XO) (Meyer and Casson, 1986). It might appear, therefore, that sex determination and dosage compensation are wholly independent processes. In fact, however, three genes have been identified whose mutants show *both* aberrant sex determination and dosage compensation.

The first of these to be characterized, initially on the basis of two hypomorphs, was *sdc-1 X*. The phenotype of XX animals homozygous for either of these *sdc-1* mutations, when derived from homozygous mothers, exhibits partial masculinization (a Tra phenotype), Egl defects, short bodies, and abnormally protruding vulvae. XO animals, in contrast, are phenotypically normal (Trent et al., 1983; Villeneuve and Meyer, 1987). Homozygotes derived from heterozygous mothers show a large degree of rescue, indicating significant maternal expression. The principal feature that differentiates these mutants from those in the sex determination pathway is that while their Tra (masculinized) phenotype can be suppressed by making the affected animals simultaneously homozygous for *her-1*, the Egl and short body phenotypes remain. The latter phenotypes apparently reflect a failure of normal dosage compensation; *sdc-1* XX animals, whether feminized by *her-1* or not, show elevated X transcript levels for genes on the X that are normally dosage-compensated (Villeneuve and

Meyer, 1987). XO *sdc-1* individuals, which do not display phenotypic abnormalities, show X chromosome gene expression typical of wild-type XO males.

Because the sexual transformation exhibited by these first *sdc-1* mutants is incomplete, the possibility existed that null mutants might show a more complete Tra phenotype. In fact, null alleles, identified as such in terms of suppressibility by amber suppressors, produce comparably incomplete masculinization (Villeneuve and Meyer, 1990). In contrast, strong lethals of the second gene in this group, *sdc-2 X*, produce XX lethality in the progeny of homozygotes, while weaker alleles produce more complete masculinization than the strongest *sdc-1* alleles (Nusbaum and Meyer, 1989). *sdc-2* is, like *sdc-1*, maternally expressed, but the extent of rescue by maternal expression of a + allele is less marked; *sdc-2* mutant effects can reflect purely zygotic deficiency for the gene. Indeed, the "stronger" effects associated with *sdc-2* mutants in general may indicate that *sdc-1* plays a helper or cofactor role with respect to *sdc-2*, similar to that of *tra-3* with respect to *tra-2* in the sex determination pathway. *sdc-1* has recently been characterized molecularly and found to encode a long (1202 amino acid) zinc-finger protein and is, apparently, therefore, a transcriptional regulator (Nonet and Meyer, 1991).

Although *sdc-1* and *sdc-2* are required for normal sex determination and dosage compensation in XX animals, the third gene, *xol-1* (where *xol* stands for *XO lethal*), is essential is 1X but not in 2X animals (Miller et al., 1988). *xol-1* is zygotic in expression and homozygous XO animals die as embryos or early L1 larvae, some of which show signs of feminization in the few sexually dimorphic traits that can be scored; homozygous XX (hermaphrodite) individuals are essentially wild-type. As with the *sdc* genes, the phenotypic effects are not a function of morphological sexual type but of X chromosome compensation. Measurements of X chromosome transcripts in XO *xol-1* dead embryos, relative to autosomal transcripts and those in XX animals, indicate that normally dosage-compensated genes are not expressed to full wild-type levels in XO animals; the finding suggests that XO *xol-1* embryos have diminished X chromosome gene expression, as if they experience hermaphrodite-specific dosage compensation. Consistent with this possibility is the finding that mutations in the X-dependent *dpy*s elevate X chromosome gene expression (as they would in XX animals) and simultaneously restore viability to XO animals (Miller et al., 1988). Nevertheless, mutations in the X-dependent *dpy*s do not suppress the feminization of XO animals but rather increase it. This effect is reminiscent of that produced by these mutants in threshold X:A ratio (2X;3A) genotypes and suggests that enhanced X chromosome expression can, at certain thresholds, trigger the female sex determination pathway, as if in response to a higher X:A ratio (Meneely, 1990). (This point will be discussed below.)

If *xol-1* acts in a genetic pathway that sets both sex type and dosage compensation, does it act before or after the *sdc* genes? The double mutant combinations, *xol-1 sdc-1* and *xol-1 sdc-2*, provide a clear answer: The *sdc* mutants are epistatic to *xol-1*—XO double mutants are fully normal males while XX animals show the Sdc phenotype (Miller et al., 1988). The result shows that *xol-1* acts "upstream" of *sdc-1* and *sdc-2*, which, in turn, act upstream of the two branches of the pathway that control sex determination and dosage compensation, respectively. The data as a whole can be summarized in a diagram which depicts the complete set of regulatory relationships (Fig. 6.20). As with the Hodgkin model of the sex determination genetic pathway, it should be emphasized that this is a formal model which does not speak to the

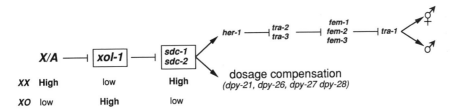

Fig. 6.20. Pathway for setting sex determination and dosage compensation. See text for discussion. (Reproduced from Miller et al., 1988, with permission of the publisher. © Cell Press.)

molecular nature of the regulatory events involved. However, the finding that *sdc-1* encodes a putative transcriptional regulator (Nonet and Meyer, 1991) supports the idea that transcriptional regulation plays a key part. The scheme, it should be noted, does not exclude the possibility that other genes and gene products are involved; indeed, the incompleteness of the transformations associated with *xol-1* indicate that other gene products are involved at the point marked by *xol-1*.

Some Questions

While the genetic investigations of the last 15 years have revealed much about the developmental pathway that determines sexual phenotype and sets dosage compensation levels in response to the X:A ratio, much remains to be elucidated.

A number of questions concern the first steps: What is the nature of the X:A signal? When does this signal act? How is this signal processed to set the appropriate levels of activity of *xol-1* (and other genes) at the start of the pathway?

The fact that there are, as yet, no clear answers to these questions may reflect a flaw in the questions themselves. The underlying assumption has been that there is *one* X:A signal and that this signal is assessed once and for all at a discrete time, thereby setting the pathway on one of either of two discrete courses. This appears to be the way that *Drosophila* does it (chapter 7) and would, indeed, be the simplest solution from an engineer's viewpoint. However, the evolution of biological processes often does not lead to the theoretical minimalist solution but to more elaborate outcomes, dictated by the (genetic and organismal) starting materials and the chance elements that often enter into a tinkering process (Jacob, 1977). With respect to sex determination in the nematode, an alternative to the once-and-for-all assessment of the X:A ratio is an early provisional assessment, which sets the first sexual divergences within the embryo, which, in turn, sets the stage for further assessments of the X:A ratio and subsequent divergences. This somewhat heterodox notion would allow for both sites on and gene products of the X chromosome to play a part in assessment of the X:A ratio (both models of Fig. 6.18).

Although this idea is also just a hypothesis, the feminizing effects of the X-dependent *dpy*s in 2X;3A genotypes, these genes which act comparatively late in the pathway (Fig. 6.20), are consistent with it (see discussions by Hodgkin, 1987b; Plenefisch et al., 1989; Meneely, 1990). While these effects are primarily maternal ones, there is also some zygotic genome contribution, presumably due to elevated expression of the X chromosome (see, for instance, Table 10 in Plenefisch et al., 1989). If comparatively late elevation of the numerator of the X:A ratio can shift sexual phenotype, then the notion of a once-and-for-all early assessment of this ratio will need to be reevaluated.

Secondly, there is the question of whether cells and regions perform the assessment, and subsequent steps, wholly independently or whether there are cooperative, interactive effects. The traditional view, derived from the phenotypic mosaicism of intersexual animals, is that there is strict autonomy (see, for instance, Hodgkin, 1988). There is, however, some contradictory evidence: When partially transformed *sdc-1* animals are scored with respect to both gonad and tail formation, statistical independence is not found (Villeneuve and Meyer, 1990). Similar findings have been reported in certain "borderline" X:A ratio situations, namely 2X;3A animals containing large X duplications (Schedin et al., 1991). These results suggest that the pathway does not operate in each cell/tissue independently but can be influenced by other cells.

If there should be a certain degree of nonautonomy in the sex determination pathway and, hence, cellular interactive events that affect sexual phenotype, it becomes of paramount interest to determine which, if any, of the identified genes in the pathway are responsible for such effects. Mosaic studies with *tra-1*, the final arbiter of somatic sexual characteristics, indicate that it does act autonomously within cells (Hunter and Wood, 1990). In contrast, other mosaic studies, employing a fusion Dp carrying *her-1+* and an autonomous nucleolar cell marker, show that *her-1*, the first autosomal gene in the sex determination hierarchy, acts nonautonomously, with effects that can spread between cells of the AB and P1 lineages (Hunter and Wood, 1992). These results indicate that the *her-1* gene product may mediate the repression of *tra-2* activity in males, indirectly or directly, by a diffusible factor.

Thirdly, turning to the various component pathways of sexual phenotype, as depicted in Figures 6.19 and 6.20, one would like to know more about the detailed mechanics of its operation; this will entail cloning and analysis of the individual regulatory genes and their products. Is most of the control at the transcriptional level? Does posttranscriptional regulation play a part? The fact that *sdc-1* has strong maternal expression yet the TSP maps to a defined segment within embryogenesis (Villeneuve and Meyer, 1990) suggests that there are *some* elements of posttranscriptional control, at least of the activity of this putative transcriptional regulator. Nevertheless, transcriptional control appears to be the predominant mode: Measurements of *her-1* transcript levels, with the cloned *her-1* gene, show sex-specific differences in these levels—high in males, low in hermaphodites—and *sdc-1* and *sdc-2* XX animals show elevated levels (Trent et al., 1991). These results accord with the idea that *her-1* is transcriptionally regulated, although the data are consistent with the possibility that it is the stability of its mRNA that is being affected.

Finally, one comes to the action of the terminal gene products in the pathway. How does *tra-1* promote hermaphrodite development (and sperm formation in males)—is it a direct transcriptional regulator of downstream sexual differentiation genes, for instance—and how do the X-dependent *dpy* genes reduce X chromosome expression in XX animals? One may anticipate that further molecular studies, in conjunction with additional genetic characterization, will help to resolve these fundamental questions concerning the cellular and molecular bases of the pathway of sexual development in the nematode.

DAUER LARVA DEVELOPMENT

The dauer larva constitutes a bypath of nematode development. It is a semidormant larval form, exhibiting greatly reduced metabolic activity and no feeding behavior, and is formed at the end of the second larval stage under conditions of starvation or overcrowding. Entry into the dauer larva state is triggered by response to the concentration of a low-molecular-weight, fatty acid-like "pheromone" that is produced throughout the life cycle. When the pheromone concentration is high relative to the food supply, as occurs under conditions of starvation or overcrowding, the animals begin dauer formation (Golden and Riddle, 1982). The sensing of the pheromone: food ratio signal is probably mediated by one or more of the sensilla of the head, the amphids, that border the mouth region (Riddle, 1988). The option of dauer development can be seen as an adaptive response favoring survival during a spell of adverse conditions.

The term "dauer larva" has a German origin and means "enduring larva." Lacking the normal pharyngeal pumping of medium that constitutes the nematodes' feeding behavior, the dauer larva does not feed and is less motile than the normal larval forms. It can persist for greatly extended periods at normal growth temperatures. In some parasitic nematodes, the dauer stage is an obligatory part of the life cycle, but in *Caenorhabditis* it is a developmental option, taken only under adverse conditions and available only at the end of the L2. At 25°C, formation of the dauer larva takes 11–12 hr following the L1 molt.

The structure of the dauer larva in *C. elegans* has been described by Cassada and Russell (1975). In form, it is slightly longer than the late L2 juvenile but only one-half as wide (Fig. 3.3). Consonant with the loss of feeding activity, the mouth becomes plugged, and in the fully formed dauer, there is no open space within the pharynx and intestine. Furthermore, these dormant larvae exhibit a higher density than the normal L2 stage, due to the shrinkage of the body, in particular the hypodermis, that takes place during the developmental sequence. The outer covering is a thickened cuticle, possessing a characteristic striated inner layer not seen in the normal cuticle. This layer is not uniformly thick but grades down and vanishes at the lateral edges of the animal, being replaced there by a fibrillar layer typical of the basal layer of the adult cuticle. In combination with the absence of pharyngeal pumping (and an occluded buccal opening), this thickened cuticle confers resistance to certain solubilizing agents added to the medium, in particular the detergent sodium dodecyl sulfate (SDS). Normal worms exposed to 1% SDS for a few minutes quickly lyse, but dauer larvae can survive such exposure for hours.

The normal life cycle is resumed upon exposure to a new food supply. In the presence of food, pharyngeal pumping begins after 1 to 2 hr and increases in rate over the next few hours; molting takes place after approximately 14 hr at 22°C. During the recovery period, animals swell to the normal width but show no detectable increase in length, presumably owing to the physical constraints of the cuticle. Bursting of the cuticle is followed by rapid growth, and after 6 hr, the normal length of L4 juveniles is attained. The recovery period thus replaces the normal L3 stage. Intriguingly, entry into or exit from the dauer state phenotypically circumvents some of the heterochronic mutant defects, restoring the match between cuticle phenotype and developmental stage in several mutants that either give reiterated juvenile cuticles or which normally skip the L2 stage (Ambros and Horvitz, 1984). Furthermore, transit through the dauer state can reduce the penetrance of several *lin* mutants that produce the Vul or Muv defects (Ferguson and Horvitz, 1985); the nature of this suppression is not understood.

Because dauer larva development is an optional stage in development, it is possible to isolate mutants that affect dauer formation yet are fully viable. Three classes of mutants might be predicted: those which enter the pathway "constitutively" (under conditions of abundant food, no overcrowding); those which cannot transform to dauers under the standard inducing conditions ("dauer defectives"); and those which cannot readily leave the dauer state ("reversal defectives"). Of the three categories, the first and third are potentially genetic lethals since animals that enter the dauer state but cannot leave it will not give rise to descendants. However, selection for hypomorphs or temperature-sensitive mutants facilitates recovery of these classes. All three categories of mutant have been found, though the distinctions between them are not absolute; many of the dauer constitutives, for instance, are slow or defective reversers. Furthermore, the selections necessarily enrich for mutants in genes whose products are preferentially or selectively employed in dauer formation. Mutants in genes required for both essential general functions and dauer formation would be less readily obtained.

The strong SDS resistance of the dauer larva permits a selective method for isolating constitutives and reversal defectives (Cassada and Russell, 1975). To isolate dauer constitutives, one exposes well-fed F_2 progeny of a mutagenized population to 1% SDS for a suitable interval (10–20 min) and then plates for survivors. (In normally growing populations, dauers are present at a frequency of 10^{-6}.) Since a survivor will be picked up only if it can subsequently emerge from the dauer state and give rise to a clone, the procedure can be modified to select for temperature-sensitive constitutives. The worms are grown at 25°C prior to the SDS treatment and then, following SDS treatment, plated at the permissive temperature (15°C), which allows reversal and production of the clones from the survivors. An analogous approach, using SDS selection, can be used to isolate slow reversers. Dauer defectives, however, never achieve SDS resistance and can only be identified by direct inspection; they are identified as worms that have not been transformed into dauers upon exposure to normal inducing conditions.

The constitutives and defectives are, in a sense, opposite classes. The constitutives clearly possess the capability for forming dauer larvae; they err in switching into dauer development when it is not called for. The most plausible explanation of constitutive behavior is that sensory defects are involved, the animal being told by its nervous system that conditions are worse than they truly are. Each of the dauer defectives, on the other hand, is missing some essential component of the developmental equipment, either for initiating the pathway or for constructing the dauer once the pathway has been entered. For any given defective, the lesion might be part of the sensory apparatus, such that a high pheromone:food ratio cannot be sensed, or the defect might be any of the components required to build the dauer once the sensory signal is received, such as enzymes required for making the dauer cuticle. The defectives would be expected to be a broader class, embracing a larger variety of defects and involving more gene functions than the constitutives.

This expectation was borne out by an extensive search for dauer formation (*daf*) mutants (Riddle, 1977; Riddle et al., 1981). It was found that mutation to dauer defectiveness occurs at a high rate, about 5%, following standard EMS mutagenesis, while mutation to constitutivity occurs with a frequency of only 0.2%. This difference reflects the different numbers of genes that can mutate to produce these phenotypes. Dauer constitutivity can be produced by mutation in 13 known genes, a number that is probably close to the maximum that can mutate to this phenotype (as judged from the frequency with which multiple alleles of the same genes have been found). In contrast, dauer defectiveness can probably be produced by mutation in any of about 100 genes (a figure based on the overall mutation rate to this phenotype divided by the standard mutation rate for single gene inactivation). Both kinds of *daf* genes are found on all linkage groups and show no obvious clustering. All of the *daf* mutants are recessive, suggesting that each mutant defect results from a deficiency of the corresponding wild-type gene product.

That functional relationships exist between dauer constitutives and dauer defectives is shown by the analysis of "revertants" of dauer constitutives, obtained by plating the temperature-sensitive constitutives at the restrictive temperature and picking survivors. The great majority of these strains are found to have a mutation in a second gene, and isolation of these second site mutations in homozygous form reveals that most produce a dauer-defective phenotype. Furthermore, a number of these new defective mutants are found by complementation to be in genes previously identified as *daf* defectives on the basis of direct isolation of this type. Thus, some of the *daf* defectives are epistatic to some of the *daf* constitutives, showing that the two mutant classes affect a common sequence of events in dauer formation.

The demonstration that the constitutives and defectives define some form of common pathway permits the delineation of that pathway, in terms of the genes involved, by a series of epistasis tests. As noted above, the simplest explanation of the constitutive response is that it involves a sensory defect, the generation of a false sensing signal in a particular cell type that thereby initiates dauer construction. Therefore, if a particular *daf*-defective mutation were found to be epistatic to the constitutive, it must act after that step, either by blocking transmission of the false signal or some other subsequent step. Conversely, if the constitutive is epistatic to a given defective, its signal must come after the *daf*-defective step or be independent of it.

When the dauer development phenotypes of the various double mutants are compared, it is indeed found that particular dauer defectives are epistatic to particular dauer constitutives while other constitutives are, in turn, epistatic to these defectives. The results as a whole can be used to construct a sequence of dauer development, in which particular steps are blocked by particular dauer defectives (Riddle et al., 1981; Swanson and Riddle, 1981).

One scheme is shown in Figure 6.21, in which the intermediate steps of the pathway are defined by the *daf* constitutives arrayed in two parallel paths; the points of blockade by the *daf* defectives are indicated by the dashed vertical lines. In this part of the pathway, which is interpreted as involving sensory processing steps leading to dauer development, the constitutives can be arranged in a linear sequence, with the exception of *daf-2*. *daf-11* and *daf-8* are suppressed by the greatest number of *daf* defectives and therefore mark the earliest constitutive steps of the pathway; *daf-4*, in contrast, susceptible to suppression by the fewest defectives, defines the end point of the sensory processing part of the pathway. That the constitutives are, in some sense, "derepressed" for the normal sensory processing part of the pathway is indicated by the hyperresponsiveness of these strains to pheromone concentration at the permis-

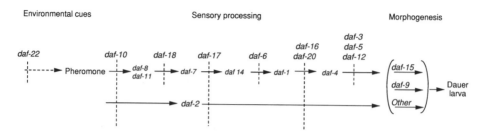

Fig. 6.21. Genetic pathway for initiation of dauer larva development by sensory signaling. See text for description. (Adapted from Riddle, 1988.)

sive temperature (Golden and Riddle, 1984). In contrast, the inability of any of the defectives to suppress *daf-2* suggests that mutations in this gene generate a novel signal that is processed independently of the normal sequence. Some independent evidence suggests that this is the case; mutants of *daf-2* are not hyperresponsive to pheromone, in contrast to the other constitutives (Golden and Riddle, 1984).

As might be expected from this interpretation, many of the *daf* defectives in the putative sensory processing part of the pathway have neuronal defects and altered chemotactic behavior. Chemotactic behavior is mediated in part by special head neurons, the amphidial neurons, and in mutants of two of the *daf*-defective genes, *daf-6* and *daf-10*, these neurons are ultrastructurally abnormal (Albert et al., 1981). Such abnormalities, and others not visible as ultrastructural defects, presumably interfere with the initial reception of the pheromone:food ratio signal or the processing of this signal.

There is beginning to be some biochemical and molecular characterization of other steps of the pathway as well. The initial event may be the synthesis of the pheromone, since without this substance, dauer development never begins in the wild-type animal. Synthesis of the pheromone requires *daf-22*, a dauer defective that makes undetectable amounts of the substance (Golden and Riddle, 1985). At the other end of the sequence, the final steps of dauer formation involve the events of morphogenesis. Most of the estimated 100 genes that can mutate to dauer defectiveness are presumably in this category. A mutant of *daf-13*, for instance, makes a dauer larva that looks normal but which is sensitive to SDS. This mutant is probably altered in some property of the cuticle which is essential for SDS resistance. When placed in a constitutive background, this *daf-13* mutation does not alter or suppress the constitutivity but causes the production of SDS-sensitive dauer larvae (Albert and Riddle, unpublished observations; cited in Riddle, 1988).

Between signal reception and morphogenesis, there must be steps of transduction and transmission of the signal, involving cellular interactions. Molecular characterization of one of the *daf* constitutives, *daf-1*, suggests that this gene may be involved in such events. The gene has been cloned and conceptual translation of the sequence indicates it to be an integral membrane protein with serine and threonine kinase activity in its cytoplasmic domain (Georgi et al., 1990). The wild-type protein might either act to suppress a false signal that was otherwise generated or might be, like *lin-12*, involved in a cell fate setting event.

The final steps of the dauer developmental pathway are those of dauer reversal. There seem to be few gene functions that are exclusively devoted to this. Selection for slow reversers yields dauer constitutives and hypomorphic dauer defectives that are affected in both entry and exit from the dauer state. From an analysis of the reversal defectives, it appears that the genes required for reversal are a subset of those required for entry (Riddle et al., 1981).

It will be apparent that many questions still remain about both the sensory and morphogenetic steps of the pathway. It should also be pointed out that the phenotypic categorization of mutants affected in dauer development is not always unambiguous and that the analysis of double mutant classes is often indirect. Such complexities can lead to alternative interpretations, and an alternative pathway to the one shown in Fig. 6.21 has, in fact, been presented by Vowels and Thomas (1992). An intriguing tangential set of issues, as noted earlier, relates to the fact that passage through the dauer state can suppress some of the heterochronic and *lin* mutant defects (Ambros and Horvitz, 1984; Ferguson and Horvitz, 1985).

CELLULAR EVENTS AND GENETIC "CONTROLS": AN OVERVIEW OF THE ISSUES

The singular and impressive feature of development in *C. elegans* is the extreme determinacy of somatic cell lineages and fates. If a cell has been identified with respect to the history of its antecedents within the embryo, its next division and the ultimate developmental outcomes of its descendants can be predicted (although, in wild-type development there are a number of instances where neighboring cells have a "choice" between alternative fates). Each somatic cell division event exhibits both characteristic cell division properties (the division plane of each mother cell and the relative sizes of the daughters) and characteristic developmental assignments of the resulting daughters. These patterns pose clear questions about both the cellular controls that produce the fixed sequence of cell division events and the relationship between assignment of cell fate and cell division.

Despite the complexity of the full set of embryonic and postembryonic cell lineages, each separate lineage, beginning with either an embryonic or postembryonic blast cell, can be regarded as a sequence of simpler, component sublineage "routines," modified in various ways (Chalfie et al., 1981; Kimble, 1981b). Fundamentally, each cell division event can be viewed as consisting of either of two kinds of cell division: symmetrical (proliferative) divisions and asymmetrical (difference-generating) divisions. The former correspond to the simpler cellular command "reproduce yourself without change," whereas the latter are the source of new cell types in development. While proliferative divisions almost always produce like-size daughters, difference-generating divisions are usually stem cell division events, in which a cell gives rise to daughters of unequal size.

Because each lineage (originating from either an embryonic or postembryonic blast cell) can be written as a tree of such cell division events, each division being of one type or the other, the fundamental "decision" between the two division types may involve a fairly simple mechanism (though one whose nature we do not yet understand). The choice between symmetric and asymmetric divisions might be activated by the same cellular switch at each division but be initiated by a large variety of different gene-encoded factors. The particular outcome in each case is probably dependent on a balance between various intrinsic, cell autonomous factors (such as the particular complement of transcription factors present) and of extrinsic factors (such as the particular contacts made with neighboring cells or the extracellular matrix or diffusible factors).

The importance of intrinsic factors is illustrated by the effects of mutations such as *unc-86*, which produce reiterations of certain basic cell division and fate patterns (Fig. 6.9) in certain specific lineages, while the role of extrinsic factors can be seen when removal of a cell's neighbors produces a dramatic effect on its cell division pattern. An example of the latter is the conversion of a relatively complex cell lineage pattern of the spermathecal precursor cell, when some of its neighbors are removed, to a reiterated stem cell division pattern (Kimble, 1981a). The fact that complete left–right symmetry reversal, with a complete set of corresponding left–right lineage patterns being reversed (Wood, 1991), can be produced by an early physical shift of a few cells suggests that the universe of cell extrinsic factors is crucial in determining the sets of intrinsic cell fate determining factors which arise during development.

At the purely cellular level, the relationships between the geometrical and directional characteristics of cell division, on the one hand, and cell fate, on the other, are highlighted by the kinds of alterations in asymmetrical division patterns seen after laser ablation of cells. These can be categorized as follows: 1) an alteration of cell *type* in a particular division; 2) reversal of *polarity* of a division, with consequent reversal of the lineage; 3) increases or decreases in cell *number* (e.g., insertion of a proliferative division); and 4) divisions that *duplicate* a sublineage through duplication of the precursor cell (Kimble, 1981b). Furthermore, all of these changes can be produced by various mutations within particular lineages of both embryonic and postembryonic development. The fact that rather specific responses can be produced in response to ablation of cell neighbors suggests that the responding cells have a limited repertoire of potentials and that fairly nonspecific signals, or the absence of specific signals that would normally originate from the ablated cell(s), can tip the balance toward one potential in preference to the others.

While the precise cause–effect relationships between a particular kind of cell division event and the activation of a particular cell fate still elude us, it is possible to imagine, from the characterized known switch genes, the kinds of regulatory events that might generate the diversity of cell fates. The traditional form of explanation envisages a key set of transcriptional regulatory genes, each activated by one or more specific signals and each activating (or repressing) a particular set of genes. While such schema have been mooted for at least 30 years (see, for instance, Monod and Jacob, 1962), the results of gene cloning experiments in the last five years have provided increasing support for this idea. We have seen, for instance, that such genes as *unc-86, mab-5,* and *lin-14* are all candidates for such putative transcriptional regulators. On the other hand, it is equally clear that other key "switch genes," such as *lin-12* and *glp-1*, exert their *effects* at the cell membrane, mediating the reception of signals there. (The ultimate effects of such signals may, and probably, do include transcriptional switches in many cases.) The clearest operational criterion for a switch gene is that its loss-of-function and gain-of-function mutational phenotypes should produce opposite effects; by this criterion, both *tra-1* and *lin-12*, whose products almost certainly act in different fashions, are both switch genes.

Much has been accomplished in nematode developmental genetics in the last quarter century; however, much remains to be done. The elucidation of the general features of the regulatory architecture that underlies development and its relationship to basic cell division events persists as a major aim of research on the development of this animal. A second, and no less important goal is the elucidation, at the molecular level, of the specific intercellular interactions and signaling systems which are so crucial to so many of the developmental events seen in *C. elegans.*

Though the findings reviewed in this chapter have concerned principally diverse features of postembryonic development, this emphasis has been chosen almost by default, because the detailed analysis of embryonic development in *C. elegans* by genetic means has hardly begun. Given that it is in embryogenesis that the basic body plan of the animal develops, and in which key events of morphogenesis take place, this is a major area for future study. One may confidently anticipate that as zygotically acting embryonic mutants are isolated and characterized, both at the cellular level, by the newly developing forms of microscopy, and at the molecular level, new features of development in the nematode will be revealed and our current understanding of the cellular and molecular events in postembryonic development will be deepened.

SEVEN

DROSOPHILA MELANOGASTER

From Blastoderm to Imago

Development is, of course, the orderly development of pattern, and therefore, after all, genes must control pattern....Most of our knowledge of pattern formation comes from experimental embryology, which is the science of analysis of pattern formation. We must now try to find out whether genetics has furnished material that permits an attack upon the problem of pattern in terms of gene action.

R. Goldschmidt (1938)

INTRODUCTION

Following the cycles of nuclear cleavage divisions and the cellularization of the surface layer of nuclei, the *Drosophila* blastoderm consists of a single layer of approximately 6000 somatic cells surrounding a yolky center, which contains a few hundred internal "yolk nuclei"; at the posterior end of the embryo, there is a distinct "polar bud" consisting of about 50 pole cells. Although the 6000 columnar somatic cells show little ultrastructural differentiation from one another, the entire somatic blastoderm cell layer comprises a mosaic of cells possessing widely differing developmental fates and capabilities. These differences originate in the actions of the maternally specified patterning systems—the anterior, posterior, terminal, and dorsoventral (d-v) gene systems—and become rapidly fixed through subsequent processes into an array of distinct determinative states (chapter 4).

In this chapter we examine the mechanisms by which the differentially distributed maternal cues are translated into the visible differences of regional and cellular phenotype apparent in the early embryo. For each of the maternal systems, the process involves the conversion of a differential spatial distribution of the key maternal substance into a differential activation of zygotic gene activities. These activities are one of the principal foci of this chapter: How were these genes identified? What are their products? How do they act? These differential patterns lay the foundations not only for the embryo but, ultimately, for the imago itself.

As background to this material, we will begin with a descriptive overview of the developmental sequence, from blastoderm to imago, and briefly review the general patterns of gene expression that accompany *Drosophila* development. We will then proceed to the genetic and molecular analyses of embryonic development.

The second subject of this chapter is the development of the imago and the genetic and molecular analyses used to explore its component developmental events. Although the formation of the embryo and the imago take place in ways that are, outwardly, markedly different, there are many connections and similarities at the genetic level, and these will be discussed. The chapter concludes with a review of what is known about the distinctive events of sex determination in *Drosophila*.

EMBRYOGENESIS AND LARVAL DEVELOPMENT

The first stage of *Drosophila* development, like that of *Caenorhabditis*, is embryogenesis, which takes place within the eggshell. It consists of the early syncytial phase, which produces the cellular blastoderm, and the postsyncytial or cellular phase, in which the cellular monolayer of the cellular blastoderm is converted into the segmented, differentiated first instar larva; the latter emerges from the egg case only 20 hr after blastoderm formation. Topographically, the problem consists of transforming a hollow ball of cells into a series of tubes within tubes. As in *Caenorhabditis*, the first steps involve multiple invaginations that remove cells from the surface and place them in characteristic positions within the embryo. Each of these invaginations alters the shape of the embryo and contributes to the formation of the three germ layers and those structures that arise within them.

The first invagination occurs almost immediately after blastoderm formation, at about 3.5 hr of development at 25°C, and takes place along the midventral line (Fig. 7.1a); the invaginated cells give rise to the mesodermal tissue of the embryo and larva. Beginning about one-sixth of the way from the anterior end and extending nearly to the posterior end, a block of cells about 50 cells long and 20 wide, approximately one-sixth of the total surface, moves inward to form a hollow tube just inside the (new) midventral line. The sides of the tube soon become apposed, obliterating the lumen, and the closed tube flattens into a layer that gradually spreads laterally and dorsally on both sides of the embryo. This process of mesodermal spreading is not complete until the 11th or 12th hr of development. By the 7th hr, the mesoderm has become divided into an inner layer, the splanchnopleure, which eventually gives rise to the visceral musculature surrounding the gut, and an outer layer, the somatopleure, which gives rise to all the other mesodermal structures of the larva (the body musculature, the circulatory system, and the larval fat bodies); it is also the source of a set of special precursor cells for muscles of the imago (Bate et al., 1991; Broadie and Bate, 1991). The midventral invagination thus lays the foundation for the development of one set of longitudinal tubular elements within the developing embryo and, ultimately, for the muscles of the adult.

The nervous system constitutes a second set of longitudinally distributed elements in the body. It arises from neuroblasts, detectable by their large size, flask-like shape, and basophilic cytoplasm. First observed in the future brain region, within the dorsolateral blastodermal layer, they become apparent along either side of the midventral invagination, within an area termed the neurogenic ectoderm, an hour after blastoderm formation. Comprising about 25% of the cells in the neurogenic ectoderm, these progenitor cells always divide asymmetrically and perpendicularly

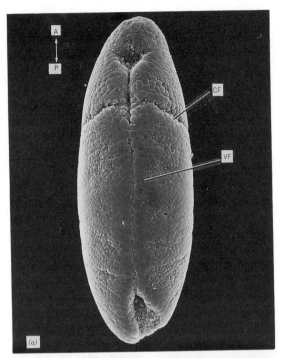

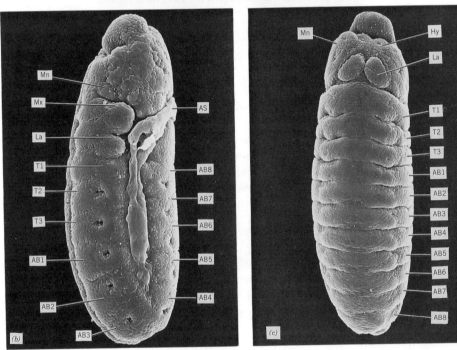

Fig. 7.1. Three stages in *Drosophila* embryogenesis. **a:** Formation of the ventral furrow. **b:** Extended germ band stage. **c:** Germ band shortening. AMG, anterior midgut invagination; CF, cephalic furrow; VF, ventral furrow; A, anterior; P, posterior; Hy, hypopharynx; Mn, mandibular segment; Mx, maxillary segment; La, labial segment; T1–T3, thoracic segments; AB1–AB8, abdominal segments; AS, amnioserosa. (Photographs courtesy of by Dr. F.R. Turner.)

to the outer surface, producing a large (outer) neuroblast and a smaller preganglion cell. The latter undergo further divisions to form the neurons of the larva; each preganglion cell ultimately generates about 18 ganglion cells during embryogenesis. Nerve fibers subsequently grow out from these ganglia, the core of the central nervous system (CNS), to generate connections with the peripheral nervous system (PNS) of the animal; this process begins in the 10th hr of development. The spatial relations of the blastodermal cell precursors of the mesoderm and body nervous system can be seen in the fate map of the embryo (Fig. 4.10).

The third major set of longitudinal elements consists of those of the intestine and gut, which arise from a number of invaginations, occurring at several distinct locations. The primary component of the intestinal system is the midgut, which is derived from two invaginations near the opposite ends of the midventral furrow, the anterior midgut (AMG) arising just anterior to the central furrow and the posterior midgut (PMG) from an invagination posterior to it. A third very slight inpocketing, immediately posterior to that of the PMG, gives rise to the terminal portion of the intestine, the hindgut and the proctodaeum. The mouth and foregut arise from a separate invagination, the stomodaeal invagination, just anterior to that of the AMG. Originating at nearly opposite ends of the embryo, the AMG and PMG grow together to form the midgut intestinal structure. The stitching together of these two cell groups occurs as part of a complicated morphogenetic movement that involves the whole surface of the embryo, in the course of which the definitive segmented form of the larva emerges.

The visibly segmented region of the embryo is termed the "germ band," and the first morphogenetic movement in its formation is an anterior movement initiated by the region of the PMG invagination; this phase is known as germ band elongation. The PMG first moves dorsally, scooping up the pole cells (Fig. 4.25a), and then anteriorly, eventually covering nearly two-thirds of the forward distance and reaching almost to the cephalic furrow. (The latter, whose position, as we have seen, serves as an indicator of the amount of *bcd* activity, demarcates the future head region, which arises in the anterior third of the embryo.) The dorsal and anterior movement of the PMG involves both the ectoderm and the underlying mesoderm. The forward extension of the germ band takes place between 4 and 6 hr of development and is followed by the formation of visible segments which persist through the remainder of development.

The first internal stages of segmentation involve the development of segmental ganglia within the ventral neural cord and the segmentation of the body muscles. Externally, between the 6th and 7th hr of development, regularly spaced clusters of large spherical cells appear. This is followed by the development of the tracheal pits, one on each side of every segment. Distinct segmental folds develop between 7 and 8 hr. The disposition of the larval segments and the appearance of the tracheal invaginations in the fully extended germ band, 8 hr embryo are shown in Figure 7.1b. The internal branches of the T-shaped tracheal invaginations soon join up and external pits disappear. The fully developed tracheal system, terminating in lateral pairs of anterior and posterior spiracles respectively at the two poles, forms the larval system of respiratory tubes.

Germ band lengthening is followed by the reverse movement, shortening. During germ band shortening, the germ band moves posteriorly and then ventrally around the caudal end. At completion, the terminal embryonic segment comes to lie at the posterior end of the animal, an arrangement of segments that is retained throughout the rest of development (Fig. 7.1c). During the process of germ band development, the

AMG and PMG extend, and as growth proceeds, the two regions come to enclose the internal, central yolk mass. The fusion of midgut extensions and the enclosure of the yolk are completed during the 11th hr of development and are quickly followed by enclosure of the gut by the visceral musculature. Externally, germ band shortening is succeeded by dorsal extension and closure of the ectodermal segments, a process that leads to the displacement and partial absorption of the amnioserosa membrane, the dorsal extraembryonic structure formed between the cephalic furrow and the PMG.

The last major morphogenetic change of embryogenesis is the involution of the head structures. The maxillary, mandibular, and labial segments of the larval head and the imaginal disc cells which give rise to the adult head structures are still external at the end of germ band shortening. These all move forward into the ventrally located stomodaeum. The consequence of head involution is that the first thoracic segment comes to occupy the anterior end of the embryo, preceded only by a region of external larval sense organs and the stomodaeum. This arrangement persists until the pupal stage. Head involution is completed between 11 and 12 hr and is followed by rapid cuticle deposition over the entire external surface.

The remaining 8 to 10 hr of embryonic development is largely devoted to the further differentiation of the internal organs and the external cuticular processes of the larva. The intermediate and final stages of development take place with very little additional cell division after 8 hr. The overall sequence of changes during embryogenesis has been described by Poulson (1950), Sonnenblick (1950), Fullilove et al. (1978), and most recently and extensively by Campos-Ortega and Hartenstein (1985).

Beyond the striking morphological and differentiative changes that occur in embryogenesis, which give rise to the larval body, are those involving the delimitation of the various groups of imaginal cells from the surrounding larval cells. The imaginal cells are distinguishable from the larval cells by their smaller size, diploid chromosomal constitution, and retention of cell division capacity. Larval cells, having lost their division capability, become progressively larger during development and develop polytene chromosomes, as described earlier. An additional difference between imaginal and larval cells is that the former show little or no tissue-specific differentiation during embryogenesis or throughout most of larval development. The imaginal disc cell clusters, and the groups of neuronal and muscle precursor cells, thus appear as islands of small, undifferentiated cells within a sea of larger, differentiated (larval) cells.

Where do the imaginal cells arise? Those that give rise to internal adult organs derive from precursor cells carried into the embryo by the first set of invaginations (although the majority of these internalized cells give rise to differentiated larval cells). However, most of the imaginal cell precursors, which will produce epidermal structures of the adult, are located on the surface of the embryo, having been set aside at characteristic positions within the sheet of embryonic cells that comprises the ectodermal surface. Imaginal disc cell clusters can be detected by their distinctive shape and by particular antibody stains as early as 9 hr of development (Bate and Martinez-Arias, 1991).

Emergence of the fully formed larva from the egg at approximately 23 hr marks the start of postembryonic development. Postembryonic development consists of three consecutive larval instars followed by a prolonged period of metamorphosis, in which the animal finally transforms into the imago. The first two larval stages each take about a day, while the third lasts nearly three. During larval development, the animal increases in size greatly, and many of the larval structures, such as the mouth hooks, trachea, and neural system, are further elaborated (Bodenstein, 1950). Accompanying these changes are continued division of the imaginal disc cells of the head and thoracic regions. In contrast, the ectodermal imaginal cells of the abdominal histoblast nests only begin their divisions at the beginning of metamorphosis (when the imaginal disc cells stop dividing), rapidly increasing in number during the first half of the pupal period.

Both the larval molts and the onset of metamorphosis are triggered by rises in the internal titer of the hormone ecdysone. Preceding metamorphosis, there is, in addition, a decline in juvenile hormone titer, at the end of the third larval instar. (For reviews of the endocrinology and the effects of these hormone changes on gene expression, respectively, see Riddiford [1985] and Ashburner [1990].) Metamorphosis itself is divisible into two broad stages. In the first, pupariation, the larval cuticle darkens and the larva shrinks away from the surrounding cuticle, secreting a new prepupal cuticle. The enclosing tanned larval cuticle is termed the puparium. At 12 hr after puparium formation, the developing prepupa begins the process of pupation proper. The larval mouth hooks and associated structures are ejected, the prepupal cuticle is shed, and the undifferentiated wing, leg, and haltere discs are everted.

The entire process of metamorphosis, lasting 4 days from pupation, is an extremely complicated one. Some larval organs are completely replaced by imaginal ones; these include the salivary glands, fat bodies, intestine, and larval muscles. Nests of imaginal cells within each of the larval organs undergo rapid divisions and replace the histolyzing larval cells. Other larval organs, including the Malpighian tubules (the excretory system), and the brain are retained, though with some "remodeling." Concomitantly, the external surface comes to be occupied by the unfolding and differentiating imaginal discs. During the final stages of the pupal period, the fully pigmented eyes and darkened wings can be seen through the puparium. At eclosion, the newly formed imago emerges through a slit in the puparium, the operculum, revealing a recognizable fly, though one with an extended, larval-like abdomen and folded wings. The mature adult shape, accompanied by some body darkening, takes form within the first 2 hr after emergence.

Gene Expression During Development: Molecular Surveys

An important question, from a molecular biologist's point of view, concerns the degree of molecular diversity that underlies, and drives, the complex pattern of phenotypic change. Are discrete stage-, tissue-, or cell type-specific diversifications produced, or accompanied, by either a small or large number of qualitative changes in gene expression? Or, are quantitative changes in gene expression equally important, involving particular thresholds for phenotypic change? Or, are both kinds of change involved and equally important?

Attempts to answer these questions have generally involved either surveys of protein synthetic capacities or the use of nucleic acid hybridization. The protein surveys have utilized either one-dimensional or two-dimensional separations of pulse-labeled polypeptides of various stages or imaginal discs. Although there are differences in details of approach and result, the general consensus is that the profile of synthesized polypeptides shows a high degree of similarity between different stages and imaginal discs (Rodgers and Shearn, 1977; Sakoyama and Okubo, 1981). In one detailed study of the changes in polypeptide protein synthetic profile during the first 8 hr of embryogenesis, Trumbly and Jarry (1983) found that the proteins synthesized by dorsalized embryos (from *dl* mothers) were nearly identical in pattern over this period, despite the absence of mesodermal and endodermal tissues from the mutant embryos. Thus, with respect to the abundant proteins at least, the early mesodermal cells do not synthesize a dramatically different set of proteins from those made in the other germ layers.

The large degree of constancy in protein synthetic pattern in *Drosophila* development, assayed by these techniques, is reminiscent of that seen in *Caenorhabditis* postembryonic development, where comparable studies reveal the same pattern of overall similarity (Johnson and Hirsh, 1979). Such apparent constancy, of course, may only reflect the fact that most proteins abundant enough to be scored on gels perform general cellular ("housekeeping") functions.

RNA hybridization experiments have inherently greater sensitivity than most of the protein separation methods and have also been used to explore general patterns of gene expression. There are several different hybridization procedures that can be used for measuring mRNA pool complexities. The most informative studies use DNA enriched specifically for the expressed genes. This is accomplished by means of the retroviral enzyme "reverse transcriptase," which copies RNA into DNA. By copying an isolated mRNA pool with this enzyme into complementary or cDNA, one effectively enriches for those gene sequences being expressed in the cells that manufacture the mRNA. By eliminating from the hybridization mixture the very large portion of DNA which is not expressed, the ratio of RNA to hybridizable DNA in the reannealing mix is substantially increased, thereby enhancing the sensitivity and accuracy of the measurements.

Estimates of expressed gene number in different *Drosophila* stages, from several studies, are summarized in Table 7.1. Most of the estimates of mRNA complexity pertain to the poly A$^+$ mRNA pool only, because this fraction is easy to isolate without heavy contamination from unprocessed nuclear transcripts (which lack the poly A tails) and are substantially more common. These results all give estimates of about 5000–7000 different mRNA species for all stages tested. However, the poly A-tailed mRNA pool may only account for a fraction of the total mRNA complexity. The measurements of total mRNA complexity shown in the table indicate a figure corresponding to 15,000–17,000 different species for the different stages tested. If the mRNA used in these experiments were substantially contaminated with nuclear RNA, these estimates would be spuriously high. However, control experiments described by Zimmerman et al. (1980) suggest that such contamination was not extensive in their experiments.

The most significant feature of the data is the high degree of shared sequence relatedness between the different stages. For the poly A$^+$ mRNA fraction, about 85–90% of the sequences are shared between embryos, larvae, pupae, and adults (Izquierdo and Bishop, 1979). For the total mRNA pool, 15,000–16,000 sequences are found in third instar larvae, pupae, and adults, with the greater proportion shared between these stages. Only a modest number, about 1600 sequences, were found exclusively in pupae and adults, and these sequences were all in the poly A-tailed pool.

It might be that this quantitative, general similarity in gene expression patterns between developmental stages masks important qualitative differences in composi-

Table 7.1. mRNA Complexities in *Drosophila*

References	mRNA fraction	Stage or structure	Number of sequences
Levy and McCarthy (1975)	Poly A$^+$	Schneider line 2	~6900
		Third instar larvae	7600–7900
Izquierdo and Bishop (1979)	Poly A$^+$	Whole embryos	3500
		Larvae	≥4900
		Pupae	~6800
		Adults	≥4900
Arthur et al. (1979)	Poly A$^+$	Embryos	~14,500
Zimmerman et al. (1980)	Total	Third instar larvae	13,000–17,400
	Poly A$^+$	Third instar larvae	~5400
Levy and Manning (1981)	Total	Third instar larvae	~15,000
	Poly A$^+$	Third instar larvae	~5,400
	Total	Pupae	~15,000
	Poly A$^+$	Pupae	~6600
	Total	Adults	~16,000
	Poly A$^+$	Pupae	~6100
	Total	Adult head	~11,000

tion between different cell types. However, one interesting result suggests otherwise. Levy and Manning (1981) measured the sequence complexity of isolated *Drosophila* heads, structures that are approximately 50% neural tissue and which lack numerous cell types found in the body (fat body cells, intestinal cells, gonadal cells, etc.), and obtained a figure of 11,700 different mRNAs. This number is approximately 70% of that determined for the entire adult body. The implication is that neural cells and probably many others will be found to have very high individual mRNA informational diversity. Indeed, complexity measurements on single vertebrate tissues reveal comparably high mRNA pool complexities (see chapter 8).

The *Drosophila* findings are puzzling and interesting in two respects. The first puzzle is that different cell types should share such a large number of expressed gene sequences, perhaps on the order of 10,000 to 15,000. The existence of such a large common pool prompts one to label all the shared sequences as housekeeping functions. Yet, if one defines housekeeping functions as those necessary for cell metabolism and reproduction, numbers greater than 2000 appear to be excessive. The bacterium *Escherichia coli*, for instance, is an organism that devotes itself solely to housekeeping—it has no known developmental program—and does very well with about 2000–3000 genes. Nor should eukaryotic cells, despite their greater complexity, be substantially more demanding. The estimated number of housekeeping functions in the sea urchin is 1000–1500 (Galau et al., 1976).

Furthermore, there is an independent genetic estimate of the number of cell-essential, housekeeping functions needed by *Drosophila*; the number is similarly small. Ripoll and Garcia-Bellido (1979) scored the survival capacity of cell clones made homozygous for individual genetic deficiencies of varying length. Any deficiency for an essential function that is cell autonomous, namely one required in the cell in which it is expressed, will be lethal when homozygous and hence unable to give rise to a surface ectodermal clone. From the distribution of chromosome bands whose deletion results in inability to form a clone, Ripoll and Garcia-Bellido estimated that only 12% of all

Drosophila bands are essential for cell viability and reproduction. With the number of essential genes in the fruit fly estimated at 6000–10,000 from genetic tests (see chapter 4), the number of essential cell autonomous functions would be 710–1000. Even taking 16,000 as the number of genes from the message complexity experiments, and assuming that essential cellular functions are distributed randomly among the 5000 different bands, a proportion of 12% still gives only 2000 cell-essential genes.

The biological function of potentially thousands of additional shared mRNA sequences thus poses something of a mystery. One possibility is that many are "fine-tuning" functions of some sort, which facilitate a developmental outcome, without being strictly essential for it (Williams and Newell, 1976). If so, does the nature of such fine-tuning consist of partial functional redundancy by many of these genes, with many collectively ensuring a particular property? Or, might many of the mRNAs represent diverged duplicates of various essential genes, the copies being sufficiently different to register as discrete mRNAs in hybridization experiments but retaining full functional equivalence? Or, are they transcripts of genes which either play defined developmental roles at discrete times and remain expressed, perhaps at reduced levels, at little or not cost to the organism? Such "sloppiness" might result from the inflexibilities of gene expression imposed by combinatorial systems of control (chapter 2). Though posing the questions does not provide the answers, simply listing them may be of some use in thinking about the phenomenon.

The second puzzle raised by the data concerns the basis of cellular phenotypic diversity. The assumption that has informed most thinking about eukaryotic differentiation during the last three decades is that cellular qualitative diversity reflects underlying qualitative differences in protein composition (Jacob and Monod, 1963). The *Drosophila* findings reviewed above, taken in conjunction with similar results in the mouse (chapter 8), suggest, on the other hand, that the qualitative patterns of gene expression between very different cell types are broadly similar.

If many of the mRNA sequences are without significant biological function, then the paradox disappears; a large number of qualitative, biologically significant differences could be hidden in the background of inessential shared mRNAs. However, if the greater part of the mRNAs are doing something useful for the cells that contain them, then qualitative cellular differences must spring either from a relatively small number of qualitatively different mRNAs between the different cell types or, primarily, from quantitative differences in gene expression or from a combination of the two sorts of difference.

Both kinds of regulation in gene expression occur during development: The mass hybridization experiments detect the former, while hybridization experiments with cloned genes reveal significant quantitative modulations for many individual genes (for a review, see Ashburner, 1989). It does not seem unlikely that many significant changes in cellular properties between successive cell generations during development are catalyzed by a small number of qualitative and quantitative changes in gene expression and that these cumulative changes in cellular phenotype over longer periods and/or numerous cell generations account for the visible and marked developmental changes we observe (Wilkins, 1984).

Such considerations serve to emphasize the importance of identifying the critical gene changes that catalyze particular developmental changes. Genetic analysis is crucial to such identification and, in the next section, we will examine how it has been applied to analyzing the development of pattern in the early *Drosophila* embryo. In this particular instance, where much of the sequence occurs rapidly while the embryo is still a syncytium, the *initial* changes triggered by the four different maternal instructional systems (chapter 4) involve relatively small numbers of qualitatively new gene functions, differentially distributed in space, whose activities set off further cascades of gene expression involving larger sets of genes.

EMBRYONIC PATTERN: GENETIC AND MOLECULAR FOUNDATIONS

Anteroposterior Patterning: Identifying and Classifying the Key Zygotic Gene Functions

The most immediately apparent feature of the *Drosophila* embryo, and one which first becomes visible at about 7 hr, is its segmental pattern (Fig. 7.1c). The processes that establish this pattern must be linked, if only indirectly, with the maternal patterning systems of the anteroposterior (a-p) axis discussed in chapter 4 (since mutations that affect those systems affect the segmental pattern), and the pattern itself is clearly a fundamental feature of the embryo's development. Not surprisingly, the analysis of the zygotic segmental functions has been a major preoccupation of *Drosophila* developmental genetics in recent years.

The principal repetitive feature of the segmental pattern is the spacing of segments. From the anterior end of the first thoracic segment (T) to the posterior end of the eighth abdominal segment (A8), the segment boundaries occur with well-spaced regularity. In addition, certain characteristic internal motifs recur. For instance, the ventral side of each segment is marked at its anterior edge by a band of small cuticular protuberances arranged in rows, the denticle hooks, and by a naked cuticle in the posterior portion, thus displaying a characteristic polarity (see, for instance, Fig. 4.18a). There is also a d-v differentiation; numerous fine hairs cover the dorsal side, spreading laterally to differing extents on the different segments.

The principal external features that distinguish segments from one another are the size of the denticle hook bands, the extent of the dorsal hair region, and the presence or absence of certain sensory organs. The broadest differentiation is that between thoracic and abdominal segments. The former possess relatively narrow denticle bands, each consisting of two to three rows, small ventral sensory structures (Keilin's organs), and small black lateral sensory organs. The prothoracic segment (T1) differs from the mesothoracic (T2) and the metathoracic segment (T3) in having an additional denticle band and pointier dorsal hairs, while T2 differs from T3 in showing a slightly wider denticle hook band. The abdominal segments differ from the thoracic in the greater width of their denticle belts and in the absence of sensory organs. The first abdominal segment (A1) appears intermediate between the thoracic and remaining abdominal segments in some respects (width of denticle band, spread of dorsal hairs) and the eighth (A8) is distinguished by the absence of a naked posterior ventral region. From the second through the seventh abdominal segments, there is a gradient of increasing width in the denticle band. The collective set of repeated and unique segment characteristics permits a clear assessment of the effects of pattern-perturbing mutations, whether global, regional, periodic, or localized to individual segments.

The first systematic search for mutants of zygotically acting genes required for the development of embryonic segmental pattern was carried out by Nüsslein-Volhard and Wieschaus (1980). Large numbers of recessive embryonic lethal mutations were isolated and the embryos of each line were screened individually on agar, using a stereomicroscope, for abnormalities in cuticular segment pattern. (For a detailed description of the embryo collection and screening procedures, see Wieschaus and Nüsslein-Volhard [1986].) Altogether, a total of 15 loci were identified in this first extensive search. Subsequent searches have approximately doubled the number, and the total set of genes identifiable in this way is probably close to saturation, at least for zygotic genes whose first major roles are in segmentation.

The first conclusion from the initial mutant phenotypes identified was that these could be ranged in three general categories: the "gap" genes, the "pair-rule" genes, and the "segment polarity" genes. Mutant embryos of the first group are deleted for large contiguous areas of the normal cuticular pattern, hence their designation as gap genes. In contrast, pair-rule mutants display a repetitive aberration throughout the germ band, the removal of integral, *alternate* segment-width areas. The third group, the segment polarity mutants, also display a repetitive deletion of pattern but, for these mutants, the pattern deletion occurs within *each* segment and is followed, for many though not all mutants in this group, by a partial mirror-image duplication of the part that remains, hence their name. All of the mutant phenotypes represent loss-of-function mutations. Representatives of each major group are depicted schematically in Figure 7.2.

Within each of the three categories, there is a distinct diversity of pattern defects, reflecting both locus-specific and allele-specific characteristics. The gap genes illustrate locus-specific differences particularly well. Thus, in *Kruppel* (*Kr*) embryos, all of the thoracic and much of the abdominal segments are missing, while in *knirps* (*kni*) embryos, the thoracic region is normal but nearly all of the abdominal segments are missing (Fig. 7.2). In complete contrast to *kni* embryos, homozygous *hunchback* (*hb*) embryos are deleted for the head segments, mesothorax, and metathorax while showing a normal abdominal segment region. Segment polarity mutants show a comparable diversity of phenotypes, as a function of locus, as will be described later.

Although the existence of three classes of zygotic segmentation mutants (and genes) is now taken for granted, it is worth emphasizing that these categories had not been expected. The most surprising result was the discovery of the pair-rule mutants. Their existence suggested that the process by which segmental pattern arises might be progressive in both time and space, with a double-segmental regional unit being an intermediary step in segment pattern development, as first suggested by Sander et al. (1980).

The discovery of the gap genes as a group was also something of a surprise. The existence of a set of genes, each of whose products is required for specific broad regions of the embryo's segment pattern, had not been predicted. Nor was the potential significance of the gap mutants much commented upon until the mid-1980s, when the three groups of maternal effect genes that are necessary for a-p patterning—the anterior, posterior, and terminal genes—were first identified and described. Comparison of several of the embryonic maternal mutant phenotypes with those of certain of the gap mutants revealed strong similarities: For instance, embryos from *bicoid* (*bcd*) mutants are similar to zygotic *hb* embryos, and embryos from mothers mutant for posterior group genes, such as *nanos* (*nos*), are similar to zygotic *kni* embryos

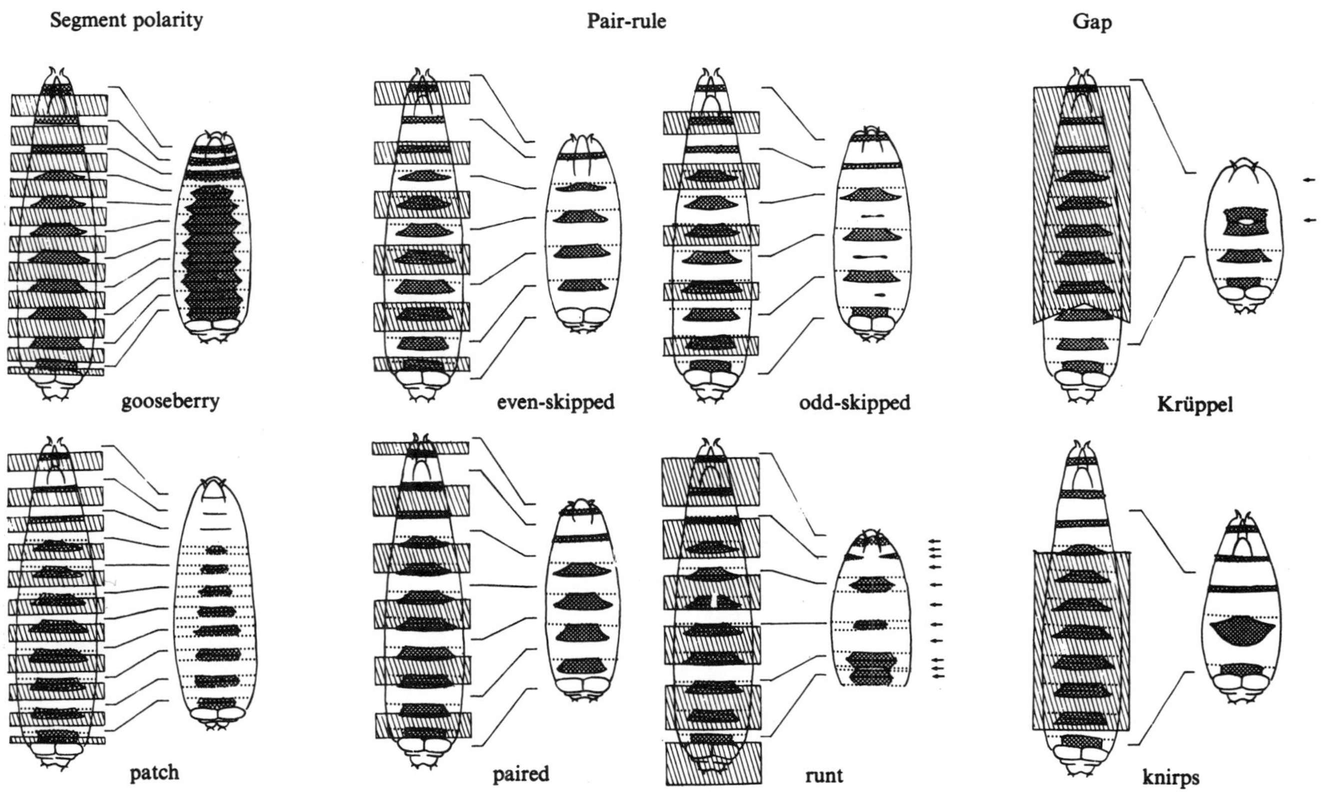

Fig. 7.2. Embryonic segment pattern defects: representative mutants. Dotted regions represent denticle bands, dotted lines show segmental boundaries, and hatched regions indicate the parts of the pattern that are missing in the mutants. Transverse lines link the corresponding regions in mutant and wild-type embryos, and the arrows, lines of polarity reversal. See text for discussion. (Reprinted by permission from *Nature*, vol. 287, p. 796. © 1980, Macmillan Journals Ltd.)

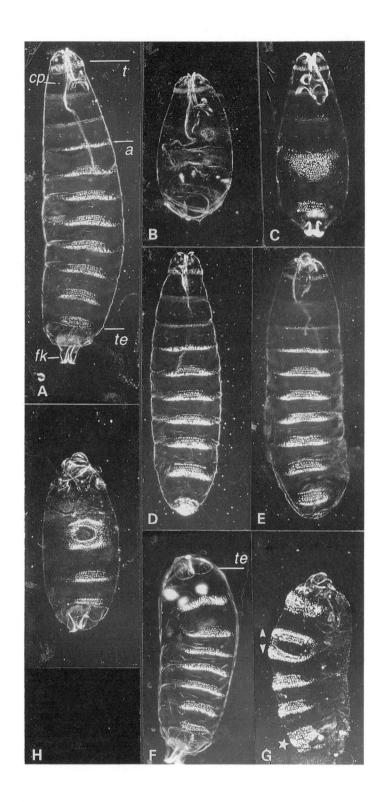

(Nüsslein-Volhard et al., 1987; Lehmann, 1988). Some of these comparisons are illustrated in Figure 7.3.

One way to explain these similarities is to posit that each of the maternal systems activates the expression of one or more specific gap genes; if such is the case, then mutational inactivation of the responding gap gene should produce a phenotype similar to that of the maternal one. For example, a chief role of *bcd* might be to activate zygotic *hb* activity in the anterior part of the embryo, while maternal terminal group activities might activate *tailless (tll)* Nüsslein-Volhard et al., 1987). That activating these particular genes could not be the sole functions of the anterior and posterior group genes was evident, however, from the fact that some differences in phenotype between the respective maternal and embryonic mutants are apparent (Lehmann, 1988).

If the gap genes are activated first, then they might in turn activate the genes whose mutant domains are the next largest in size, the pair-rule mutants (Meinhardt, 1986). The pair-rule genes, in turn, might then activate the genes whose mutant domains are the smallest, the segment polarity genes. In effect, under this hypothesis, development of segmental pattern entails a sequence of progressively finer regional specifications through a hierarchical sequence of gene expression controls. This scheme is diagramed in Figure 7.4.

More than a decade of studies following isolation of the zygotic segmentation mutants has provided support for this hypothesis. It has also become clear, however, that the development of segmental pattern in *Drosophila* entails many complexities that are not encompassed by it. In the following sections, we will examine these second-generation studies of the zygotic segmentation genes in some detail to see where these complexities lie.

Gap Genes and Their Interactions With the bcd Gradient and Each Other

Gap gene expression is now known to be very early, occurring in the first wave of zygotic transcription, well before blastoderm formation. For instance, *in situ* hybridization detects *Kr* transcription much earlier, beginning just after the 11th nuclear cleavage division (Knipple et al., 1985). Similarly, zygotic *hb* is also detected at this time (Tautz, 1988). Thus, both *Kr* and *hb* are transcribed from the zygotic genome during the first period of zygotic genome activation. (*hb* is one of the few gap genes to be maternally expressed as well, and the significance of its maternal transcripts has been discussed earlier.)

The domains of expression of six gap genes, as measured by *in situ* hybridization, are illustrated schematically in Figure 7.5, which also depicts the defective regions in embryos lacking these gene activities. It can be seen that there is only an approximate correlation between transcript domain and area of phenotypic defect. For *Kr* and *kni*,

---

Fig. 7.3. Comparison of zygotic gap gene versus maternal system defects, for loss-of-function mutations. **A:** Wild-type embryo from wild-type mother. **B:** Embryo from *nos* mother (a posterior maternal group gene). **C:** Embryo that is homozygous for a *kni* mutation. **D:** Embryo from mother homozygous for a *tor* mutation (a terminal maternal system gene). **E:** Embryo that is homozygous mutant for *tll*. **F:** Embryo derived from *bcd*-deficient mother (loss of anterior maternal system function). **G:** Embryo that is homozygous for an *hb* mutation (derived from a maternal germ line clone also lacking *hb*). **H:** Embryo that is homozygous for a *Kr* mutation. (Reproduced from Lehmann, 1988, with permission of the Company of Biologists, Ltd.)

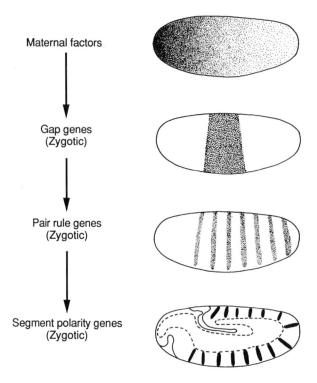

Maternal factors

Gap genes
(Zygotic)

Pair rule genes
(Zygotic)

Segment polarity genes
(Zygotic)

Fig. 7.4. Segmentation gene hierarchy in *Drosophila*.

the area of defectiveness is considerably broader than that of the expression domain; for the other genes, the expression domain is larger than the region of defect (indicating, in the latter case, that the activities are not of primary importance in certain parts of the expression domain).

Furthermore, for these genes the pattern of expression is a dynamic one and most are expressed in more than one area of the embryo. The patterns of change are shown in Figure 7.6 for five of these genes. For *hb* and *Kr*, at least, these two domains do not appear simultaneously; rather, the anterior domain appears slightly before the posterior one (Jäckle et al., 1986; Tautz et al., 1987). Further changes follow during gastrulation.

The inference, from mutant phenotypes, that the gap gene activities are regulated by the three a-p maternal gene systems is substantiated by molecular tests. Thus, in embryos whose mothers are deficient for either the anterior or posterior gene systems,

Fig. 7.5. Gap gene expression domains and regions of defectiveness in the respective mutants. The defects are indicated with respect to the larval segments affected but projected onto the blastoderm fate map. Black bars, regions where defects are shown; cross-hatched bars, early expression domains; open bars, late expression domains (many of which may not relate to initial gap gene action in segmentation). (Reproduced from Hülskamp and Tautz, 1991, with permission of the publisher © ICSU Press.)

Fig. 7.6. Early and late gap gene expression domains for five gap genes, as shown by *in situ* hybridization in whole mounts. (Reproduced from Hülskamp and Tautz, 1991, with permission of the publisher. © ICSU Press.)

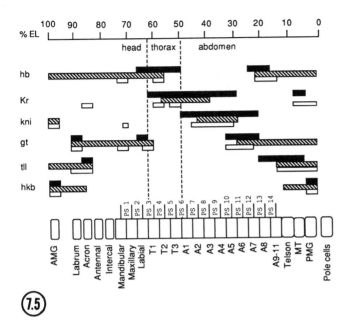

(7.5)

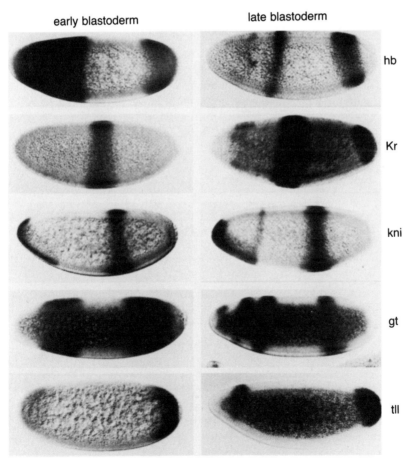

early blastoderm   late blastoderm

hb

Kr

kni

gt

tll

(7.6)

277

the sizes and relative positions of the initial gap gene transcript domains are significantly altered. For instance, while the anterior border of *Kr* transcripts in embryos from wild-type mothers is 54% EL (egg length, as measured from posterior end), in embryos from *bcd*⁻ mothers, this border is shifted to 69–77% EL (the precise position of the anterior border depending on the *bcd* allele used), while embryos from mothers deficient in the posterior gene group system (*osk, pum, vas, stau, tud*) show a pronounced posterior shift of the posterior *Kr* transcript border (from 39% in wild-type to 25–30% EL) (Gaul and Jäckle, 1987). Similarly, *hb* zygotic expression depends on the presence of maternal Bcd protein; in embryos from *bcd*⁻ mothers, the anterior domain of zygotically encoded *hb* protein does not appear, reflecting a failure of activation (Tautz, 1988; Driever et al., 1989). *Kr* and *hb* are, evidently, regulated by and, hence, "downstream" of the maternal genes that organize the anterior and posterior segmental regions.

Direct evidence for transcriptional control by Bcd protein of *hb* has been produced in *in vitro* experiments. Purified Bcd protein was incubated with fragments of the *hb* promoter and regions with binding sites were identified and then further characterized by "footprinting" experiments (in which the specific sequences that bind protein are protected from nuclease digestion). These initial experiments (Driever and Nüsslein-Volhard, 1989) identified three strong Bcd binding sites within 300 bp 5′ to the transcription start site.

In vivo tests utilizing "reporter genes" were then carried out. In these experiments, parts of the *hb* promoter are fused to a coding sequence whose product is easily analyzed (such as the bacterial chloramphenicol transacetylase, or CAT, protein), and these constructs are then used in P-element germ line transformation to create strains carrying the reporter gene and which are susceptible to those factors that normally regulate the *hb* promoter. The results showed that the binding sites identified in the footprinting experiments are indeed required for *hb* activation in *bcd*⁺ embryos. Furthermore, in embryos from *bcd*⁻ mothers, which would lack *bcd* activity, none of the engineered versions of the *hb* gene are activated (Driever and Nüsslein-Volhard, 1989).

Given the evidence of a Bcd protein concentration gradient in the embryo (Fig. 4.20), the crucial question is whether the *hb* gene can respond *differentially* to different concentrations of this protein. Employing *in vitro*-modified *hb* gene constructs, containing various amounts of 5′ flanking sequence, Driever et al. (1989) and Struhl et al. (1989) showed that the Bcd-dependent expression of *hb* is, indeed, a function of the number of strong and weak binding sites 5′ to the promoter and that *hb* can be activated at Bcd concentrations present in the anterior half of the embryo; the degree of response is a function of the number of binding sites and their relative strengths.

In addition to its activation effect on *hb* transcription, high concentrations of Bcd protein, as found in the anterior half of the embryo, evidently repress *Kr* expression, and presumably its transcription. However, the molecular details of this repression have not yet been worked out and, as described below, low concentrations may actually help to activate *Kr*, in conjunction with Hb protein (Hülskamp et al., 1990).

A comparable, though less direct, regulation of transcription pertains between the key gene of the maternal terminal system, *torso (tor)*, and certain zygotic gap genes for the nonsegmented terminal regions of the embryo. At the anterior end of the embryo, the phenotypic relationships between maternal terminal and zygotic gap genes are less obvious than the *bcd-hb* relationship because the morphological features at the anterior tip are a composite of the activities of the anterior and terminal systems (Nüsslein-Volhard et al., 1987). However, at the posterior end, matters are clearer.

In particular, the zygotic gene *tll*, which maps near the end of the right arm of the third chromosome (100A5–B2), is one of the principal gap genes activated by maternal *tor* activity. One indication is that the posterior phenotype of *tll* embryos (Fig. 7.3E) is broadly similar to that of embryos from *tor* mothers (Fig. 7.3D). More compellingly, the phenotype produced from maternal *tor* gain-of-function mutants (in which segmental pattern of the embryos is abolished) is suppressed by zygotic homozygosity for *tll* (Klingler et al., 1988; Strecker et al., 1989). Evidently, the maternal mutant effect operates through the zygotic *tll* gene, and activation of *tll* expression must be a major consequence of *tor* activity.

Yet *tll* is not the only zygotic gene activated by maternal *tor* activity. While maternal *tor* deficiency abolishes the PMG rudiment (the invagination associated with the PMG) and germ band elongation, *tll* embryos show a rudimentary PMG and exhibit germ band elongation (Strecker et al., 1986). However, deficiency for a second gene, *huckebein* (*hkb*), in combination with *tll,* produces the full *tor* phenotype (Casanova, 1990). If *tor* activity is strongest, in wild-type, at the extreme terminus and decreases in a short gradient from the end (Casanova and Struhl, 1989), it seems probable that the highest concentration activates *hkb* at the terminus while *tll* is activated at the somewhat lower *tor* activities just inside the termini (Casanova, 1990).

The activation of *tll* is itself part of a transcriptional cascade. The *tll* gene product is a member of the steroid receptor gene family and has DNA binding activity (Pignoni et al., 1990). It is undoubtedly a transcriptional regulator, whose level of activity influences the limits of the terminal *ftz* stripe (Casanova, 1990), though this influence may be indirect, involving other gap gene products (see below). *tll* also regulates other genes directly involved in terminal pattern formation. One of these is the homeotic gene *forkhead* (*fkh*) (3–95; 98D2,3) whose mutants cause homeotic transformations (Jurgens and Weigel, 1988; Casanova, 1990). *fkh* is also a DNA binding protein and putative transcriptional regulator and therefore continues the transcriptional cascade (Weigel et al., 1990). Thus, if one were to trace the sequence of events originating with the maternal terminal activities, one would obtain a scheme similar to that shown in Figure 7.7: *torsolike (tsl)* activity at the termini activates *tor* locally, which in turn phosphorylates one or more target proteins that activate the transcription of *tll* (and *hkb*), whose activities lead to further transcriptional activations.

Gap Gene Interactions and the Posterior Pattern

In the first analyses of the maternal and zygotic patterning systems, all of the early evidence supported the notion of a hierarchical system of gene control that runs from key maternal gene products to particular gap genes and from the latter to the pair-rule genes (Fig. 7.4). The cloning and characterization of three of the gap genes, *hb, Kr,* and *kni*, revealed them to have sequences characteristic of transcripitional regulators, of the so-called "zinc fingers" class, suggesting further that the regulatory hierarchy involved a direct transcriptional cascade, from maternal factors to pair-rule genes. (For the characterization of *Kr*, see Rosenburg et al. [1987]; for *hb*, Tautz et al. [1987];

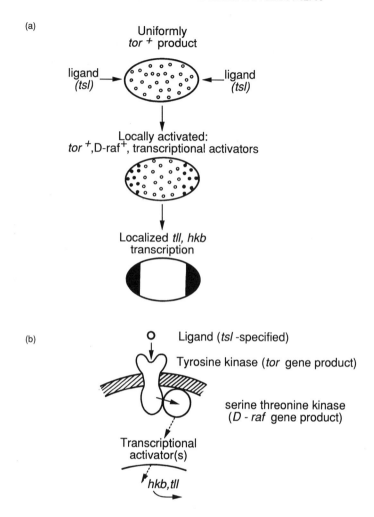

Fig. 7.7. The relationship between the maternal terminal system and the activation of zygotic transcriptional regulators at the termini. **a:** Uniformly distributed, maternally specified *tor* gene product is activated by the ligand, whose synthesis is under the control of *tsl,* producing a localized activation of expression of *tll* and *hkb.* **b:** The presumptive signaling cascade. See text for description. (Adapted from Pignoni et al., 1990.)

and for *kni,* whose product has distinct sequence relatedness to the steroid receptor superfamily, see Nauber et al. [1988].) Yet, other, early molecular studies soon revealed that the situation is more complex than this kind of simple hierarchy.

The principal class of events not predicted by the model consists of various *regulatory interactions between pairs of gap genes.* To take one example, in *hb* embryos, the *Kr* domain shifts anteriorly into the region that would normally be occupied by *hb* transcripts (Jäckle et al., 1986). These observations, based on transcript domains, are confirmed when *Kr* expression is monitored by assaying *Kr* protein with immunological techniques. Thus, the central *Kr* protein domain is shifted anteriorly in embryos lacking zygotic *hb* activity (Gaul and Jäckle, 1987). These results suggest that *Kr* activity is repressed, directly or indirectly, by either high levels of *hb* activity or *kni* activity.

In addition to repression events, there are distinct enhancement events. For instance, the expression of *kni,* whose domain is contiguous to that of both *tll* and *Kr,* is affected in embryos deficient for either gene activity. While *kni* expression is repressed by *tll* activity (in *tll* mutants, the *kni* transcript domain expands posteriorly), it is enhanced throughout the posterior domain by *Kr* activity (its expression in this region being measurably less strong in *Kr* mutants) (Pankratz et al., 1989). The interactions between gap gene activities thus embrace instances of positive control as well as repression. Furthermore, these effects extend beyond the readily detectable transcript domains (Gaul and Jäckle, 1989), presumably through diffusion of the encoded proteins. Indeed, careful analysis of the shifts in expression suggest that regulatory effects occur at concentrations lower than those detectable by current methods (Hülskamp and Tautz, 1991).

Such effects can help to explain the spread of *Kr* and *kni* defects well beyond the measured transcript domains (Fig. 7.5). They also bear on a model that was significant in early thinking about gap gene action. Meinhardt (1986), noting that the defective regions for *Kr* and *kni* are roughly double the extent of the transcript domains for these genes in the wild-type embryos, proposed that it is not the gap gene domains *per se* that are important but the borders between these domains; these boundary lines, in the model, act as demarcation points for the subsequent patterned expression of the pair-rule genes. The findings outlined above, indicating that gap gene products are active well beyond the transcript domains, suggest that the situation is considerably more complex, involving interactions well within the gap gene domains that have no obvious reference to the borders defined by the transcript domains.

The idea that the gap genes exert regulatory interactions on each other can explain, in principle, the puzzle of how the posterior segmental pattern is initially specified. In chapter 4, we reviewed the evidence that the segmental function of the posterior gene group is to inhibit translation of maternal *hb* transcript. The absence of maternal factors that actively "instruct" abdominal segmental pattern, in contrast to the *bcd* gradient in the anterior half, leaves open the question of how the first cues for pattern formation in the posterior half arise. If, however, the gap genes of the middle and posterior regions (*Kr, kni, gt,* and *tll*) exert regulatory effects on each other, then, in principle, one need only an initial differential regulation of gap genes at the anterior end (as provided by the *bcd* gradient) to initiate the sequence of gap gene expression events throughout the embryo; their regulatory interactions in the posterior half of the embryo would furnish the basis of specification in that region (Hülskamp and Tautz, 1991).

In addition to these regulatory interactions between the gap genes, there are unexpected additive effects between maternally specified gene products that affect gap gene regulation. For instance, embryos derived from maternal germ line cells that are simultaneously *bcd* and homozygous for *hb* fail to activate the central domain of *Kr* expression (Hülskamp et al., 1990). While Bcd protein at high concentrations represses *Kr,* it would seem that at lower concentrations it activates *Kr* additively with Hb protein.

The picture of gap gene action that has emerged is thus a complex one. On the one hand, differential concentrations of maternal factors along the a-p axis, in particular Bcd, help to produce a differential activation of at least two gap genes, *hb* and *Kr,* while regulatory interactions between gap genes whose protein domains overlap involve both enhancement and repression. The critical first events may involve the differential settings by maternal factors in the anterior region of the embryo, followed by consequent activation and delineation of gap gene domains in the posterior domains and sharpening of these domains throughout the embryo by the various cross-regulatory effects. The terminal regions also involve the differential activation of transcriptional regulators but the process there probably is less direct, involving *tor*-mediated phosphorylation events to achieve the activation.

Pair-Rule Genes: Establishing Initial Periodicity

The gap gene domains are notably broad and aperiodic, yet, as discussed below, they initiate the first periodic gene expression patterns, those of the eight pair-rule genes. The direct regulation of the pair-rule genes by the gap genes was first inferred from the early and sharp alterations in pair-rule gene expression patterns in the embryos of gap mutants (reviewed in Ingham and Gergen, 1988) and later, directly, by molecular tests, involving DNA binding experiments of gap gene products to promoter regions of pair-rule genes (e.g., Stanojevic et al., 1989; Pankratz et al., 1990).

It is the pair-rule gene activities that provide the first molecular signs of the ultimate (morphological) periodic pattern, the segments themselves. The archetypal pair-rule gene expression pattern, the seven-stripe *ftz* array, the first to be visualized (Hafen et al., 1984b), illustrates this periodicity strikingly and is shown in Figure 7.8.

Although the canonical stripe pattern of pair-rule genes is the first evidence of segmental periodicity, comparison of the stripe patterns produced by different pair-rule genes reveals that few are precisely overlapping. Rather, most are slightly offset with respect to each other, as was indeed first indicated by their various mutant

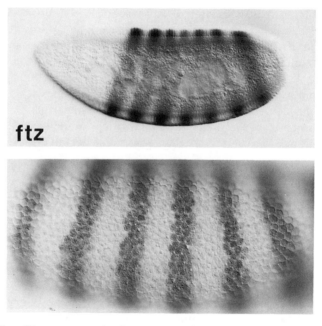

Fig. 7.8. The wild-type, seven-stripe *ftz* pattern, seen laterally (top) and ventrally (bottom). (Photograph courtesy of Dr. P. Ingham.)

phenotypes (Nüsslein-Volhard and Wieschaus, 1980). Thus, pair-rule gene expression patterns exhibit a characteristic *polarity*, with respect to the a-p axis, as well as periodicity. Furthermore, these patterns strongly suggest that few or no pair-rule genes individually specify particular sets of segments or segment regions. Rather, the visible pattern of segments derives from a complicated set of regulatory interactions involving the pair-rule genes.

These regulatory events begin with induction by the gap genes of certain pair-rule genes, during the syncytial blastoderm stage, and proceed by a sequence of both one-way hierarchical control events of some pair-rule genes by others and more complicated regulatory interactions by individual pair-rule genes on other members of this gene set. In its general features, therefore, the network of regulatory events governing the ensemble of pair-rule gene activities bears a strong resemblance to that of the gap genes.

The patterns of expression of *hairy* (*h*) and *ftz*, the first two pair-rule genes to be cloned, illustrate some of the fundamental behaviors of this class of segmentation gene. When studied by *in situ* hybridization, their expression is found to begin in cleavage stage 12, shortly after the onset of gap gene expression, as broad, diffuse transcript domains throughout the embryo. These transcript domains then progressively resolve into the definitive striped patterns that characterize pair-rule genes, by late syncytial blastoderm, just prior to cellularization. In the case of *h*, this consists of seven stripes at cellular blastoderm, which arise in a characteristic temporal sequence (later, an eighth, posteriormost stripe arises during gastrulation) (Howard, 1988; Pankratz et al., 1990). The *ftz* seven-stripe pattern also resolves itself rapidly during cellularization from an initially diffuse distribution (Edgar et al., 1987), the stripes being partially offset with respect to the *h* stripes (Ingham et al., 1985).

Despite the fact that the transcript stripes of *h* and *ftz* occupy largely discrete regions, these two gene activities are not independent of one another, but exhibit a hierarchical relationship, with *h* (partially) controlling *ftz* expression. This becomes apparent when embryos mutant for either gene are examined for the transcript patterns of the other. It is found that *h* expression strongly influences *ftz* expression but that *ftz* mutants exhibit normal *h* expression. The principal alteration in *h* embryos is an enlargement of all *ftz* stripes, while in *ftz* mutants, *h* stripes are unchanged (Carroll and Scott, 1986).

Furthermore, one can test the effect of *h* on *ftz* by artificially producing uniform *h* expression throughout the embryo at cellular blastoderm and then monitoring the effects on *ftz* expression. The procedure is to put the *h* coding region under the promoter of a heat shock gene, transforming flies with this gene construct using P element-mediated germ line transformation, and then heat shocking embryos of this strain during early cycle 14 (at the beginnings of cellularization). The result is an extinction of *ftz* expression (Ish-Horowicz and Pinchin, 1987). Clearly, *h* exerts a strong repressive effect on *ftz* expression and the results suggest that the existence of largely nonoverlapping *ftz* and *h* domains reflects the repression of the former by the latter, at least in part. (The existence of *some* degree of overlap, however, indicates that the repression of *ftz* by *h* is normally conditional to some degree upon the presence of one or more other factors.)

The regulatory relationship between *h* and *ftz* is not unique to these genes among the pair-rule set. *runt,* for instance, appears to have a similar repressive effect on *even-skipped (eve),* accounting for the fact that *eve* activities are offset with respect to *runt* and that in *runt⁻* embryos, *eve* is expressed broadly, with only slight periodicity throughout the germ band (Ingham and Gergen, 1988). On the other hand, *ftz* expression is not affected by every other pair-rule gene; embryos that are zygotically deficient for *odd-skipped (odd)*, *paired (prd)*, *odd-paired (odd-prd)*, or *sloppy-paired (slp)* exhibit normal *ftz* transcription (Carroll and Scott, 1986).

The finding that some pair-rule genes appear to regulate other pair-rule genes simplifies the problem of the generation of periodic stripes in response to an aperiodic (gap gene product) pattern in one respect, in that it is no longer necessary to imagine that *all* pair-rule genes develop periodic expression independently of one another. Rather, a three-stage process may be envisioned. In the first, a slight biasing toward periodicity of an initial small set of pair-rule genes is generated by a set of preexisting gene product domains (those of the gap genes). In the second stage, these initial minor periodicities are enhanced, the stripes becoming sharper, through a set of mutually repressive interactions, probably involving cooperative binding events of pair-rule gene (and other gene?) products (Edgar et al., 1989). Then, in the third phase, these first sharp pair-rule stripe patterns "template" the subsequent development of other pair-rule gene periodicities (Howard, 1988).

If this general scheme is correct, then special attention should be focused on the earliest pair-rule gene activities, as the precipitators of the overall pair-rule patterns, and on the regulation of these "primary pair-rule" genes (Ingham and Gergen, 1988) by gap gene products. *h,* as one of the primary pair-rule genes, has been intensively analyzed in this regard. The findings indicate that the promoter structure of *h* is complex, with different regions of the promoter being responsible for different groups of stripes (Howard, 1988) and that the development of each *h* stripe is a function of the relative concentrations of the regulatory gap gene products within the region that the stripe appears (Pankratz et al., 1990). The results of one such analysis, the regulation of *h* stripe 6 by the joint (repressive) action of *Kr* and the (positive control) activity of *kni,* are shown in schematic form in Figure 7.9. It appears that, just as the expression of individual gap genes involves combinations of sensitive, concentration-dependent repression and activation events, the setting of the primary pair-rule activities by the gap gene products involves similar combinatorial elements.

Segment Polarity Genes: Setting Up the Morphological Pattern

The third tier of segmentation genes is that of the segment polarity genes. The designation derives from the appearance of the embryos of the first identified mutants in this group. These display an intrasegmental region of reversed polarity in which part of each segment is deleted and replaced by a mirror image of the remaining part (Fig. 7.2) (Nüsslein-Volhard and Wieschaus, 1980). Today, a gene is defined as belonging to this group if mutant homozygotes show an abnormality in each and every segment (as opposed to the regional or alternate-segment abnormalities seen in the gap and pair-rule gene mutants, respectively).

Altogether, four classes of phenotypic abnormality are seen, and amorphs, or putative amorphs, of each of the 14 known segment polarity genes fall into one of these classes (Table 7.2). In contrast to the gap genes and the pair-rule genes, which are required transiently for the establishment of segmental pattern (though some of the pair-rule genes are later employed for neural development), the segment polarity genes are required either continuously or over extensive periods for maintenance of

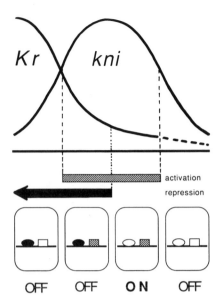

Fig. 7.9. Model for the activation of the sixth *h* stripe, as a function of overlapping gap gene activities. The sixth stripe is activated by the combination of high concentrations of *kni* activity and low *Kr* activity; the concentration gradients of the two gap gene products and the structure of the *h* promoter ensure that this activation occurs only in a narrow region, to generate the sixth stripe. The other *h* stripes are presumably generated in similar or comparable fashions by the combined activating/repressing actions of other overlapping gap gene domains. (Reproduced from Pankratz et al., 1990, with permission of the publisher. © Cell Press.)

segmental pattern. In addition, most or all are required for maintenance of pattern imaginal tissues; the latter requirements are indicated by the phenotypic abnormalities associated with homozygous clones induced late in development (Table 7.2).

The defining characteristic of the segmental polarity genes is their segmental periodicity. Comparably, when the expression patterns of the segment polarity genes in the early embryo are examined by *in situ* hybridization, several are found to show a distinct periodicity of pattern, one stripe per segment anlage. These genes are *engrailed (en), wingless (wg), patched (ptc), cubitis interruptis-Dominant (ci-D), gooseberry (gsb),* and *hedgehog (hh);* the *en* pattern is shown, as an example, in Figure 7.10. The two questions that such spatial arrays immediately raise concern, first, how the periodicities of the transcript patterns arise, and, second, how they relate to the ultimate morphological patterns.

Although it is clear that the pair-rule genes are essential, directly or indirectly, for the initial establishment of these patterns (see, for instance, Carroll and Scott, 1986; DiNardo and O'Farrell, 1987; Martinez-Arias et al., 1988), the precise pathway by which this is achieved is not known. However, two kinds of explanation have been proposed. The first is that the overlapping patterns of particular pair-rule genes create a repeated sequence of cell states. In this view, each row of blastodermal cells is specified differently from its immediate neighboring rows, though the patterns of overlap produce repeated cell states (Gergen et al., 1986). Each cell state, in this model, translates into a particular pattern of activities of specific segment polarity

Table 7.2 Segment Polarity Genes

Gene	Map position	Embryonic phenotype (References)	Embryonic domain of transcript (References)	Imaginal disc domain of transcript (References)	Clonal defects (References)	Molecular Characterization (References)
engrailed (en)	2–62	Pairwise fusion of adjacent segments (Kornberg, 1981a)	Posterior region of segment (Kornberg et al., 1985)	Expressed nonuniformity within posterior compartment (Brower, 1986)	Defects in posterior compartments associated with clonal overgrowth at AP and DV boundaries (Lawrence and Morata, 1976)	Contains homeobox (Fjose et al, 1985)
naked (nkd)	3–47	Missing anterior region of segment (Jurgens et al., 1984) (Perrimon et al., 1989) (Nüsslein-Volhard et al., 1984)	—	—	—	—
lethal (1)3ba (l(1)3ba) lines (lin)	1–1.1 2–59		—	—	—	—
patched (ptc)	2–59	Narrow denticle belts within double segment borders; mirror-image duplication of anterior region including segmental border (Nüsslein-Volhard et al., 1984) (Simpson and Grau, 1987)	Anterior region of segment (Hooper and Scott, 1989; Nakano et al., 1989)	Anterior region overlapping A-P order (Phillips et al., 1990)	Anterior regions of embryonic segments show "domineering" nonautonomy (Roberts et al., 1989; Phillips et al., 1990)	Membrane-spanning protein (Hooper and Scott, 1989; Nakano et al., 1989)
costal-2* (cos-2)	2–57		—	—	Nonautonomous (but see Simpson and Grau, 1987)	—
wingless (wg)	2–30	Duplication of denticles with mirror-image reversal in posterior segmental regions	Anterior. Within ptc domain and anterior to en domain (Baker, 1987, 1988a)	A and P, in wings and halteres, ventral-anterior in legs (Baker, 1987, 1988a)	Nonautonomous (Lawrence and Morata, 1977; but see Baker, 1988b)	Secreted factor homologous to int-1 (Rijsewijk et al., 1987; van den Heurel et al., 1989)

Gene	Map	Reference	Expression		Cell autonomy	Molecular nature
cubitus interruptus-D (*ci-D*)	4–0	(Nüsslein-Volhard and Wieschaus, 1980)	Anterior within *ptc* domain (Eaton and Kornberg, 1990; Orenic et al., 1990)	—	—	Zinc-finger protein (Orenic et al., 1990)
fused (*fu*)	1–59.5	(Nüsslein-Volhard and Wieschaus, 1980; Martinez-Arias and Lawrence, 1985)	—	—	Largely cell autonomous (Martinez-Arias and Lawrence, 1985; but see Gergen and Wieschaus, 1985)	Membrane protein kinase (Préat et al., 1990)
*armadillo** (*arm*)	1–1.0	(Nüsslein-Volhard and Wieschaus, 1980; Klingensmith et al., 1989)	Ubiquitous (Riggleman et al., 1989)	Uniform (Riggleman et al., 1989)	—	Plakoglobin homolog (Peifer and Wieschaus, 1990)
*dishevelled** (*dsh*)	1–34	(Nüsslein-Volhard and Wieschaus, 1980; Perrimon and Mahowald, 1987)	—	—	—	—
*porcupine** (*porc*)	3–59	(Nüsslein-Volhard and Wieschaus, 1980; Peifer and Wieschaus, 1990)	—	—	—	—
hedgehog (*hh*)	3–90	(Nüsslein-Volhard and Wieschaus, 1980)	Posterior (Hidalgo and Ingham, 1990)	—	Posterior/distal "domineering" nonautonomy (Mohler, 1988)	—
gooseberry (*gsb*)	2–104	(Nüsslein-Volhard and Wieschaus, 1980)	Posteriormost region of segment overlapping *en*-expressing cells (Baumgartner et al., 1987)	—	—	Contains homeobox (Baumgartner et al., 1987)

Source: Adapted from Wilkins and Gubb (1991).

Genes are grouped according to embryonic phenotype of mutant alleles. The embryonic region affected in mutants is often larger than the domain in which transcripts are expressed in wild-type embryos. The severity and extent of segmental defects in mutant embryos are often correlated with the degree of residual gene activity. Genes marked with an asterisk show a maternal effect segment polarity phenotype with mutant alleles; homozygous mutant progeny from heterozygous mutant mothers show reduced segment polarity defects or give wild-type embryos. The embryonic phenotypes of *fu*, *arm*, *dsh*, *porc*, *hh*, and *gsb* are associated with cell death in posterior regions.

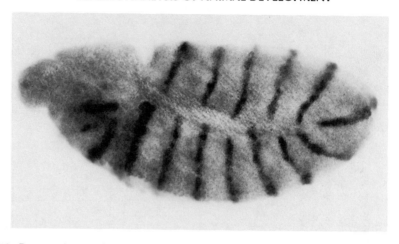

Fig. 7.10. Pattern of expression of a segment polarity gene, *en,* at the extended germ band stage, as measured by *in situ* hybridization to the *en* transcript. Anterior is to the left, dorsal is at the top. (Reproduced from Wilkins and Gubb, 1991, with permission of the publisher.)

genes, which in turn, gives rise to a particular aspect of the pattern within each segment.

The second kind of hypothesis is that particular gene activities establish crucial boundaries that act as "reference points" for the subsequent organization of gene activities, and pattern, between these boundaries (Meinhardt, 1982, 1986; Lawrence et al., 1987). In the particular version of this hypothesis that has been applied explicitly to the relationship between pair-rule genes and segment polarity genes, the boundaries are those of the regions termed "parasegments." Parasegments are metameric units that are out of register with respect to the morphological segments, and which consist of the posterior portion of each segment plus the anterior portion of the next segment (Martinez-Arias and Lawrence, 1985) (Fig. 7.11). The parasegment has a transient existence as a morphological domain, with its own borders, in the early embryo and is also the domain of expression of certain homeotic genes that establish regional identities (Martinez-Arias and Lawrence, 1985).

The principal reason for postulating a role of parasegmental borders in setting subsequent patterning events is that two pair-rule genes, *eve* and *ftz,* are expressed in alternate parasegments and exhibit a pronounced sharpening at their anterior parasegment borders, immediately after blastoderm formation (Lawrence et al., 1987; Lawrence and Johnston, 1989). This sharpening process is accompanied by a graded diminution of the protein products posteriorly within the parasegments and is not observed in several mutants that derange segmental patterning (Lawrence and Johnston, 1989). In the wild-type embryos, these anterior borders demarcate the boundaries of the *wg* and *en* expression domains.

The difficulty in evaluating this hypothesis, however, is a practical one. Parasegments are, like segments, only 3.5 cell diameters in width, on average, at the time of *ftz* and *eve* border sharpening (Sullivan, 1987; Lawrence, 1989), while the *en-* and *ftz*-expressing borders are themselves one to two cells wide. With regions of these relative dimensions, it is difficult, if not impossible, to design experiments that discriminate effects that arise secondarily after formation of a boundary from those

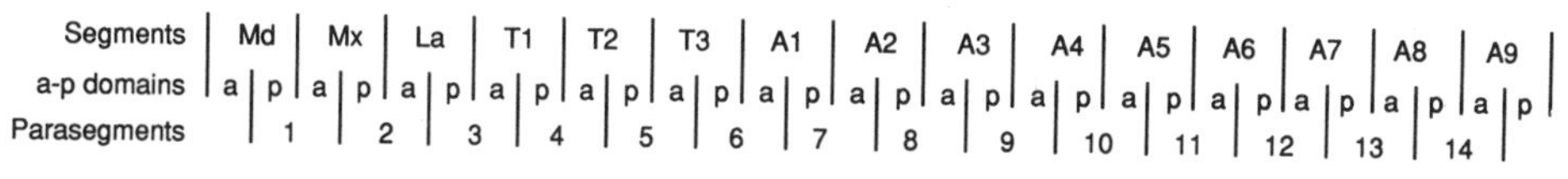

Segments	Md		Mx		La		T1		T2		T3		A1		A2		A3		A4		A5		A6		A7		A8		A9	
a-p domains	a	p	a	p	a	p	a	p	a	p	a	p	a	p	a	p	a	p	a	p	a	p	a	p	a	p	a	p	a	p
Parasegments		1		2		3		4		5		6		7		8		9		10		11		12		13		14	

Fig. 7.11. Relative disposition of segments and parasegmental domains. The anterior and posterior regions of the segments are indicated. The posterior segmental regions are defined by their expression of *en*. See text for further discussion of the parasegmental domain.

that appear virtually simultaneously through cell-cell contact interactions within the domain.

Furthermore, it is not apparent how one might distinguish the kind of proposed boundary effect, involving an interaction between neighboring cells, from a simple threshold effect for particular gene products within one of the cell populations at the boundary. Such a threshold effect could then trigger secondary events within those cells and, hence, produce other cell states. In effect, key boundary models reduce, in this instance at least, to one class of cell state models, in which the relative timing of the acquisition of particular cell states is given particular emphasis.

Leaving the question of how the periodic patterns are initially established, one comes to the more crucial one of function: How do these gene products organize or contribute to the formation of segmental pattern? Many of the segment polarity genes have now been cloned and sequenced and the group has been found to be a biochemically diverse one. Thus, at least three, namely *en, gsb,* and *ci*^{D}, encode protein motifs characteristic of transcriptional regulators (Fjose et al., 1985; Baumgartner et al., 1987; Orenic et al., 1990), while *ptc* is a membrane-spanning protein (Nakano et al., 1989; Hooper and Scott, 1989), *fused (fu)* specifies a protein kinase (Préat et al., 1990), and *wg* encodes a secretable factor, though one that does not travel far from the cells that produce it (Rijsewijk et al., 1987; van den Heuvel et al., 1989). Finally, *armadillo (arm)* encodes a protein homologous to mammalian plakoglobin, a component of adhesive junctions of epithelial cells (Peifer and Wieschaus, 1990).

In considering how the products of these genes act to produce a repeated segmental (and subsegmental) pattern, it is useful to divide them first into two broad expression classes: those that are expressed maternally as well as zygotically and those that are expressed solely from the zygotic genome (Table 7.2). Among the strict zygotics are those that are expressed in a periodic pattern, and it is these that would seem to have the most obvious role in establishing the morphological periodicity of the segmental pattern.

The first of these genes to be cloned and studied was *en* (Fig. 7.10); it was found to be expressed in narrow (one to two cell diameter) bands at cellular blastoderm in the region that becomes the posterior part of every segment (and the anterior portion of each parasegment). Similarly, the transcripts of the *gsb* locus are also expressed in the posterior regions of the segmental primordia, being slightly broader than, and overlapping, the *en* stripes. In contrast, *ptc* is expressed in broad bands in the anterior region of each future segment (Hooper and Scott, 1989; Nakano et al., 1989), and *wg* is also expressed in the anterior domain, in bands that are slightly narrower than *ptc* and which are immediately adjacent to the *en* stripes (Baker 1987, 1988a). *en* and *gsb* are expressed in regions that give rise to naked cuticle, while *ptc* and *wg* are expressed in the regions that give rise to the denticle belts.

These subsegmental domains, however, are not independent of each other but are highly interdependent. For instance, while *en* and *wg* are expressed in different cells within each segment anlage, the continued expression of each is necessary for the maintenance of the other. Furthermore, in *ptc* embryos, unusually broad bands of *wg* expression form and these become secondarily flanked by new bands of *en* expression (DiNardo et al., 1988; Martinez-Arias et al., 1988). Since *en* and *wg* are expressed in different cell populations, these effects are clearly produced across cell boundaries and reflect close-range cell-cell interactions. (While the *wg* gene product is itself a short-range diffusible factor, the effects of *en* deficiency must involve secondary consequences, given that it encodes a transcriptional regulator.)

One conceptual framework for these results is to view the segment polarity gene activities as determining a linear sequence of repeated positional values within each segment (Meinhardt, 1984). Indeed, both regenerative responses to surgical removal of parts of insect segments (Lawrence, 1981) and the regulatory responses to deficiency for particular gene activities are highly reminiscent of the polar coordinate model (Wilkins and Gubb, 1991). In the polar coordinate model, it will be recalled, there are constant geographical relationships of "positional values," and removal of certain positional values produces a response leading either to regeneration or duplication of pattern elements.

In this view, the expression of certain key segment polarity genes within particular spatial domains produces characteristic cell states or positional values, while removal of one or more of these positional values leads to the "intercalation" of the missing values (Martinez-Arias et al., 1988; Wilkins and Gubb, 1991). The canonical segment polarity phenotype, namely mirror-image duplication of the anterior part of each segment, is, of course, highly reminiscent of the duplicative response engendered by removing a majority of the positional values, in the polar coordinate model.

For the periodically expressed segment polarity genes, the application of the idea seems straightforward, but where would it leave those segment polarity genes that are expressed maternally and whose products, therefore, might be distributed ubiquitously throughout the embryo initially? One possibility is that that some may mediate signals that are themselves localized; a ubiquitous distribution could thereby be translated into a localized effect.

This explanation seems highly applicable to, at least, one well-studied example, that of *arm*, whose transcripts are distributed uniformly throughout the embryo and, later, in imaginal tissues (Riggleman et al., 1989). Despite the ubiquity of the transcript, the *arm* protein comes to be distributed in a distinct segmental pattern by early germ band extension (Riggleman et al., 1990). This segmental pattern, indeed, seems to "track" *wg* expression, with the strong *arm* stripes developing slightly after *wg* transcript patterns appear and occupying virtually the same locations. Furthermore, in the absence of *wg*, the striped pattern does not appear. These findings suggest that *wg* activity produces a posttranscriptional accumulation of *arm* protein in these regions in some manner (Riggleman et al., 1990).

Yet the relationship to *wg* has a second important feature. *arm* is homologous to the intermediate junction protein plakoglobin, found in vertebrates, and is found in association with actin, preferentially at cell membrane surfaces (Peifer and Wieschaus, 1990). Taken in conjunction with the similarity of *arm* and *wg* phenotypes, these biochemical facts suggest that *arm* protein may help to mediate the *wg* signal. Given that *porcupine* (*porc*) and *dishevelled* (*dsh*) mutations, like *wg*, also prevent the initial segmental patterning of *arm* (mutants of the other segment polarity genes do not prevent its appearance), Peifer and Wieschaus (1990) have suggested that these gene products might also be part of the same signaling pathway.

The *arm-wg* relationship shows that a ubiquitously transcribed gene can give rise to a product that, via its interactions with localized signals, produces a localized response. On the other hand, the existence of a localized expression pattern, in the first place, does not necessarily signify that the gene product *must* be localized in order to exert its effect. When *ptc* expression is made ubiquitous by means of heat shock induction of *ptc* (produced by means of a *hsp70* promoter–*ptc* construct), embryonic segmental development is unaffected (Sampedro and Guerrero, 1991). *ptc* is, nevertheless, biologically active when so induced because the treatment can rescue *ptc* defects in mutant embryos (Sampedro and Guerrero, 1991).

If ubiquitous expression need not imply ubiquitous developmental function, and if localized expression does not necessarily imply that expression *must* be localized, then, clearly, expression patterns are, at best, an unreliable guide to ascertaining where gene products either must function or be prevented from functioning. In contrast, clonal analysis can provide information on the essential site of action of a gene product, its *focus* (Hotta and Benzer, 1973). Although clonal analysis in the embryo is difficult, because of the small size of clones that are obtained, such analyses of segment polarity gene requirements have been carried out in imaginal tissues and have been informative about the foci of action of several segment polarity genes. The results provide some valuable information on their roles in pattern formation in imaginal tissues and some hints about their roles as positional value genes; these findings will be discussed later in this chapter.

The Genetic Basis of Segment Individuation

As noted earlier, the segmental pattern of *Drosophila* has a dual character. On the one hand, it features a repetitive motif, involving both the regular spacing of the metameres and certain pattern elements within the segments, and, on the other, it displays individuating aspects of segment character, which serve to distinguish segments from one another. In the material discussed in the preceding sections, we concentrated on the genes whose activities create the repetitive, or periodic, aspects of the pattern. We will now turn to those genes, generally designated as the "homeotics," whose mutations leave the fundamental periodic pattern untouched but which transform individual or groups of segments (or parasegments) into other, recognizable segment types. We will also briefly examine the regulatory network that governs the expression of these genes. This network includes certain of the segmentation genes, which participate in the initiation of the patterns, and a group of distinct genes, which specify general chromatin or transcriptional proteins and which function in maintaining the initial expression patterns (and, thus, segmental phenotype).

Although mutations in many *Drosophila* genes can affect segment phenotypic identity, two gene clusters in particular are especially important. Both gene groups were detected through their mutant homoeotic effects in the adult fruit fly; only subsequently were the genes discovered to have profound effects on larval segment identity and structure. The first such cluster, the biothorax complex or BX-C, designated as such by Lewis (1978), is named after the original mutant mapped to this region, a viable mutant that produces a partial transformation of the haltere (the small balancer organ found on the third thoracic segment) into wing, a second thoracic structure. The BX-C is located on the right arm of the third chromosome at 3–58.8 (bands 89E–E4) and is required for normal segmental development from the second thoracic segment (T2) (with principal effects in T3 and A1) to the eighth abdominal segment (A8).

The second gene cluster involved in the control of larval segment identity is the Antennapedia complex or ANT-C (Kaufman et al., 1980). The *Antennapedia* (*Antp*) mutants, which give the complex its name, are dominant homoeotics that show degrees of transformation of the antennal structures of the adult fly into leg structures; the typical *Antp* transformation has been mentioned earlier and was shown in Figure 2.5. ANT-C is also located on the right arm of the third chromosome, but many bands proximal of BX-C (toward the centromere), at 3–48 (bands 84B1–B2); it is required for development of the head segments and the first two thoracic segments (T1 and T2).

Though not of immediate significance for what follows, an evolutionary point of interest is that while the BX-C and ANT-C are physically separate in *Drosophila,* their apparent homologs are part of a single cluster in the more primitive insect, the red flour beetle, *Tribolium* (Beeman, 1987). As will be discussed in chapter 8, the mammalian homologs of the BX-C and ANT-C are also found grouped together. Where the genes of the BX-C and ANT-C need to be referred to collectively, they will be designated here as the HOM genes (Holland, 1990) or, alternatively, as "homeotic selectors." (The origins of the term "selectors" will become apparent later in this chapter.)

The BX-C: Organization and Domains of Expression

Although both gene complexes appear to play central roles in the delineation of segmental phenotype, the BX-C has been more extensively characterized to date. The first description of segmental transformations in the embryo produced by mutations of the BX-C was given by Lewis (1978) and was subsequently elaborated by Struhl (1981). These first-generation studies relied on the morphological transformations in the adult wrought by genes of the BX-C (see Fig. 2.3). An alternative approach, which ushered in a second generation of studies focused on defining the number of vital genes in the BX-C, was pioneered by Sanchez-Herrero et al. (1985) and Tiong et al. (1985).

From these two different approaches arose two different interpretations, which are depicted in Figure 7.12. In the first view, the BX-C consists of a sequence of genes, *in the same proximodistal (p-d) order on the chromosome as the a-p sequence of the segments,* each required for the specific phenotypic individuation of a given segment. For the abdominal segments, these putative genes were designated as *infra-abdominal* genes and consisted of *infra-abdominal-2* (*iab-2*), required for A2 identity, to *infra-abdominal 8* (*iab-8*), required for A8 identity.

In the second model, there are only three genes—*Ultrabithorax* (*Ubx*), *abdominal-A* (*abd-A*), and *Abdominal-B* (*Abd-B*)—and each has what may be termed "primary responsibility" for a particular suprasegmental region. In this scheme, each gene is a complex locus of some kind and differences between segments arise from the complexities of expression or regulation of each gene. The conflict of interpreta-

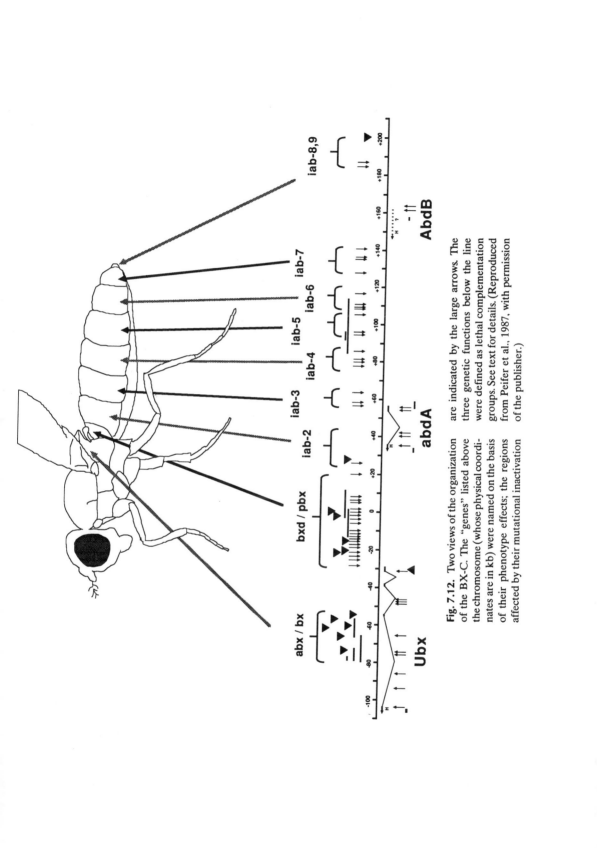

Fig. 7.12. Two views of the organization of the BX-C. The "genes" listed above the chromosome (whose physical coordinates are in kb) were named on the basis of their phenotype effects; the regions affected by their mutational inactivation are indicated by the large arrows. The three genetic functions below the line were defined as lethal complementation groups. See text for details. (Reproduced from Peifer et al., 1987, with permission of the publisher.)

tions represented by these two models has now been resolved but the initial difference in views provides a useful illustration of how conclusions about genetic organization can be influenced by the particular approach taken.

The first interpretation of the BX-C, as a sequence of genes, with individual genes required for individual segments, stemmed from approaches that emphasized the identification of genetic changes affecting external (cuticular) morphology. This involved, on the one hand, the scoring for effects in the embryo of mutations that were known to alter imaginal phenotype and, on the other, the examination of embryos deleted for various regions of the BX-C.

Two classes of imaginal mutations found to produce embryonic defects are the *Ubx* and *bithoraxoid* (*bxd*) mutations. The *Ubx* mutants are haplo-insufficient dominant mutants, distinguished in adults by a slight enlargement and partial transformation of the halteres to a wing-like state. In late embryos, *Ubx* homozygotes show both AB1 and T3 transformed to a T2-type phenotype. Thus, removal of this gene function, which happens to be the most proximal gene in the BX-C, "drives" development to a T2-like state. In contrast, the *bxd* imaginal phenotype is a partial transformation of the A1 segment to a T3-like phenotype, and embryos homozygous for *bxd*, which is just distal of *Ubx*, show a partial transformation of AB1 to a T3-like phenotype. These observations indicate that there may be separate genetic functions for individual segmental phenotypes and suggest that the BX-C serves to "raise" development "above" a T2-like state (Lewis, 1964, 1978). This inference about BX-C general function is confirmed by homozygotes for the deletion *P9*, which deletes the entire BX-C. These embryos show normal head segments and the first two thoracic segments (the domain of the ANT-C) but exhibit a transformation of all segments, from T2 to A8, to a T2-like state (Fig. 7.13).

To this picture were added the results of analysis of partial deletions of the BX-C. This procedure employed various translocations that break the complex into two parts. To create the requisite translocations, flies were treated with X-rays and progeny were examined for dominant mutant phenotypes involving T2, T3, or the abdominal segments, such mutations signifying probable breaks within the BX-C. Flies showing segmental transformations in the BX-C spatial domain were then tested to see which mutations involved translocations that split the complex into two parts, in which one part of the BX-C was transposed to another chromosome. The positions of the breaks are, of course, not under the control of the experimenter, but by generating enough translocations, one can often obtain the separations desired.

For many translocations, possession of both halves of the translocation by the animal permits viability. From these viable parents, the two halves can be segregated into progeny and analyzed for the embryonic phenotypic effects associated with removal of the other half of the translocation. By placing the deleted third chromosome over a complete deficiency for the region, *Df(3)P9*, one can determine the effects of that deficiency. The results of these studies delimited a proximal portion of the BX-C responsible for normal T3 development and a large region required for abdominal development, in which the addition of successively larger, more distal pieces promoted progressively more wild-type abdominal development (Lewis, 1978; Struhl, 1981b).

In general, deficiencies and loss-of-function mutations produce transformations toward a more anterior segmental phenotype, the extreme being the row of T2-like segments in *Df(3)P9* embryos. In contrast, a number of dominant mutants, identified

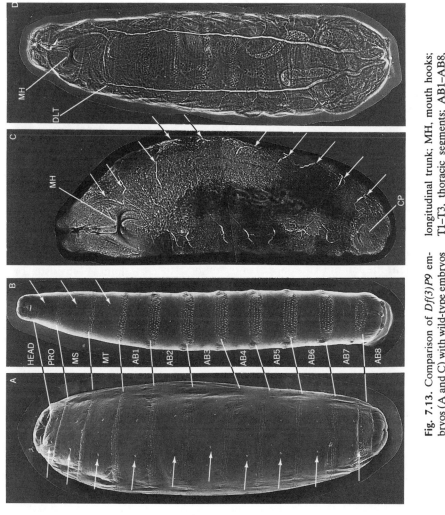

Fig. 7.13. Comparison of *Df(3)P9* embryos (A and C) with wild-type embryos (B and D). **A, B:** Surface cuticular patterns. **C, D:** Internal tracheal patterns. In A, arrows indicate Keilin's organs (a thoracic structure); in C, arrows indicate separate tracheal sections. DLT, Dorsal longitudinal trunk; MH, mouth hooks; T1–T3, thoracic segments; AB1–AB8, abdominal segments. Magnifications: A, 160×; B, 60×; C and D, 120×. (Reprinted by permission from *Nature*, vol. 276, p. 569, © 1978 Macmillan Journals Ltd.)

on the basis of their effects in the adult, do the reverse, promoting transformation of particular segments to that of more posterior ones. Thus, for instance, *Hyperabdominal* (*Hab*) causes both T3 and A1 to take on an A2 phenotype, and *Miscadestral pigmentation* (*Mcp*) transforms AB4 to AB5 (readily observed in adult males because the fifth abdominal segment is darkly pigmented while the fourth in wild-type males is not) (Lewis, 1978). Such dominant mutations are most readily interpreted as neomorphs, individually expressing a particular segment gene function in one or two segments anterior to the "appropriate" one. Conversely, recessive loss-of-function mutations that cause individual segments to assume the appearance of the next most anterior one, such as the *iab-2* mutant (Kuhn et al., 1981), can be viewed as inactivating such segment-specific genes. Finally, the first molecular mapping of some of these mutations within the "abdominal" part of the BX-C supported the idea of a sequence of segment-specific abdominal genes, mapping in a p-d order that mirrors the a-p order of the segments (Karch et al., 1985).

In sum, these findings indicated that each segment requires the expression of both a particular BX-C function and of all the genes proximal to it (though AB8 apparently did not require *Ubx* function for normal morphological development) (Lewis, 1978, 1981). Thus, in going from T3 to A8 within the embryo and larva, there would be a roughly corresponding step-like gradient of expression within BX-C such that with each additional posterior step (segment), an additional BX-C function is activated. In this construct, a given segmental phenotype results from the *summed* expression of the activated BX-C functions. The Lewis model, whose first, prescient formulation was made in the 1960s on the basis of several imaginal mutant phenotypes (Lewis, 1964), is diagrammed in Figure 7.14. In it, the basis of this differential activation was proposed to be an a-p gradient of "repressor" molecules of some sort, running from a high point near the center (T2 anlagen) of the embryo to a low point at the posterior end, along with a p-d gradient of "operator strengths" along the chromosome such that the most proximal genes had the weakest affinity and the most distal (those governing A6 and A7), the strongest. These two conditions would suffice to explain the apparent pattern of increasing activation of BX-C functions from T3 to A8. The subsequent discovery that many BX-C transformations take place within parasegmental (Fig. 7.11), rather than segmental, units (Martinez-Arias and Lawrence, 1985) does not alter the fundamental tenets of the Lewis model.

While the hypothesis that each segment or parasegment requires its own gene, at least in the abdominal regions, was based primarily on the search for genetic changes that produce morphological alteration, the alternative three-gene model of the BX-C stemmed from the search for lethal mutations within the BX-C (Sanchez-Herrero et al., 1985; Tiong et al., 1985). Individually mutagenized third chromosomes were placed over complete or partial deletions of the BX-C and those that failed to give viable larvae were provisionally identified as harboring lethal alleles. Different lethals were then intercrossed to test for complementation and the number of complementation groups were ascertained. Finally, test lethals for each complementation group identified as falling within the region of the BX-C (by the criterion of failure to complement with *Df(3)P9*) were examined for their morphological effects. Two ways of scoring for such morphological approaches were used: the phenotype of rare adult "escapers" (individuals surviving the normally lethal block) were examined or, where no escapers were obtained, mitotically recombinant clones were induced (involving induced loss of a Dp

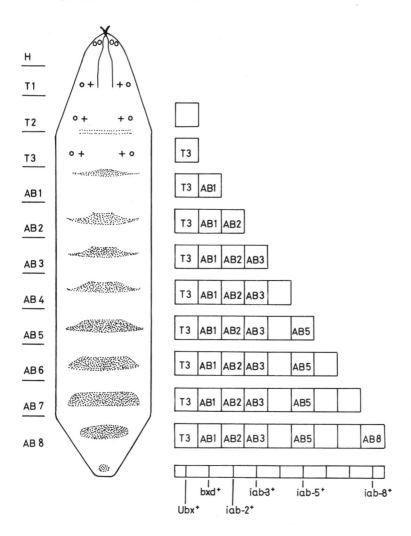

Fig. 7.14. The Lewis model of BX-C expression from T2 through AB8. The hypothesis posits a gradient of BX-C expression, which increases posteriorly, each segment expressing a unique BX-C expression plus all those expressed in more anterior segments (and proximally, with respect to the chromosome). (Adapted from Lewis, 1981.)

containing the wild-type BX-C) and scored for the nature and location of any segmental transformation associated with them.

This approach showed that the BX-C contains three vital genes (and two immediately flanking vital loci). Furthermore, lethal mutations in these three genes were found to be associated with identifiable segmental transformations. The first and most proximal of these genes is identical with *Ubx* by all morphological criteria; heterozygotes for lethal mutations in this gene produce a *Ubx*-type haplo-insufficient transformation of haltere to wing, while embryos show the characteristic *Ubx* transformation of T3p–A1a to T2–3a (parasegment [ps] 6 to ps5) and lesser degrees of transformation in the more posterior parasegments.

Mutations in the central gene are fully recessive, but homozygotes are lethal, producing larvae with a strong transformation of A2, A3, and A4 (or ps7, 8, and 9) toward A1 and lesser degrees of transformation toward an A1-type morphology in A5–A8; all segments anterior to A1a are normal, however. This gene is designated *abd-A*. Finally, mutations in the third vital gene are slightly halpo-insufficient, producing mild segmental transformations in segments A5–A8 in adults and a strong transformation in homozygous larvae, or rare escaper adults, toward A4; these larvae are normal, however, in the head, thoracic segments, and first four abdominal segments. This third, and centromere distal, gene has been designated *Abd-B* (the uppercase A signifying its slight dominant effect, in contrast to the full recessivity of *abd-A*).

The results support the notion that there are only three functional units in the BX-C: *Ubx*, with its strongest requirement in ps6 (T3p–A1a) but which is expressed and required to a lesser extent in all posterior parasegments; *abd-A*, required especially in ps7–ps9 (posterior A1 to anterior T4) and to a lesser extent in more posterior parasegments; and *Abd-B*, required in ps10 (posterior A4 and anterior A5) and all more posterior parasegments. Lethal mutations producing transformations in single abdominal segments or parasegments were not obtained. How might one reconcile this picture with that obtained from the earlier studies?

The answer is implicit in the results from the molecular mapping of various mutations, both point mutations and chromosomal rearrangements, as summarized in Figure 7.12. Of the 300 kb of DNA sequence that comprises the BX-C, only a relatively small portion is directly dedicated to coding, and, in fact, only three coding regions exist in the BX-C, corresponding to the genetic units *Ubx*, *abd-A*, and *Abd-B*. Mutations that destroy the functioning of these units, either point mutations in essential exons or rearrangements that break them up, result in lethal mutations, and the complete set of lethals define the three lethal complementation groups identified by Sanchez-Herrero et al. (1985) and Tiong et al. (1985). In contrast, mutations that do not destroy any of these complementation groups and which lie either outside a coding region or, in certain cases, within introns (such as the *abx* or *bx* mutations within *Ubx*) can produce segment-specific (or parasegment-specific modulations of the distribution of one of the three protein products, resulting in segment- or parasegment-specific transformations of phenotype (Peifer et al., 1987). Thus, "genes" such as *iab-2* or *iab-7* are, in fact, *cis*-acting regulatory regions rather than coding regions in themselves, and much of the BX-C consists of such sequences, dedicated to regulating the transcription of a comparatively small number of coding regions (Beachy et al., 1985; Peifer et al., 1987). The regions designated as *iab* genes appear, in fact, to be enhancer regions that are activated in specific parasegments, promoting expression of individual coding regions in those parasegments. One, and possibly two, of these enhancer-like regions may be shared between *abd-A* and *Abd-B* (Celniker et al., 1990).

All of the effects described so far involve the outer cuticular surface, but genes of the BX-C are actively expressed in internal tissues as well, including large regions of the nervous system and the mesoderm. The evidence indicates that this expression is essential for normal development of at least some of these tissues and structures, in particular the CNS. To examine this requirement, Jiminez and Campos-Ortega (1981) studied the ventral neural cord patterns in several lethal embryonic BX-C genotypes. Although the various segmental ganglia, or neuromeres, fuse ("condense") in late embryonic development, they remain distinguishable by two structures. Each neuromere contains a distinctive medial cell column and the thoracic neuromeres are distinguished by an intracortical band of densely staining cells at the anterior edge of each neuromere. The number of these bands varies directly with the cuticular transformations produced in the mutants. Thus, wild-type embryos show three intracortical bands in the thoracic region; lethal *Ubx* embryos, which convert AB1 to T2, have four; and *Df(3)P9* embryos, which have all segments posterior to T2 converted to a T2 phenotype, have 11 (Jiminez and Campos-Ortega, 1981).

In addition to these embryonic transformations, BX-C expression is also required in the nervous system of the imago. For instance, Teugels and Ghysen (1983) determined the numbers of leg ganglia in adult *Hab* and *bxd* mutants, both of which show variable penetrance, and found that either reduced or additional numbers of leg ganglia could be produced by the two mutations respectively. Furthermore, these transformations could be produced independently of the occurrence of the cuticular transformation. The results show that the CNS can be transformed autonomously by mutation in the BX-C, indicating that the gene complex is expressed in and necessary for normal CNS development. Expression of *Ubx*, visualized by immunostaining directed against Ubx protein, is shown in Figure 7.15.

As in the ectoderm, the domains of expression for both *Ubx* and *abd-A* in the visceral mesoderm of the embryo are parasegmental, with their anterior boundaries shifted one parasegment posterior, relative to the ectoderm (Bienz and Tremml, 1988). The results indicate that the visceral mesoderm is not directly "instructed" by the ectoderm to follow its expression pattern but instead exhibits some independence in its regulation of BX-C genes. Nevertheless, the different patterns of expression of

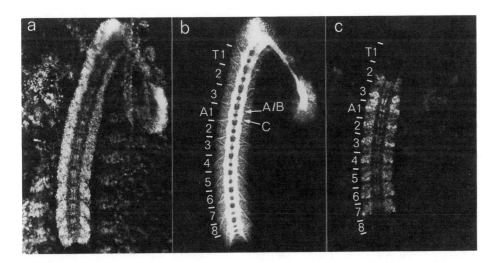

Fig. 7.15. Expression of *Ubx* in the ventral nervous system of the embryo. **a:** Embryo stained with Hoechst 33258, to reveal the nuclei. **b:** Embryo stained with flourescein-conjugated anti-HRP, which stains neural tissue and reveals the commissures (transverse nerve bundles) (the commissures of abdominal segment 1 are indicated). **c:** The pattern of expression of *Ubx*, with strongest staining in ps6, but staining showing heterogeneous expression both in this parasegment and in more posterior ones. (Reproduced from White and Wilcox, 1984, with permission of the publisher. © Cell Press.)

BX-C genes in different tissues are not wholly independent of one another. Expression of *Ubx* and *abd-A* in the visceral mesoderm has an inductive effect in the neighboring endoderm, in evoking the expression of a third homeotic gene, *labial* (*lab*), and this effect is mediated by diffusible signals (Immergluck et al., 1990).

ANT-C: Organization and Domains of Expression

ANT-C is less well characterized than the BX-C but plays a comparable role to that of the BX-C in the anterior half of the animal, from the mesothorax, T2, through the gnathocephalic segments (the mandibular [Mn], maxillary [Mx], and labial [La] segments). The first work on the ANT-C was inspired by the early work on the BX-C and involved a determination of the number of genes and lethal complementation groups in the vicinity of the *Antp* gene. Comparably, the initial characterization of its mutants was in terms of adult phenotypic defects and, subsequently, of embryonic and larval abnormalities. While this work has shown that there is a parallel between the sequence of those genes that affect anterior segments and their relative positions on the chromosome, as exists in the BX-C, it has also revealed some departures from the BX-C organizational pattern.

The first mutants of the ANT-C were dominant *Antp* neomorphs, many associated with chromosomal rearrangements, that show various degrees of transformation of antenna to mesothoracic leg (Fig. 2.5). Because such mutational changes are difficult to analyze, both in terms of the complexities of the aberrations themselves and because it is difficult to work back from a neomorphic phenotype to wild-type function, new mutants in *Antp* and its vicinity were sought. This work was first carried out by Thomas Kaufman and his colleagues and involves two stages.

In the first stage, nonhomeotic "revertants" of the dominant mutants were isolated, by screening heterozygotes for loss of the dominant phenotype. Many of these revertants will have a second mutation that inactivates or deletes the gene responsible for the neomorphic phenotype and, hence, convert the neomorphic mutations to loss-of-function ones. As anticipated, many of the inactivating mutations proved to be deletions, which were viable over the wild-type third chromosome but which were lethal when homozygous. In the second step of the procedure, flies carrying EMS-mutagenized third chromosomes were crossed to flies carrying these deletions and the heterozygotes were screened for new phenotypes. Chromosomes giving inviable, semiviable, or homeotic transformations must have a mutation in the same region as that spanned by the deletion. These new mutations were then tested in pairs to determine whether or not they complemented. Finally, the mutations were mapped with respect to one another to generate a map of the various complementation groups in the region (Lewis et al., 1980a,b). With the later placement of the *bcd* gene as a genetic function contained within the ANT-C (identified and mapped but not named as *bcd* by Frigerio et al., 1986), the complete genetic map of the ANT-C was obtained (Fig. 7.16).

Analysis and comparison of the mutant phenotypes produced by ANT-C mutations reveals the first interesting difference in its organization to that of the BX-C. While each of the three BX-C genes is associated with clear homeotic phenotypes, loss-of-function mutations in only three of the genes of the ANT-C, taking the leftmost (*lab*) and rightmost (*Antp*) homeotic genes as its boundaries, give marked segmental homeotic transformations in either imaginal tissues or in the embryo. These genes are *proboscipedia* (*pb*), *Sex combs reduced* (*Scr*), and *Antp* itself (Lewis et al., 1980 a,b;

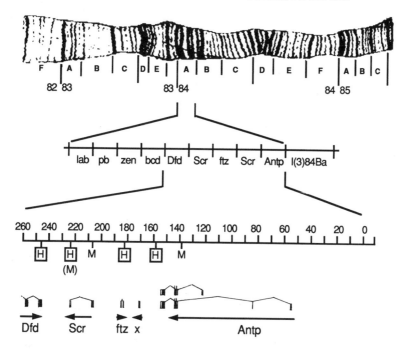

Fig. 7.16. The map of the ANT-C. The top portion indicates the cytogenetic region of the ANT-C; the lower portion is an expanded view of part of the ANT-C, showing the orientation of four genes, their homeoboxes (signified by a boxed H), and the approximate locations of the repeat sequence, "opa" (M). (Adapted from Gehring, 1985.)

Wakimoto and Kaufman, 1981). Loss-of-function mutations in a fourth gene, *Deformed* (*Dfd*), which was first identified on the basis of a viable, dominant neomorphic mutation affecting eye size, may also give partial transformations of dorsal, posterior cuticular structures on the head to those of thorax (Merrill et al., 1987). Homozygotes for loss-of-function mutations in *Dfd* show abnormalities in the three gnathocephalic segments but no evident signs of homeotic transformation (Merrill et al., 1987). *lab* embryos are associated with slight changes that have been interpreted as mild homeotic changes.

Mutants in the remaining genes of the ANT-C show a variety of other developmental phenotypes. Thus, *zerknult* (*zen*) is required for the development of embryonic dorsal tissues (Doyle et al., 1986), while *ftz,* as we have seen, is a pair-rule gene involved in establishing the periodic segmental pattern, and *bcd,* as discussed in chapter 4, specifies a key maternal protein that carries out the first step of creating the embryonic a-p axis. (Embryos from *bcd* mothers, it will be recalled, do show a homeotic transformation of their acrons.) The inclusion of genes dedicated to roles other than segment identity specification marks a difference in ANT-C organization from that of the BX-C. (However, as discussed below, all the genes of the ANT-C, like those of the BX-C, contain homeoboxes and act as transcriptional regulators in development [Mahaffey et al., 1989].)

A second difference between the two complexes concerns the nature of the genetic units. While the ANT-C as a whole is comparable in size to the BX-C (more than 400 kb vs. 300 kb), the latter consists of three very large genes, when one includes the *cis*-acting sequences for each (Fig. 7.12). However, there is only one "giant" gene in the ANT-C, and that is *Antp* itself, with an extent of approximately 100 kb (Scott et al., 1983); the other genes of the complex are small to average-size eukaryotic genes (Fig. 7.16). Consistent with its size and complex internal organization, there are multiple *Antp* transcripts (four have been identified) but, in contrast to the genes of the BX-C (see below), they have identical open reading frames and, hence, encode the same protein (Schneuwly et al., 1986). The multiplicity of transcripts derives, in part, from two different promoters, which determine two somewhat different spatial patterns of appearance of the *Antp* protein (Jorgensen and Garber, 1987).

The principal organizational similarity between the two gene complexes lies in the general correspondence between a-p position of the segments affected and the linear order of the genes required for the development of those segments. Thus, the map order of five of the genes expressed in particular anterior segments is *lab-pb-Dfd-Scr-Antp*, while the order of regions and anterior segments is procephalon-Mn-Mx-La-T1-T2.

Although the patterns of several of the proteins are both dynamic and spatially complex, principal domains of expression can be detected by immunostaining, and these domains in the surface ectoderm are as follows: The *lab* protein is expressed primarily in the procephalon (anterior to the gnathocephalic segments); *Dfd* is expressed posteriorly of the procephalon, principally in the Mn and Mx segments; and *Scr* protein appears primarily in La and T1 (Mahaffey et al., 1989). The protein domain of the rightmost gene in the complex, *Antp*, initially overlaps both the *Scr* and primary *Ubx* domain (extending from posterior La through T1, T2, and T3 to anterior A1 in the ectoderm at the extended germ band stage), but the principal area of staining then becomes concentrated in T1, T2, and T3, thereby overlapping while extending posteriorly from the *Scr* epidermal domain. *pb* expression, however, does not fit this overall pattern; though located just to the left of *lab* on the genetic map, its spatial domain is not between *lab* and *Dfd*, as one might predict from the map position/a-p position correlation, but partially overlaps both *Dfd* and *Scr* (Mahaffey et al., 1989).

Indeed, the overall expression patterns show both dynamic change and complexity of pattern. Thus, while the patterns noted above describe the surface ectodermal domains, there are often intriguing differences of expression in neighboring mesodermal regions. Furthermore, the patterns of expression in neural tissue show complex modulations in space and time that do not match the changes in the overlying surface ectoderm. A final complexity to be noted is that the precise metameric register of expression is a function of both the gene and tissue or regional position. Parasegmental registers are neither universal for homeotic genes, as was predicted by Martinez-Arias and Lawrence (1985), nor even necessarily an invariant feature for those genes that show parasegmental registers in some positions. Thus, while the metameric frame of expression for both *Dfd* and *Scr* is parasegmental in the ventral part of the embryo, it is a segmental pattern in dorsal regions, the shift occurring at or near the boundary between the dorsal epidermis and the neurogenic region (Mahaffey et al., 1989). As is the case for the BX-C, the regulation of expression of genes of the ANT-C is clearly complex and not reducible, at least at present, to a few simple rules.

Pattern Formation and the Regulation of the Genes of the BX-C and the ANT-C

The large diversity of segmental phenotypes in the *Drosophila* embryo (and of the imago), from the head segments to the most posterior abdominal segments, is created by the activities of a relatively small number of genes, the homeotic selectors of the BX-C and the ANT-C, collectively the HOM genes. The beginning of an understanding of the relationships between the overt pattern and the activities of these genes lies in an appreciation of the fact that these genes are part of a transcriptional cascade. They are themselves transcriptionally regulated by "upstream" genes and, in turn, regulate a variety of "downstream" genes. In order to encompass the regulatory complexities, this material will be covered in the following four sections.

Initiation. What mechanisms ensure that particular homeotic genes are always expressed in their appropriate locations? In principle, only one or a small number need be initially localized, because subsequent regulatory interactions can then create or sharpen the boundaries of the expression domains of others. In fact, the patterns of transcripts of both *Ubx* and *Antp* are very broad initially, at the blastoderm stage, within that part of the embryo that will give rise to the germ band (Akam, 1983; Levine et al., 1983) and subsequently sharpen to narrower domains. These dynamics suggest that other gene products distributed anisotropically along the a-p axis are involved. One group of such substances is, of course, the segmentation gene products, and, as first pointed out by Gubb (1985a), one or more of these are likely to be involved in refining the initial homeotic gene patterns. Several lines of evidence implicate one gap gene, *hb,* and one pair-rule gene, *ftz,* as especially important in defining the domains of BX-C products.

Both genetic and molecular data support a specific regulatory role for *hb* in BX-C regulation. In the course of characterizing 17 different *hb* alleles (isolated on the basis of various noncomplementation searches with known *hb* alleles), Lehmann and Nüsslein-Volhard (1987b) identified three alleles that give a distinct homeotic transformation of head and thoracic segments to abdominal segments, tending to resemble either A1 or A8 depending upon the allele and the position of the transformed segment. Simultaneous deletion of the BX-C suppresses the transformation (cited in Lehmann and Nüsslein-Volhard, 1987b), showing that they depend on BX-C activity and implying that the effects involve an ectopic activation of *abd-A* or *Abd-B* in the head and thoracic regions of these embryos. That there is indeed some misregulation of BX-C functions in *hb* embryos is indicated by observations on the expression patterns of *Ubx.* While all *hb* mutant embryos show broadened *Ubx* expression both anteriorly and posteriorly, the most marked anterior extensions are produced by the homeotic *hb* alleles (White and Lehmann, 1986). These homeotic effects, it should be stressed, cannot be an indirect consequence of simply reducing levels of *hb* activity because only a subset of all *hb* alleles, and these by no means the most deficient, yield these transformations. Evidently, the *hb* gene product itself participates in the molecular interactions that regulate and affect BX-C gene expression.

A similar line of reasoning suggests that *ftz* also plays a role in defining the limits of the *Ubx* domain. As discussed by Kaufman (1983), Laughan and Scott (1984), and Duncan (1986), certain viable dominant alleles of *ftz* give homeotic transformations in the thorax and abdomen. These alleles are not null mutants but rather neomorphs.

Perhaps the most intriguing of these mutants are those described by Duncan (1986), the so-called *Ultra-abdominal like* (*Ual*) mutants, whose principal transformation in adult flies is the partial conversion of A1 to A3. That the mutations are in *ftz* was indicated first by mapping, which placed the mutations at the location of *ftz* (3–47.7), and then by screening for revertants of the dominant phenotype; homozygous embryos of these revertants show the typical *ftz* mutant phenotype. Although *ftz* is expressed with equal strength in seven stripes, the transformation seen in the dominants is highly localized (there are lower frequencies of transformation in other structures, in particular, partial transformation of A3 to A5). Evidently, in whatever manner *ftz* regulates BX-C activity, there must be other factors as well which serve to spatially restrict the effects.

Indeed, the actions of *hb* and *ftz* in defining the *Ubx* domain are probably connected, with *hb* exerting its effect through *ftz*. *hb* is expressed before *ftz*, is known to be part of the regulatory circuitry of *ftz*, and, perhaps most significantly, the homeotic *hb* alleles have been shown to give a distinctive reduction in *ftz* stripes 3 and 4, an effect not seen with the *hb* nulls (Carroll and Scott, 1986). Stripe 3, in fact, corresponds to ps6, which is the principal domain of *Ubx* expression. It seems likely that other gap/pair-rule gene combinations are involved in initiating some of the other homeotic gene expression patterns.

Interactions and fine-tuning. Once *some* differences in regional expression have been established, other factors come into play in defining and sharpening the boundaries; we have seen the operation of this principle in the regulation of the pair-rule genes. In the case of the HOM gene set, particular members begin to regulate other members of the set. This regulation is at the transcriptional level: All three genes of the BX-C (*Ubx, abd-A,* and *Abd-B*) and *Antp, Scr,* and *Dfd* are homeodomain proteins and act as transcriptional control proteins (reviewed in Gehring, 1985; Peifer et al., 1987). It was, in fact, the sequence relatedness between portions of *Ubx* and *Antp* that led to the discovery of the homeobox (McGinnis et al., 1984a) and the provisional identification of these genes as transcriptional regulators (Laughan and Scott, 1984).

The first evidence for regulatory interactions between HOM genes was genetic. Mitotically recombinant clones in T2 legs that had been made homozygous for *Ubx* at blastoderm were found to show transformations to a T1 leg phenotype (Morata and Kerridge, 1981). If, however, these clones are also *Scr⁻*, the transformation is abolished (Struhl, 1982b). The simplest interpretation is that early *Ubx* activity is required in the mesothorax for normal (T2) leg development specifically to prevent an *Scr*-dependent transformation to T1.

With the cloning of the various HOM genes, molecular techniques could be brought to bear on such interactions. One of the first findings was that *Antp* expression is repressed by one or more genes of the BX-C: In *Df(3)P9* embryos, lacking the entire BX-C, *Antp* transcript abundance increases dramatically throughout ps6–ps13 (Hafen et al., 1984b). The conversion of posterior segments to T2-like segments in the absence of the BX-C, as seen in *Df(3)P9* embryos (Fig. 7.13), reflects this derepression of *Antp* activity.

Other interactions involve those between members of the same complex. Thus, embryos lacking either *abd-A* or *Abd-B,* or both, show elevated expression of *Ubx* proteins throughout ps7–ps12 (Struhl and White, 1985), signifying a degree of repression of *Ubx* by those two genes within their domains. This increase in expression seems to involve both an elevation in *Ubx* proteins within individual cells and in the total number of cells expressing the protein within the abdominal region. These effects are illustrated schematically in Figure 7.17. The cellular pattern of *Ubx* expression within each parasegment is determined by a combination of the particular *cis*-acting elements within the complex, the presence or absence of *abd-A* and *Abd-B*, and, possibly, regional-specific *trans*-activators that help to initiate transcription of these genes (Peifer et al., 1987). The morphological landscape of each parasegment within the BX-C domain is, thus, in part a reflection of the cellular mosaic expression pattern of Ubx proteins, mediated in part by *abd-A* and *Abd-B*. In general, HOM genes, whose main expression domains are posterior to those of others, tend to repress the more anteriorly expressed ones.

It seems probable that the relative times at which these proteins appear is important, since the gene products of the ANT-C and BX-C are all homeodomain-containing transcriptional regulators and a particular transcriptional outcome may be, in part, a "balance of forces," turning on which products are present and in what concentrations. The fact that *Antp* and *Ubx* have large introns may be relevant in this respect. The net effect of such total gene size is a lag time, on the order of an hour for a gene of 70

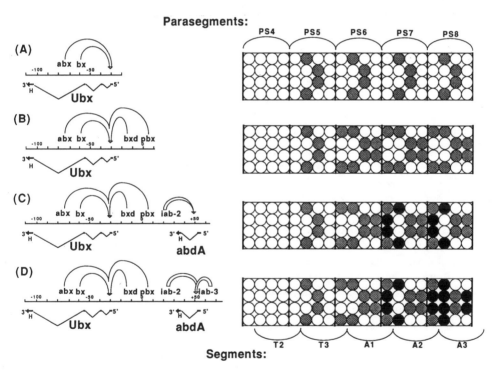

Fig. 7.17. Cellular mosaic patterns of expression of *Ubx* and *abd-A,* in various deletion mutants that remove chromosomal material from the right. Schematic, generalized pattern of *Ubx* expression in cells indicated by grey circles, pattern of *abd-A* expression in cells indicated by black. Adding *cis*-regulatory sequences from the right (**B** vs. **A**) increases the number of cells expressing *Ubx* in ps6 and in more posterior parasegments; expression of *abd-A* decreases *Ubx* expression in *abd-A* domain. (Reproduced from Peifer et al., 1987, with permission of the publisher.)

kb, from the start of its transcription to the completion of the first transcripts. Such "intron delay" may serve to retard, within particular regions, the synthesis of the encoded homeotic gene products, whose otherwise premature appearance might alter the overall transcriptional balance (Gubb, 1986). It may be significant, in this connection, that *Ubx*, which represses *Antp*, is the shorter of the two and that both *abd-A* and *Abd-B*, which repress *Ubx*, are considerably shorter than *Ubx*.

In addition to gene length, the intron-exon organization of each gene may contribute to phenotypic diversity through differential splicing. Where such splicing takes place within the coding regions, it can give rise to variant proteins for each gene, which might, in turn, have different functions. Differential splicing has been documented, from the existence of variant cDNAs, for both *Ubx* and *Abd-B*.

The first demonstration of differential splicing for *Ubx* was obtained by Beachy et al. (1985) from the analysis of *Ubx* cDNAs, which were found to differ in their composition with respect to certain very small internal exons ("microexons"). This description was completed by Kornsfield et al. (1989), who demonstrated that all *Ubx* mRNAs encode identical amino-terminal and carboxy-terminal regions of 247 and 99 amino acids, respectively, but differ in a splice site for the 5' exon and in the two microexons. The variable splice site and the microexons generate three "optional elements," encoding 9, 17, and 17 additional amino acids, which are designated, respectively, the b, I, and II elements.

The best evidence that different Ubx proteins have different roles comes from experiments in which different encoded forms are not differentially removed or inactivated but, rather, added back, singly, to the genome. This can be done by making cDNAs to *Ubx* mRNAs, differing in their optional element content, placing them under the *hsp70* promoter, and doing germ line transformation with these constructs. The developmental effects of brief heat shocks, which serve to induce expression of the proteins, can then be assayed. When otherwise wild-type strains are prepared and tested in this way, and constructs encoding different Ubx proteins are compared, it is found that all three forms transform ps 1–ps5 into ps6-like regions in the outer ectoderm with apparently equal success (Mann and Hogness, 1990). The result indicates that *Ubx* activity is sufficient to transform the anterior parasegments into the ps6 state. However, when the embryonic PNS is examined, it is found that the cDNA lacking only element b can create an abdominal-cell type transformation in ps3, 4, and 5, following brief heat shock, but that the cDNA lacking all three cannot. This result is strong evidence that optional elements I and II confer certain developmental abilities on Ubx proteins.

Various genetic and molecular findings indicate that *Abd-B* also undergoes differential splicing. In a study of 27 mutations identified as *Abd-B* mutations on the basis that they fail to complement either of two deletions that overlap only in *Abd-B*, Casanova et al. (1986) identified two distinct classes on the basis of their effects (plus a third which seems to combine both types of effect). In the first, only abdominal segments A5–A8 (ps10–ps13) are affected, while the terminalia are normal; in the second class, the principal effect is the appearance of an extra denticle belt posterior to A8, while A5–A8 are substantially normal. An analysis of *Abd-B* transcripts confirmed that the two regions distinguished by the mutational classes are, in fact, occupied by different *Abd-B* transcripts (Sanchez-Herrero and Crosby, 1988). Studies of the distribution of *Abd-B* proteins in different mutants confirm the existence of two different protein distributions (DeLorenzi and Bienz, 1990).

The net effect of all the regulational interplay is to establish characteristic cellular mosaics of expression for the different homeotic genes within each parasegment. When examined at the cellular level by immunostaining, there are highly significant differences in the precise cellular arrays *within segments* in which the various proteins of the BX-C appear. Thus, while the highest levels of staining are produced in ps6 (T3p + A1a), with slight staining in ps5 (T2p + T3a) and somewhat stronger staining in ps7–ps12, and ps13 (A7p + A8a) showing very weak staining (a pattern consistent with the extent of transformation in the mutants), these patterns are not uniform within parasegments (White and Wilcox, 1984; Beachy et al., 1985). Rather, each stained parasegment shows a characteristic mosaic ("pepper-and-salt") pattern of staining: Only a portion of the cells within each parasegment show *Ubx* protein and the particular array of stained cells appears characteristic of the individual parasegment (Fig. 7.17).

Maintenance. Although the expression patterns of the homeotic genes are not static, there is a general, region-specific maintenance of these patterns. One element in such stability may be self-activation by these genes of their own activity. Thus, for instance, once genes such as *Ubx* and *Antp* are turned on in their respective domains, they may continue to autoregulate.

There is both genetic and molecular evidence to support this idea. To study the control of *Ubx* expression, Bienz and Tremml (1988) constructed a fusion gene, consisting of the *E. coli* β-galactosidase coding region under the control of the *Ubx* promoter. Germ line transformation was then carried out to obtain a strain in which β-galactosidase expression can be monitored to track activity of the *Ubx* promoter. They found that the expression in visceral mesoderm was dependent on an intact (endogenous) *Ubx* gene; in *Ubx⁻* embryos containing this fusion gene, there was no β-galactosidase synthesis in the visceral mesoderm. A comparable approach has also demonstrated autoregulation of the *Dfd* gene (Kuziora and McGinnis, 1988). In this instance, the coding region of the homeotic gene was placed under a *hsp70* promoter and germ line-transformed flies were heat shocked. A novel pattern of *Dfd* expression in the ventral ectoderm was obtained, which was dependent on the endogenous *Dfd* gene. The results indicate that the pulsed induction of the construct promoted, in turn, the activation and autoregulation of the endogenous *Dfd* gene.

Molecular analyses support the genetic inferences. It has been shown that purified *Ubx* protein binds to its own promoter and that of *Antp* at specific DNA binding sites (Beachy et al., 1988). Self-binding activity reflects *Ubx* autoregulatory capacity, while binding to the *Antp* promoter is connected to the ability of *Ubx* to repress *Antp* expression. An even more direct demonstration of *Ubx* transcriptional control of itself and of *Antp* utilizes cotransfection, in *Drosophila* cells in culture, of *Ubx* with different reporter gene constructs, containing *Ubx* or *Antp* promoters. It has been found, using this approach, that *Ubx* activity can repress expression from one of the *Antp* promoters (P1) and activate expression from the *Ubx* promoter (Krasnow et al., 1989).

Nevertheless, even for genes, such as *Ubx* where autoregulation has been demonstrated, it is not an invariant property. In the ectoderm, in contrast to the visceral mesoderm, *Ubx* does not autoactivate its own expression (Bienz and Tremml, 1988; Mann and Hogness, 1990). Taken as a whole, the entire set of results indicate that *Ubx* has *potential* autoregulatory capability but that this property is itself dependent on what may be termed "transcriptional context," presumably the presence or absence of other transcriptional factors. This molecular context, in turn, must reflect the particular developmental state of the cell.

In those instances where a HOM gene both activates its own synthesis and represses that of another, the former activity can, in principle, serve to maintain the state of repression of the other gene. Thus, autoregulation can ensure not only continued regulation of the various downstream genes under its own control but the repression of those regulators that are subject to its control and their downstream genes.

Apart from cross-regulation and autoactivation of the genes of the BX-C and ANT-C, many other gene products are involved in maintenance of the activities of these two gene complexes. In particular, there is a large set that keeps the BX-C and ANT-C *off* in cells and regions where they are normally not expressed. Many of these were first identified from dominant but relatively slight imaginal homeotic mutant effects, produced by haplo-insufficiency in heterozygotes.

The first of these genes to be characterized was *Polycomb* (Pc) (at 3–47.1). *Pc* was initially identified as a dominant mutation in adult male flies that causes the appearance of extra sex combs on the meso- and metathoracic legs (in the wild-type, the sex comb appears only on the prothoracic legs of the male). This phenotype, however, is not a sex-limited transformation but a reflection of a more general conversion of meso- and metathoracic legs toward prothoracic legs in both sexes. *Pc* mutants are, in fact, pleiotropic, exhibiting partial transformation of abdominal segment structures to those of more posterior segments and, under certain conditions, provoking a partial transformation of antenna toward pro- or mesothoracic legs (Duncan and Lewis, 1982). Gene dosage studies reveal that these phenotypes reflect reduced dosage for the wild-type gene; *Pc* is thus, like *Ubx* (but not many other genes in the *Drosophila* genome), a haplo-insufficient locus (Denell, 1978).

The most dramatic transformation produced by *Pc,* however, is seen in embryos homozygous for strong loss-of-function alleles. Homozygosity had long been known to be a lethal condition, but when the embryos were examined, they were found to have extensive transformations, with reduced head structures, and extensive partial transformations throughout the thoracic and abdominal regions toward a posterior abdominal segment phenotype. These transformations require at least one dose of the wild-type BX-C in the genome (embryos homozygous for *Pc* and simultaneously deficient for the BX-C look like *Df P9* homozygotes), while additional copies of the BX-C in *Pc* homozygotes produce a more striking transformation in thoracic and A1–A7 segments toward A8. Furthermore, simultaneous deficiency for the BX-C produces a string of T2-like segments (Lewis, 1978). Thus, BX-C genes are necessary for the *Pc* loss-induced transformation to A8.

The interpretation initially made by Lewis (1978) was that *Pc* encodes the repressor of the putative repressor gradient proposed in his model of BX-C activation. In the absence of *Pc+* activity, there will be a complete "derepression" of BX-C genes, producing a repeated sequence of A8-like segments. In this model, therefore, *Pc* plays an essential role in *establishing* the initial states of BX-C activities (Lewis, 1978).

We now know, however, that *Pc* activity is essential for *maintaining* states of BX-C expression rather than setting the initial differentials. This possibility was first suggested by Denell (1978), who argued that the multiple homeotic transformations produced by a simple halving of gene dosage are more likely to reflect indirect and secondary consequences on gene expression rather than a highly specific role in initiating a particular pattern of homeotic gene expression. Furthermore, if *Pc* is an establishment function rather than one of maintenance, then removal of wild-type *Pc* activity by mitotic recombination after establishment of the segmental pattern should have no phenotypic effect. In fact, homozygous clones of *Pc* cells, induced late in embryogenesis, exhibit homeotic transformations in imaginal tissue (Struhl, 1981b). Clearly, *Pc* activity helps to maintain particular states of BX-C activity.

Within a short time, it became evident that *Pc* is not alone in this role but that there are many other genes required for such maintenance (Struhl, 1981b; Duncan, and Lewis 1982; Jurgens, 1985). Not all, when individually mutant, can produce as dramatic a transformation as *Pc,* but several do so when present in pairs; altogether, as estimated by Jurgens (1985), there may be as many as 40 or so *Pc*-type genes (classified as such in terms of their mutant phenotype) scattered throughout the genome. Furthermore, careful examination of the embryos reveals that the transformations involve both anteriorward and posteriorward effects and that the former require at least the *Scr* gene (Sato and Denell, 1985). Evidently, some, at least, of the *Pc* class of genes are involved in regulating the expression not just of the BX-C but of the ANT-C as well.

Molecular evidence confirms the genetic inference of a maintenance role performed by genes of this group. Thus, in *extra sex combs* (*esc*) embryos (zygotically deficient for *esc* and derived from mothers lacking this activity), *Ubx* expression is essentially normal in early embryogenesis, showing highest expression in ps6, and remains so until the extended germ band stage. Its expression then becomes ubiquitous in both ectoderm and mesoderm throughout the germ band, ps 1–ps14 (Struhl and Akam, 1985). A similar, late "derepression" of *Ubx* expression (Beachy et al., 1985) and of *Dfd* (Kuziora and McGinnis, 1988) occurs in *Pc* embryos themselves. What might the wild-type functions of the gene products of the *Pc* "family" be? The simplest hypothesis is that these genes, rather than encoding specific regulators of the BX-C and ANT-C, specify chromatin proteins of some kind that are required to maintain transcriptional inactivity of various genes, including those of the BX-C and ANT-C. These genes need not be identical, but they might have a measure of functional redundancy and a degree of interchangeability, as suggested by the synergistic interactions of several specific pairs (Jurgens, 1985).

Some evidence in support of this idea is available. The *Pc* gene has been cloned and its product shown to be a nuclear protein that binds to approximately 60 sites on polytene chromosomes, including ANT-C, BX-C, and *Pc* itself (Zink and Paro, 1989). Furthermore, sequence analysis shows it to have homology, over a 37 amino acid residue stretch, with a known heterochromatin-localized protein, HP1 (Paro and Hogness, 1991), which itself has been implicated as an influence in transcriptional control. This involvement comes from the discovery of a mutant of the HP1-coding gene that acts as a suppressor of the form of transcriptional inhibition termed "position effect variegation" or PEV (James and Elgin, 1986).

PEV is a form of somatically heritable transcriptional inactivity which is observed when virtually any euchromatic gene is repositioned, by chromosome rearrangement, to the vicinity of centromeric heterochromatin, where in polytene chromosomes, the genes frequently exhibit "heterochromatinization" (Ananiev and Gvozdev, 1974). Though the precise nature of the transcriptional inhibition in PEV, in molecular terms, has long been obscure, recent years have witnessed something of a breakthrough. This has consisted of the identification of a large number of genes whose mutants either suppress or enhance the inhibition of expression of the PEV-affected genes. Analysis has shown that, for many of these genes, the type of modulating activity, whether suppressing or enhancing, is a function of the gene's dosage (Locke et al., 1988). Such dosage-sensitive behavior is precisely that expected for chromatin proteins that are required in stoichiometric amounts to stabilize chromatin states (Locke et al., 1988; reviewed in Eissenberg, 1989).

It seems probable that many members of the *Pc* gene group will prove to be genes for general chromatin proteins and, simultaneously, members of the gene set identified as modifiers of PEV. The identification of the HP1-encoding gene as a gene whose loss-of-function mutations can suppress PEV, and its sequence relatedness to *Pc,* shows that the qualitative distinctions between euchromatin and heterochromatin proteins are not absolute. The findings also validate the long-standing contention by many of those who have worked on PEV that the phenomenon holds a key to understanding the stability of determinative states in general (see review by Baker, 1968). The specific involvement of the *Pc* group of genes in this stabilization of chromatin states in determination has been discussed by Paro (1990).

Target genes of the BX-C and ANT-C: Toward a molecular biology of the phenotype. The genes of the homeotic selector complexes, the BX-C and ANT-C, must exert their developmental effects, ultimately, through the transcriptional regulatory activities of their products on downstream genes whose products directly affect cell phenotypes. If one is to understand how the activities of BX-C and ANT-C dictate segmental pattern, the identification of the downstream genes is essential. Although progress in this area has been comparatively slow, beginnings have been made. Gould et al. (1991) have reported experiments involving immunoprecipitation of chromatin with anti-UBX antibody and the cloning of the gene fragments purified in this manner. They describe four genes identified in this manner, of which two are transcriptionally regulated *in vivo* (directly or indirectly) by *Ubx.* One of these genes, designated "35," shows elevated transcript levels in *Ubx*-deficient embryos, with an especially strong effect in ps6; evidently, "35" is normally repressed by *Ubx* (and, as shown by even higher levels in *Df(3)P9* embryos in ps7–ps14, by *abd-A* and probably *Abd-B* as well.) In contrast, transcripts of gene "48" are reduced in the absence of *Ubx* activity in ps5 and 6; clearly, "48" is activated by *Ubx* expression. The biological functions of these genes are not yet known but the isolation of mutants should help to clarify their roles. For the present, the primary significance of the work lies in the demonstration that both positively and negatively regulated downstream target genes of the homeotic selector genes can be isolated and characterized.

Setting the Dorsoventral Embryonic Pattern

In the preceding sections, we have examined the genetic activities that generate pattern along the a-p axis of the embryo. A comparable, though less well-delineated, series of events establishes pattern along the second major axis of the blastoderm embryo, the d-v axis. The primary maternal "instruction" along the d-v axis, as discussed in chapter 4, is the nuclear gradient of *dorsal* (*dl*) protein, a presumptive transcriptional activator. Its concentration is at a maximum along the ventral midline of the embryo and diminishes symmetrically, and sharply, on both sides of the embryo as one moves dorsally. It is this initial chemical difference along the d-v axis that lays the foundation for specification of the basic tissue and cell types in the embryo, in longitudinal bands along the blastoderm surface, as depicted in the fate map (Fig. 4.10). These specification events, from mesoderm at the ventral side to the amnioserosa at the dorsalmost position, require the activation and action of particular genes from the zygotic genome.

Specifying the Mesoderm

Among the first of these zygotic genes to be discovered were two that are needed to specify the mesoderm, which forms from the band of cells that invaginate along the ventral midline. These two genes, *twist* (*twi*) (at 2–100) and *snail* (*sna*) (at 2–51), were discovered in an extensive mutant hunt of second chromosome zygotic mutants that give altered embryonic cuticular patterns (Nüsslein-Volhard et al., 1984). The *twi* and *sna* phenotypes are very similar to those of embryos produced by mothers mutant for *dl* or other members of the dorsal group, the first visible developmental defect being the failure to form the ventral furrow and consequent absence of mesodermal tube formation.

While the phenotypic similarities between the embryos derived from dorsal group maternal mutants and those homozygous for *twi* or *sna* could, in principle, be coincidental, some early genetic evidence, involving synergistic interactions, indicated that the wild-type *dl* and *twi* and *sna* genes are part of the same functional network (Simpson, 1983). The test involved an examination for interaction between reduced maternal dosage of dl^+ activity, which creates a temperature-sensitive partial dorsalization of embryos, and reduced zygotic genome dosage for *twi* or *sna*. Although *twi* heterozygotes from homozygous dl^+ mothers are fully viable at temperatures from 18°C to 29°C, such progeny from heterozygous *dl/+* mothers show enhanced temperature-sensitive lethality and dorsalization; their +/+ siblings do not show the effect. A smaller degree of synergism is found for *sna* heterozygotes. The effect is not a function of particular alleles but of reduced maternal *dl* and zygotic *twi* and *sna* dosage.

In effect, deficiency for *twi* activity, and to a lesser extent that of *sna*, enhances the *dl* dominant effect, suggesting that all three genes affect the same process in d-v pattern setting at the midventral position. (Furthermore, increased *zygotic* dosage of dl^+ can produce some measure of rescue of *twi* heterozygotes from *dl/+* females, showing that *dl* is zygotically expressed, despite the absence of measurable paternal rescue of $dl/^+$ embryos [Simpson, 1983]. This illustrates the point made in chapter 3 that paternal rescue tests provide only a rough measure of whether or not early zygotic expression of a gene occurs.)

Although the results establish that the three genes are linked in a common process, they do not reveal the precise nature of the functional connection, whether the effects are on mesoderm formation or on gastrulation (ventral furrow formation) *per se*, since both processes are affected concomitantly. Some molecular evidence, however, indicates that the primary effect is on mesoderm formation. Both *twi* and *sna* have been cloned and their sequences and nuclear location indicate that they too, like *dl,* are transcriptional activators (Boulay et al., 1988; Thisse et al., 1988). By *in situ* hybrid-ization with the cloned sequence to the embryo, Thisse et al. (1987) showed that *twi* is expressed in the invaginating cells of the ventral furrow and later in the differenti-ated mesoderm. These investigators also established the nature of the functional link between *dl* and *twi* indicated by Simpson's experiments: In the absence of *dl* activity, *twi* is not transcribed, showing that *dl* is necessary for the transcriptional activation of *twi*.

With the cloned sequences for both *twi* and *sna*, Leptin and Grunewald (1990) investigated the relationship between the expression patterns of these two genes and the pattern of gastrulation in various genetic backgrounds. The *twi* pattern was shown earlier in this book (Fig. 2.1). The key observations are that while *twi* and *sna* are required for mesoderm formation, their expression can be uncoupled from the events of gastrulation in several respects. Thus, for instance, both *twi* and *sna* embryos exhibit partial ventral furrow formation though no mesodermal tissue is formed.

In addition, ventralized embryos from dominant *Toll* mothers show uniform *twi* expression (consistent with the ventralization phenotype) around the circumference of the blastoderm embryo but still undergo a localized ventral furrow formation (though the furrow is more extensive) in the "appropriate" place. Evidently, there is a localized predisposition for ventral furrow formation that is independent of *twi* and *sna* formation; these two genes appear to be required for mesoderm formation specifically and not for the morphogenetic events that accompany mesoderm forma-tion. Furthermore, embryos that are homozygous mutant for both *twi* and *sna* differ in their cellular behavior in several respects from the singly mutant embryos. Thus, while *twi* is expressed prior to *sna* (first appearing during late syncytial blastoderm), it does not simply serve to regulate *sna* expression. *sna*, in fact, is not restricted to expression in mesoderm, as is *twi*, but is expressed later in embryogenesis in other tissues (Grau et al., 1984; Alberga et al., 1991).

The Neuroectoderm

The invagination of the cells that give rise to the mesoderm leaves a thin strip of mesectoderm cells (which also initially express *twi*) at the ventral midline, bordered by two broad lateral bands of neuroectoderm, totaling approximately 2000 cells. It is in this region that the *dl* nuclear gradient is particularly steep, diminishing to zero near its dorsal edge (Rushlow et al., 1989).

Within the neuroectoderm, the second major observable event in the differentia-tion of embryonic d-v pattern takes place, the separation of neuroblasts from surface ectoderm, the hypoderm. Altogether, one-quarter of the neuroectoderm, or about 500 cells, will become neuroblasts, these cells ultimately giving rise to the CNS. The process by which a particular cell is "chosen" to become a neuroblast appears to be quasi-random, but once the initial choice is made, that cell appears to inhibit its immediate neighbors from becoming neuroblasts.

The principal evidence for such "lateral inhibition" of neuroblast formation comes from two sources: laser beam ablation experiments in the grasshopper embryo, in which elimination of a newly initiated neuroblast is soon followed by replacement of one of its neighbors (Doe and Goodman, 1985), and cell transplantation experiments in *Drosophila*, where removal of individual cells from the neuroectoderm to a more dorsal position leads to their development as neural cells at a significantly higher frequency than 25% (Technau and Campos-Ortega, 1986a).

Not surprisingly, both populations of cells in the neuroectoderm, the epidermoblasts and the neuroblasts, have specific genetic requirements. Deficient genetic activity for a particular group of genes, termed the "neurogenics," leads to excessive numbers of neuroblasts at the expense of epidermal cells; these genes are required for epidermoblast development. In contrast, inactivation of a smaller, less well-defined gene set, the "proneurals," leads to a reduction in number of the neuroblasts and these genes must be required for neuroblast formation.

Neurogenic genes. The neurogenic genes were the first to be categorized as a group. Indeed, observations on one of these genes, the X chromosomal locus *Notch* (*N*), were among the earliest in *Drosophila* developmental genetics. In the late 1930s, Donald Poulson reported that in homozygous embryos of *N* (named *Notch* for its heterozygous dominant wing phenotype in adults), considerably more neuroblasts are formed in the neurogenic region, with a consequent hyperplasia of the CNS (Poulson, 1937). As pointed out by Wright (1970), this neurogenic phenotype involves a recruitment of cells, which would otherwise give rise to epidermis, to the neural developmental pathway rather than stimulated proliferation of the neuroblasts. The neurogenic phenotype is illustrated in Figure 7.18. Originally, a group of seven genes whose mutants give the neurogenic phenotype were identified: *N* (1–3.0), *big brain* (*bib*) (2–34.7), *master mind* (*mam*) (2–70.3), *neuralize* (*neu*) (3–50), *Delta* (*Dl*) (3–66.2), *almondex* (*amx*) (1–27.7), and *Enhancer of split* (*E(spl)*) (3–89.1) (Lehmann et al., 1981, 1983; Jiminez and Campos-Ortega, 1982). The neurogenic phenotype for all seven genes reflects the loss of wild-type gene activity rather than neomorphic effects. Differences in the "strength" of the neurogenic phenotype among them, even among amorphs, have been shown to reflect differing degrees of maternal expression (Jiminez and Campos-Ortega, 1982).

The first clues to the wild-type functions of these genes were provided by the discovery of lateral inhibition in the neuroectoderm, which implicates the existence of some form of cell signaling process by means of which cells "talk" to each other and negotiate their fates. Signaling systems necessarily involve both genetic machinery for producing the signal(s) and receptors for responding to it.

The molecular characterization of two of the genes suggested the possibility that they might encode part of the reception apparatus. The first to be cloned was *N*. Like *lin-12* in *C. elegans*, the *N* gene encodes a large (2703 residue) transmembrane protein with multiple repeats of an EGF (epidermal growth factor)-like motif. There are 36 of these repeats altogether, and they are located in the extracytoplasmic domain of the protein (Wharton et al., 1985). The structure of *N* is, in its overall organization, similar to *lin-12* and *glp-1* of *C. elegans* (Fig. 6.17).

Although the resemblance to EGF suggests that the *N* might encode a diffusible factor, a possibility if the extracellular domain is cleavable from the rest of the protein, mosaic analysis argues against this idea. Clonal analysis, using induced mitotic recombination, in adult flies (Dietrich and Campos-Ortega, 1984) and gynandromorph analysis in *Drosophila* embryos (Hoppe and Greenspan, 1986) shows that *N* cells display distinct cellular autonomy; *N* clonal patches show no signs of rescue by neighboring wild-type tissue to develop as wild-type, as one might expect if *N* encoded a freely diffusible factor. The results suggest, therefore, that the *N* protein is localized

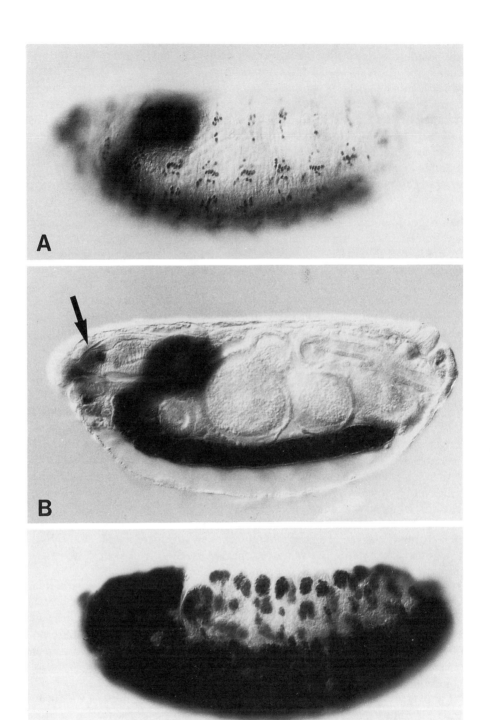

Fig. 7.18. Expression of a neural marker in wild-type embryos (**A,** lateral view; **B,** sagittal view) and in a neurogenic embryo (**C**). (Reproduced from Knust and Campos-Ortega, 1989, with permission of the publisher. © ICSU Press.)

to the cell surface and there participates as part of a receptor apparatus in an intercellular signaling process that mediates lateral inhibition. The receptor function may be shared in some manner: *Dl* similarly encodes a transmembrane protein, though a smaller one (approximately one-third the size of *N*), with an extracellular domain containing nine EGF-like repeats (Vassin et al., 1987).

Indeed, *N* and *Dl* can interact at the cell surface to produce cell adhesion, a reaction that may be part of the signal transmission process. The evidence comes from mixing cells transfected with either *N* or *Dl* and immunostaining for both expressed proteins. It is found that the two populations of cells can adhere together (Fehon et al., 1990). Furthermore, under these conditions, *N*-expressing cells do not exhibit self-adhesion while *Dl*-expressing cells do, suggesting the possibility that, in some fashion, *N* and *Dl* may compete with each other at the cell surface.

Though *N* did not exhibit self-adhesive (homophilic) behavior under these conditions, there is some genetic evidence that *N* transmembrane molecules within the same cell can interact. This comes from the finding of "negative complementation" between certain so-called *Abruptex* (*Ax*) mutations in the locus (Welshons, 1971). Unlike the canonical *N* mutations, *Ax* mutations of *N* produce gapped wing veins and some bristle defects. While many are viable as homozygotes and as hemizygotes, certain heterozygotic combinations are lethal, the phenomenon of negative complementation. The simplest explanation is that their gene products produce a nonfunctional dimer (or higher-order aggregate) and, by implication, the wild-type products also form dimers or oligomers. (Negative complementation, it should be mentioned, is neither unique to *N* nor to *Drosophila;* a comparable situation, involving the *W* gene of the mouse, will be described in chapter 8.)

Xu et al. (1990) employed negative complementation, involving two *Ax* mutations in the EGF region, to determine which genes specify products that either interact with *N* or which act downstream of it, mediating the lethal *Ax* signal. They found that loss-of-function mutations in only two genes would act as suppressors; the genes are two of the previously identified neurogenics, *Dl* and *mam*. The identification of *Dl* confirms the importance *in vivo* of the *in vitro* adhesion results while the identification of *mam* indicates that its product is part of the immediate interactive machinery.

Additional suppression results reported by Xu et al. (1990) implicate the cytoplasmic domain of *N* as involved in the lethal *Ax* interaction, presumably as the portion of the molecule that transmits the signal. One of the interactions detected involves a point mutation in *E(spl)* and this portion of the molecule. (*spl* mutations are recessive viable *N* mutations and the *E(spl)* locus was initially detected because of the enhancement by one of its dominant mutations of the *spl* phenotype.)

E(spl) is a good candidate for a downstream function from *N* and *Dl*. It is a particularly complex gene and contains 10 nonoverlapping transcription units within 35 kb. Some of these transcripts show homology to the c-*myc* class of DNA binding proteins (reviewed in Knust and Campos-Ortega, 1989) while others show homology to a so-called G protein, one involved in the transduction of signals received at the cell surface (Hartley et al., 1988). The molecular evidence suggests that *E(spl)* is involved in several aspects of the transduction of the epidermis-promoting signal, whose precise interrelationships are still unclear.

These findings define part of the putative signaling pathway involving the neurogenic genes, which begins with a cell adhesion event, produced by *N* and *Dl*, which, in turn, generates a signal mediated by one or more products of *E(spl)*. Yet the functions of the other neurogenic genes (*bib, amx, neu*) remain obscure. However, another genetic approach suggests that *neu* and *amx*, though not *bib*, are part of the *N-Dl-mam-E(spl)* pathway.

The approach makes use of the fact that the purely zygotic effects of the mutants, except for *Dl*, produce less than complete conversion of the neuroectoderm. For genes whose products are part of the same pathway, increasing the wild-type gene dosage for a gene that is downstream of another gene in this sequence should, in principle, either enhance or reduce the extent of neurogenesis. In contrast, modulating the dosage of a wild-type gene that normally acts before (upstream) the gene whose mutant affect is being assayed should have no effect on the extent of neurogenesis (de La Concha et al., 1988). (To test *Dl*, whose amorphic zygotic phenotype produces nearly complete neurogenesis, *Dl* hypomorphs were used.)

The results show a distinct pattern of nonreciprocal pairwise effects, which can be assembled into a pathway of putative actions (Fig. 7.19). The scheme, it should be emphasized, is purely a formal one; such tests cannot reveal whether the steps involve alterations in synthesis of the gene products (a formal possibility) or modulations of protein activity, or combinations of both kinds of effect. The significant feature is the demonstration of a sequence of interactions, with *bib* acting independently of the other genes in the generation of the signal. The elucidation of its full meaning in molecular terms awaits future studies. Yet, the outcome of the operation of the pathway in wild-type development is clear, namely the suppression of neural development. This probably involves the inhibition of activity of one or more of the proneural genes, a group to which we now turn.

Proneural genes. The proneural genes are defined as those necessary for neuroblast development (Garcia-Bellido, 1979; Campos-Ortega and Hartenstein, 1985; Caudy et al., 1988). Mutants of these genes show a reduction in the numbers of neural cells in either the CNS or PNS, or, generally, both. The best characterized of these proneural genes, and the first to be identified, are those encoded by the so-called achaete-scute complex (AS-C) (Campuzano et al., 1985). These genes were initially named for hypomorphic mutants affecting particular bristle patterns on the mesotho-

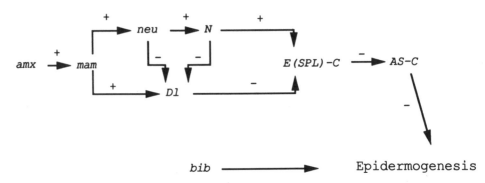

Fig. 7.19. The genetic pathway in neurogenesis, as deduced from gene dosage effects. A plus sign indicates an inferred activation of product (or expression), a minus sign, an inferred inhibition. See text for discussion. (Figure courtesy of Dr. J. Campos-Ortega.)

rax (Dubinin, 1929). (Bristles are innervated structures and the genes of the AS-C are now known to be required for development of the PNS, in both the embryo and imago. These requirements in the imaginal PNS, particularly with respect to bristle development, will be discussed in chapter 9.)

Following the first studies on bristle pattern effects, a gene closely linked to *achaete* and *scute,* which is now known to be part of the AS-C, was shown to encode one or more vital functions (Muller and Prokofyeva, 1935). Embryos that are completely deficient for this locus, named *lethal of scute (l'sc),* show greatly reduced development of both the CNS and PNS of the embryo (Garcia-Bellido and Santamaria, 1978; Dambly-Chaudière and Ghysen, 1987).

These early effects on neural development involve the neuroblasts directly rather than their descendant cells. Elimination of the AS-C complex suppresses the formation of certain neuroblasts in the neuroectoderm of otherwise wild-type embryos (Cabrera et al., 1987) and also suppresses the phenotype of most neurogenic mutants (with the exception of *bib* mutants) (Campos-Ortega and Brand, 1988). Interestingly, however, AS-C deficiency does not eliminate all neuroblast formation nor completely suppress the neurogenic phenotype, suggesting that there is some degree of functional redundancy between genes of the AS-C and other proneural genes.

Some evidence suggests that the AS-C has a role in early maintenance rather than in the initial step of neuroblast determination. Campos-Ortega and Brand (1988) report that in embryos doubly deficient for a neurogenic gene and the AS-C, all cells in the neuroectoderm initiate neural development (as shown by antibody staining of a neural antigen) but that many cells subsequently lose their neural character. It is possible, of course, that AS-C functions are required for both initiating and maintaining neural development. Several of the transcripts encode a homeodomain protein homologous to the vertebrate proto-oncogene, c-*myc,* and the *Drosophila* genes are now known to be transcriptional regulators (reviewed in Garrell and Campuzano, 1991).

From the genetic evidence, indeed, it seems probable that the neurogenic gene activity serves to inhibit AS-C activity in those cells that become epidermal precursors. The genetic data, however, do not define the molecular nature of this regulatory system. Cabrera (1990) has found that the T3 transcript of the AS-C, the transcript encoded by *l'sc,* is made within the presumptive neuroectoderm, well before the synthesis of its protein begins. However, only in those cells that become neuroblasts is T3 protein made, showing that the "decision" to become a neuroblast involves release from a block to translation of this RNA. In contrast, in embryos homozygous for *N* or *Dl,* synthesis of T3 protein is seen in many more cells within the neuroectoderm, beginning at the time the translational block is lifted in wild-type, and these are the cells that become the additional neuroblasts in the neurogenic embryos. Evidently, at least one important output of the neurogenic gene pathway (Fig. 7.19) is to block translation of AS-C gene transcripts (and, possibly, of transcripts of other proneural genes as well).

The Dorsalmost Tissues and the Problem of Integrating the Dorsoventral Pattern

Dorsal to the neurogenic ectoderm are regions that give rise to epidermis and, at the dorsalmost position, the extraembryonic membranes or amnioserosa. Isolation of various zygotic mutants that show a partial or complete "ventralization" of the embryo, with a concomitant reduction or elimination of these dorsal tissues, has identified several zygotic genes required for the specification of these tissues.

The *zen* gene is one such gene. Discovered as a lethal complementation group within the ANT-C (Fig. 7.16) and initially classified phenotypically as a gastrulation defective incapable of normal germ band elongation (Wakimoto et al., 1984), the primary developmental lesion was subsequently recognized as a failure of dorsal tissue specification. *In situ* hybridization experiments with the cloned gene demonstrate that it is expressed only in the dorsal region of the embryo, along its full length, starting in syncytial blastoderm, with expression becoming localized to the presumptive amnioserosa by gastrulation (Doyle et al., 1986). In embryos from mothers deficient in activity for genes of the *dl* group, however, *zen* is expressed ubiquitously along the d-v axis (Rushlow et al., 1987). Evidently, *zen* is repressed by the *dl* group genes, and, indeed, the *zen* promoter has a high affinity for *dl* protein (Ip et al., 1991). These findings suggest that *dl* directly represses *zen* transcription and that expression of *zen* in the dorsal region of wild-type embryos reflects the lack of nuclear *dl* protein in those cells.

In fact, *zen* is probably part of a transcriptional cascade, being repressed by *dl* but regulating the transcription of various downstream genes. The *zen* locus consists of two closely linked, homeobox-containing genes, of which only one, Z1, is required to supply *zen*+ function (as shown by germ line transformation and rescue of *zen* phenotype by this gene) (Rushlow et al., 1987). Presumably, the downstream genes that *zen* regulates are required for amnioserosal formation.

A number of other zygotic genes are expressed and required in the dorsal region of the embryo but the most unusual, in terms of its genetics and developmental functions, is *decapentaplegic* (*dpp*). Originally thought to be a gene complex and named the decapentaplegic complex or DPP-C on the basis of its imaginal mutant phenotypes (in which defects were found in 15 of the 19 imaginal discs) (Spencer et al., 1982), *dpp* is now known to consist of a single coding region that possesses extensive 5' and 3' *cis*-acting regulatory regions (St. Johnston and Gelbart, 1987).

While mutations in the *cis*-acting regions are now known to be responsible for the imaginal phenotypes (which we will return to), loss-of-function mutations in the coding region cause a dramatic alteration of embryonic phenotype along the d-v axis. Amorphs for this region, which encodes a 3.5 kb mRNA, produce a haplo-lethal condition (an unusual characteristic in itself), termed dpp^{Hin} (for *ddp*-haplo-insufficient). Heterozygous embryos show deficiencies for dorsal tissues while homozygotes are completely ventralized, with abdominal denticle belts running circumferentially around the embryo (Irish and Gelbart, 1987). Furthermore, there is no detectable maternal expression; embryos derived from homozygous germ line clones appear identical to those experiencing simply a zygotic deficiency (Irish and Gelbart, 1987).

The results indicate that the *dpp*-encoded function is required in two doses zygotically for normal dorsal and lateral d-v patterning. *In situ* hybridization results are consistent with this inference; the transcript appears in a wide dorsal band in late syncytial blastoderm along the full length of the embryo and then becomes localized to the future dorsal epidermis. Subsequently, the pattern breaks up into two stripes, one just ventral to the amnioserosal region, the other just above the neuroectoderm, with other areas of expression appearing in parts of the visceral mesoderm and the fore- and hindgut (St. Johnston and Gelbart, 1987).

In embryos derived from *dl* mothers and carrying fewer than two zygotic doses of *dpp*[Hin], the embryos appear neither fully dorsalized (as genetically wild-type embryos from *dl* mothers would) nor fully ventralized (if deficient for *dpp*[Hin] but from wild-type mothers) but appear to show circumferential lateral pattern elements (Irish and Gelbart, 1987). This observation suggests that zygotic *dpp* function is regulated, directly or indirectly, by the maternal dorsal group genes and that there must be genes other than *dpp* which the nuclear *dl* gradient inhibits, again directly or indirectly, in the dorsal or lateral regions of the embryo.

These results focus attention on an as yet unresolved question about patterning along the d-v axis in the embryo. It is clear that the maternal nuclear *dl* gradient differentially regulates a number of key zygotic genes along the d-v axis, involving both specific repression and activation events. Given that *dl* nuclear concentration becomes vanishingly small in regions dorsal to the neurogenic ectoderm, it seems unlikely that all of these localized regulatory outcomes are direct responses to nuclear *dl* protein concentration. Is some other mechanism involved for the lateral and dorsal regions? One possibility is that there are only a small number of discrete responses, perhaps at the extremes of the axis (the ventral and dorsalmost positions) with a great deal of filling in, in the intermediate positions. Such filling in, if it occurs, would presumably involve a sequence of cellular interactions, akin to the creation of pattern in segmental anlagen mediated by the segment polarity genes.

IMAGINAL DISC DEVELOPMENT AND PATTERNING

Larval and Imaginal Cells: How Different Are They?

The imaginal disc cells are markedly different from all the other cells of the embryo, which go on to form the body cells of the larva. First distinguishable as clusters of cells in the newly hatched first instar larva, the cells of the imaginal discs are small, diploid, and essentially undifferentiated until the pupal period; only the eye imaginal disc, revealing the presence of ommatidia in late third instar, shows signs of its future role. In contrast to the imaginal cells, most of the larval cells are large, have big nuclei with polytene chromosomes, and are differentiated with respect to one another from early or mid-embryogenesis onward. Although most cells in the larval stages can be grouped in these categories, a small proportion cannot: Some of the larval cells, such as those of the larval ganglia, remain diploid and continue to divide, while the histoblast cells, which give rise to the imaginal abdominal integument, are larger than disc cells and although diploid, are partially differentiated, participating in the secretion of larval cuticle (Madhavan and Schneiderman, 1977).

One hypothesis to explain the broad differences between imaginal disc cells and larval cells is that there are distinctly different imaginal and larval developmental pathways, involving the expression of unique sets of genes. If imaginal and larval development differ in this manner, then it should prove possible to isolate mutants that form normal larvae but which are deficient in the formation or structure of their imaginal discs. (The hypothetical complementary class of mutants, defective in presumptive larval-specific genes, would die in embryogenesis and therefore be classified as embryogenesis defectives.) Recessive mutants of imaginal cell- or disc-specific functions should be viable as homozygotes throughout larval development but would die at or near the time of pupation, as the imaginal disc program commenced. A precedent for such imaginal disc mutants has long been known: The recessive *lethal giant larva* (*lgl*), described by Hadorn (1961), forms functional larvae possessing severely defective discs. These larvae never pupate but reach abnormally large sizes before dying. (Their large size reflects their impaired synthesis of ecdysone, which during normal development is secreted by the imaginal ring gland.)

To examine the genetic basis of imaginal disc development, Shearn et al. (1971) carried out a large screen for third chromosome late larval lethal mutants, following chemical mutagenesis. The search produced 134 mutant strains, out of a total of 3167 mutagenized lines. Of these, 66 showed one or more disc types absent or defective in third instar larvae. One small class of mutant consisted of those defective in particular pairs of disc types. However, the great majority, 64%, proved defective in all their imaginal discs (Shearn, 1977), either lacking discs entirely (discless) or showing rudimentary discs (the small disc class). Both kinds of mutants can also be found in X chromosome mutant searches (Stewart et al., 1972). Indeed, the genome is probably liberally sprinkled with genes that can mutate to give a seemingly specific disc-defective condition. Shearn and Garen (1974) calculated that there are a total of 1000 genes specifically required for general imaginal disc development and inessential for larval development.

Although these results seemed to support the notion of distinct imaginal disc and larval developmental pathways, closer inspection of the mutants and subsequent genetic results have undermined this notion. In the first place, many of the presumptive disc-specific defectives have larval defects. A simple criterion of larval defectiveness involves time of death; any mutant strain which dies before puparium formation must be deficient in some larval-essential function. By this measure, 38% of the original 66 third chromosome imaginal defectives are also larval defectives (Shearn, 1977).

In the second place, whether or not a mutant appears to be a disc defective or a more general lethal is largely, and perhaps entirely, a function of the mutant allele rather than of the genetic locus itself. Different mutant alleles of the same gene can give either the discless or small disc mutant phenotype (Shearn et al., 1971) or the small disc phenotype or that of a general lethal, the difference reflecting different degrees of hypomorphism (Shearn et al., 1978a). Furthermore, some or many of the genes identified by single mutants as lethals may also mutate to alleles that give the disc-defective condition (Shearn et al., 1978b). Clearly, disc defectiveness does not rigorously define a group of genes expressed solely in discs.

It is nevertheless of interest to know what the characteristics of the late larval lethals are and why it is so comparatively easy to obtain mutants that seem specific for imaginal disc development. One clue came from an unexpected source, an investigation of mutants involved in DNA repair mechanisms. In *Drosophila,* as in several other organisms, one can isolate viable mutants of such repair functions, these mutants being characterized by high sensitivity to certain forms of induced DNA damage. Severely defective mutants of two such repair gene functions, first identified by hypomorphic viable mutants which exhibit a mutagen-sensitive (*mus*) phenotype, were found to give a late larval phenotype (Baker et al., 1982). Examination of the cytological defects in these lethals revealed a sufficiently high frequency of chromosome breaks and aberrations in their dividing larval ganglion cells to account for the death of all dividing cells, whether imaginal or diploid larval cells.

Furthermore, examination of a larger set of late larval lethals showed that a high percentage exhibit mitotic abnormalities (cited in Baker et al., 1982). Finally, Szabad and Bryant (1982) showed that in a group of five discless mutants, the mutants show a strong inhibition of division in all cells that normally divide, both larval and imaginal. The larval cell defects show up as division blocks in the larval ganglia and reduced numbers of blood cells in the lymph glands. Careful inspection of the mutants shows that all, despite their initial classification as discless, possess vestigial discs, containing the same cell numbers as those present in hatching wild-type L1 larvae. (One mutant, however, the X chromosome mutant *dl-1,* seems to lack even rudimentary wing and haltere discs, though possessing vestigial discs of the other kinds.) Thus, even in the most extreme class of imaginal disc defectives, the discless mutants, imaginal discs are normally established in nearly all instances but fail to undergo normal cell proliferation. The fact that such division-defective mutants can carry out cell division during embryogenesis probably reflects maternal storage of essential division components in the egg, provided by the heterozygous mother.

In conclusion, the imaginal disc cells as a group do not require the expression of a gene set very different from that employed by larval cells as a whole. And, indeed, many imaginal cells cease to divide and develop polytene chromosomes later in development, in this sense acquiring a larval cell phenotype. The essential difference between the two cell types may therefore be one of timing in the shutdown of normal division and the onset of polyteny. One implication of this view of the imaginal/larval cell distinction is the expectation that contiguous larval and imaginal cells will often have more in common with one another than do cells within each category that come from different blastodermal regions. This idea has been borne out in two different ways: The blastodermal fate maps of larval and imaginal precursors are well matched with respect to placement of the anlagen for the same segments and, secondly, the larval and imaginal pathways within segments are governed by many of the same genes (Lewis, 1978).

Characterizing the Imago by Clonal Analysis: Fate Mapping With Gynandromorphs and Analyzing Imaginal Growth by Mitotic Recombination

Because the imaginal cells are undifferentiated and exist within a sea of larval cells, their characterization during embryogenesis and larval growth presents a challenge. Much of this characterization, in fact, has been indirect and employed the techniques of clonal analysis. These methods have been those of gynandromorph fate mapping and induced mitotic recombination. The principal application of gynandromorph fate mapping has been to determine the relative locations of the imaginal precursor cells on the blastoderm-stage embryo. In contrast, clonal analysis via induced mitotic recombination has concentrated on characterizing the growth of imaginal anlagen following the blastoderm stage and searching for developmental restrictions of developmental capacity during the phase of imaginal growth. The finding that such restrictions do take place has raised some questions about their genetic basis and significance.

Gynandromorph Fate Mapping

Gynandromorph fate mapping was invented by A.H. Sturtevant (1929) for localizing the precursor cells of imaginal structures, but the method lay unutilized for 40 years until revived by Garcia-Bellido and Merriam (1969), who analyzed Sturtevant's data in detail and refined the method. Fate mapping the surface of the adult onto the blastoderm embryo involves deducing the mapping transformation between two different surfaces. This is accomplished by positioning the anlagen of structures initially two at a time, then triangulating the primordium for a third structure with respect to the first two, then adding a fourth by the same method and so on.

The data for such a mapping are provided by a set of adult gynandromorphs—the larger the set the better—in which a recessive marker on the "exposed" X chromosome in the XO tissue can be clearly discerned in the adult gynanders (Fig. 4.12). In the imago, there is a distinctly stronger tendency for mosaic boundaries to follow segmental boundaries and the longitudinal midline than in the larva, an effect of the separate development and subsequent suturing together of the different parts of the imago, relative to larval gynandromorphs (compare Figs. 4.12 and 4.13). As in the larval mapping, however, the distance between each pair of structures is measured in sturts, the percentage of gynandromorph sides in which the structures are of different X chromosomal composition.

To position the fate map with respect to the physical surface of the embryo, one needs some external reference point. A convenient one is distance to the dorsal midline for the various imaginal primordia, obtained by dividing the sturt distance between equivalent structures on left and right sides by two. Anteroposterior locations are obtained by normalizing distances along the longitudinal axis to the d-v distances (Garcia-Bellido and Merriam, 1969). A typical gynandromorph map is shown in Figure 7.20. Mesodermal and CNS anlagen are not shown on the map because there are not yet appropriate markers for these tissues on the X chromosome.

The placement of some primordia in circles emphasizes the statistical nature of the gynandromorph map. Because restrictions on potency and assignments of developmental fate continue right up to the pupal stage with an ever-expanding imaginal cell population, it follows that the map itself cannot be a strict one-to-one localization of anlagen to invariant blastodermal cell positions. Each point on the gynandromorph fate map, or each center of a circle (for large primordia), is the *high point of the probability distribution* for the cells that give rise to that structure or cell type (Janning et al., 1979).

To interpret the biological significance of the fate map, one needs to know which stage of embryonic development is being mapped. All of the evidence points to the cellular blastoderm as the stage being described. In the first place, the map closely resembles the fate map of the embryo derived from direct observation, in which the external ectodermal structures are lateral and dorsal, the neural structures ventrolateral, and the mesodermal-derived structures ventralmost. In the second place, and most tellingly, it is possible to map the germ line precursors (the pole cells) relative to the gonadal mesoderm, and these results place the pole cells at the posterior pole

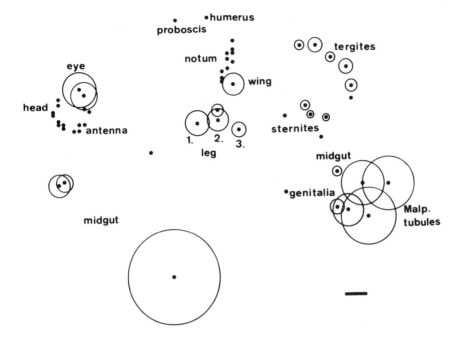

Fig. 7.20. Gynandromorph fate map of imaginal primordia. The diameter of the circle is a measure of the size of the primordia at blastoderm. The relative positioning of the anlagen largely prefigures that of the adult structures. Constructed by means of an empirically derived mapping function, which corrects for the apparent shrinkage of distance between more distant landmarks. (Reproduced from Janning et al., 1979, with permission of the publisher.)

of the embryo. The only cellular stage at which this holds true is blastoderm; immediately following blastoderm, the pole cells move dorsally and then internally. Gehring (1976) was able to map the germ line, starting from the premise that gynandromorphs lacking mature cells possess a gonadal mesoderm of different sex composition from the germ line itself (see van Deusen, 1976). The mapping placed the germ line precursors at the posterior pole, as expected; a similar result, employing a different mapping strategy, was obtained by Nissani (1977). The results confirm that the blastoderm embryo is the stage whose fate map is described by sturt distances measured from adult gynandromorphs.

In addition to plotting locations of imaginal precursor cells, gynandromorph mapping can be used to estimate relative numbers of cells in each primordium. Absolute numbers of precursor cells can be estimated from the inverse of the smallest markable fraction, for a given structure or region, from a set of gynandromorphs but such estimates are only approximate ones (Merriam, 1978). Undoubtedly, however, the most useful application of gynandromorph mapping is in ascertaining the "focus" of action of mutations on the X chromosome. The focus is the specific location in which deficiency for the gene activity produces the phenotype. In mapping the focus, one is, in effect, determining where the wild-type allele must be expressed in order to produce wild-type development.

The method was first used to ascertain the tissue site of action of various behavioral mutants (Hotta and Benzer, 1973). However, the method is also useful for determining the focus of certain developmental mutations. An example is that of mutations in

the *disconnected* (*disco*) gene (mapping at 1–53) whose phenotype involves the failure
of eye photoreceptor neuron axons to reach the developing optic ganglia; the consequence is degeneration of these ganglia. Gynandromorph fate mapping showed the
focus of *disco* to be far from the precursor cells of both the optic ganglia and the eye
disc itself (Steller et al., 1987). Subsequent examination of late larvae indicated that
the initial disruption was in a pioneer nerve, Bolwig's nerve, which provides the initial
connection between the eye discs and the optic ganglia (Steller et al., 1987).

Mitotic Recombination: Characterizing Growth Parameters

Gynandromorph fate mapping provides a description of imaginal cell precursors
at the blastoderm stage, but cannot be used for characterizing the precursor cell
numbers or positions at later (postblastodermal) stages of development. For such
characterizations, induced mitotic recombination has proved versatile and informative. For instance, by inducing marked clones at various stages of development, one
can ascertain the rates of cell proliferation of different imaginal cell groups.

For such experiments, one employs genetic markers that alter a visible cell phenotype without affecting the underlying cellular growth pattern. Suitable "growth
neutral" markers include those that affect general cuticular pigmentation, such as
yellow (*y*), bristle morphology, for example, *singed* (*sn*) or *forked* (*f*), or hair morphology, such as *multiple wing hairs* (*mwh*). (Bristles, as noted earlier, are innervated
structures while hairs are formed by single cells.) Since most areas of the *Drosophila*
imaginal cuticle are distinguished by their pigmentation or bristle or hair patterns,
almost any region can be marked and its cell composition and proliferation dynamics
inferred from study of the clonal patterns obtained (Merriam, 1978; Postlethwait,
1978).

The estimates of imaginal cell numbers, using mitotic recombination, are based on
the principle illustrated in Figure 7.21. One induces homozygosity in cells, by X-ray

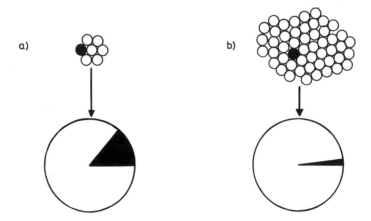

Fig. 7.21. Clone size as a function of cell number. **a:** One out of seven progenitor cells is homozygous for
the marker; in the fully developed structure, approximately one-seventh of the total area shows the marker.
b: One out of 50 cells, in a later stage of the same primordium, is homozygous for the marker; only
one-fiftieth of the surface of the developed structure displays the marker phenotype.

treatment, for scorable markers and then measures the relative sizes of the resulting marked clones in the adult. The basic premise is that the proportion of marked cells in a single clone at the end of development is an inverse measure of the number of precursor cells present at the time of recombinant cell induction (Fig. 7.22). If both daughters become homozygous for different markers (Fig. 4.23), the result is a "twin spot" and the reciprocal of the fraction of all marked cells provides the estimate directly; if one is scoring only one homozygous genotype, one multiplies the reciprocal by one-half.

To measure growth kinetics, one induces clones at successive stages of development and takes the reciprocal of the measured average area marked by the clonal patches at each point, the growth curve then being constructed from these estimated cell numbers. In making the calculations, one needs to take into account whether twin spots (representing both daughters) or just single patches (one daughter only) are being used. The fact that the cells in which mitotic recombination can be induced are in a restricted part of the cell cycle (the G2 phase) must also be taken into account (see Postlethwait, 1978, for the details of this correction).

Growth curves for several imaginal primordia, constructed in this way, are shown in Figure 7.22. In each case, the estimate of final cell number was obtained either by counting, as is possible in the wing where every cell produces a hair (which can be counted directly), or from the relative areas of disc surface and clone surface, using an average cell density measured over a small area. The curves represent average rates of growth throughout the primordium. However, regional differences in precursor cell growth can, in principle, be measured in the same fashion.

The figure shows that (with the apparent exception of the leg) growth is exponential in the imaginal discs from the end of embryogenesis to the end of the third larval instar, with typical doubling times being 8–10 hr. The histoblast nests follow a different pattern: After a brief initial division period in embryogenesis, they do not divide again until the end of larval development. Although the leg appears to follow a third pattern, with an initial period of exponential growth followed by a later tailing off, this plot reflects the fact that only marked bristles were scored in the experiment; the asymptote of 440 reflects the average number of bristle cells in the legs.

These measurements were obtained before there were direct counts of cell numbers in the different discs but the two sets of measurements have been found to be in generally good agreement (Madhavan and Schneiderman, 1977). Nevertheless, the genetic estimates consistently give 1.5 to 4 times fewer estimated cells than the direct measurements. There are at least two reasons for the quantitative discrepancy. The first is that X-rays, used to induce the recombination events, produce some cell mortality. The second and more significant factor is that the genetic experiments score only the outer cuticular cells yet the disc possesses other cell types including tracheal and mesodermal cells. Direct cell counts in discs include these cells whereas clonal analysis of the cuticle perforce misses them. The overall agreement, however, between the genetic estimates and the direct cell counts gives confidence in the genetic method, with respect to its usefulness in situations where direct cell counts cannot be obtained. What the method provides, in the absence of the more laborious direct cytological counts, are good relative measures of cell number and accurate determinations of average or regional growth rates.

The shapes of the clones can also reveal much about the growth characteristics of imaginal cells. First, for most imaginal disc clones, the marked cells comprise a single

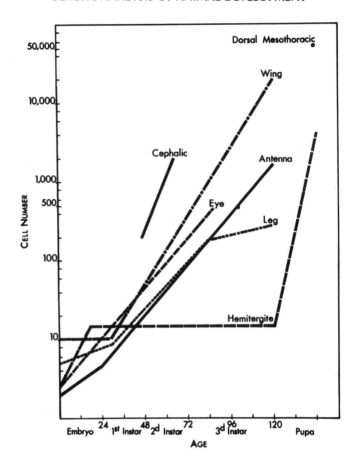

Fig. 7.22. Growth kinetics in imaginal primordia, estimated from mitotic recombination experiments. Detailed references are given in Postlethwait (1978). (Reproduced from Postlethwait, 1978, with permission of the publisher.)

patch, showing that there is little cell dispersion during clonal growth. In contrast to the disc clones, however, those induced in the abdominal histoblasts are often broken up into several patches. This difference between histoblast and imaginal disc clones reflects their different growth patterns. The histoblast cells, unlike those of the imaginal discs, only begin their period of multiplication during the pupal period— rapid growth in this phase tends to disperse members of single descendant clones (Madhavan and Schneiderman, 1977). (The histoblast cells apparently use the larval segment as a "template" of some sort, however; prior injury to the larval segments yields defects in imaginal abdominal segments.)

Second, clones tend to have a shape characteristic of the region in which they arise. In the mesonotum (the proximal body portion of the mesothorax), clones are usually as wide as they are long. In the appendages—antennae, legs, and wings—clones are generally elongate, stretching along the p-d axis. When clones are induced early, these elongate patches can extend over a large part of the appendage, frequently over many contiguous segments in the legs and antennae. In general, such clones are continuous, but in certain cases, reproducible discontinuities are observed for clones in particular regions. Such discontinuities reflect morphogenetic changes during growth. Thus, antennal clones present on both the second and third antennal segments always show a displacement of about 90°, reflecting a quarter-turn rotation between these segments during development of the antenna.

Third, within the limits posed by these regional characteristics, clonal growth has a strong element of indeterminacy. Collections of clones induced at the same time in different individuals and spanning the same region often differ in their precise shape and size. This is illustrated in Figure 7.23. Thus, single progenitor cells within a primordium can give rise to somewhat variable numbers of descendants and specific delimited regions and structures can arise from any of several possible progenitor cells in the early primordium. (The observation of overlapping but different mosaic patterns within imaginal structures in gynandromorphs reflects the same fact.)

It is this degree of indeterminacy in clonal growth that makes the strongest contrast to cell behavior in *C. elegans,* where every division is determined with respect to orientation and, except in the relatively few instances of interchangeable roles, every cell has a fixed number of assigned divisions among its descendants. In *Drosophila,* which possesses many more cells, there is considerably more latitude in cell division

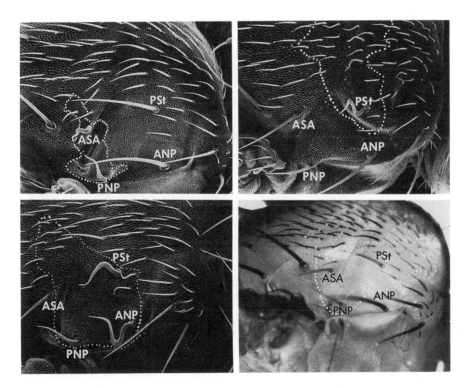

Fig. 7.23. Indeterminate growth of clones on the mesonotum; dotted areas indicate clonal extent. Bristle landmarks; ANP, anterior notopleural bristle; ASA, anterior supra-alar; PNP, posterior notopleural bristle; PSt, posterior sternopleural bristle. (Reproduced from Postlethwait, 1978, with permission of the publisher.)

behavior. Yet there are *some* restrictions on clonal growth and it was mitotic recombination experiments that revealed these.

Restrictions of Cell Fate: Segments and Compartments

The first of the cell fate restrictions to be discovered involved those limiting participation of clones in neighboring discs and segments. To ascertain whether there are such restrictions, one needs to determine the existence and extent of cross-clonal participation between discs whose precursors lie relatively close together. In a detailed gynandromorph fate mapping of four pairs of thoracic discs, the three leg discs and the wing disc, Wieschaus and Gehring (1976a) estimated the size and relative location of their primordia, using specific bristle landmarks within each. The results showed that all four primordia are contiguous, with many interdisc distances being less than many intradisc distances. The precursor cells for one pair of wing and leg bristles were found to be particularly close. The anterior notopleural bristles (aNP) of the wing and the edge bristles of the second leg (IIEB) are only 6.2 sturts apart. As can be seen in Figure 7.24, these bristles are not especially close in the adult (nor are the wing and leg discs contiguous in the larva). If cell assignments are not disc specific at blastoderm, then clones induced at this stage might cover both structures.

In a companion study, Wieschaus and Gehring (1976b) found just this. In 13 clones marked at blastoderm which covered either the edge bristle or the preapical bristle, sever were found to cover part of the wing and then only in the region predicted from the fate mapping, embracing the aNP. Since only 1–2% of all wings and legs had any clones in these areas, the overlap could not have been produced by independently induced clones. By 7 hr of development, however, clones covering either of the respective wing or leg regions do not spread into the other disc. Evidently, some restrictive event intervenes between blastoderm (3 hr) and the 7th hr of embryonic development. Because clone size also decreases by a factor of 2 between these two time points, a cell division is inferred to take place between 3 and 7 hr. However, by

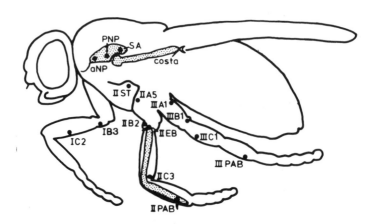

Fig. 7.24. Regions of early clonal overlap between wing and leg. The key bristles illustrating shared wing-leg ancestry are IIEB, edge bristle of the second leg; IIPAB, preapical bristle on the second leg; aNP and pNP, anterior and posterior notopleural bristles of the wing. (Reproduced from Wieschaus and Gehring, 1976b, with permission of the publisher.)

itself, this decrease in clone size is not sufficient to explain the change. Evidently, either a disc-specific restriction in capacity or a disc-specific restriction in cell cross-over between disc progenitors, perhaps reflecting a morphogenetic movement of precursors, must occur.

The study also revealed that possession of common ancestral cells by discs is not simply a function of primordium distance. The closest precursor distance between landmarks of the mesothoracic and metathoracic legs is essentially the same as that between the closest wing and second leg landmarks. However, mitotically induced clones are never observed to overlap both legs. The evidence shows that while the cells of imaginal disc primordia on discs in different, albeit neighboring, segment anlage are prevented at blastoderm from intermingling, imaginal precursor cells *within* a segment are not necessarily limited in this fashion at blastoderm, though subsequently some restrictions on clonal contributions to different discs within segments take place.

As development proceeds, induced clones cover fewer structures and occupy smaller areas. For example, from the blastoderm stage until the middle of the first instar, clones marking bristles on the leg extend from the femur to the fifth tarsal segment, thus stretching nearly the full length of the leg; by the third instar, a given clone is restricted to only one leg segment (Bryant and Schneiderman, 1969). Similarly, in the wing, some clones induced before the beginning of the first larval instar extend onto both the dorsal and ventral surfaces; from the first instar onward, any clone is restricted to one surface or the other (Bryant, 1970).

Cumulatively, such results present a picture of progressive restriction of developmental capacity. However, if there are such restrictions in developmental capacity, they are not, until late in development, to *specific cell types* but rather to particular regions since clones often embrace different cell types. It might be, however, that this progressive narrowing of clonal scope does not reflect an inherent restriction in cellular developmental capacity but only the trivial fact that later clones, being smaller, cannot include as many cells. If the distance between the two structures embraced by the large early clones is greater than the diameter of later clones, then no single clone can extend to both structures. Smaller clones are also more likely to be truncated by any morphogenetic process such as infolding or regionalized cell death; larger early clones are more likely to extend through such local interruptions.

To determine whether the restrictions on clonal extent reflect genuine restrictions in potential or the trivial constraints imposed by limited division capacity, one needs to uncouple growth restrictions from the developmental program. This can be done by generating clones which grow faster, by virtue of some feature in their genetic makeup, than the surrounding cells. If there are fundamental restrictions in developmental capacity that accompany normal imaginal development, these should show up as common boundaries for some fraction of the the fast-growing clones. If, on the other hand, the apparent restrictions on clonal extent observed in the earlier experiments are a direct consequence of growth limits, then no such boundaries should be apparent. The so-called Minute technique permits precisely this kind of experiment to be performed (Garcia-Bellido et al., 1973; Morata and Ripoll, 1975).

The method depends on the fact that a class of dominant mutants, termed *Minutes,* show a slower cellular growth rate, when the mutations are heterozygous, than wild-type. For the whole organism, this manifests itself in a slower growth rate during postembryonic development. Altogether there are about 60 *Minute* loci scattered around the genome. Homozygosis for any *Minute* mutation is a lethal condition, while all strains heterozygous for a *Minute* (the genotype being symbolized as M/M^+) exhibit delayed development (though the extent of delay varies between different mutants). Minute adults are characterized by shortened thinner bristles and, in a few cases, by minor morphogenetic abnormalities. The biochemical basis of the mutant phenotype is obscure but may involve general defects in protein synthesis, possibly involving faulty ribosomal proteins.

When embryos or larvae heterozygous for M mutations are X-irradiated to induce mitotic recombination, two kinds of homozygous daughter cells are produced: M/M, which are cell-lethal and leave no progeny, and M^+/M^+, which are wild-type and grow at the wild-type cell rate. If the M^+ chromosome carries an autonomous cell marker, such as y or mwh, homozygosis for M^+ simultaneously produces cells homozygous for that marker and these cells produce a distinctively marked clone against a wild-type background. The essential point is that because the clone has a wild-type growth rate, while the surrounding cells grow only at the Minute rate, the marked clone has a growth advantage, producing a disproportionately large patch (Morata and Ripoll, 1975). The genetic scheme is illustrated and contrasted with that of standard twin spot analysis in Figure 7.25.

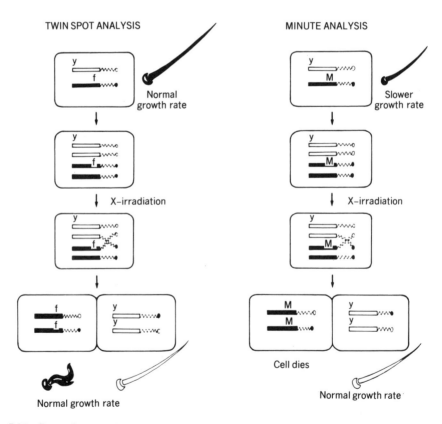

Fig. 7.25. Comparison of mitotic recombination in non-Minute (M^+/M^+) background (left) and Minute background (M/M^+) (right). In the latter, the non-Minute clone (marked with *yellow* (y) outgrows the surrounding nonrecombinant cells, while its homozygous Minute sister dies. (Reproduced from Wieschaus, 1978, with permission of the publisher.)

The Minute technique was initially used to generate large clones in the metathorax; these first results were striking. The induced M^+/M^+ clones were found never to overlap a certain demarcation line (Garcia-Bellido et al., 1973). From the earliest times of clone induction around the blastoderm stage, the marked area can occupy either the anterior part of the mesonotum (the body portion of the mesothorax) and the anterior part of the wing *or* the respective posterior sections of the notum and wing. The boundary line in the wing is perfectly straight for large clones and can extend for hundreds of cells (Fig. 7.26). Interestingly, this line of clonal restriction does not correspond to any obvious morphological feature of the wing (although it may have a special functional significance as the point of wing flexure during flight; see discussion by Weis-Fogh in Lawrence and Morata [1976b]). Somewhat later in development, a second line of restriction appears: New clones are restricted either to the dorsal or ventral surface of the wing. Thus, clones induced after this point are restricted to one of four quadrants, defined by the a-p and d-v axes. Within a demarcated region, early M^+/M^+ clones can occupy up to 60–90% of that region and later clones as much as 30–50% (Garcia-Bellido et al., 1976). (Two further putative restriction events of this kind in the mesothorax are mentioned by Garcia-Bellido et al. [1973], but these are now suspected to reflect morphogenetic separations of cells rather than inherent restrictions imposed between adjacent cells.)

The groups of cells which give rise to one of these demarcated regions have been termed "compartments," and the process by which groups of clones become so restricted, "compartmentalization" (Garcia-Bellido et al., 1973; Garcia-Bellido, 1975). From the fact that single clones are invariably restricted to the anterior (A) or posterior (P) compartment, even when clonal induction occurs at blastoderm, it was originally concluded that this first separation occurs no later than the first division after blastoderm—this division being required to generate the M^+/M^+ cell (Garcia-Bellido, 1975). The development stages at which the dorso-ventral restriction appears to occur in M^+ clones in Minute flies is during the second larval instar (Lawrence and Morata, 1977).

The discovery of compartmentalization had a major impact on *Drosophila* workers. The phenomenon was a clear departure from traditionally described assignments of cell fate and capacity and had been discovered by means of a purely genetic technique. In the classical view, based upon various approaches and observations in amphibia, ascidians, and nematodes (before the cell lineages of the latter had been fully traced), development was viewed as proceeding by a series of specifications of cell *function,* usually with respect to a particular structure or tissue type. In contrast,

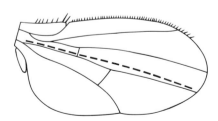

Fig. 7.26. Anterior-posterior compartment border in the wing. (Figure courtesy of Dr. D. Gubb.)

compartmentalization consists of assignments to future *geographical area* without respect to either precise cellular function or prior cell lineage.

The possible significance of compartmentalization was enhanced by the discovery that three recessive, viable homeotic mutations—two in the BX-C (*bithorax* [*bx*] and *postbithorax* [*pbx*]) and the third in a gene known as *engrailed* (*en*)—cause apparently compartment-specific transformations (Garcia-Bellido, 1975). The *en* mutation was of special interest because it transformed posterior wing compartments, incompletely but recognizably, into anterior wing compartments. The effects of this mutation, en^1, are, indeed, general for P compartments; it produces varying degrees of transformation to the corresponding A compartments in the leg and eye-antenna discs, as well as the wing disc (Morata and Lawrence, 1975, 1979a; Lawrence and Morata, 1976a). Furthermore, the transformations are seen in clones: Large homozygous en^1 clones, generated by the Minute technique, in the A compartment of the wing are morphologically normal and "respect" the A-P compartment boundary in the wing where they touch it for hundreds of cells, but clones arising in the P compartment are morphologically abnormal and where they touch the boundary, transgress it (Morata and Lawrence, 1975). From such results, it was concluded that *en* has no effect in the A compartment but is expressed in the P compartment, where it is required both for the normal development of that compartment and the maintenance of the compartment boundary, the latter property perhaps being a consequence of failure of P-type cells to mix with A-type cells (Morata and Lawrence, 1975).

The existence of compartment-specific homeotic transformations led to the formulation of the influential selector gene hypothesis of Garcia-Bellido (1975). A selector gene was defined as one that sets a particular course of development within a compartment by governing the activities of numerous downstream "realizator" genes; it is the activities of the latter that create the phenotype of that compartment (as discussed earlier in connection with the search for the "target genes" of the BX-C).

There were three main elements to the selector gene hypothesis, as originally formulated: 1) that the development of *Drosophila* involves an early, and perhaps progressive, blocking out of groups of cells, namely compartments, to form particular areas of the animal—the embryo develops, in effect, as an increasingly detailed mosaic of compartments; 2) that each compartmentalization event is accompanied by the activation in that compartment of (at least one) selector gene; and 3) the initial activation of particular selector genes is caused by the products of another set of differentially distributed gene products, those of "activator genes" within the early embryo.

The selector gene hypothesis marked a dividing line between the traditional emphasis in *Drosophila* developmental genetics on the study of individual gene and mutant effects and the present-day field that centers on attempts to understand global pattern formation mechanisms in terms of its genetic basis. The presentation of the idea, following the discovery of compartmentalization itself, was catalytic in producing this change of emphasis in the field. More than 15 years after its formulation, however, it is pertinent to ask how well the original hypothesis has held up.

Its key postulate was the universality of compartmentalization as an event in pattern formation. The realities have proved more complex, however. On the one hand, compartment formation *does* appear to be a general event within the cell groups that give rise to external ectodermal structures. Thus, A-P compartments have been found, by means of the Minute technique in the leg discs (Steiner, 1976), the eye-antennal disc (Morata and Lawrence, 1979a), the proboscis, which is formed by the labial disc (Struhl, 1977, 1981a), the haltere (Hayes, 1982), and the genital discs of male and female (Dubendorfer and Nöthiger, 1982).

On the other hand, each of these disc types undergoes compartmentalization in its own way. While in the wing the initial event appears to be the formation of A and P compartments early in embryogenesis, perhaps just after blastoderm formation (a point that we will return to), an A-P separation only occurs in the eye-antennal disc in the second instar, and thus much later than the comparable event in the thorax (Morata and Lawrence, 1979a). The proboscis, like the wing, is divided into A and P compartments along a smooth boundary that follows no obvious morphological discontinuity but further subdivisions could not be detected (Struhl, 1981a). In the legs, there is an early separation into A and P compartments, which may be followed by a further subdivision of the A compartment alone into D (dorsal) and V (ventral) compartments (Steiner, 1976). However, the precise position of the A-P compartment border in both the leg and the antenna has a certain indeterminacy, unlike that of the wing (Morata and Lawrence, 1979; Lawrence and Morata, 1979). In fact, the only "universal" feature is the occurrence of A-P compartmentalization for those cell groups that give rise to the outer surface of the head and thorax of the fly. Progressive compartmentalization events, postulated as a means of specifying increasingly fine regional patterns (Garcia-Bellido et al., 1973; Lawrence and Morata, 1976), have conspicuously failed to emerge.

Furthermore, the search for compartments in internal structures or tissues has had little success. The observations of mesodermal and neuronal tissue, using a variety of clonal analyses and histochemically detectable gene activities, form a complex pattern, but none of these tissues shows a clear A-P compartment separation, nor any effect of en^1 (the original *en* allele) or *en*-lethal alleles, either of transformation or in clone frequency (Ferrus and Kankel, 1981; Lawrence, 1982). Using *maroon-like* (*mal*) as an internal tissue marker, Janning et al. (1986) could not find any sign of compartments in the Malpighian tubules.

A detailed study of the muscles of the thorax does reveal an early separation into dorsal and ventral lineages. Using the Minute technique and succinate dehydrogenase inactivity (Sdh^-) as a marker, Lawrence (1982) showed that early (blastodermal) clones can mark both the dorsal and ventral muscles of the mesothorax. By mid-first instar, however, clones can mark only dorsal or ventral muscles within any of the three thoracic segments. It is not clear whether this separation of lineages reflects an imposed restriction on cell mixing of primordia which continue to lie adjacent to one another or to a physical separation of the primordia. No A-P compartmentalization events were found.

The second major element of the selector gene hypothesis is that compartments comprise the fundamental domains of expression of homeotic genes. A corollary is that the unique properties of individual compartments are produced by the expression of selector genes unique to those compartments. We now know that there are many exceptions to both rules. There is, for instance, much evidence that parasegments (Fig. 7.11) rather than compartments are the domains of expression of genes of the BX-C (Lawrence and Morata, 1983; Hayes et al., 1984; Struhl, 1984). In addition, two of the "genes" whose activities were inferred to represent compartment-specific domains, namely *bx* and *pbx,* are now known not to be genes but regulatory elements for the *Ubx* proteins (Fig. 7.12) (Peifer et al., 1987) and the activities of these elements, in fact, embrace more than single compartments (Cole and Palka, 1982; Peifer and

Bender, 1986). Correspondingly, the transformations observed in various homeotics do not "obey" compartmental boundaries (Struhl, 1982a; Peifer and Bender, 1986; Jorgensen and Garber, 1987). In sum, compartments are not the sole domains for either expression of selector genes or of homeotic transformation nor do single homeotic genes confer unique compartmental properties.

Perhaps the strongest claims for such specification of compartment properties by a single gene were those in connection with *en* of P compartments. Thus, the original observations on *en*[1] phenotypic effects were interpreted to mean that the difference between A and P compartments "depends critically on the selector gene *engrailed* which, when active, selects the posterior developmental pathway" (Lawrence and Morata, 1979a). In accordance with the early genetic inferences, the *en* protein, detected by immunochemical methods, has been found to be expressed in the P compartment of the wing imaginal disc, in second and third instar larvae, and not in the A compartment.

Yet, if *en* is the "selector" for P compartment development, then the incompleteness of the transformations associated with *en*[1] can only be explained by the ostensible hypomorphic condition of the allele (Garcia-Bellido and Santamaria, 1972). In fact, however, it is now known that *en*[1] is not a typical loss-of-function allele of the gene but has some of the character of an antimorphic mutation, albeit one recessive to wild-type (Eberlein and Russell, 1983; Gubb, 1985b). Indeed, presumptive amorphs produce a less substantial P-to-A transformation than does *en*[1] (Kornberg, 1981a; Lawrence and Struhl, 1982; Eberlein and Russell, 1983). The *en* gene may be one of the determinants of A-P compartment differences but it is not the *sole* determinant of those differences.

The third element of the selector gene hypothesis was the postulated existence of activator genes, genes whose products directly and differentially turn on the selector genes. The *bcd* gene does form a gradient but it activates and represses certain of the gap genes, as we have seen, rather than directly switching on the activities of particular homeotic genes. No other candidates for the role of activator genes have come forward and it now seems that the expression domains of homeotic selector genes are set by the activities of the gap genes and pair-rule genes rather than by the maternal systems directly.

Thus, in its principal tenets, the original hypothesis has not been validated. Compartments are not universal building blocks in *Drosophila* development, progressive compartmentalization is not a general feature of *Drosophila* development (though a few discs experience a later D-V restriction of clones), and homeotic selector genes are not activated in single compartments to confer specific compartment identities. Yet, from the perspective of the 1990s, the hypothesis should not be viewed as having "failed." Rather, it was a highly useful idea: It made specific testable predictions and it stimulated a great deal of further characterization of clonal behavior and homeotic gene expression in *Drosophila*. Perhaps its most significant contribution was to focus attention on the roles of cell lineage and homeotic gene expression in the setting of regional patterns, in a way that was novel and stimulating.

Furthermore, the compartmentalization phenomenon remains an interesting one, raising questions about both its biological role(s) and the mechanism(s) by which it takes place. One feature that has seemed to give compartmentalization special significance is that the A-P separation in the wing is an early postblastodermal event. Indeed, because no single clones spanning the A-P boundary have been found (a few events interpreted as doubles have been detected), it has been argued that the separation must be no later than the first division following blastoderm formation (Lawrence and Morata, 1977). Though this is a seemingly plausible argument, the separation into A and P compartments may, in fact, take place several cell divisions later (Gubb, 1985a). It has long been known that clonal growth in the wing is greatly biased along the p-d axis; indeed, clones induced at 3 hr (blastoderm), in a non-Minute background, can extend the whole p-d axis of the wing (Bryant, 1970). If the first non-Minute imaginal cells, generated in a Minute background, are comparably biased in their division pattern, the frequency of boundary crossing by early clones could be effectively zero (Gubb, 1985a).

Whatever the mechanism of the early A-P separation in the wing, it appears different in kind from that of the later d-v clonal restriction. The event that separates dorsal from ventral clones in the wing during the third larval instar appears to be coincident with the formation of a furrow along the D-V boundary consisting of nondividing cells (O'Brochta and Bryant, 1985). Given the relatively small number of cells in the imaginal disc anlagen during embryogenesis, it seems unlikely that the A-P separation occurs by a comparable mechanism.

In addition, even the A-P separations, which take place in all discs, may not involve the same mechanism. Thus, in the wing, as we have seen, it occurs during embryogenesis within relatively small groups of cells, while in the eye-antennal disc, it occurs late during the second instar, when the disc comprises several thousand cells. Thus, hypotheses such as that of Kauffman et al. (1978), which impute compartmental separations to certain constants of mass and growth dynamics, cannot be true, at least in their simplest form.

Irrespective of the mechanism(s) involved, it might be argued that the relatively early A-P separations are essential for the subsequent construction of the imaginal structures. There is some evidence, however, that the timing of the event, with respect to cell numbers in the imaginal anlagen, can be altered without seriously impeding the construction of a normal structure. The key observation is that of Morata and Kerridge (1981). They determined the time of the A-P segregation in a leg formed in the head region, in flies mutant for a strong *Antp* allele. They found that the "cephalic leg" formed an A-P boundary at the stage *typical of the eye-antennal disc.* Morata and Lawrence (1979) and Struhl (1982a) have made similar observations using the less complete *spineless-aristapedia* (*ss*[a]) transformation of antenna to leg. Evidently, the stage at which A-P separation occurs is not critical for the construction of a leg.

The most important question, however, is not of the timing or mechanism but of biological function. There are at least two broad categories of hypothesis. One category concerns roles in growth control, the other posits special "instructional" aspects of compartments or their boundaries. The first idea to be proposed was that compartments serve as "units for the control of shape and size" (Crick and Lawrence, 1975). The differences in compartmentalization patterns, in this view, might reflect the different requirements of or constraints on growth and morphogenesis in the imaginal discs.

If this idea has validity, then neighboring compartments might also interact to guide each other's growth to some degree. Indeed, Lawrence and Morata (1976) observed that *en*[1] clones at the margin of the posterior dorsal subcompartment of the wing cause a local enlargement of the wing that extends beyond the clone on both dorsal and ventral surfaces. They concluded that the ventral surface/compartment may be "modeled," in part, on the dorsal compartment, suggesting the existence of an interaction between compartments despite their clonal independence.

Comparably, it is possible that separation into A and P regions is needed for the control of growth of imaginal cells in general. In the abdominal region, where the cells of the imaginal histoblast nests do not undergo division until the pupal stage, the genetic evidence suggests that there are no early clonal restrictions between the anterior and posterior parts of each segment. Although the P regions are marked by *en* expression (Kornberg et al., 1985) and only the posterior histoblast nests seem to be affected by *en* mutations (Kornberg, 1981b), the patterns of gynandromorph boundary lines in the larva show no discontinuities between the *en*-expressing (P regions) and non-*en*-expressing regions (A regions), as would be expected if these regions had the developmental independence of compartments (Szabad et al., 1979). Finally, in the cephalic region, where the imaginal segments are much less clearly defined, the A-P compartmental separation happens late, as we have seen. If the A-P separation, in general, contributes to strengthening segmental separations, the differences in clonal behavior between thoracic, abdominal, and cephalic segments become more comprehensible.

The other class of hypothesis is that some feature of compartments, most probably the compartmental boundary, produces a developmental signal of some kind that is important for future development. In this view, the compartmental borders function not as barriers to restrain growth but as the source of further developmental cues. This possibility was first suggested by some tests of regeneration capacity in imaginal discs. The results indicated that fragments possessing part of the A-P boundary are most capable of regenerating distal structures in both wing (Wilcox and Smith, 1980; Karlsson, 1981a) and leg discs (Schubiger and Schubiger, 1978).

One particular hypothesis about the instructional capacity of compartmental boundaries is that of Meinhardt (1983, 1986). He has suggested that the intersection of the A-P boundary of each thoracic disc with an additional, hypothetical D-V boundary of some kind generates a gradient specifying relative position along the p-d axis. In this proposal, the boundary intersection may activate the production of a morphogen, whose high point, at the intersection, specifies the extreme distal point and whose lower concentrations specify progressively more proximal ones.

Although the notion of intersecting boundaries has not been supported by experimental evidence, recent findings lend some dramatic support to the notion that the A-P boundary is directly important for specification of positional values along the p-d axis. These involve the *dpp* gene, whose activity in the imago is known to be necessary for distal structure specification (see Spencer et al., 1982, and discussion later in this chapter). *dpp* has been found to be transcribed in a narrow zone stretching along the A-P border in third instar wing discs (Posakony et al., 1991), slightly overlapping the region of *en* expression (Raftery et al., 1991). Most significantly, *dpp*-deficient clones on the anterior side of the A-P boundary, and only clones located there, are associated with distal defects (Posakony et al., 1991), indicating that expression of *dpp* from this region *must* take place if distal structures are to develop.

These results suggest that the A-P boundary is necessary for *dpp* expression, which, in turn, is necessary for development of distal disc-derived structures. Although *dpp* is transcribed in a localized fashion, it encodes a diffusible product, which, it will be recalled, is a member of the TGF-β family of secreted factors. Altogether, the findings on *dpp* not only confirm a crucial element of the Meinhardt hypothesis, the "instructional value" of the A-P boundary, but provide a molecular basis for understanding the earlier results from the regeneration experiments cited above. If the A-P boundary is the site of synthesis of an essential distalizing substance (Dpp protein), the results of those experiments make perfect sense. It will be interesting to see whether *dpp* expression proves to be a general feature of A-P boundaries in other discs; preliminary results (cited in Posakony et al., 1991) suggest that it will be.

Homeotic Genes: Initiating and Maintaining the Pattern Differences Between Discs

In contrast to the larva, whose basic segmental pattern can be discerned in early embryogenesis, the form of the imago only comes into being late in development, during the final events of morphogenesis. As the imaginal discs evert, differentiate, and join, the characteristic head and thoracic segmental pattern emerges. In the abdominal region, the segments of the imago arise through a replacement of the outer surface of the larval segments, as the histoblasts grow and differentiate. Much evidence, from fate mapping experiments, indicates that the precursor cells of the imaginal primordia arise in the same process that delimits the corresponding embryonic primordia and are stamped at the outset with properties of the embryonic/larval segment that contains them (Lohs-Schardin et al., 1979a,b; Szabad et al., 1979).

The genetic analysis of imaginal homoeotic transformations produced by BX-C and ANT-C mutations confirms this relationship; particular mutations produce comparable phenotypic changes in imaginal primordia and embryonic segment anlagen. The imaginal precursor cells in the early embryo therefore undoubtedly have their initial patterns of BX-C and ANT-C gene activity established in parallel with the cells that surround them in the embryonic segment anlagen. However, genetic and molecular analyses of homeotic gene effects in the imago have revealed additional facts about the regulation of these genes, and, in particular, about the factors that maintain and modulate their expression over long periods of developmental time.

As in the embryo, the domain of action of the BX-C in the imaginal anlagen is found to be from the mesothorax (T2) posteriorly to the last abdominal segments (A8 and the terminal, vestigial segments) and that of the ANT-C from mesothorax (T2) anteriorly to the cephalic imaginal segments. Indeed, as noted earlier, a number of dominant mutant effects, involving genes in both complexes, were first detected on the basis of the transformations elicited in the adult and then subsequently found to affect the corresponding segment(s) in the embryo.

Furthermore, while certain homeotic selector gene activities are needed only briefly early in embryogenesis for subsequent imaginal development (in particular *Ubx* activity in the first 7 hr of embryogenesis to repress *Scr* in T2), most, if not all, of these genes remain "on" in their primary imaginal domains, whether by autoinduction or other means, throughout the greater part, or all, of larval development. Thus, for instance, the *Antp* locus continues to be expressed in the leg imaginal discs of all three thoracic segments as shown by both *in situ* hybridization to transcripts and immunostaining of proteins (Levine et al., 1983; Wirz et al., 1986).

Expression patterns, however, only provide a clue to requirements. Thus, while *Antp* is expressed in all leg discs, genetic analysis indicates that it is only required in the mesothorax and metathorax leg discs. Struhl (1981c) induced homozygous, loss-of-function *Antp⁻* clones at blastoderm and found that such clones are associated, ventrally, with transformation of leg tissue to antennal tissue while, dorsally, such clones in T2 and T3 provoke a transformation to T1 structures. In contrast, *Antp⁻* clones in antennae are perfectly normal, showing that the gene is not required for antennal development. However, if *Antp* is required for leg development but not for that of antennae, it follows that the dominant *Antp* transformation (antennae into T2 legs) reflects a neomorphic expression of *Antp⁺* activity in the antennal discs in these mutants.

Molecular evidence supports this inference about the dominant *Antp* mutants. In wild-type development, *Antp* protein is expressed in all the thoracic imaginal discs but is not found in the eye-antennal disc (Wirz et al., 1986). In contrast, in one of the strong dominant *Antp* mutants, produced by an inversion that breaks the *Antp* gene between the promoter and the first translated exon, the *Antp*-coding sequence is found to be under the control of a new promoter (Schneuwly et al., 1987) and the gene is transcribed in the antennal disc (Schneuwly et al., 1987).

Indeed, placing the *Antp*-coding sequence under the control of a heat shock promoter and causing it to be expressed throughout the developing animal, by giving a brief heat shock in third larval instar, produces adults with antennae partially transformed to mesothoracic (T2) legs (Schneuwly et al., 1987) (Fig. 7.27). Although *Ubx* expression antagonizes and reduces *Antp* expression in the wild-type metatho-

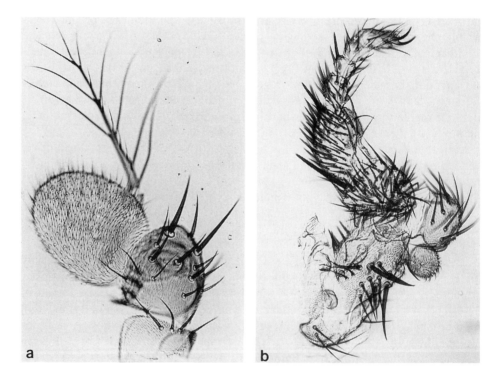

a   b

Fig. 7.27. Ectopic antenna-to-leg transformation produced by heat shock in third larval instar to transgenic animals carrying *Antp* under the control of an *hsp70* promoter. **a:** Control, wild-type following development after larval heat shock. **b:** Transgenic fly following heat shock and subsequent development. (Photographs courtesy of Dr. W.J. Gehring.)

rax, an ectopic expression of *Antp* in the eye-antennal disc would take place there unhindered by the presence of *Ubx*. In contrast, if *Ubx* expression is briefly induced in the eye-antennal disc at the end of the second larval instar (produced by a short heat shock to *Ubx* under the *hsp70* promoter), a leg is produced in a high fraction of the surviving flies but the leg that develops is a metathoracic (T3) leg (Mann and Hogness, 1990). Evidently, the expression stage of antennal disc cells is some kind of "ground state" for leg development; the latter can be triggered, and its direction determined, by excess production of particular proteins of either the ANT-C or the BX-C.

Not only are many of the homeotic selector genes active until late in development, but these late expression patterns are needed to ensure appropriate imaginal development. One of the earliest genetic observations to suggest this was of clones made homozygous for a strong *bx* mutation late in the third instar; despite the late time of their induction, such clones can develop mesothoracic (T2) bristles in the metathorax (T3), indicating the need for Ubx proteins at or subsequent to this time and the irrelevance of earlier segmental identity (Lewis, 1964). Another kind of evidence comes from temperature shifts of temperature-sensitive (ts) mutants. Thus, temperature pulses at the restrictive temperature to embryos homozygous for a ts *Dfd* mutation revealed ts periods in embryogenesis, the larval stages and the pupal stages, with the greatest effect on imaginal structures being produced in the pupal stages (Merrill et al., 1987).

Whatever the inherent capacity for self-activation of such genes, or the role of more complicated regulatory circuits in keeping such genes "on," their expression appears sensitive to events mediated by cell-cell interactions. A significant fact about partial homeotic transformations caused by hypomorphic or neomorphic mutations in the BX-C is that the transformed cells need have no relationship through cell lineage; induced, marked clones can overlap the areas of transformation to various degrees (Duncan and Lewis, 1982). Apparently, these transformations take place in neighboring groups of cells that are defined solely by their relative proximity and not by shared ancestry. Botas et al. (1988), comparing the patterns of observed transformation with the regions of *Ubx* expression in such strains, have proposed that cell neighbors must influence each other in setting their levels of *Ubx* expression, either enhancing or extinguishing such autocatalytic expression in successive cell divisions, when the protein has only partial activity. In the wild-type, in contrast, the full activity of the protein would tend to guarantee continual reinforcement in the "correct" cells. Indeed, wild-type imaginal cells are remarkably stable in those epigenetic states that confer disc-specific characteristics. This stability was first demonstrated by Hadorn (1965), who showed that disc cells can retain their characteristic developmental fate during prolonged culture. By this most stringent of criteria, discs are highly determined pieces of tissue.

The standard culture procedure for discs is diagramed in Figure 7.28. Discs are first removed from larvae (usually third instar larvae, which have the largest discs) and either cut into fragments or partially dissociated and reaggregated, the latter procedure promoting cell differentiation during culture. Test pieces are then cultured as in the Chan-Gehring experiment (see chapter 4). At the end of this first culture period (or transfer generation, Trg), the implant is removed, cut into several pieces, and the procedure is repeated. At each stage, a small piece can be assayed for developmental state by placing it within a mature host larva; the implant, exposed to the ecdysone

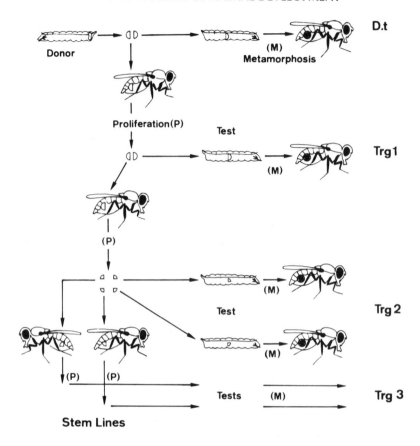

Fig. 7.28. In vivo culture procedure for imaginal disc fragments. See text for discussion. D.t, direct test of imaginal disc developmental state; Trg 1–3, transfer generations 1 to 3. (Reproduced from Hadorn, 1978, with permission of the publisher.)

of the metamorphosing host, differentiates along with its host and can be removed from the abdomen after emergence of the imago.

The adult structures generated by the tested fragment reveal its developmental capacity before differentiation. Thus, at each Trg, the determinative state of replicate implants can be tested. The experiment reveals that disc-specific states of determination can be reproduced with great faithfulness. In one culture line started from a male genital disc, characteristic genital disc structures could be produced for a period of up to 55 Trg (Gehring, 1978).

Despite this great stability of disc-determined states during culture, it is not perfect; changes in state can and do occur. Yet the changes in state are for the most part to recognizable *alternative* disc types. (Only after very extensive subculturing do truly abnormal developmental states, shown by abnormal differentiated patterns, appear.) Such changes of state are termed "transdeterminations" (Hadorn, 1965). If one designates the original state as the "autotypic" state and the new one as an "allotypic" state, the striking fact is that the allotypic disc states show both the stability and types of (further) change (see below) that are characteristic of the disc type whose phenotype they mimic. By these criteria, the allotypic cells have acquired determined states indistinguishable from those of the discs they resemble.

Transdeterminative changes are genuine changes of developmental state, rather than mutational events. They occur at too high a frequency to be conventional mutations and many are reversible, at high frequency. Furthermore, a transdeterminative change can simultaneously entrain cells from different clones, as shown by the induction of marked clones, by mitotic recombination, in transdetermined tissue (Gehring, 1967). In this latter aspect, of cell cooperativity or entrainment, transdetermination resembles the action of certain homeotic mutants. An example is the change from antennal to homeotic leg growth within the antennal disc of an *Antp* mutant. Although this switch probably occurs in the early third instar stage, genetically marked cells induced before this period never solely occupy the homeotically transformed patch, as should occur some of the time if the *Antp* switch takes place in individual cells (Postlethwait and Schneiderman, 1969, 1971). Such homeotic changes within appendages must also involve two or more cells at a time.

Indeed, developmental assignments in *Drosophila* may always involve small groups of cells. Compartmentalization always occurs in cell groups, termed "polyclones" by Crick and Lawrence (1975); a marked M^+ clone, no matter how large, never completely fills either an A or P compartment. And, as shown by many clonal analyses, imaginal primordia always arise from more than one blastodermal cell; the completed structure of an imaginal disc is never composed wholly of tissue derived from a single marked cell (reviewed in Merriam, 1978). In the case of transdetermination, it seems probable that some, at least, of these changes involve cell interactive mediated switches in homeotic genes, in events not dissimilar to the reinforcement extinction processes described by Botas et al. (1988).

It seems likely, in fact, that many, perhaps all, transdeterminative changes involve switches in activity of one or a few homeotic genes. This supposition seems probable from the single most striking fact about the phenomenon of transdetermination, that the entire sequence of observed changes comprises a single recognizable pattern (Fig. 7.29). Each disc type undergoes primary changes to one or a few other specific disc types, with a characteristic frequency and probability. With further growth, these changes are succeeded by characteristic secondary switches. Furthermore, autotypic and allotypic implants of the same disc type experience essentially identical patterns of change (Hadorn, 1978).

Many, though not all, of these changes mimic known homeotic changes produced by mutations in known genes, including genes of the BX-C or ANT-C. In the instances that find no parallel among the known homeotics, e.g., genital to leg, there may be a coordinated set of changes in activities of several homeotic changes or a cascade sequence of such changes. Yet, despite the fact that the entire pattern can be described in terms of switches in state of a small number of "control circuits" (i.e., homeotic genes) (Kauffman, 1973), the experimental system has remained too cumbersome to put this idea to the test and it remains, today, more than 25 years after the discovery of transdetermination, only an interesting speculation.

Nevertheless, the phenomenon of transdetermination remains of interest because of its basis in cellular interactions and its dependence on cell proliferation; in these respects, it appears linked to normal pattern formation within discs, as described by the polar coordinate model (Bryant et al., 1981; reviewed in Bryant and Simpson, 1984). For transdetermination, active cell growth is equally essential, with treatments

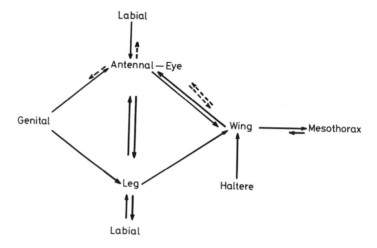

Fig. 7.29. Global pattern of transdeterminative changes. Dotted arrows indicate rare or suspected changes. (Adapted from Hadorn, 1978.)

such as cell dissociation and reaggregation that stimulate cell division actively promoting transdetermination (Hadorn, 1978). Furthermore, the cells that are most likely to undergo transdetermination are probably not a random sample of the cells of a disc but probably come from particular regions. When different parts of the foreleg disc are cultured under various conditions, there is an approximate correlation between capacity for regeneration of distal structures and for transdetermination (Strub, 1977). Such results suggest that the stimulus for transdeterminative change is not dissimilar to that which promotes normal pattern formation and, in particular, distal regeneration. In an earlier section, we looked at some of the possible factors involved in distal regeneration; in the following section, some possible genetic bases of the rules of the polar coordinate model will be discussed.

Segment Polarity Genes: Do Some Encode the Positional Values of the Polar Coordinate Model?

The central postulate of the polar coordinate model is that there are distinct positional values associated with both circumferential and distal locations in discs and that the completion of these sets of positional values during growth arises through the shortest intercalation and distalization rules (see chapter 2). Yet, while the polar coordinate model provides a view of pattern formation in terms of cellular behavior, it signally lacked, in its original version, a basis in genetics.

There are some grounds for believing that this lacuna in the hypothesis is about to be filled, in that there are now candidate genes for the specifiers of both circumferential and distal positional values (see review by Wilkins and Gubb, 1991). The genes that may be important for the creation of, at least some of, the circumferential positional values are a subset of the segment polarity genes, those that are expressed in position-specific stripes in the embryo. The grounds for thinking so are that these genes seem to be required for expression in particular regions of all or most discs and that *en, wg, ptc, hh,* and *ci* are expressed *similarly with respect to the a-p axis in both embryo segments and imaginal discs.* Thus, *en* is transcribed in posterior regions of segments and discs, while *wg, ptc,* and *gsb* are expressed in anterior regions of segments and discs.

The continued requirement for these genes, and their expression in particular regions of discs, has been shown by clonal analysis (Table 7.2). One can test for such requirements by inducing mitotic recombination in heterozygotes and then scoring the phenotype of homozygous clones that are detected by means of a linked but independently expressed cuticular marker. Three segment polarity genes that have been tested in this fashion, namely *en, hh,* and *ptc,* all show disturbances of development, if induced either early or late in development, consistent with either a continuous or late requirement for these gene products (Kornberg, 1981a; Lawrence and Struhl, 1982; Mohler, 1988; Phillips et al., 1990).

Most significantly, the particular abnormalities that develop are a function of the position of these clones in the disc. Thus, *en* clones induced in the P compartment of discs but not in the A compartment are associated with abnormal wing development (Lawrence and Morata, 1975; Morata and Lawrence, 1975; Lawrence and Struhl, 1982). In contrast, removal of *ptc* from the A, but not the P, compartment leads to abnormal clonal development in the wing (Phillips et al., 1990).

Particularly interesting is the case of *hh.* Clonal defects are predominantly associated with clones induced in the posterior compartments of discs and the defects extend *beyond* the region of mutant tissue. Such effects have been termed ones of "domineering nonautonomy" (Mohler, 1988). In particular, the effects are predominantly in distal structures; in the distalization rule of the polar coordinate model, it will be recalled, the development of distal positional values is a consequence of completing circumferential values. Other domineering nonautonomy effects have been reported for *ptc* clones in the anterior compartment (Phillips et al., 1990).

Two segment polarity genes whose hypomorphs are associated with wing pattern defects, *wg* and *cos-2,* do not give defects in clones (Simpson and Grau, 1987; Baker, 1988b), but in the case of *wg,* this is believed to reflect nonautonomy of action of the mutant (the ability of surrounding wild-type cells to repair the developmental defect), and, similarly, for *cos-2,* there is some suggestive evidence for nonautonomy. As noted earlier, *wg* encodes a secreted protein, a property that can explain, in principle, the correction of a mutant deficiency by surrounding wild-type cells.

The possibility that some of these segment polarity genes specify comparable positional values in embryos and imaginal discs is strengthened by the finding of similar alterations in expression patterns of these genes in embryos and discs, as a result of mutation in other genes of the set. Thus, in both embryos and discs homozygous for *en* mutations (for the discs, viable *en* mutations were used), *ci-D* transcription appears in posterior regions (Eaton and Kornberg, 1990). Similarly, in *ptc* embryos, both *wg* and *en* expression appear in the regions that would normally express *ptc* but not *wg* or *en* (DiNardo et al., 1988; Martinez-Arias et al., 1988) and similar ectopic expression of *wg* and *en* takes place in parts of *ptc* discs that do not express these genes in wild-type (Simcox et al., 1989).

Although there are some differences in expressional patterns of some of these genes between embryos and discs with respect to the a-p axis (see, for instance, Baker, 1988b), and some involve differences in regulatory relationships (see Phillips et al., 1990), the similarities are sufficient to warrant consideration of the idea that some of the segment polarity genes are among the "missing genes" of the polar coordinate hypothesis, either directly encoding or regulating those molecules that constitute the positional values. Some, like *wg*, may involve short-range, diffusible signals, while others might be localized transmembrane proteins. Others, like *en, gsb,* and *ci-D,* would be more likely to be regulators of such cell surface markers. Not all of the positional values, however, are necessarily encoded by distinct genes. Some, perhaps most, are created by combinations of various gene products and others by quantitative differences in particular molecules.

Interpreting some of the segment polarity genes as positional value genes helps to explain some observations. For instance, the observation that *ci-D* is expressed only in anterior regions of embryonic segments and discs yet its mutant defects are predominantly in posterior regions (Eaton and Kornberg, 1990) makes more sense within this framework than in the alternative one that *ci-D* specifies segmental features in the region in which it is expressed.

There are three predictions that this hypothesis makes. The first is that, if different imaginal discs are examined for their expression patterns of the segment polarity genes of interest, they will show broadly similar patterns, both temporally and spatially, for these genes. The second prediction is that even brief ectopic expression of any of these genes, which are normally expressed in particular disc regions, should produce abnormalities in the form of pattern duplications. (For the particular case of *ptc,* this has already been shown *not* to be the case [Sampedro and Guerrero, 1991].) The third prediction is that extirpation of particular regions of imaginal discs should provoke characteristic reexpression of certain segmental polarity genes. The specific patterns should be a function of the region extirpated, but similar patterns should be seen in different discs as a response to removal of geographically comparable regions.

Genes Required for Distal Structures: Another Link to the Polar Coordinate Model?

While certain of the segment polarity genes may account for some of the circumferential positional values in imaginal discs, there is another group of genes whose products are required for creating distal positional values. These three genes were not identified as a group, or in the course of massive mutant screenings to identify genes involved in distal specification, but were identified individually and serendipitously on the basis of mutant phenotypes involving defects in distal imaginal structures. They are listed with their map positions and associated mutant phenotypes in Table 7.3.

The first gene is one that we have already encountered, namely *dpp,* which encodes a diffusible product, a member of the TGF-β family of proteins (Padgett et al., 1987). In contrast to the embryo, where *dpp* is required for formation of dorsal tissues and structures, the imaginal requirement for *dpp* is in the specification of distal structures. While the null allele is, as we have seen, an embryonic lethal, producing extreme ventralization of the embryo, mutations that create less severe reductions in *dpp* protein cause deletions or aberrancies in distal structures in all imaginal discs. These mutations can, indeed, be arranged in a phenotypic series, leading from mild to severe defects in distal structures in discs to the near elimination of the discs themselves (Spencer et al. 1982).

As discussed earlier, the expression pattern in discs is interesting, occurring in a stripe along the A-P boundary, along the p-d axis, but required only within the A

Table 7.3. Genes Required for Distal Specification

Gene name	Cytogenetic (map) position	Imaginal phenotypes	References
*Brista (Ba)*or *distalless (dll)*	60F5–6 (2–107.6)	Distal antenna to leg; antennal distal defects	Cohen and Jurgens (1989); Sunkel and Whittle (1987); Cohen et al. (1989)
rotund (rn)	84D3	Defects of distal structure in proboscis, leg, antenna, wing	Cavener et al. (1986); Kerridge and Thomas-Cavallin (1988); Agnel et al. (1989)
decapentaplegic (dpp)	22E3–F2 (2–4.0)	Defects/abnormalities in distal structures for 15 of 19 discs	Spencer et al. (1982); Posakony et al. (1991)

compartment (Posakony et al., 1991; Raftery et al., 1991). The structure of the gene is itself highly unusual: Mutations producing the different categories of disc defectiveness map not in the coding region nor in the 5′ flanking region but in discrete regions within the 3′ flanking region. The entire 3′ flanking region, which is transcribed little or not at all, apparently consists of a set of *cis*-acting sequences that regulate transcription from the 5′ side; chromosome breaks that interrupt this region inhibit function, and, apparently, the fewer of these regions that are juxtaposed to the coding sequence, the more severe the imaginal defects (St. Johnston et al., 1990). The simplest explanation for the phenotypic series of mutants is that there is a graded requirement for this protein along the p-d axis, the most distal positions requiring the most *dpp* product.

The second gene required for development of distal imaginal structures was originally named *Brista (Ba)* (and is now usually designated as *distalless [dll]*). Its first mutants were identified as homeotic dominants causing transformation of distal antennal structures (but never the most proximal antennal segment, AI) to the comparable (distal) leg structures (Sunkel and Whittle, 1987). This transformation was identified by the Mullerian test as a haplo-insufficient phenotype. More complete reductions of gene activity, in certain *Ba* heterozygotes, result in deletions or abnormalities of distal leg structures, while still further reductions produce embryonic lethality, with the dead embryos lacking various sense organs believed to be the larval equivalents of various distal imaginal structures (Cohen and Jurgens, 1989; Sunkel and Whittle, 1987). Clonal analysis shows that the gene acts autonomously within cells, while temperature-shift experiments on ts mutants indicate that it is required from embryogenesis through to puparium formation, a period coextensive with that of cell proliferation in the leg discs (Sunkel and Whittle, 1987).

dll encodes a homeodomain protein (Cohen et al., 1989). It seems likely, therefore, that the gene is a transcriptional activator required for the expression of genes needed for distal structures. The fact that reduction of dosage of the wild-type allele is sufficient to trigger a homeotic change (antenna to leg) illustrates, once again, the intimate connections between cell proliferation, development of positional values, and the setting (or resetting) of homeotic gene activities in response to positional values.

The third gene, *rotund* (*rn*), was identified on the basis of pleiotropic defects in distal leg, proboscis, and antennal and wing structures, though, interestingly, the most distal wing and antennal structures are less affected than those immediately proximal. *rn*, unlike *Ba/dll*, can be completely deleted and still give viable adults, though ones that are sterile (Cavener et al., 1986; Kerridge and Thomas-Cavallin, 1988). Clones homozygous for an *rn* null in distal imaginal regions, induced during early larval development (24–48 hr AEL), are largely autonomous though pattern disturbances outside the clones ("domineering nonautonomy") are also seen (Kerridge and Thomas-Cavallin, 1988). Other clonal analyses show that the gene is required for development of distal structures as late as post-third larval instar (cited in Agnel et al., 1989). In contrast, clones found in proximal leg regions were without developmental effect, consistent with the phenotype of *rn* animals. *In situ* hybridization experiments with the cloned gene confirm this regional pattern; the transcripts in late imaginal discs are found only in the distal regions of wing, haltere, and antennal discs (plus a stripe in the genital disc and in some of the adepithelial, muscle precursor cells) (Agnel et al., 1989).

The discovery that three genes are required for development of distal structures is intriguing but much remains to be done. The functional relationships between these genes, if any, have yet to be determined and it is not known whether other genes are required for development of distal structures. The particular element of progress in pattern formation studies that the analysis of these genes represents is that the phrase "distal positional values" has begun to acquire specific molecular correlates. From the phenomenological characterization of distal regeneration, which led to the formulation of the "distalization rule" (Bryant et al., 1981), this is an important advance, indeed.

The Eye as a Case Study: Cell Contact Interactions and Global Factors in Disc Development

While the polar coordinate model presents a valuable framework for understanding imaginal disc development in general, there is one disc in which pattern formation operates, quite possibly, by other rules. The development of the eye disc differs visibly from that of the other imaginal discs in two respects. First, its final differentiated pattern begins to emerge during the third larval instar, rather than during the pupal period, in contrast to all the other discs. Second, the main epithelium of the disc develops into a mosaic of essentially identical units, the facets or ommatidia. Although these units, develop in close juxtaposition to one another, there is evidence, to be discussed below, that each ommatidium is developmentally independent of the others. This absence of obvious "intercalative" interactions, taken in conjunction with the essentially repetitive nature of the pattern, suggests that pattern formation in the eye involves novel aspects.

Yet, despite its idiosyncratic aspects, development of the *Drosophila* eye also exhibits three kinds of influence that are seen generally in development: short-range, contact-mediated cellular interactions (which create the basic structural unit, the ommatidium); regional factors (which affect initiation of eye development and its symmetry properties); and "developmental noise" (which affects the number of units present in each eye and produces occasional small variations in the unit module).

Genetic analysis has been particularly instructive in analyzing the formation of the basic unit structure, and we will concentrate on these findings. However, a few facts about longer-range influences and developmental noise will also be described.

Structure of the Eye

Each eye of the adult *Drosophila* consists of approximately 700–750 facets (females, with larger eyes, have more facets), which are packed in a hexagonal, honeycomb pattern to produce the characteristic insect compound eye (Fig. 7.30). The mature ommatidium is a 22-cell structure, which consists of 8 photoreceptor cells and 14 accessory cells, of various types (Fig. 7.31). The photoreceptor cells comprise the

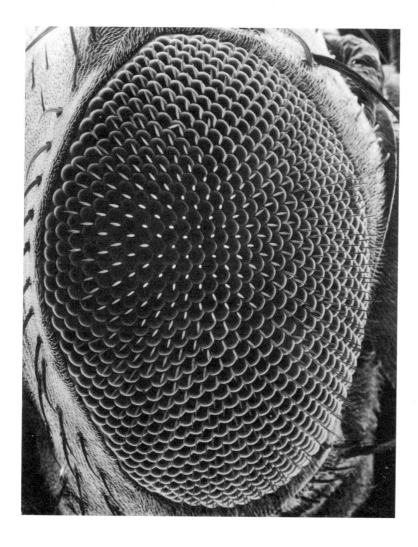

Fig. 7.30. The *Drosophila* compound eye. (Reproduced from Tomlinson, 1988, with permission of the Company of Biologists, Ltd.)

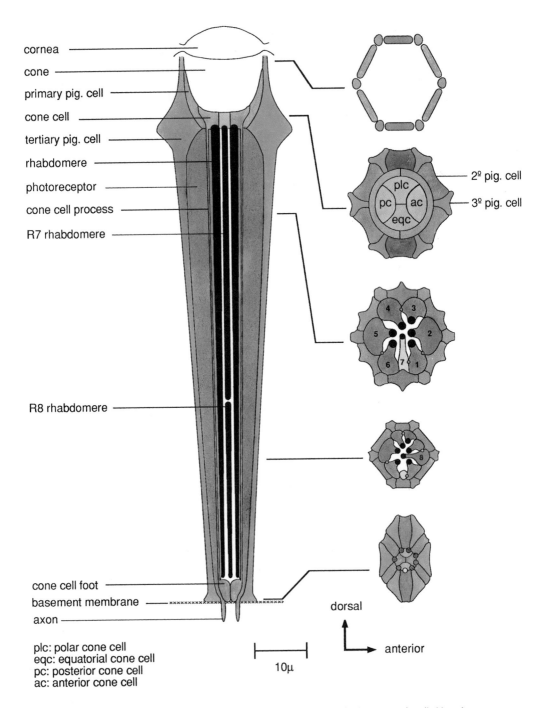

cornea

cone

primary pig. cell

cone cell

tertiary pig. cell

rhabdomere

photoreceptor

cone cell process

R7 rhabdomere

R8 rhabdomere

cone cell foot

basement membrane

axon

plc

pc   ac

eqc

2° pig. cell

3° pig. cell

plc: polar cone cell
eqc: equatorial cone cell
pc: posterior cone cell
ac: anterior cone cell

10µ

dorsal

anterior

Fig. 7.31. The structure of the *Drosophila* ommatidium, in longitudinal cross section (left) and transverse sections at several positions (right). The placement of the various accessory cells and photoreceptor cells (indicated by number, 1 to 8) are as shown. (Reproduced from Ready, 1989, with permission of the publisher.)

core of each unit: Cells R1–R6 essentially enclose the two central photoreceptors, R7 and R8, while above the photoreceptor cells are four cells, termed cone cells, that secrete the lens of the ommatidium. The cone cells are surrounded by a collar composed of the two so-called primary pigment cells, while other pigment-containing cells, the secondary and tertiary pigment cells, serve to enclose the primaries, the cone cells and the photoreceptors. The final cell complement is the four-cell unit that makes up the bristle organule at the alternate vertex of each ommatidium.

The highly stereotyped pattern of the ommatidia suggests the necessity for having cells of fixed function in specific places for ommatidial function. Thus, the cone cells secrete the secondary lens, the pseudocone, which lies just underneath the corneal lens of each facet, while the pigment cells of each ommatidium provide optical shielding of the photoreceptor cells within each ommatidium. The primary pigment cells may also be involved in secreting part of the lens.

Perhaps, somewhat more unexpectedly, the eight photoreceptor cells can be divided into at least three physiological groups. These groups are R1 through R6 (the peripheral photoreceptor cells) and the single member groups, R8 and R7. While R1–R6 send their axons to the first optic ganglion, both R7 and R8 send theirs to the second optic ganglion. Each photoreceptor group has a characteristic spectral sensitivity profile (Harris et al., 1976), apparently produced by a characteristic rhodopsin complement. (Rhodopsins are the essentially universal visual pigments found in the animal world, and consist of a chromophore, usually 11-*cis* retinal, coupled to a transmembrane protein of the opsin family.) The principal visual pigment of the R1–R6 group was identified by O'Tousa et al. (1985) and two R7-specific pigments have been reported (Zuker et al., 1985; Fryxell and Meyerowitz, 1987; Montell et al., 1987). The precise basis of the spectral sensitivity profile in R8 is not yet known. While all eight photoreceptor cells show differing responses to visible light, R7 is the only one capable of detecting long-wavelength UV.

Despite the invariance of fundamental ommatidial structure, there are some variations in less critical features of the eye, the aspects that reflect developmental noise. Thus, both the numbers of rows of facets and the numbers of facets per row can differ from eye to eye; in addition, some of the bristles are displaced from their anterior vertices (Ready et al., 1976). Evidently, the constancy of unit facet development has been more highly selected in evolution than either the specific number of facets or the precise placement of the associated bristles.

Ommatidial Development and Its Cell Biology

The ommatidium is such a well-defined structural unit that it is tempting to speculate that it, or at least its set of eight photoreceptor cells, arises from a founder cell, comparably to neurons from a ganglion mother cell or a bristle organule from a bristle mother cell (chapter 9). Indeed, an early suggestion in the literature was that the eight photoreceptor cells of the ommatidium arise in a sequence of three cell divisions from a single founder cell.

This idea can be tested genetically by clonal analysis: One induces clones, which can be detected by some readily visible marker, well before ommatidia form in the third larval instar, and then scores individual facets, and their groups of photoreceptors, within single clones. If every facet, or its group of photoreceptors, is genetically homogeneous, then these cell groups are almost certainly of clonal origin, while if mosaic ommatidia are found, they cannot be. The first detailed analysis of this kind was reported by Ready et al. (1976). They induced homozygous *white* clones in first larval instar, when the eye primordium contains an estimated 20 cells, and scored single clonal patches in the adult flies. (Because they are induced early, the events are comparatively rare, but each one, marking the descendants of one of only 20 precursor cells, is large.) The results showed widespread mosaicism both within individual ommatidia and within the photoreceptor cell groups of ommatidia. Furthermore, the collective set of results showed no preferred clonal boundaries among any cells of the ommatidium; evidently, there are no fixed relationships by descent among any of the component cells of the ommatidium.

Lawrence and Green (1979) extended these results by examining clonal relationships of cells dividing during late third instar. To ensure that only single events were being scored, they looked for rare red (wild-type) spots in a white background, by selecting for intragenic recombination between two different white alleles. The resultant clones again failed to reveal any obligate clonal relationships. For instance, in a number of two-cell clones, one cell was a photoreceptor (wild-type photoreceptor cells have some pigment) and its sister, a pigment cell. The results show that neither individual ommatidia nor their sets of eight photoreceptor cells need be clones. Evidently, ommatidia are assembled from cells that share propinquity rather than ancestry.

The first detailed observations of ommatidial construction confirmed the polyclonal origins of ommatidia, showing that three of the photoreceptor cells (R1, R6, and R7) arise from mitotic events, involving distinct precursors, that come after those that generate the other five R cells (Ready et al., 1976). Ommatidia are found to arise from a seemingly uniform epithelial sheet during the third larval instar, following the passage across the eye disc epithelium of the "morphogenetic furrow," a groove that visibly traverses the disc from its anterior edge to its posterior edge. Ahead of the furrow, the epithelial sheet remains undifferentiated, with many cells proliferating. As the groove passes over the cells, they sink into it, and cease proliferating. Behind the groove, six- to seven-cell clusters form, each the nucleus of a future ommatidium, consisting of the precursors of R2, R3, R4, R5, and R8 plus one or two "mystery cells," the latter subsequently leaving the group.

Following passage of the furrow, there is a further wave of mitoses immediately in its wake, and subsequently, additional cells generated in this zone, including the cells that will become R1, R6, and R7, join the newly forming unit. The precise sequence of recruitment has been mapped by Tomlinson and Ready (1987), who used two different antibodies, both specific for *Drosophila* neural cells, to detect the photoreceptor cells as they form. The first cell to acquire the specific antigen is the future R8 cell; it is followed by the cells that will become R2 and R5; R3 and R4 then follow; these are succeeded by R1 and R6; and, lastly, R7 joins. (R7 is formed in the same mitotic wave as R1 and R6, but expresses neural antigen 8–10 hr later, presumably as a function of its making the appropriate contacts upon joining.)

These symmetrical additions create a bilaterally symmetrical structure (as seen from the apical surface) whose symmetry is only broken after R7 has joined (Tomlinson, 1985; Tomlinson and Ready, 1986). Though all the cells of the future ommatidium exist initially in a monolayer, there is a characteristic, highly choreographed movement, between the apical surface and the basement membrane, of nuclei within the participating cells as the sequence unfolds. A portion of the process is diagrammed, in a lateral view, in Figure 7.32. The accessory cells, in particular the future cone cells, also join during this sequence to create first a two cone cell stage, then the four cone cell stage. Further events shape the ommatidium into its final form and differentiated state (including the formation within the photoreceptor cells of the stacks of photosensing membranes, the rhabdomeres, that carry the rhodopsins). The period of construction of the ommatidial rows lasts from mid-third instar to more than 12 hr past the white prepupa stage (Campos-Ortega and Hofbauer, 1977; Basler and Hafen, 1989).

The net effect of the progression of the furrow is the elaboration of rows of developing ommatidia, each row forming approximately once every 70 min on average (at 25°C), though the progression in the posterior part is slower than that in the anteriormost region (Basler and Hafen, 1989). As the furrow moves on, each row behind it can be seen to be at a characteristic state of development (with respect to the final ommatidial structure and the time elapsed since passage of the furrow) (Tomlinson, 1985; Tomlinson and Ready, 1987). This progression of rows led Ready et al. (1976) to propose, initially, that the whole process could be likened to crystalline growth, in which each newly forming layer uses the previously formed layer as a template.

Subsequent experiments have disproved this idea, however, and shown that ommatidia can self-assemble in the absence of neighboring ommatidia. For instance, when small pieces of the undifferentiated eye disc epithelium, which the morphogenetic furrow has not yet reached, are put into larval hosts and allowed to go through metamorphosis, they can develop ommatidia (Lebowitz and Ready, 1986). Other experiments show that the cut itself does not induce a new morphogenetic furrow and indicate that the sequence of ommatidial development, from posterior to anterior, is intrinsic to the epithelium (Lebowitz and Ready, 1986). Study of eye development in the mutant *Ellipse* (*Elp*) (see below) confirms the idea that ommatidia can form in the absence of a template of other ommatidia. In this mutant, far fewer facets develop and these are spaced relatively far apart, yet the structure of individual ommatidia is normal (Baker and Rubin, 1989).

If, as we've seen, the cells of the ommatidium have no fixed lineage relationships, and if each ommatidium can develop in isolation, then the "rules" of ommatidial construction must be written in the sequential cellular interactions that take place. In particular, following the initial "nucleation event" in the morphogenetic furrow, specific contacts in the developing ommatidium may serve as localized induction events, each involving particular genetic activities, with each collective set of particular contacts specifying future cell identity (Tomlinson and Ready, 1987). If this hypothesis is correct, then the elimination of either an individual signal or its reception by the target cell(s) should prevent the development of specific cells, or subsets of cells, within the ommatidium. The existence of a number of eye mutants has permitted tests of this hypothesis and provided both striking confirmation of the idea and additional information about the first events in ommatidium formation.

Genetic and Molecular Analysis

Because the eye is not essential for viability, it is particularly favorable material for genetic analysis—null mutants in genes required only for the eye can be directly isolated, while the role(s) of previously identified vital genes suspected of a role in

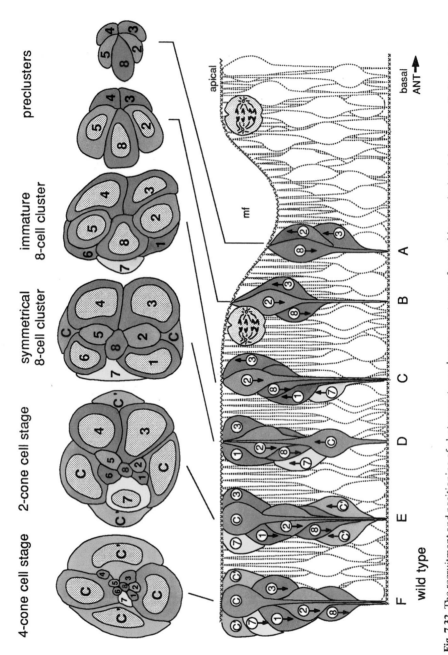

Fig. 7.32. The recruitment and positioning of photoreceptor and accessory cells to an ommatidium as a function of time, following passage of the morphogenetic furrow. The direction of passage of the furrow is from posterior to anterior (left to right in the diagram) and successive stages of assembly are depicted from A to F. See text for general description and reviews by Tomlinson (1988) and Ready (1989) for details of the recruitment and construction process. (Reproduced from Ready, 1989, with permission of the publisher.)

development of the eye can be tested by means of clonal analysis. A description of some of these genes and their roles will be given here, while a more extensive review can be found in Basler and Hafen (1991).

One group of genes whose mutants produce a variety of eye defects are the neurogenic genes, as seen both in clonal analysis and by the study of defects in ts mutants subjected to heat pulses during development. As in the embryo, a common phenotype associated with mutations in these genes is an overproduction of neural cells, specifically the photoreceptor cells in the case of the eye. Thus, in homozygous clones for certain alleles of several of these genes (*N, mam, neu,* and *Dl*) there are excessive numbers of R cells (Dietrich and Campos-Ortega, 1984). Homozygous clones for amorphic alleles of *Dl* and *E(spl),* however, were not obtained, indicating that these are vital functions in the eye epithelium.

The most thoroughly characterized of the neurogenic genes in eye development is *N,* whose roles in eye development have been explored by means of heat pulses during development to a ts *N* mutant. In the first study of this kind, Foster and Suzuki (1970) showed that successive heat pulses to different batches of larvae during third larval instar produced a line of defective ommatidia, whose position moved from posterior to anterior as a function of time. These defects were not characterized in detail but, in retrospect, can be seen to "track" the progress of the morphogenetic furrow. The most complete study of this kind was carried out by Cagan and Ready (1989), who gave heat pulses of varying duration from late larval to mid-pupal stages. The results indicate that *N* is required throughout this period and that heat pulses produce different but characteristic sets of cell conversions and losses in both neural (R) and nonneural cells, as a function of the time of the pulse. During late third larval instar, the primary defect is an excessive recruitment of cells to be photoreceptors; heat pulses at later times produce losses, and sometimes gains, of pigment and cone cells. Evidently, *N* is needed at multiple points in eye development in several cell types.

Its role may be to provide a needed signal at these times, as suggested in the neurogenesis work, or alternatively, to promote cell adhesion, as a precondition for the more specific cell contacts needed to drive development (Hoppe and Greenspan, 1986; Cagan and Ready, 1989). *N* is also required at more than one point in bristle development, and the same questions about its precise function(s) arise there; we will return to this matter in chapter 9.

A second set of genes, whose effects are much more specific, and narrowly timed, are those that affect R8 development. R8, it will be recalled, is the first pre-R cell to be selected for the ommatidial unit, and whatever the nature of the cellular interactions involved in its "choice," they are bound to be different from those involving the other cells, which join a preformed cluster or cluster nucleus. Indeed, it appears likely that, in contrast to the cell contact interactions that shape the subsequent development of the ommatidium, there are mechanisms which "space" the R8 precursor cells and unique genetic requirements that set it apart (Basler and Hafen, 1991).

In the dominant *Elp* mutant, for instance, as mentioned earlier, the preommatidial cell clusters are relatively few and far between. This gene is now known to encode the *Drosophila* homolog of the EGF receptor (DER) and to be identical to the *torpedo* (*top*) gene (at 2–99), the maternal, somatically expressed gene whose mutants produce a ventralized eggshell and embryo (Price et al., 1989). The *Elp* mutant is a gain-of-function mutation in this gene (Baker and Rubin, 1989) and the gene has been shown, by clonal analysis, to function only in R8. If overexpression of a putative receptor reduces the number of ommatidial groups, the mechanism might involve titration of an activator.

Mutants in another second chromosome gene, *sca* (at 49 D1,3), may identify a gene with the opposite function. In eyes of *sca* flies, the ommatidial preclusters appear more crowded, not less as in *Elp* eyes (Mlodzik et al., 1990). *sca* encodes a protein that is homologous to fibrinogen and which, from its structure, appears to be secreted, like fibrinogen (Baker et al., 1990). Significantly, in the emerging precluster, it is the R8 cells that express this gene most strongly. It seems likely that *sca* encodes an inhibitor, which functions in the process of lateral inhibition, to help maintain the spacing of preclusters. As with *N*, *sca* also functions in bristle development and may be part of the mechanism that enforces the spacing of bristles (chapter 9) (Mlodzik et al., 1990).

However, of all the known eye mutants, it is those that affect R7, the last photoreceptor cell to be recruited, whose effects are best understood. Of this group, the most thoroughly characterized are those of the *sevenless* (*sev*) gene, on the X chromosome (map position 1–33.2). As its name implies, *sev* flies lack the R7 photoreceptor specifically and exclusively; in all other respects, their eyes are normal. The first *sev* mutant was isolated by Benzer (1967) as one of a number of mutants defective in phototaxis. Wild-type *Drosophila* move toward the light and Benzer's procedure was designed to maximize recovery of flies capable of movement but who showed no preferential movement toward light. The first *sev* mutant was one of these and subsequent tests showed that R7 is the only photoreceptor capable of responding to long-range UV light (Harris et al., 1976).

A further analysis of *sev* action in flies mosaic for this gene showed that the gene is autonomous: In gynandromorphs, only R7 cells that carry the wild-type *sev* allele develop as such and no R7 cells that carry the *sev⁻* allele can do so (Harris et al., 1976). Observation of the course of development of ommatidia in the eye discs of animals homozygous for *sev* show that the cell that initiates the R7 type of contacts and movements develops instead into one of the four cone cells (Tomlinson and Ready, 1986). In effect, *sev* can be regarded as a cell homeotic.

If one views the elaborate orchestrated movements of cells during ommatidial development as producing a particular sequence of cell contact-mediated signals, then the *sev* defect can be seen as interfering with either the production or reception of one of these signals necessary for R7 development. As in the comparable case of neurogenesis, the finding of genetic autonomy provides a provisional assessment: The fact that the *sev* defect cannot be rescued by wild-type neighbors strongly suggests that the defect is in reception of the signal. The sequencing of the cloned gene bears this out. Conceptual translation of the *sev* protein sequence reveals the gene product to be a transmembrane protein with a cytoplasmic tyrosine kinase domain, in common with a large number of well-characterized membrane receptors (Hafen et al., 1987). The *sev* protein evidently serves to transmit a signal, presumably originating in one or more of its neighbors, and this signal is necessary to ensure R7 development.

Though *sev* expression is clearly necessary for R7 development, expression of the protein is not restricted to R7. Both Tomlinson et al. (1987) and Banerjee et al. (1987), using immunostaining, found abundant *sev* expression on other cells of the developing ommatidium (though the two studies differ in the reported patterns of expression, perhaps reflecting differences in the antisera used). Evidently, expression of *sev* is not sufficient for specifying "R7-ness," since other cells express the gene strongly without becoming R7 cells. Furthermore, the results suggest that the unique assignment of R7 fate requires the presentation of a *sev*-specific ligand at the time the R7 candidate cell joins the developing ommatidium. The expression of *sev* at other times, and in other tissues (as has been found), would be without effect if not simultaneously accompanied by presentation of the ligand.

Indeed, the current evidence suggests that specification of R7 involves a timed and spatially restricted presentation of ligand. The *sev* ligand has been provisionally identified as the product of a gene expressed by the R8 cell, which comes into close contact with the R7 precursor cell early in the latter's recruitment. The gene, named *bride-of-sevenless* (*boss*) (located on the third chromosome), was first identified by a mutant that, like *sev* mutants, is missing R7 specifically. However, unlike *sev*, *boss* expression was found, by clonal analysis, to be needed only in R8, and not in R7 (Reinke and Zipursky, 1988). The result establishes the importance of the interaction of R8 with R7 for the latter's development and *boss* is an excellent candidate for the ligand-presenting gene for the *sev* protein. Whether or not *boss* plays this role, however, it seems certain that transient activation of the *sev* protein is the specifying event for R7. When a constitutively active *sev* protein is engineered, and germ line transformation is carried out, the resulting flies display many cells in their ommatidia with R7 characteristics (Basler et al., 1991).

Another gene required for expression specifically in some but not all R cells is *rough* (*ro*), whose original mutation, described by H.J. Muller in 1913, was one of the first eye mutants isolated. As its name suggests, the *rough* eye exhibits a disordered facet array, reflecting abnormally formed ommatidia, possessing variable numbers of photoreceptor cells. Examination of the pattern of ommatidial development in this strain reveals that in the five-cell precluster, R8 and then R2/R5 show normal neuronal staining but that the process then falters; no R3/R4 pair joins and exhibits neuronal antigen, as seen in wild-type (Tomlinson et al., 1988). The failure of this recruitment step then leads to further aberrancies but with a variable set of outcomes. To determine the site of action of *ro,* marked clones were induced by mitotic recombination and the *ro* patches were analyzed. In 78 mosaic but normal ommatidia scored, the R2 and R5 cells were always wild-type, while all other R cells, including R3 and R4, were mutant in some of the ommatidia (Tomlinson et al., 1988). The results show that normal development requires that the wild-type *ro* allele be expressed in R2 and R5 in order to permit subsequent development involving the R3 and R4 cells (which make extensive contact with R2 and R5, respectively). *ro* encodes a homeobox-containing protein and, hence, a putative transcription factor (Tomlinson et al., 1988). It seems likely that the mutant phenotype in *ro* eyes is a homeotic switch, which changes R2 and R5 into R cells resembling the other members of the R1–R6 group. In the absence of the normal contacts of the R2 and R5 cells, R3 and R4 develop awry.

The idea that ommatidial development involves not only a specific sequence of cell contacts but of specific, resulting gene inductions has thus received some important confirmation, but much remains to be done. While *sev* and *ro* seem to be required specifically in the eye, many of the other relevant molecules may be utilized in multiple places in development. Some of the pleiotropic mutants that affect the eye may yet prove unexpectedly informative in further tests of the hypothesis.

The story of *Drosophila* eye development involves more, however, than just the rules of ommatidial construction. One aspect, which involves large regional properties, is the fact that the dorsal and ventral halves are essentially mirror images of one another. As ommatidia begin to form within the furrow, they all, in effect, point posteriorly; after the furrow passes, they each turn 90°. Yet the turning in the dorsal and ventral hemispheres is in opposite directions, to produce the mirror-image pattern (Tomlinson, 1988; Ready, 1989). The processes which produce this chirality are unknown. The only clue to date is that the boundary between the dorsal and ventral halves is preceded by a slight equatorial groove (Ready et al., 1976). Somehow, newly forming ommatidial clusters on either side of the groove "know" which way to turn. It is not impossible that the first turnings on either side propagate rapidly through the respective hemispheres. Equally possible, however, is some form of prior bias differentially "imprinted" in the epithelium. There is a precedent for such imprinting in the posteroanterior sequence of ommatidial development, which exists as a propensity within the epithelium prior to passage of the morphogenetic furrow, as shown with excised and cultured pieces (Lebowitz and Ready, 1986).

Nor does the eye develop in isolation; its development has effects on neighboring structures. Thus, as shown with the mutant *disco,* in which neuronal fibers from the eye do not travel, as in wild-type, along a pioneer larval nerve (Bolweg's nerve) to the optic lobes of the brain, the consequence is degeneration of these brain structures. As noted earlier, the defect in this mutant appears, from gynandromorph mapping, to be within the pioneer nerve itself rather than in the optic lobes or the eye disc (Steller et al., 1987). A similar dependence of optic lobe development on correct innervation by axons from the eye had earlier been shown in a clonal analysis of several eye mutants, performed by Meyerowitz and Kankel (1978). This observation highlights how much is yet to be learned, not only about the details of ommatidial development but the global factors that shape the eye and which, in turn, influence other developmental processes as a result of eye development.

SEX DETERMINATION IN *DROSOPHILA*

A Brief History of the Problem

Sex has a twofold significance for geneticists. It is both the biological phenomenon that makes genetic analysis possible and is, itself, a subject of intrinsic biological interest. For those animal species in which sex is specified genetically, sex determination recommends itself as a genetic subject both because of the 1:1 segregation of the sexes, which indicates Mendelian inheritance, and the clear phenotypic differences that separate males and females. Not surprisingly, the determination of sex was one of the first problems to be investigated by geneticists in the early 1900s. *Drosophila* emerged as the premier model system for its study in the second decade of the century.

In *Drosophila,* the difference in sex is correlated with, and caused by, the difference in sex chromosome composition: X/X animals are female, X/Y individuals are male. The morphological signs of sexual dimorphism between males and females involve three parts of the body and can be briefly listed. These are the sex comb of the male foreleg, a group of 10–13 thick black bristles on the most proximal segment of the tarsus, the basitarsus (Fig. 7.33); the fifth and sixth tergites in the abdomen, which are uniformly darkly pigmented in males but which only show a posterior band of pigment in the females—in addition, females possess a seventh tergite while the abdomen of the male is both smaller and more rounded (Fig. 7.34); and the terminalia (Fig. 7.35), formed from the genital discs of the two sexes.

All of these structures derive from precursor cells that exist in both sexes but which take different developmental routes as a result of the sex determination "decision." For instance, the sex comb derives from a group of bristle-secreting cells that, in the female, develop into the most distal row of transverse bristles in the basitarsus. In the male, these bristle precursor cells rotate 90° and develop into the thick "teeth" of the sex comb. The terminalia have a more complex set of cellular origins. They consist of the anal plates, which derive from a common set of precursors in both sexes, and the

Fig. 7.33. Sex comb on the foreleg of male *Drosophila*. (Photograph courtesy of Dr. D. Gubb.)

Fig. 7.34. Abdomens of wild-type female (**a**) and male (**b**) *D. melanogaster*. (Reproduced from Baker and Ridge, 1980, with permission of the publisher.)

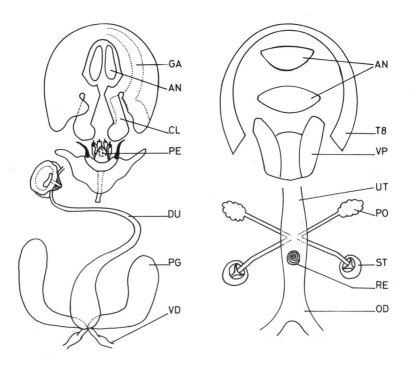

Fig. 7.35. External structures of the terminalia of male (left) and female (right) *D. melanogaster*. Male structures: GA, genital arch; AN, anal plates; CL, claspers; PE, penis; DU, ejaculatory duct; PG, paragonium; VD, vas deferens. Female structures: AN, anal plates; T8, eighth abdominal tergite; VP, vaginal plate; UT, uterus; PO, parovarium; ST, spermatheca; RE, seminal vesicle; OD, oviduct. Dotted lines are lines of clonal restriction. (Adapted from Dubendorfer and Nöthiger, 1982.)

genital structures, which derive from two distinct sets of primordia, initially present in both sexes, but only one developing, depending upon the sex chromosome composition of the cells (Nöthiger et al., 1977; Wieschaus and Nöthiger, 1982).

The genetic basis of sex in fruit flies was first elucidated by T.H. Morgan and his colleagues, in the course of the work that synthesized Mendelian genetics and the chromosomal theory of inheritance. The central individual in these investigations was Calvin Bridges, who established that the genetic basis of sex determination in *Drosophila* is the ratio of X chromosomes to autosome sets (the X:A ratio), rather than the presence or absence of the Y chromosome or the difference in the absolute number of X chromosomes between females and males. His first important findings were that X/O animals are male and X/X/Y animals are female, establishing that the Y does not confer maleness (Bridges, 1916). Then, in a landmark series of experiments, involving various triploid genotypes, Bridges demonstrated the critical importance of the X:A ratio (Bridges, 1921).

His approach was the model for the comparable approach used much later in *C. elegans* for analyzing X:A ratio effects in that organism (described in chapter 6). He first obtained some rare, spontaneously occurring triploid females and then crossed these with normal males; the various surviving progeny were found to have a variety of sexual phenotypes which could be correlated with their X chromosome and autosomal compositions. (Because of the need for chromosomal balance, all survivors had either diploid or triploid autosomal sets, except for some variability in the presence or absence of the tiny fourth chromosome, monosomy for the fourth having little effect on viability.) Bridges found that low X:A ratios favor maleness and high X:A ratios, femaleness. The results also included clonal patches where the X:A ratio could be scored. From the earlier studies on gynandromorphs (Morgan and Bridges, 1919), it was clear that sexual phenotype was set autonomously within clones.

The observations, for the entire range of X:A ratios scored, are summarized in Table 7.4. The results show that animals or tissue patches with an X:A ratio of 1.0 or more are phenotypically female, while animals or clones with an X:A ratio of 0.5 or less are male. For individuals with an X:A ratio between the male-determining value of 0.5 and the female-determining value of 1.0, the phenotypic outcome is intersexuality. Thus, an X:A ratio of 0.67 (2X:3A) yields an intersexual phenotype, with sexually dimorphic regions being a mosaic of female and male cells and structures.

Following Bridges' work, the problem of sex determination seems to have receded in perceived importance as a research subject for Drosophilists for many decades. Yet

Table 7.4. X:A Ratio and Sex Determination in *Drosophila melanogaster*

Sexual phenotype	X chromosomes	Autosomes	XA ratio
Metafemales	3	2	1.5
Female (triploid)	3	3	1.0
Female (diploid)	2	2	1.0
Haploid[a]	1	1	1.0
Intersex	2	3	0.67
Male	1	2	0.5
Metamale	1	3	0.33

two important advances were made in the period between 1921 and the 1970s. The first was the discovery, by Herman Muller, of the phenomenon of dosage compensation, the equalization of expression of (most) X chromosome genes between females and males despite the twofold greater dose of X genes in the former (Muller, 1932, 1950). The function of dosage compensation, as Muller realized, is to preserve relative genetic balance between the products of X chromosome and autosomal genes in the two sexes, despite their difference in X chromosome number.

The second set of discoveries involved the isolation of various autosomal mutants that produce a phenotypic sexual transformation. The first of these to be described in the literature, by Morgan's group in 1943, was the *intersex* (*ix*) mutation, which produces a partial masculinization of X/X animals but is without visible effect in X/Y animals. The second was the *transformer* (*tra*) mutation, which produces an even more complete phenotypic transformation of chromosomal females into phenotypic males ("pseudomales") albeit sterile ones (Sturtevant, 1945). A subsequently identified sex-transforming gene was *tra-2* (now *tra2*), whose mutants show a similar masculinizing effect to that of *tra* (Watanabe, 1975). The fourth in this set is *double sex* (*dsx*), whose first mutant, a dominant masculinizing one, was isolated in 1940, but not described in the literature until much later. Subsequently, it was shown that *dsx* can mutate to recessive forms that produce intersexes in both X/X and X/Y animals (Hildreth, 1965) and, also, less frequently, to recessive alleles that affect one sex or the other, converting either XX animals to intersexes or XY individuals to intersexes (Baker and Ridge, 1980; Nöthiger et al., 1987).

Despite their dramatic effects on phenotypic sexual characteristics, these genes play no role in dosage compensation; the expression of X-linked genes in the phenotypically transformed individuals behaves according to X chromosome composition, not phenotypic sex (Muller, 1950; Smith and Lucchesi, 1969). Although phenotypic sex and dosage compensation properties are associated in normal development, they can, evidently, be uncoupled genetically.

Despite the intrinsic interest of these findings, the problem of *Drosophila* sex determination remained relatively dormant as a research subject until the late 1970s, when the pace of work accelerated and two important discoveries were made. Indeed, much of our present understanding of the genetic basis of sexual development stems directly from these discoveries. In the first, the existence of a master regulatory gene for sexual phenotype, the X chromosomal gene *Sex lethal* (*Sxl*), was delineated (Cline, 1978). *Sxl* is expressed in and essential for female sex determination, being switched on in response to a high X:A ratio, while the active gene product is not produced in males, and, in fact, must remain unexpressed to permit male viability. In the second important investigation, the sex-transforming autosomal mutants were analyzed and found to comprise a genetic pathway downstream of the X:A ratio (Baker and Ridge, 1980) (and, later, of *Sxl* itself).

In reviewing what is known about sexual development in *Drosophila*, we will begin with *Sxl* and then examine what is known about the separate genetic mechanisms governing somatic sex determination and dosage compensation, respectively. Finally, the nature of germ line determination, which involves some distinctive features not shown by the somatic sex determination pathway, will be discussed. The literature on the subject of *Drosophila* sexual development is large and, particularly with respect to the genetic analyses, complex. What follows therefore is a general overview rather than a detailed account. Readers wishing to see more complete reviews should consult

Steinmann-Zwicky et al. (1990) for a detailed discussion of the sex determination mechanism, Lucchesi and Manning (1987) for the background and details of the phenomenon of dosage compensation, and Pauli and Mahowald (1990) for a discussion of the special aspects of germ line determination.

Initiating the Choice of Sex: From the X:A Ratio to *Sxl*

One of the most fundamental questions about sex determination in *Drosophila* is how the X:A ratio sets this "decision." This question breaks down into two parts: How is the ratio first sensed by cells? How, once sensed, does it translate into the choice between male and female pathways? The answer to the second question—how the choice is made—was discovered first and centers on the X-linked gene *Sxl* (at 1–19.2). As noted above, it is *Sxl* activity that is the crucial switch in setting the embryo on one path of sexual development or the other. An X:A ratio of 1.0 sets *Sxl* "on," promoting femaleness, while an X:A ratio of 0.5 leads to *Sxl* being "off," with consequent development of maleness. Understanding the role of *Sxl* is central to understanding what the X:A ratio does and how it is sensed.

The first indication of a key role of *Sxl* was obtained during an investigation of an autosomal gene, *daughterless* (*da*; 2–41.5). The original *da* mutant was a ts maternal effect sex-specific lethal; X/X embryos from homozygous *da* mothers at 25°C die while X/Y embryos are fully viable. At 18°C, a small proportion of daughters survive. During the course of an investigation on the *da* mutant, a spontaneously arising mutant in the stock was found to nearly completely restore the viability of daughters (Cline, 1978). The mutation suppressing the *da* maternal effect was determined to be X linked and, intriguingly, lethal to males; the mutant allele was named *Sex-lethal, Male-specific* (*Sxl^{M1}*). To rescue, *Sxl^{M1}* must be present in the genome of the daughters themselves. *Sxl^{M1}* females are fully viable whether or not they come from *da* mothers.

The potential significance of the *Sxl^{M1}* mutation was highlighted by the fact that this mutation maps to virtually the same position (1–19.2) as a previously isolated female-specific lethal (formerly *Female-lethal* [*Fl*], now designated *Sxl^{f1}*). This mapping suggests that the two mutations are allelic but produce opposite phenotypic effects. Cline (1978) characterized these two mutations by the Mullerian test and found that *Sxl^{f1}* is a hypomorph and *Sxl^{M1}* a neomorph, exhibiting (uncharacteristic) constitutive expression in males.

The opposing consequences for viability of the two mutant types in the two sexes suggest a simple hypothesis: that *Sxl* activity is normally expressed in and required in females, while *Sxl* is normally "off" in X/Y animals but if expressed in them, causes male-specific lethality. In this interpretation, the *Sxl^M* mutations are male-lethal precisely because they trigger inappropriate expression of *Sxl* activity. In contrast, female-lethal *Sxl* mutants, of which there are now many alleles, make effectively less *Sxl^+* product, a condition that reduces or eliminates female viability. In terms of this hypothesis, the function of maternal *da^+* is to help activate *Sxl* in X/X zygotes, without which these embryos would die; correspondingly, the *Sxl^{M1}* mutation rescues X/X embryos from the effects of deficiency for maternal *da* activity by constitutive activation of *Sxl* (Cline, 1978).

Although the effects on viability are sex specific for the two classes of alleles, they do not depend on phenotypic sex *per se;* lethal genotypes of either class when made homozygous for sex-transforming mutations to the opposite sex are not rescued by the transformation (Cline, 1978). Rather, the two kinds of lethality are associated with particular X:A ratios; homozygous Sxl^f mutations are lethal to X/X animals, regardless of phenotypic sex, while Sxl^M mutations are lethal to X/Y animals, similarly without respect to phenotypic sex.

The simplest explanation for these lethal states is that they involve inappropriate, that is reverse, settings of dosage compensation with respect to X chromosome composition. Dosage compensation, like phenotypic sex, is set by the X:A ratio and would be expected to be lethal if set incorrectly. As will be discussed below, dosage compensation in fruit flies involves a regulated hypertranscription of the single X in males to produce twice as much as either X in females. Thus, if dosage compensation in X/Y embryos were to be set in the female mode, they would transcribe their single X at a lower than normal rate, relative to the autosomes, creating a lethal genetic imbalance. Similarly, if the two Xs of the female were wrongly set in the male hypertranscription mode, they would make twice as much X chromosomal gene products as normal females, which would also be expected to be a lethal condition. In effect, Sxl activity might be necessary in females to prevent X chromosome hypertranscription, while inappropriate Sxl^M activity in X/Y animals might set X chromosome transcription at too low a rate for the single X condition (Cline, 1978). Indeed, Lucchesi and Skripsky (1981) found just such X chromosome hypertranscription in the polytene chromosomes of larvae heterozygous for partially complementing Sxl^f alleles (this condition permitting survival until the larval stage). (The mechanism of dosage compensation in fruit flies, it will be noted, appears rather different from that employed in *C. elegans* [chapter 6], although the result, equalization of X chromosome activity between the two sexes, is the same.)

However, Sxl is involved in more than setting the appropriate X chromosome transcription level in the two sexes; its activities influence sex determination as well. The link between dosage compensation and sex determination became apparent when small patches of mutant lethal Sxl tissue were examined in mosaics. These clones were obviously viable (unlike the whole organism homozygotes), but most significantly, they were observed to be sexually transformed. In XX//XO gynandromorphs, where the single X in the XO tissue carries Sxl^M, sexually dimorphic XO regions are found to be phenotypically female (Cline, 1979). Conversely, in X/X clones homozygous for a female-lethal Sxl allele, the transformation is to male tissue (Sanchéz and Nöthiger, 1982). The fact that clones of X/X tissue with male-level expression of the X are viable shows that hyperexpression of X is tolerable in some regions, at least, if not in the whole organism.

Although it might be that the clonal sex transformation effects are an indirect consequence of the inappropriate X chromosome expression caused by the mutations, some genetic results indicate that dosage compensation and sex determination are, in fact, regulated independently by the gene. Cline (1984) first selected viable "revertants" of Sxl^{M1} in males, in other words, second-site mutants that permit survival of the Sxl chromosome in X/Y animals through some reduction of expression of Sxl activity, and then examined these revertant chromosomes in females. Two such chromosomes were found to be partially viable in females, when heterozygous, indicating that, while hypomorphic for Sxl, they possess sufficient dosage compensation activity to allow some survival. However, surprisingly, these animals were not phenotypically female, as expected from their chromosomal sex, but were phenotypic males. Evidently, in these genotypes, there is enough *Sxl* activity to regulate dosage compensation, permitting survival, but insufficient activity to switch on those genes essential for female somatic sexual characteristics. An additional mutant, one directly selected as a female lethal, but which proved to be hypomorphic for *Sxl* function, exhibited similar behavior (Cline, 1984). The results show that *Sxl* is not only essential for dosage compensation but for somatic sex determination and that these two capacities can be uncoupled in certain genetic circumstances.

These experiments also provided the first clue to another important property of *Sxl,* namely that once a wild-type allele is switched on, it can maintain its own activity, by some form of autoregulation (Cline, 1984). Although no *Sxl+* X/X animals from *da* homozygous mothers survive at the restrictive temperature, viable X/X heterozygotes for the two revertants described above can be obtained from such mothers, albeit developing as phenotypic males. However, if these heterozygotes also carry an *Sxl+* allele, these individuals are phenotypically female, though sterile (Cline, 1984). The frequency of rescue is a function of the dosage of the revertant *Sxl* alleles, while it is insensitive to wild-type allele *Sxl* dosage. The results strongly suggest that the revertant alleles retain some of their constitutive *Sxl* activity (not requiring maternal *da* for activity) and that this is sufficient to activate the wild-type *Sxl* activity, which then continues to function, promoting female development.

Other results support the existence of autoregulation for *Sxl* and the separation of genetic elements for the initiation of such activation from those for subsequent maintenance of the gene's activity (Maine et al., 1985a). Furthermore, genetic analysis, based on P element excisions from *Sxl,* which produce a variety of deletions, show that the gene is a complex one, with separate regions for initiation of *Sxl* activity, maintenance of this activity, and for the somatic sex determination functions (Salz et al., 1987). There is also a distinct region required for *Sxl* function in the germ line (Salz et al., 1987). Perrimon et al. (1986) had shown earlier that *Sxl* can mutate to a form that allows viability of females but confers sterility, while Schüpbach (1985) had shown, by means of pole cell transplantation, that *Sxl+* activity is required in the germ line for oogenesis. All of these findings illustrate both the complexity of *Sxl* itself, as a genetic entity, and the manifold requirements for its activity in different aspects of sexual development.

If it is *Sxl* activity that sets the course of female development, while its absence triggers male development, a critical question concerns how *Sxl* is activated in response to a high X:A ratio, which, in turn, is related to the question of how that ratio is sensed in the first place. Molecular studies, made possible by the cloning of the *Sxl* gene, have shown that it is expressed very early in X/X embryos. *Sxl* protein is detectable by cleavage division 12 and thus distinctly before cellular blastoderm (Parkhurst et al., 1990).

As in the comparable case of *Caenorhabditis,* the critical question about X:A ratio sensing is whether it involves: 1) titration of autosomal "repressor(s)" by multiple sites on the X or 2) the excess production of critical X-linked genes in 2X genotypes relative to 1X genotypes (Fig. 6.18). If the former is the case, then the X "numerator" elements are binding sites and do not themselves produce a product; if the latter explanation holds, then the X numerator elements themselves encode products.

Genetic identification of the X numerators should, in principle, be able to decide the issue. The key criteria for identifying such elements, as listed by Cline (1988), are: 1) a decrease in their dosage should tend to kill females, as a result of failure to activate

Sxl, while an increase should kill males through inappropriate activation of *Sxl;* 2) female lethality should be suppressed by *Sxl*[M1], given the constitutive, neomorphic nature of this mutation, while male lethality should be suppressed by loss-of-function *Sxl* mutations; 3) there should be synergistic interactions between maternal *da* dosage and numerator element dosage, and, again, female-specific lethality should be suppressible by *Sxl*[M1]; 4) altering numerator element dosage in either direction should shift triploid intersexual phenotype.

Several candidate genes have been identified by screens for new female-lethals and searches among X chromosome duplications that can shift choice of sex pathway. The two principal candidates for the X numerator are the *sis-a* (at 1–34.3) and *sis-b* (1–0.0) genes (Cline, 1986, 1988). Both are strictly required in the zygotic genome and both show the kinds of dosage effects expected for X numerator elements that activate *Sxl* (Cline, 1988).

Molecular evidence concerning both *da* and *sis-b* suggests that the sensing of the X:A ratio, and the subsequent activation of *Sxl* in 2X embryos, involves transcriptional activation. In the first place, the initial regulation of *Sxl* by *da* is almost certainly transcriptional. *da* is a member of the so-called helix-loop-helix (HLH) group of proteins and also of the *myc* homology group (Caudy et al., 1988). Furthermore, beyond its resemblance to *myc*, it is also closely related in sequence to certain known mammalian transcription factors (Murre et al., 1989). (The amorphic condition for *da*, not surprisingly, is a lethal zygotic condition, indicating that the transcriptional factor it encodes is essential for development [Cronmiller and Cline, 1987].)

Furthermore, *sis-b* also has the HLH motif and is related to *myc* by homology. It is one of the earliest-expressed zygotic genes, making it a suitable switch element for *Sxl* (Torres and Sanchéz, 1989). Most unexpectedly, *sis-b* proved to be one of the genes of the AS-C with *myc* homology, *scute-alpha* (Torres and Sanchéz, 1989). (This is a particularly good illustration of how misleading gene names, based on initial mutant phenotypes, can be with respect to their biochemical functions.)

The significance of the HLH motif in *da* and *sis-b* is that it confers on them, potentially, the capacity for forming heterodimers, which possess novel transcriptional activity. This supposition is strengthened by an unexpected finding of Parkhurst et al. (1990), who discovered that early ectopic expression of the *h* gene under the influence of the *hb* promoter produced female-specific lethality and that this effect followed from an inhibition of *Sxl* expression in the domain of induced *h* expression (the anterior *hb* domain). *h*, like *da* and *sis-b*, is an HLH protein and therefore capable, in principle, of forming heterodimers with the *da* or *sis-b* gene products.

Parkhurst et al. (1990) suggest that *Sxl* transcription is initially activated by the formation of specific HLH heterodimers between *da* and *sis-b* (and possibly with other numerator element products as well), which would form in sufficient concentration when the latter is present in two doses (X/X embryos) but not one (X/Y embryos). Under this hypothesis, the inhibition of *Sxl* expression during ectopic *h* expression involves the titration of *sis-b* monomers by *h* monomers, thus lowering the concentration of *da/sis-b* heterodimers below the requisite concentration for *Sxl* activation.

Altogether, the results provide the first plausible molecular explanation, based on experimental evidence, of how the X:A ratio "signal" might work. Given the identification of one X numerator element as a gene that encodes a product, it seems probable that other X numerator elements will also be found to encode products;

discoveries of this kind would strengthen the hypothesis that high X:A ratios turn on *Sxl* through excess production of X chromosome-encoded activators rather than by titration by the X of autosomal "repressors." Nevertheless, the issue is far from definitively resolved and it is possible that X numerator elements include both genes encoding activators and sites for binding autosomal products. Two different kinds of repetitive sequence that are unique to the X have been identified (Pardue et al., 1987; Waring and Pollack, 1987) and they could be the hypothetical binding sites.

Much still remains unclear. In particular, we still do not know the identity of the "A denominator" elements or even if discrete A denominator elements exist (Steinmann-Zwicky et al., 1990); genetic searches for autosomal elements whose wild-type alleles masculinize X/X animals have not been successful, to date. If, on the one hand, the X numerator elements are products, then the putative A denominator elements may be DNA binding sites for these X-encoded products. Equally possible, however, is that the A denominator elements are genes that encode gene products, for instance HLH proteins capable of forming inactive heterodimers with the X numerator elements or *da,* in a fashion similar to that of *h* protein. Whatever the precise nature of the A denominator elements, their genetic identification will remain an uncertain prospect if there is a large number of them, each possessing a small effect but whose collective activities are additive.

Furthermore, there are questions about the exact number and identity of all the X numerators (Cline, 1988) and, indeed, of other X chromosome genes that may play a part in *Sxl* activation. One such additional X chromosomal gene is *liz,* whose maternal activity contributes to the activation of *liz* in the zygotic genome, which in turn contributes to *Sxl* activation (Steinmann-Zwicky, 1988). If additional X chromosome genes contribute in comparable fashions, then *Sxl* activation might well involve the additive, incremental effects of such interactions, being highly favored in 2X, relative to 1X, genomes (Steinmann-Zwicky, 1988). Indeed, such a mechanism does not seem improbable. Were the activation of *Sxl* a function of just a single gene product, varying only twofold in dose, the chances of error would be much higher. In contrast, if there are synergistic, dose-dependent interactions between such products, analogous to those of certain chromatin proteins involved in maintaining states of chromatin inactivity (Locke et al., 1988), then the initial twofold difference in dosage for each becomes amplified, to achieve a large differential effect, one that sets *Sxl* activity reliably and exclusively in 2X genotypes. The complexities and remaining conundrums of the X:A ratio are discussed in Steinmann-Zwicky et al. (1990).

A final point about *Sxl* regulation is that a key element in control of its activity is a posttranscriptional RNA processing step. This aspect, in turn, is probably intimately connected to the gene's autoregulatory properties. Despite the essentiality of *Sxl* activity in females and its apparent nonexpression in males, the gene itself was found to be transcribed in both males and females (Maine et al., 1985a). In fact, a complex pattern of 10 transcripts has been found (Salz et al., 1989) and, as noted earlier, transcription begins in the syncytial blastoderm (Parkhurst et al., 1990). However, when the sequence of later-stage male and female transcripts is analyzed, the male transcripts are found to differ from those in females in a key respect: They are spliced differentially to include an exon with a stop codon (Bell et al., 1988).

The consequence is that male-specific transcripts of this kind produce only partial and, hence, inactive *Sxl* protein products. Furthermore, the *Sxl* gene itself has distinct homology to several RNA binding proteins and may well contribute to its own splicing

(Bell et al., 1988). (This aspect of *Sxl* activity is discussed below.) Thus, the initial transcription of active *Sxl* in females could lead to continued production of active *Sxl* in this sex, while in males, this autoregulatory process would not even begin. These findings provide a specific molecular explanation for the *Sxl* autoregulatory property (described earlier), which was initially deduced by Cline (1984) on the basis of genetic experiments.

However, these observations do not explain how the *initial* female-male difference in *Sxl* activity is established. A clue may lie in the fact that of the 10 different *Sxl* transcripts, a few forms unique to early embryos, and possessing unique 5′ ends, have been identified (Salz et al., 1989). In view of this, and the involvement of transcription factors in *Sxl* activation, it seems reasonable to posit that the initial activation of *Sxl* is at the transcriptional level in females, from specific promoters and in response to the X:A ratio, and that these first transcripts encode active Sxl protein (Salz et al., 1989). Later transcriptional activation events that occur in male embryos would be independent of the X:A ratio, would involve other promoters, and would produce transcripts that encode inactive Sxl.

Somatic Sex Determination

Long before the discovery of *Sxl,* several autosomal sex-transforming mutants had been reported and their genes mapped. Of the four identified autosomal genes, loss-of-function mutations in three (*tra, tra2,* and *ix*) affect somatic sex determination only in females, producing female-to-male transformation (reviewed in Baker and Ridge, 1980). In contrast, most of the loss-of-function mutations in the fourth gene (*dsx*) produce a mutant phenotype of intersexuality in both 1X and 2X animals, while minority classes transform either X/Y animals to intersexes or X/X animals to intersexes or, in the case of a few dominant alleles, to males (Baker and Ridge, 1980; Nöthiger et al., 1987). These four loci are listed and briefly described in Table 7.5.

All four autosomal genes affect all aspects of somatic sexual phenotype, including internal sex organs and behavior. Their roles are, therefore, in the determination of general sexual phenotype rather than in the specification of particular sexually dimorphic structures. Furthermore, with the exception of *tra2,* they are without substantial effect in the sex not transformed; *tra2,* however, is required in males for the completion of spermatogenesis (Belote and Baker, 1982; Schüpbach, 1982).

Though somatic sexual transformation can be complete in those mutants that do not produce interesexes, there is far less frequent gametic transformation and those

Table 7.5. Autosomal Sex Determination Mutants in *Drosophila melanogaster*

Locus	Map position	Allele	Phenotype
transformer	3–45	tra	Transforms females into males; males normal
transformer-2	2–70	tra2	Transforms females into males; males sterile
intersex	2–60.5	ix	Transforms females into intersexes; males normal
doublesex	3–48.1	dsx	Transforms males and females into intersexes
		dsx^D	Transforms females into intersexes; males normal
		dsx^{Mas}	Like dsx^D
		dsx^{136}	Transforms males into intersexes; females normal

germ cells that exhibit some degree of transformation undergo abortive development (Nöthiger et al., 1989). Thus, *tra-* and *tra2*-transformed X/X animals are male in appearance but sterile, lacking functional sperm. The absence of sperm stems from an incompatibility between the transformed mesodermal tissues surrounding the germ cells and the untransformed germ cells themselves (see below). Furthermore, as noted earlier, the transformations do not alter states of dosage compensation, which remain set according to chromosomal sex. Indeed, were this not so, phenotypic sex transformation would be a lethal event, as evidenced by the *Sxl* mutants.

An important question about any putative regulatory loci for a developmental pathway is whether it is required to initiate the pathway or rather, primarily, to maintain it. Like the homeotic selector genes of the ANT-C and BX-C, the expression of the autosomal sex determining genes is found to be required late, as well as early, in development; they are, therefore, not solely establishment functions. To ascertain the temporal requirements, as in the case of the BX-C and ANT-C functions, one removes the wild-type allele from heterozygotes by mitotic recombination at particular times in development and then scores the emergent adults for the presence or absence of the homeotic phenotype. For all four genes, the retention of wild-type sexual phenotype within individual clones is dependent on the continued presence of the wild-type allele until late in the third instar, or, in the case of the abdominal histoblasts, the pupal period (Baker and Ridge, 1980; Wieschaus and Nöthiger, 1982). In effect, all four genes (*tra, tra2, ix,* and *dsx*) are required in the sex in which they act until the last few cell divisions in all of the sexually dimorphic structures.

These genes are probably also required early, given the sexual differentiation in the larvae of the gonads and genital discs, but not the latest, since these gene products may "perdure" in the various tissues even after recombinational removal of the genes. In fact, *tra2* activity is required even in adult females for continued maintenance of, at least, one female-specific trait, namely yolk protein (YP) synthesis. Thus, temperature-shift experiments with a ts *tra2* mutant reveal a requirement for active gene product in the fat body cells, one of the two major sources of YPs of the adult female, to ensure their continued synthesis (Belote et al., 1985). YPs are also made in the follicle cells of the egg chamber, but, in these cells, *tra2* activity is not needed in adults to maintain YP synthesis (Bownes et al., 1990). (The nature of the difference between the requirement for *tra2* activity in fat body and follicle cells will be described later in this section.)

Of the four sex determining loci, *dsx* is the only one required in both sexes for somatic sex determination. As noted above, it can, unlike the other three, mutate to produce at four very different phenotypes: the amorphic and recessive *dsx* allelic type, which transforms both X/X and X/Y animals into intersexes; dominant mutants that masculinize females; and recessive mutants of two different classes, each of which transforms one sex but not the other into intersexes (Nöthiger et al., 1987). The simplest extrapolation from these genetic findings is that the wild-type *dsx* locus is a bifunctional gene, possessing distinct masculinizing, dsx^m, and femininizing, dsx^f, activities. In wild-type X/Y animals, only the former is expressed, while in wild-type X/X, only the latter is. The different mutations either abolish both activities (the majority class), leading to intersexuality in both sexes, or "set" one activity or the other inappropriately with respect to chromosomal sex. The dominant masculinizing alleles, under this interpretation, are sufficiently "strong" to override the dsx^f activity of the dsx^+ activity in heterozygotes.

Unlike *ix*, which does not affect males, *dsx* activity is needed in both sexes for the development of normal sexual phenotype (Hildreth, 1965). In both *dsx*-transformed X/X and X/Y intersexes, for instance, the sex comb develops into a structure almost exactly intermediate, in orientation and bristle morphology, between the normal sex comb and the homologous transverse bristle row of the female. This phenotype is thus different from that of the 2X:3A intersexes, where each cell assesses the X:A ratio and makes a "clean" decision as to sexual phenotype. If the setting of *dsx* activity, however, is part of the execution process, following that decision, then an intermediate phenotype in *dsx* flies would not be unexpected.

Whenever a group of genes is identified as being important for a particular phenotype, it becomes necessary to determine if their actions are independent of one another or part of a genetic pathway. In the latter case, there should be a clear pattern of epistatic relationships discernible between those mutants differing in phenotype. The appropriate pairwise tests were carried out by Baker and Ridge (1980) and, indeed, such a pattern was found. The results demonstrate the existence of a pathway involving these genes. The amorphic *dsx* allele is epistatic to *tra*, *tra2*, and *ix*. In turn, *tra* and *tra2* are epistatic to *ix*. The relative positions of *tra* and *tra2* cannot be tested because of the similarity of their phenotype.

The existence of a definite epistatic hierarchy for several of these genes shows that they, at least, are part of a single pathway of development. If the pathway is viewed provisionally as a sequence of determinative decisions (rather than a sequence of "substrate" transformations), then the suppression of the three other mutant phenotypes by *dsx* indicates that *dsx* is the final activity in the pathway.

Taking into account the nature of *dsx* as a bifunctional activity, and the role of *Sxl*, these findings permit the following interpretation: that a high X:A ratio sets *Sxl* "on" in females, that this activity in turn activates *tra*, *tra2,* and *ix*, and that these gene products set dsx^f "on." The latter would then either switch on the necessary suite of downstream genes necessary for female sexual development or repress male-specific differentiation genes, the consequence in either case being the promotion of female development. In contrast, a low X:A ratio would fail to trigger these sequential activations and, by default, dsx^m would be expressed, this activity either activating the set of male differentiation genes or simply repressing the female-specific differentiation genes, the net effect, in either case, being the promotion of male development. The proposed female and male pathways are illustrated in schematic form in Figure 7.36.

This model, originally proposed by Baker and Ridge (1980) (without the explicit stipulation of the role of *Sxl*), has been tested and largely confirmed by subsequent molecular analysis. Not surprisingly, it has also been extended and modified by the findings, particularly with respect to the role of *ix*, which was placed, on the basis of the epistasis results, upstream of *tra* and *tra2* and now appears to be an activity that modifies *dsx* (Nagoshi et al., 1988).

The aspect, however, that was completely unanticipated when the model was formulated is that much of the sequence involves not differential control of transcription but a sequence of differential splicing events to produce sex-specific mRNAs. (A pathway based on epistasis results, of course, reveals nothing about underlying molecular mechanisms, but the expectation, at the time, was that a determinative pathway would involve primarily transcriptional controls.) At the end of this chain of events, there is produced in females a set of female-specific *dsx* transcripts, and in

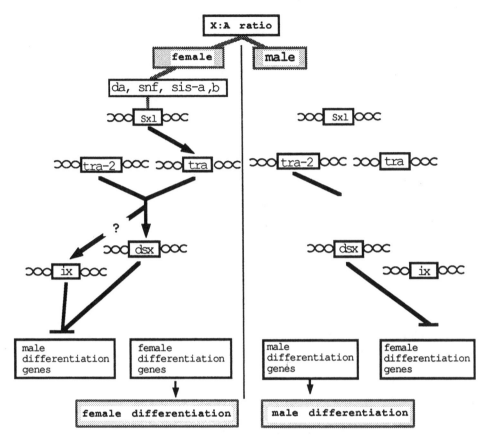

Fig. 7.36. Pathways for somatic sexual development in *Drosophila.* Female development is triggered by a high X:A ratio "sensed" through the indicated X numerator elements, which activate *Sxl* expression and the sequence of gene activities, as shown, culminating in *dsx*[f] activity, which represses male differentiation. A low X:A ratio leads to the "default" male pathway, culminating in the production of *dsx*[m] activity and the repression of female differentiation. See text for gene names and discussion. Note: *snf* is *sans fille,* one of the X numerators, also known as *liz* (Steinmann-Zwicky, 1988). (Reprinted by permission from *Nature,* vol. 340, p. 52X, © 1989 Macmillan Magazines Ltd.)

males, a set of male-specific *dsx* transcripts (Baker and Wolfner, 1988). These two transcript groups (or individual members of each) correspond to the *dsx*[m] and *dsx*[f] activities that were originally predicted by the Baker-Ridge model.

The complete sequence of events can now be summarized: In animals with an X:A ratio of 1.0 (females), there is an activation of *Sxl,* which leads to the production of active Sxl protein. Sxl protein, related to several known RNA binding proteins (Bell et al., 1988), is required for, and presumably directly participates, in the production of female-specific *tra* RNA (Boggs et al., 1987). The active Tra protein, in concert with the protein encoded by *tra2,* which itself is related to several RNA binding proteins (Amrein et al., 1988), then splices *dsx* RNA to produce *dsx*[f] transcripts (Nagoshi et al., 1988). *tra2* is transcribed in both sexes and its known RNAs are, in contrast to *Sxl* and *tra,* identical in males and females; the *tra*-encoded product made in females presumably acts as a cofactor for *tra2* in somatic cells to permit the splicing of *dsx* and the generation of *dsx*[f] RNAs (Amrein et al., 1988).

In animals with an X:A ratio of 0.5 (males), the default pattern of splicing events occurs. First, a truncated Sxl protein is made (Bell et al., 1988). In the absence of functional Sxl protein, alternative splicing produces *tra* transcripts lacking long open reading frames (Boggs et al., 1987). (These transcripts are, in fact, not specific to males; females also produce them, but, in addition, make transcripts that are capable of encoding functional product [Boggs et al., 1987].) In the absence of functional Tra product, *dsx* transcripts undergo default splicing to produce *dsx*[m] transcripts, which have long open reading frames but differ in their exonic composition (Burtis and Baker, 1989). The net result, in males, is the production of Dsx proteins with different activities from those of Dsx proteins in females. Some of the salient features of the splicing events are diagramed in Figure 7.37; for more detailed discussions, the reader should consult the reviews of Baker (1989) and Steinmann-Zwicky et al. (1990).

Given that the end result of these two different molecular pathways in the two sexes is the production of two different *dsx* activities, the crucial question is, Just how do *dsx*[f] and *dsx*[m] create the different somatic sexual phenotypes? Though *dsx* does not bear obvious sequence relatedness to known transcriptional factors, its encoded protein forms almost certainly act in that capacity, with *dsx*[m] repressing female differentiation functions and, possibly, activating male-specific genes, while *dsx*[f] does the reverse. The existence of recessive mutant forms which convert only one sex to intersexes is most readily explained on the basis that *dsx*[f] and *dsx*[m] simply act by repressing, respectively, the differentiation functions of the opposite sex. The most frequent mutant class, whose members produces intersexuality in both X/Y and X/X animals, would, under this hypothesis, simply reflect general failure of repression, with concomitant derepression of male- and female-specific differentiation functions in both chromosomal sexes.

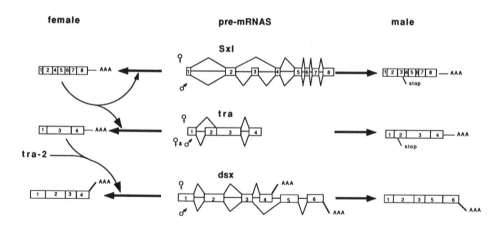

Fig. 7.37. Some of the transcript splicing patterns in the sex determination gene hierarchy. Alternative splicing leads to functional gene products for *Sxl* and *tra* in females and transcripts encoding truncated proteins in males; for *dsx*, alternative functional products, differing in their exon composition are generated. See text for discussion and Baker (1989) for details. (Reprinted by permission from *Nature*, vol. 340, p.52X. © 1989 Macmillan Magazine Ltd.)

It is also possible that *dsx* does not perform these actions directly but controls a battery of region- or cell-specific transcriptional regulators. If such were the case, then mutants of such genes should produce localized homeotic sex transformations. However, such mutants have not been found and this alternative hypothesis of *dsx* action, therefore, appears unlikely. Furthermore, both Dsx^f and Dsx^m protein forms have recently been found to bind specifically to the enhancers of two YP genes (Burtis et al., 1991). The binding of the Dsx^m form presumably reflects its activity as a repressor of the YP genes, while the binding of the Dsx^f form either reflects its activity as a positive control transcriptional factor or the absence of some condition *in vitro* that prevents Dsx^f protein from acting as a repressor *in vivo*.

Assuming that the Dsx proteins are, indeed, the direct regulators of sex-specific downstream genes, two kinds of regulation can be envisaged (reviewed by Steinmann-Zwicky et al., 1990). The first would be determination of cell types that are specific to one sex or the other, such as the different kinds of gonadal and genital tissue, and the second would involve regulation of sex-specific genes in tissues shared by both sexes. A prime example of the latter would be the YP genes that are active only in the fat body of the female, though this is a tissue common to both males and females.

For the determination of sex-specific cell types, one might predict that *dsx* activity would be dispensable after those cells have differentiated, and, indeed, this has been found to be the case in two instances. Thus, temperature inactivation of the *tra2* gene product in *tra2* ts mutants does not stop production of male-specific transcripts in the male accessory glands, the paragonia (Chapman and Wolfner, 1988) nor YP synthesis in the ovarian follicle cells (Bownes et al., 1990). In contrast, YP synthesis in the fat body cells of females is brought to a halt by such temperature shift (Belote et al., 1985). Even more dramatically, YP genes commence transcription when X/X; *tra2*^{ts}/*tra2*^{ts} pseudomales, raised at high temperature, are shifted to the permissive temperature (Belote et al., 1985). Since the activity of *tra2* is to generate *dsx*^{f} transcripts, the implication is that, in the fat body, these transcripts, which must exhibit rapid turnover, continue to be made and required.

To generalize these findings, *dsx* can act as a homeotic gene, in the setting of (sex-specific) cell fates, and becomes dispensable after those cells have been formed, while, in other circumstances, it participates as an ongoing transcriptional regulator in tissues that are common to both sexes but which have sex-specific activities.

Dosage Compensation

Dosage compensation in *Drosophila,* the second major element of sexual character, involves a doubling of the rate of expression of X chromosome genes in males relative to females; this doubling cancels the effect of halving X chromosome gene dose in males relative to females. Compensation serves to equalize X chromosome transcription between the sexes with respect to the autosomes and, as we have seen, the alternative dosage compensation states are a function of the presence or absence of *Sxl* activity. Although the majority of X chromosome genes are dosage compensated in larvae and adults, a number of exceptions are known (reviewed in Lucchesi and Manning, 1987). Prior to *Sxl* activation, there is presumably no dosage compensation, a condition which must presumably pertain to those X numerator elements that encode gene products. Failure of normal dosage compensation in females or males, as seen in both Sxl^f and Sxl^M mutants, leads to lethality.

Dosage compensation in *Drosophila* operates at the level of transcription, as first shown with autoradiographic measurements on salivary gland chromosomes (Mukherjee and Beerman, 1965; Mukherjee, 1966), The equalization of transcript levels has also been demonstrated for several individual X chromosome genes, using quantitative hybridization methods (e.g., Breen and Lucchesi, 1986). In principle, this equalization could take place either by reduction of X chromosome transcription in females or by preferential boosting of transcription in the single X of the male. Much evidence, beginning with the autoradiographic studies (Mukherjee and Beeman, 1965), shows that it involves the second mechanism, the enhanced transcription of the X in males. Indeed, as mentioned earlier, the insufficiency of *Sxl* activity in 2X larvae leads to a male-type response, namely increased transcription of the X chromosomes relative to the autosomes (Lucchesi and Skripsky, 1981). The result indicates that *Sxl* activity in females possesses some repressive activity for certain male-specific functions, though whether it is direct or indirect is not known.

Just as *Sxl* exerts its effects on somatic sex determination by regulating a downstream gene (*tra*), its effects on dosage compensation also involve a set of downstream genes. Four genes required for the male-specific hypertranscription of the X chromosome were first identified in searches for autosomal (second and third chromosome) sex-specific lethals (Belote and Lucchesi, 1980a,b) This gene set is referred to as the male-specific lethal (MSL) group and consists of *msl-1* (2–53.3), *msl-2* (2–9.0), *msl-3* (3–26), and *maleless* (*mle*) (2–55.8). In contrast to the early lethal effects of *Sxl* deficiency in 2X animals, the time of death ascertained for several of the male-lethal mutants is the late larval–early pupal stage (Belote and Lucchesi, 1980a).

The lethal effects of the loss-of-function mutations in these genes are strictly a function of the 1X condition and are not dependent on either the presence of the Y chromosome or on the expression of phenotypic maleness (Belote and Lucchesi, 1980a). Thus, for instance, *tra*-transformed X/X; *msl-1/msl-1* individuals, though phenotypically male, retain normal viability. In the homozygous male larvae, three X-linked enzyme activities were found to be depressed to about 60% of normal and X chromosome transcription in homozygous *mle^ts* larvae was reduced to about 65% of normal at the restrictive temperature for this ts mutant (Belote and Lucchesi, 1980b).

If the lethal effects of homozygosity for Sxl^f mutations in 2X animals are caused by hyperactivation of the MSL loci listed above, then mutations in these loci should rescue Sxl^f females. Somewhat surprisingly, this prediction has not been fulfilled (Skripsky and Lucchesi, 1982). One possible explanation has been offered by Cline (1984), who has speculated that the known MSL genes may not be the only ones required for X hypertranscription. While the four identified genes are required for male larval survival (and imaginal development), there might be others required for initial or early X hypertranscription in the embryo.

The phenomenon also necessitates that there be something special about genes on the X such that they will recognize the factor(s) when present and respond accordingly. To date, no such *cis*-acting sequence has been identified unambiguously, although Pardue et al. (1987) have identified a short, multiply repeated sequence that is located throughout the X and is present on the X in higher concentration than on the autosomes.

One approach to the identification of such sequences has involved the construction of transgenic animals, for either X-linked or autosomal genes, by P-mediated transformation. These lines are then tested, respectively, for the retention of dosage compensation by X-linked genes in autosomal locations or for the acquisition of dosage compensation by autosomal genes integrated into the X. These experiments have indicated that responder sequences are liberally scattered throughout the X chromosome and can confer dosage compensation on autosomal genes (Scholnick et al., 1983; Spradling and Rubin, 1983). Conversely, X-linked genes when transferred by germ line transformation to autosomal sites can maintain their dosage compensation, and the essential regulatory sequences have been shown to be closely linked (Lewis et al., 1985; Pirotta et al., 1985). It will be interesting to see if the presence of the sequence reported by Pardue et al. (1987) proves to correlate with retention of dosage compensation.

The *mle* gene has been cloned and found to share homology with a number of known RNA and DNA helicases (unwinding proteins) (Kuroda et al., 1991). The noteworthy, and completely unexpected, property of this gene is that it is expressed equally in males and females but binds strongly to hundreds of sites on the X chromosome of males and only weakly to the X chromosomes of females or to the autosomes of either sex (Kuroda et al., 1991). Its precise role in dosage compensation remains to be elucidated. However, the simplest interpretation of its X chromosome pattern is that it recognizes and binds to another protein involved in dosage compensation, perhaps one encoded by one of the three other male-specific lethal genes, which is expressed in males but not females and that recognizes X chromosome sites specifically.

Germ Line Development

The final aspect of sexual development concerns the genetic basis of germ line development. Some of the earliest observations seemed to indicate that development of the gametes was a cell autonomous function of chromosomal composition. The experiments involved transplanting pole cells from chromosomally marked embryos into donor embryos; survivors of the operation were then mated to appropriately marked strains. The experiments are done "blind," that is, without respect to the sex chromosome composition of recipient or donor embryos. However, given the appropriate chromosomal markers, any progeny derived from the donor pole cells can be identified as such. If either "heterosexual" combination, that is, X/Y pole cells transplanted into X/X embryos or the reverse, can produce maturation into the "wrong" gamete type (in terms of sex chromosome composition) under the influence of the recipient's soma, that change of fate will be detectable if the X and Y chromosomes of the donor pole cells are marked appropriately. The results of such experiments failed to reveal any switch of developmental fate of transplanted pole cells (van Deusen, 1976). Since only germ cells finding themselves within a gonadal soma of compatible sex chromosome composition could give rise to functional gametes, it was assumed that mismatched germ line cells simply followed the normal gametic developmental path associated with their X:A ratio, irrespective of the surrounding soma (van Deusen, 1976).

It is possible, however, that they fail to be detected, not because of autonomous development down the wrong path, but because they are out-competed by the recipient's own germ line cells (Schüpbach, 1985). Another explanation is that X/X germ line cells in an X/Y soma and X/Y germ line cells in an X/X environment may begin development but fail to complete it.

To resolve the issue, Steinmann-Zwicky et al. (1989) did pole cell transplants into recipients lacking pole cells (from *osk* mothers); any gametes detected would have to be from the donor cells. The results showed that, contrary to the earlier inferences, some "inappropriate" germ line development can be induced by a soma of the opposite sex chromosome composition. In particular, X/X cells in an X/Y soma were found to undergo development into spermatocytes (though not into mature sperm) in a substantial fraction of the hosts. Evidently, development of oocytes in X/X animals is not an autonomous property dictated by the X:A ratio of the germ line cells. In contrast, X/Y cells in an X/X soma are not induced to become oocytes but proceed to become spermatocytes, irrespective of the somatic environment.

These results show that germ line development involves a combination of autonomous properties and of inductive influences. The cell autonomous element(s) of germ line development are seen in the failure of X/Y germ line cells to become oogenetic in an X/X soma. The nature of the inductive signal is less clear but could be, in principle, of either of two types. It could be a male-determining signal from an X/Y soma, which is effective in initially channeling X/X germ line cells toward spermatocyte development. Alternatively, it could be a female-determining signal originating in an X/X gonadal soma, which is sufficient to prompt X/X but not X/Y germ line cells to begin oogenesis and in whose absence 2X germ line cells adopt a "default state" and commence spermatogenesis. These two possibilities cannot yet be experimentally distinguished.

The influence of surrounding cells in female germ line development contrasts sharply with the strictly autonomous character of sexual phenotype in somatic clones (Baker and Ridge, 1980). A further difference is that the somatic sex determination genes, downstream of *Sxl*, are not required for female germ line determination. This was first shown for *tra*, which was shown to be without sex-transforming potential in female germ line cells. Marsh and Wieschaus (1978) transplanted marked pole cells from X/X; *tra/tra* embryos into recipient female hosts and found that progeny derived from these pole cells gave rise to as many functional egg cells (as measured by progeny counts) in female recipients as do control +/*tra* pole cells (Marsh and Wieschaus, 1978). Similar tests, and the use of mitotic recombination to generate germ line clones homozygous for *dsx, ix,* or *tra2,* have shown that these sex determining loci are also without transforming effect in the germ line (Schüpbach, 1982).

In sharp contrast, *Sxl* is required autonomously in female germ line cells. The *Sxl* requirement was first shown by Schüpbach (1985), who marked homozygous Sxl^f pole cells with mal^+ (both *Sxl* and *mal* are on the X) and transplanted them into $Sxl^+ mal^-$ hosts and examined their development in ovarioles. The stained cells were invariably abnormal in several respects. Thus, despite a surrounding wild-type soma, *Sxl*-deficient X/X germ line cells show autonomous cell defects. Furthermore, X/X pole cells lacking *Sxl* activity have been found to undergo partial development into spermatocytes in agametic females and, conversely, agametic males receiving Sxl^M X/XY pole cells exhibited some oogenetic structures (Steinmann-Zwicky et al., 1989). The results may provide an explanation for the nonautonomous behavior of X/X pole cells in an X/Y soma; it could be that such a soma prevents activation of *Sxl* activity in the X/X germ line cells and that the propensity of wild-type 2X germ line cells to develop into

spermatocytes in an X/Y soma, mentioned earlier, involves specifically the failure of
Sxl^+ activation in these cells.

The most important resemblance of germ line development to somatic sexual
development is in the requirement for Sxl activity in X/X cells. The significant
difference between the two pathways is that Sxl does not seem to be needed to activate
the genes required for somatic sexual phenotype, including dsx. This raises the
question of the identity of the downstream genes in germ line development and, to
date, they have proved elusive. At least two genes are known whose mutants cause a
transformation of sexual phenotype in the germ line, ovo and liz, but both act
upstream of Sxl, to activate it within the germ line cells (Oliver et al., 1988; Steinmann-
Zwicky, 1988). The identification of the genes downstream from Sxl in the female
germ line is a task for the future.

Overview

The progress of the last 15 years in the analysis of sex determination in *Drosophila*
has been dramatic. The pathway is now one of the best understood, in genetic terms,
in all of animal biology. It is summarized schematically in Figure 7.38.

However, there remain many puzzling features and unsolved questions. Though
the somatic sex determination part of the pathway, involving a splicing cascade, is
well understood, the need for its complexity is not (Baker, 1989). Here, as in the
development of the posterior abdomen of *Drosophila,* one is inevitably drawn into
evolutionary questions and speculations. Furthermore, though some aspects of the
X:A ratio "signal" have been elucidated, it seems probable that other X numerator
elements exist while the numbers and modes of action of the A denominator
elements are completely unknown. Finally, and not least, the mode of regulation by
the Dsx^f and Dsx^m proteins and the particular downstream genes need to be
delineated.

Dosage compensation also presents some poorly understood features. In its reli-

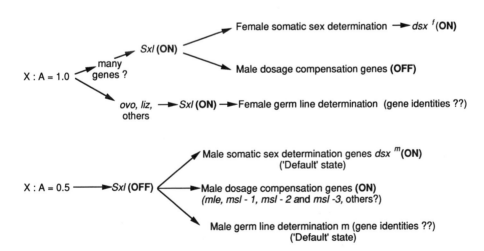

Fig. 7.38. Summary of sex determination pathway in D. *melanogaster,* showing the different consequences
of *Sxl* activity states for somatic sexual development, dosage compensation, and germ line development in
female and males.

ance on hyperactivation of the single X, it presents some marked contrasts to the other two well-characterized systems, those of *C. elegans* (chapter 6) and of mammals (chapter 8). With respect to the details, several questions need to be answered. Are there other MSL genes beyond the ones that have been identified? Does *Sxl* activity repress the activity of the MSL genes by differential splicing? Furthermore, does it act directly or through an intermediary? How does the Mle protein coat the X chromosome of the male and how does it stimulate transcription there?

Finally, there are questions about germ line determination. Although it shows a similar requirement for autonomous *Sxl* action, the pathway itself seems to involve a different set of downstream genes for germ line determination. These genes have not yet been identified. Thus, while "maleness" also appears to be a default state in germ line development, its molecular basis needs to be elucidated. Furthermore, intriguingly, there appear to be cell–cell interactions, involving the soma and germ line, which do not appear to feature in somatic sex determination. In the existence of such interactions, this aspect of *Drosophila* sex determination bears more formal similarity to mammalian sex determination (see chapter 8) than do its other aspects.

GENETIC REGULATION AND *DROSOPHILA* DEVELOPMENT: SOME PRELIMINARY CONCLUSIONS

The past dozen years have been an impressive period in the genetic exploration of *Drosophila*. To list some of the landmark findings evokes what has been accomplished: the discovery of the *bcd* gradient and how it works; the analysis of segmental pattern development in the early embryo; the elucidation of a complex molecular pathway in "neurogenesis" and an even more complex cellular and molecular sequence in eye development; and the discovery that the sex determination regulatory hierarchy involves a sequence of differential splicing events. Yet the successes should not lead one to minimize how little we still understand in certain respects. Much of what has been learned concerns the *regulatory machinery* behind the visible events, while a great deal remains obscure in how that machinery produces its effects. We still do not know how a segment is constructed, or the precise basis of pattern formation in imaginal discs, or even the nature of the "signal" that prevents an epidermoblast from turning into a neuroblast.

Nevertheless, there is sufficient information about the nature of "regulatory" events in *Drosophila* development to allow one to draw certain general conclusions. These generalizations are significant and could not have been stated with confidence even five years ago. They will be listed here.

The maternal "instructions" are neither in the form of a pair of maternal gradients nor a mosaic of tissue- or structure-specific determinants laid down in the egg but a combination of gradients and localized signals that act as initiators of gene expression sequences. Thus, *Drosophila* is organized along the a-p axis by localized signals at the anterior and posterior ends (the terminal system), the maternal *bcd* gradient which governs pattern in the future thoracic and anterior abdominal segments, and a localized abdominal signal that antagonizes the translation of a maternal mRNA. For each maternal component, the ultimate consequence is an initial differential regulation of other genes. However, these first-tier target genes are neither histo- nor organ-specific ones; the ultimate specification of recognizable phenotypic features is considerably "downstream" in each case. For both the terminal and d-v systems, the maternal components appear to be ligand-based signaling systems. Similarly, ligand-receptor interactions between blastomeres are an essential feature of much conditional specification in embryos of echinoderms and vertebrates (Davidson, 1990). It is noteworthy that such interaction play key parts in the first steps of *Drosophila* embryonic development as well, but, at this stage, involving ligands presented by either the surrounding follicle cells or the perivitelline fluid.

The other aspect of early development which deserves note is the relative importance of early interactions between zygotic genes in setting the initial pattern. Indeed, the whole posterior of the embryo, apart from the future telson, may be specified by interactions between zygotically expressed gap genes. The function of maternal "determinants" for this region may be solely to establish the initial activities at the boundaries of this region, *Kr* at the anterior boundary and *tll* and other genes at the terminus. These boundary activities then start a cascade process that fills in the middle.

Understanding developmental processes in detail will involve a much more extensive knowledge both of the key surface molecules and of the signaling events they set in motion. Although, in the late 1960s and throughout the 1970s, it was fashionable to believe that development could be "explained" in terms of transcriptional switches, that viewpoint had, clearly, lost most of its adherents by the mid-1980s. If one wants to understand how, for instance, a cell in the neuroectoderm becomes either an epidermoblast or a neuroblast, it will be necessary to understand both the nature of the signal(s) elaborated by the neurogenic genes and the sequence of events that follow upon the reception of and transduction event(s) or their failure to occur. Many more such examples could be enumerated.

Multiple use of single gene products, partial functional redundancy, and the complexity of gene regulatory networks are connected features of development. The first part of this statement—the multiple use of individual gene products—is simply the conclusion drawn from the fact of pleiotropy and might therefore appear not worth remarking upon. What has given it new point is the characterization at the molecular level of many genes known, and named for, their effects on particular aspects of development. Thus, both *ftz* and *eve* were first isolated on the basis of their pair-rule phenotypes, yet both are expressed in the nervous system and required there for certain sets of neurons (Doe et al., 1988). The interesting point about *ftz* utilization is that if *ftz* is prevented selectively from expression in the nervous system, the neural path defects that take place are somewhat variable (Goodman et al., 1988), reminiscent of the variable expressivity effects associated with amorphs in *C. elegans* for certain key gene functions. Evidently, other gene products serve as backup for *ftz,* and vice versa, in the nervous system.

A secondary inference is that much of the complexity of gene regulation probably exists in order to provide sufficient backup functions for each requisite gene process. The rigidities imposed by combinatorial control (chapter 2) are undoubtedly relieved by the existence of related gene products, encoded by gene families whose members are dispersed throughout the genome and that respond to different sets of regulatory molecules.

If some, or perhaps much, of the complexity of the regulatory circuitry has evolved to overcome the inherent restraints of combinatorial control, it becomes easier to understand the seemingly unnecessary antagonistic relationships between gene products that are occasionally found. To take one example, embryos that are zygotically deficient for either of the pair-rule genes *odd* or *eve* show a pair-rule mutant phenotype (though with different alternate parasegments missing), yet double mutant embryos are nearly normal (Coulter and Wieschaus, 1988). In effect, such situations provide a cautionary tale for the modeler of gene networks: The structure of these networks may often embody successive evolutionary contrivances, the "bricolage" ("tinkering") of natural selection referred to by Jacob (1977) rather than necessarily representing the optimal engineering design. We will return to this point in the final chapter.

Despite the triumphs in this field of the last dozen years, the most difficult tasks may lie ahead, namely understanding the complete chain of events between the regulatory machinery and the visible cellular phenotypes. Sex determination provides an example. The evidence shows that determination of somatic sexual phenotype in *Drosophila* females is governed by the production of the female-specific form of *dsx* mRNA. It may not be long before it is understood how the *dsx*[f] protein, encoded by the female-specific mRNA, turns on, let us say, vitellogenin production in the fat body and the follicle cells. However, how does the *dsx*[m] form of the protein direct those cells in the foreleg imaginal disc that would make a normal transverse bristle row in females to make a sex comb? With that kind of question, one returns to what has been the main subtext of this chapter, the problem of pattern formation in *Drosophila*.

It is in pattern formation that one can trace a subtle development in the evolution of ideas, namely a move away from hypotheses that posit control by gradients (apart from the syncytial stage of the embryo) to mechanisms involving short-range cell–cell interactions, the majority probably involving cell contacts, and a minority, short-range diffusible substances. It appears likely that pattern formation both in the formation of segments and in imaginal discs (aside from the special case of the eye disc) involves similar mechanisms and an at least partially overlapping set of shared molecules, those of (some of) the segment polarity genes. The challenge here, as in all cases of pattern formation, will be to understand the connections between the organizing events and molecules and the ultimate visible expression of pattern. As more "downstream" genes are identified and their products and expression patterns studied, it is not unreasonable to believe that, in the near future, we may begin to move from general statements about pattern formation to fairly precise pictures of how particular spatial cellular arrays are organized.

EIGHT

MOUSE POSTIMPLANTATION DEVELOPMENT

The mouse has been a slower starter than many other organisms in the race to unravel the genetic control of embryonic development. Recent cloning of putative developmental genes combined with new approaches to manipulating the mouse genome seem set, however, to allow the mammalian embryo to move towards the front of the field.

J. Rossant and A.L. Joyner (1989)

INTRODUCTION

For many years, it seemed to many developmental geneticists that the first would, indeed, be the last. Though the mouse was the initial animal model system in genetics, and was and is one of the most thoroughly genetically characterized organisms, its study was seemingly eclipsed in the 1970s and 1980s by the rapid advances in the developmental genetics of *Drosophila* and *Caenorhabditis*.

That situation, however, began to change in the late 1980s, as the result of several key technical developments. These included the discovery of ways of creating transgenic mice, the improvement of methods for obtaining new mutations, the identification of mouse homologs of key *Drosophila* genes, and the application of methods developed in *Drosophila* for identifying and isolating genes with new patterns of gene expression ("enhancer trap" techniques). Finally, and perhaps most significantly, strategies for replacing wild-type genes with engineered mutant homologs ("gene targeting" methods) were devised and brought to fruition. As a result of these innovations and discoveries, it is beginning to be possible to approach the genetic characterization of developmental events and processes in the mouse with the same degree of rigor that has been applied to the fruit fly and the nematode. Furthermore, a major legacy from the long period of classical mouse developmental genetics, the many mutants collected and provisionally characterized, is proving to be an important element in this new phase of exploration of mouse development. Increasing numbers of the corresponding wild-type genes of these mutants are being cloned and analyzed and the findings are illuminating the nature of the mutant defects.

This chapter examines the genetic analysis of postimplantation development in the mouse. We will begin with a brief summary of murine genetics and a descriptive overview of postimplantation development and its molecular biology. The role of clonal analysis in examining events in postimplantation development will then be described. In the second part of the chapter, a description of some of the contemporary methods that are transforming the analytical capabilities will be presented, along with some of the significant findings that have resulted.

GENETICS

The diploid chromosome number of *Mus musculus* is 40. Of the 20 chromosome pairs, 19 are autosomal and one pair consists of the sex chromosomes, either two Xs (females) or an X and a Y (males). Through the X-Y system is cytogenetically similar to that of *Drosophila,* there is a distinct difference in sex determination mechanisms between the mouse and the fruit fly; in mammals, sex is determined by the presence or absence of the Y rather than by the X:A ratio. Although the Y chromosome is similar in size to the smallest autosome pair, it is comparatively gene poor, containing few genes apart from the locus for the H-Y antigen (a surface male-specific histocompatibility antigen), several male fertility genes, and, most significantly, the critical locus involved in sex determination (which will be described later in this chapter).

Mouse genetic nomenclature is similar to that of other animal systems (for review, see Lyon, 1989). Genes are designated by italicized symbols which are usually abbreviations of their names. In general, genes first identified by dominant mutations are given abbreviations that start with a capital letter while the names of genes first identified by recessive mutations begin with a lowercase letter.

For many decades, new mutants could only be mapped with respect to other markers and placed in formal linkage groups, whose chromosomal identity was unknown. The development of chromosome banding techniques, G-banding and Q-banding, which distinguish specific homologous chromosome pairs, made possible the assignment of genes to particular chromosomes in the karyotype. This method uses translocations that move known genetic markers from one chromosome to another, in combination with an analysis of the changes in karyotype structure produced by the translocations, as detected by the banding techniques (reviewed in Miller and Miller, 1975). All chromosomes are acrocentric and numbered from largest to smallest. The nomenclature for chromosomal rearrangements is similar to that of *Drosophila.*

The molecular organization of the mouse genome is essentially comparable to that of the nematode and the fruit fly but the genome itself is much bigger: At 3×10^9 bp, the haploid mouse genome is nearly 20 times larger than that of *Drosophila.* Of this, 8–10% consists of highly repeated satellite DNA, principally a 140 bp adenine-thymine (AT)–rich repeat, reiterated 10^6 times in the centromeric heterochromatin. The single-copy fraction is approximately 60–76% and the mid-repetitive fraction is 15–25% of the genome, as measured under standard renaturation conditions (Church and Schultz, 1974; Ginelli et al., 1977.) In one rodent genome, that of the hamster, these mid-repetitive sequences are predominantly in long blocks, which, in turn, consist predominantly of "scrambled clusters" of shorter (300 bp) blocks (Moyzis et al., 1981), as indeed is also found for the long blocks of these sequences in the fruit fly genome (Wensink et al., 1979). As in other organisms, no distinctive functions have been found for the mid-repetitive sequences.

The number of functional genes in the mouse is not known with certainty, but might be anywhere in the range of 5000 to 40,000. The lower estimates, of 5000–10,000 genes, are based on mutation rates in essential genes within defined chromosomal regions, which are then extrapolated to the whole genome (Shedlovsky et al., 1986; Dove, 1987). However, there may be many genes whose mutational inactivation does not produce inviability or even a detectable phenotype, perhaps because they are effectively supplemented by other gene functions. In *Drosophila,* as we have seen, the number of different mRNA transcripts is approximately three to four times that of the vital genes. If similar proportions hold for the mouse, the total gene number is likely to be in the range of 15,000 to 40,000.

One of the chief genetic virtues of the mouse is its large number of well-characterized inbred lines, numbering more than 300 and each differing from the others in one or several morphological, immunological, or biochemical characteristics. By crossing two such inbred strains, and then doing successive sib matings, one obtains sets of strains possessing new combinations of the parental characteristics; the co-inheritance patterns of particular genes, compared within and between different sets, provide mapping information. Recently, such mapping techniques have been enriched by the creation of interspecies *M. musculus–M. spretus* hybrids (reviewed in Avner et al., 1988), whose distinctive DNA polymorphisms greatly facilitate new molecular mapping methods (Cox and Lehrach, 1991). To date, more than 1300 genetic loci in the laboratory mouse have been identified and mapped (Green, 1989; Copeland and Jenkins, 1991) and newly identified molecular markers are rapidly being added, filling in the spaces of the total genome map. The process is progressive; as more genes are placed on the genome map, it becomes easier to place, and isolate, newly identified mutations or gene clones. The acquisition and identification of new molecular and genetic markers is transforming studies of later stages of mouse development.

POSTIMPLANTATION DEVELOPMENT

Description of the Events

The stages of postimplantation development have been described by Snell and Stevens (1966) and are summarized in Figure 8.1. At the time of implantation, at 4.5 days postconceptus (pc), the blastocyst consists of just four cell types: the polar and mural trophectoderm cells, the primary (or primitive) ectoderm, and primary endoderm, arranged in an enclosed cup-shaped, radially symmetrical structure (Fig. 8.1A). During the following 2 days, the primitive ectoderm or "epiblast" will proliferate inside its stocking of endoderm, which in turn spreads to cover the inside of the blastocoele wall; it is the epiblast which will later give rise to the embryo proper (the fetus). During this first phase, while the endoderm is coming to line the inside of the blastocoele (Fig. 8. 1B), the epiblast with its covering endodermal layer projects progressively further into the blastocoele cavity; this structure is now referred to as the egg cylinder and the blastocoele at this point is designated the yolk cavity (in the avian embryo this space is, in fact, filled with yolk).

(The early invagination of embryo precursor material into the yolk cavity, it should be mentioned, is peculiar to mice and rats, and does not take place in other placental

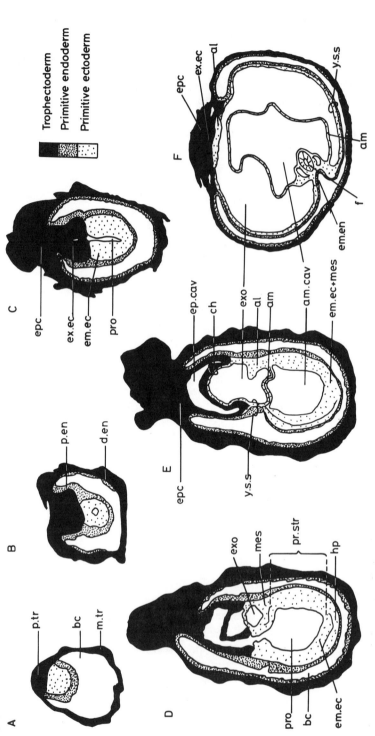

Fig. 8.1. Early postimplantation development in the mouse. **A:** 4 days, 5 hr pc. **B:** 5 to 6 day egg cylinder stage. **C:** 5 days, 12 hr. **D:** 7 days, 1 hr. **E:** 7 days, 6 hr. **F:** 8 days, 11 hr. (Not drawn to scale.) Structures listed in order of appearance: bc, blastocoele; m.tr, mural trophectoderm; p.tr, polar trophectoderm; p.en, proximal endoderm; d.en, distal endoderm; epc, ectoplacental cone; ex.ec, extraembryonic ectoderm; em.ec, embryonic ectoderm; pro, proamnion; exo, exocoelom; mes, mesoderm; pr.str, primitive streak; hp, head process; y.s.s, yolk sac splanchnopleure; ep.cav, ectoplacental cavity; ch, chorion; al, allantois; am, amnion; am.cav, amniotic cavity; em.en, embryonic endoderm; f, fetus. (Adapted from Gardner, 1978.)

Legend:
■ Trophectoderm
▨ Primitive endoderm
⣿ Primitive ectoderm

mammals, even more primitive rodents. The typical mammalian pattern is initially similar to the avian one; the ectoderm is a disc sitting atop the endodermal layer and only invaginates at a much later stage. The murine arrangement, in which ectoderm comes to be enclosed by endoderm, is referred to in classical embryological texts as the "inversion of the germ layers.")

The primary endoderm now consists of two distinct cell layers. That layer which covers the inside of the yolk cavity is referred to as the "distal" (or "parietal") endoderm, while that covering the egg cylinder is termed the "proximal" (or "visceral") endoderm. The proximal endoderm blankets both the epiblast or embryonic ectoderm and the extraembryonic cells, the latter being derived from the polar trophectodermal cells, which retain their capacity for cell division. Capping the extraembryonic ectoderm in the 5.0 day embryo (Fig. 8.1B) is the ectoplacental cone, also derived from the polar trophectoderm and a major precursor of the placenta. At this stage, a small cavity within the embryonic ectoderm can be discerned, the proamniotic cavity, which is the precursor of the amniotic cavity, the fluid-filled sac containing the fetus.

The entire structure enclosed by the visceral endoderm is termed the yolk sac. By 5.5 days (Fig. 8.1C), the embryonic ectoderm has began to delaminate the extraembryonic mesoderm, in what will come to be the posterior region. Concurrently, the cavities within the embryonic and extraembryonic portions of the egg cylinder have enlarged and temporarily joined.

By 6.5 days, the epiblast consists of about 600 cells. At this point, a thickening, termed the "primitive streak," is detectable on the proximal posterior side of the epiblast, and marks the future posterior end of the embryo proper. From this point forward, the embryogenesis of the mouse, and mammals in general, bears many similarities to that of avian embryos, despite the simpler geometry and greater accessibility of the blastodisc of the latter.

The primitive streak not only presages the future anteroposterior (a-p) axis of the embryo but is the entity that organizes gastrulation in the embryo. (In avian embryos, a primitive streak region within the blastodisc plays a similar role.) The streak both recruits cells from large lateral portions of the rest of the epiblast and expands anteriorly, through active cell prolfieration.

Although delineating the cellular origins of the different tissues and structures of the embryo proper is considerably more difficult in the cup-shaped epiblast of the mouse than in the essentially flat blastodisc of the chick, fate mapping has been carried out by means of both grafting and cell labeling and tracking experiments (Tam, 1989; Lawson et al., 1991). In experiments with 6.7 day pc embryos (at the prestreak and early streak stages), single cell injections with horseradish peroxidase (HRP) and rhodamine-dextran were carried out and development was allowed to proceed, in culture, for a further 22 hr (to late streak and neural plate stages) (Lawson et al., 1991). A key finding was that many single cells are pluripotent with respect to germ layer fate; more than 40% give rise to descendants in two different germ layers (the ectoderm, endoderm, mesoderm, and extraembryonic mesoderm) (Lawson et al., 1991).

Nevertheless, despite this developmental plasticity, cells in particular regions of the epiblast tend to give rise to cells of particular germ layers, which permits the construction of fate maps. From the grafting experiments, for instance, the cranial neural structures and the epidermis were found to derive from a broad wedge of the epiblast, while mesodermal structures all arise from cells of the primitive streak. The fate map of the brain, for instance, has been particularly clearly delineated, with the fore-, mid-, and hindbrain precursor regions in a proximodistal (p-d) sequence along the epiblast, anterior to and facing the primitive streak (Tam, 1989).

A picture of the general mouse, chick, and amphibian (*Ambystoma*) fate maps derived from the cell marking experiments (Lawson et al., 1991) is shown in Figure 8.2. Although differences in relative area and shape for the different precursor regions are apparent, the general dispositions of the different precursor regions with respect to one another are similar, as is particularly clear for the mouse and chick embryos.

During the development of the primitive streak, the mesodermal structures arise through a complex series of cellular movements accompanied by extensive growth within the primitive streak. The process begins with the production and proliferation of mesodermal progenitor cells at the proximal (allantoic) end of the primitive streak; this position marks the future caudal end of the fetus. (Proximal and distal orientations are with respect to the developing ectoplacental cone, the site of implantation.) As ectodermal cells migrate through the primitive streak, they move both laterally and distally toward the future cranial end of the embryo, extending the primitive streak toward the distal tip.

The a-p orientation of the embryo is thus further delineated by the laying down of the mesoderm in a caudocranial direction. As this process continues, mesodermal cells progressively occupy the space between the visceral endoderm and embryonic ectoderm, excepting the narrow central section that goes to form the spinal region of the fetus. At 7.0 days pc, the so-called head process has taken shape at the most distal end

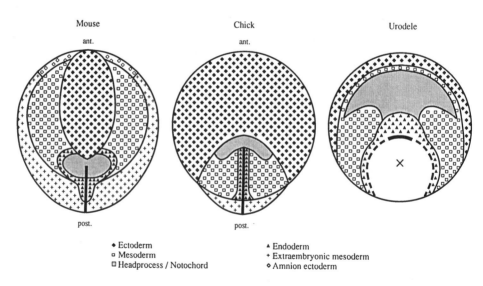

Mouse   Chick   Urodele

ant.   ant.

post.   post.

♦ Ectoderm   ▲ Endoderm
□ Mesoderm   + Extraembryonic mesoderm
▣ Headprocess / Notochord   ◇ Amnion ectoderm

Fig. 8.2. Fate maps of gastrula-stage embryos of the mouse, the chick, and the urodele, *Ambystoma*. The cup-shaped epiblast of the mouse embryo has been flattened for depiction of the fate map and the regions of overlap between neighboring areas have been eliminated for simplicity; the fate map is for the early primitive streak stage. The primitive streak of the mouse and chick embryos are represented by a bar, as is the dorsal lip of the blastopore of the urodele embryo. ♦, ectoderm; □, mesoderm; ◇, amnion ectoderm; ▲, endoderm; +, extraembryonic mesoderm; stippled regions, notochord. (Reproduced from Lawson et al., 1991, with permission of the Company of Biologists, Ltd.)

of the egg cylinder (Fig. 8.1D). The head process gives rise to the notochord and also contributes to part of the endodermal lining of the gut. The head process grows proximally and laterally; by 7.5 days pc the growing head process and the mesodermal sheets have met. Throughout most of the embryo, ectoderm is separated from primary endoderm by the mesodermal layer.

Cavities have also begun to appear within the extraembryonic mesoderm; these unite to form the beginnings of the exocoelom, which by 7.25 days (Fig. 8.1E) occupies the space between the ectoplacental and amniotic cavities. At this stage, the blood islands of the extraembryonic mesoderm, in the yolk sac, have formed and begun to produce blood cells. This is the initial source of blood for the embryo until approximately 11.5 days, when the fetal liver takes over this role. The exocoelom is separated from the amniotic cavity by the amnion, a membrane composed of a layer of mesoderm (on the exocoelomic side) and a layer of ectoderm.

By 8.5 days, the exocoelom has surrounded the amniotic cavity, except for the region occupied by the fetus (Fig. 8.1F). The ectoplacental cone has fused with the chorion (the membrane bounding the exocoelom opposite the amnion) and the allantois, a structure derived from the mesoderm that serves as a connection between the maternal blood supply and the fetus. The chorioallantoic membrane is connected to the fetus through the richly vascularized membrane that bounds the exocoelom, the yolk sac splanchnopleure.

During this period (6.5–8.5 days), the embryo proper has been developing extensively, with the digestive tract and the axial system of the embryo first forming. The gut begins as two invaginations of the yolk sac on opposite sides, near the junctures of the embryonic and extraembryonic ectoderm. The foregut precursor pushes in at the cranial end, beginning at 7.0 days, and by 7.75 days has become a pronounced inpocketing; the hindgut has just begun to form at this point. The blind ends of the foregut and hindgut invaginations eventually break through to the outer surface of the embryo, giving rise to the mouth and anus, respectively. The invagination of the foregut also lays the foundation of two other key organs, the heart and the brain. The dorsalmost region of ectoderm that is pushed in is the head fold structure, which develops into the brain. The heart arises from the mesoderm lying between the head fold and the outer endoderm.

The longitudinal axial system comes into being from the lateral sheets of mesoderm that border the notochord, the so-called paraxial mesoderm. As in other vertebrates, the axial system arises in a strict, linearized sequence of somite formation that begins near the cranial end of the embryo, just posterior to the embryonic brain, and progresses to the caudal end. Somitogenesis therefore proceeds in the opposite direction to that of mesoderm formation.

The delimitation of somites is shown in Figure 8.3. The first pair of somites arises at about 8 days. Initially, pairs of somites are laid down at the rate of one per hour; subsequently the process slows to one pair every 2 to 3 hr (Tam, 1981). Somitogenesis is complete by about 13 days of development, when 65 pairs of somites have been formed. Of these, about four or five pairs will contribute to formation of the head (including parts of the skull), 25 pairs will form the skeleton, musculature, and dermis of the body (neck, thorax, and lower back), and 35 pairs will form these elements in the tail. Each newly formed somite shows a trace of a boundary between anterior and posterior halves and each vertebra arises from cells of the posterior half of one somite and the anterior half of the next. Differentiation of the somites is accompanied by

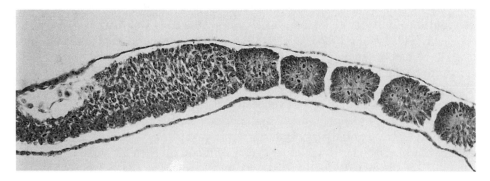

Fig. 8.3. Somitogenesis. Presomitic mesoderm and somites in a seven-somite mouse embryo. (Reproduced from Tam, 1981, with permission of the Company of Biologists, Ltd.)

dispersal of many of the mesenchymal cells that compose them, leading to an obliteration of the intersomitic boundaries.

The formation of the neural tube, the other major element of the axial system, is also closely correlated with the stages of somitogenesis. The neural tube develops from the closure of the neural plate area just above the notochord and between the sheets of paraxial mesoderm. The strict correlation of somite stage with other developmental events in normal development permits one to stage embryos according to their somite number, as in avian embryos.

By 10 days pc, the major elements of the circulatory system, also formed from the paraxial mesoderm, have developed. At this time, the embryo makes a characteristic turn, which serves to complete the formation of the gut, by generating the midgut and joining the fore- and hindguts. During the succeeding 10 days of development *in utero,* the major and minor organ systems undergo progressive differentiation. A photograph of a 26-somite embryo and a diagram of the 13.5 day embryo, showing the major organ rudiments, are shown in Figure 8.4. A timetable of some of the major events in postimplantation development is given in Table 8.1.

Gene Expression During Postimplantation Development

The events of implantation mark a point between the early but limited cell diversification of preimplantation development and a much more extensive set of cellular differentiations. Molecular surveys show that, as in the fruit fly, there is a broad base of shared gene expression between different cell types and stages during mouse development. At the same time, they show that large-scale quantitative shifts in gene expression or gene product accumulation occur during development.

Although numerous studies of polypeptide synthetic patterns have been performed, they necessarily emphasize the patterns of synthesis of the abundant proteins. RNA-DNA hybridization analyses have provided, on the whole, a better measure of the complete set of expressed genes. Several estimates of mRNA sequence diversity in mouse cells have been made, from the kinetics of hybridization of poly A^+ mRNA to cDNA made from such RNAs. As noted in chapter 7, this method neglects nonpolyadenylated mRNA species and may fail to detect rare poly A^+ mRNAs that

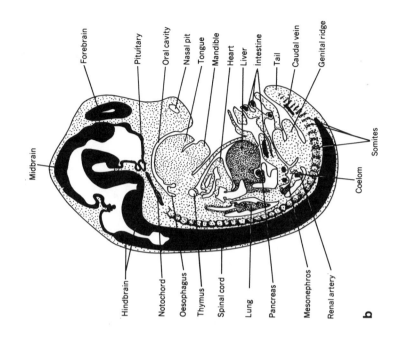

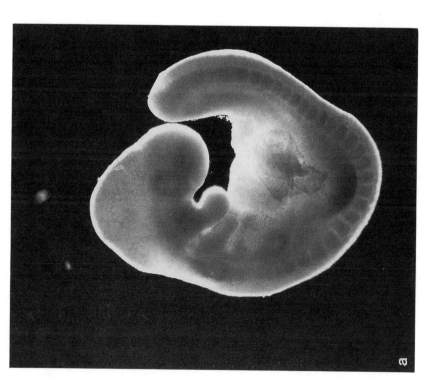

Fig. 8.4. Two stages of development of the mouse fetus. **a:** A 26-somite embryo (Reproduced from Tam, 1981, with permission of the publisher.) **b:** Schematic longitudinal section of a 13.5 day embryo. (Adapted from Rugh, 1968.)

Table 8.1 Developmental Landmarks in Postimplantation Development[a]

Event	Days p c
Implantation	4.5
Proamniotic cavity appears	5.0–5.5
Primitive streak detectable	6.5
Foregut appears	7.0
Neural plate forms	7.75
Somite formation begins	8.0
Chorioallantoic placenta forms	9.0–10.0
Primordial germ cells appear at the base of the allantois	7.75–8.0
Primordial germ cells arrive in genital ridges	10.5–11.5
Melanoblasts leave the neural crest	8.5–9.5
Neural tube development completed	10.0
Forelimb buds appear	9.2
Hindlimb buds appear	10.0
Somite formation completed	13.0
Hematopoiesis begins in the fetal liver	9.5–10.0
Hematopoiesis begins in the fetal bone marrow	16–17
Birth	19–20

Source. Adapted from McLaren (1976a) and Snow et al. (1981).

[a]Times cited are approximate; inbred mouse strains frequently show a delayed timetable.

are present in some but not all cell classes. Nevertheless, if the presence or absence of polyadenylation is characteristic of a particular mRNA sequence, the measurement of sequence complexities in different cellular poly A$^+$ mRNA pools provides useful comparisons between different cell types.

In two studies that measured mRNA complexities of the poly A$^+$ pool for different adult tissues, the range of different mRNA species was found to be 10,000 to 15,000 (Hastie and Bishop, 1976; Young et al., 1976) (although Young et al. reported a higher complexity for brain tissue, 25,000 different mRNA types, a figure that has not been supported by other measurements). By performing cross-hybridizations to near saturation in order to detect sequence overlap, Young et al. found a large measure of qualitative similarity between the mRNA pools of adult liver and brain. However, the rates of cross-hybridization, which primarily reflect the more abundant species, were lower than in the homologous hybridizations. The measurements therefore indicate that the most abundant species for each of the two tissues were less prevalent in the other.

In the comparable study by Hastie and Bishop (1976), analysis of the mRNA populations of adult liver, brain, and kidney produced estimates of 13,000, 11,600, and 11,500 different structural gene sequences in total represented in the three messenger pools, respectively, each with a relatively small number of highly abundant species and a large number of rare species. Of the 11,500 rare poly A$^+$ mRNA sequences of the kidney, 9000 to 10,000 were estimated to be shared by liver and brain. As in the study of Young et al., the abundant mRNA species for each tissue were also detected in the mRNA pools of the other two but were in the intermediate or rare class. Altogether, a total of 15,000 to 16,000 mRNA species were found within the three tissues. The main qualitative differences were in the rare mRNA classes, though quantitative differences of the abundant and intermediate classes were a marked feature.

In the third study, the mRNA complexities were estimated in several pluripotent neoplastic cell lines. These embryonal carcinoma (EC) cell lines can have broad differentiative capabilities and may, in some sense, be an analog of the early embryonic ectodermal contribution. Affara et al. (1977) measured EC mRNA diversity in two pluripotent cell lines and two specialized, EC-derived myoblast (muscle precursor) cell lines. For the pluripotent EC lines, a transcript diversity of 7700 different poly A^+ mRNAs was estimated, and for the myoblast lines, a transcript complexity equivalent to 13,200 poly A^+ mRNA species. By cross-hybridization tests, all of the EC mRNA sequences were found in the transcript pools of the myoblast cell lines, but the latter cells contained a substantial number of rare mRNA species not found in the EC cell line. As in the other studies, the abundant and intermediate sequence classes that characterize one cell type are less abundant in other cell types.

The noticeably similar feature of these studies is the high degree of sequence overlap between the pools from the different sources. While qualitative differences are undoubtedly of importance in generating different phenotypes, there is a very large shared base of expression, and a large part of the differences involves quantitative shifts between the different classes. Although the presence or absence of certain rare species probably reflects posttranscriptional processing (reviewed by Davidson and Britten, 1979), a very large body of evidence shows that the major shifts in abundance reflect primarily changes in transcriptional rates.

Altogether, the picture presented by the molecular studies on postimplantation development seems similar to that presented by *Drosophila:* Development of different cell and tissue types is accompanied by a relatively modest number of qualitative changes in gene expression that takes place on a foundation, as it were, of a large base of shared gene expression. While the switching on of expression of new gene functions in mouse development must be crucial for the acquisition of new phenotypes, quantitative changes of synthesis of mRNAs and proteins constitute an important component of the processes of cellular phenotypic change.

Such general surveys, however, will fail to reveal any differential parental imprinting that may have taken place. As noted in chapter 5, most of the phenotypic effects of such imprinting only appear during postimplantation development. This suggests that some element of the postimplantation gene expression pattern reflects this process, if only in a relatively small number of genes.

One way to approach this question is to make chimeras between normal, fertilized (F) early embryos and either diploid androgenetic (AG) or diploid gynogenetic/parthenogenetic (PG) embryos and to trace the time course of contribution by the AG or PG cells to different tissues or regions of the chimeras. To do so, one can use either an indirect marker, such as GPI-1 isozyme difference (followed by dissection and scoring of component tissues) or a direct, *in situ* marker. A summary of the combined results of several such studies is presented in Figure 8.5.

The results indicate several points of interest, of which perhaps the most significant is differential contribution of AG cells to mesodermal derivatives and of PG cells to ectodermal derivatives. A second difference is that at about 15 days, PG cells begin to experience a relative decline, both with respect to the surrounding normal cells and with respect to AG cells (in the other set of chimeras). This does not appear to reflect the onset of cell death in the PG component but rather some block to their continued

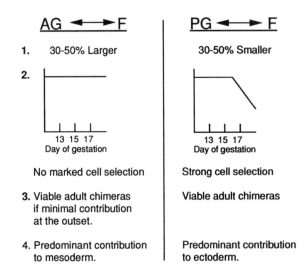

Fig. 8.5. Contrasting development of androgenetic (AG) and parthenogenetic (PG) cells in chimeric fetuses. See text for further discussion. (Reproduced from Surani et al., 1990, with permission of the Company of Biologists, Ltd.)

proliferation (Surani et al., 1990). A further difference, one not shown in the figure, is a distinct elongation of the a-p axis in AG↔F chimeras, suggesting that paternally imprinted genes can influence body patterning in certain, relatively subtle fashions.

Since it is possible that differentially imprinted genes may modify each other's activities when both genome sets are present in the same cell (as in normal cells), the results fall short of a definitive proof that differentially imprinted genes play independently different roles during normal postimplantation development. Nevertheless, the results are highly suggestive of such functional differentiation. Furthermore, the observation that triploid embryos containing one maternal and two paternal genomes develop to a considerably later stage (Kaufman et al., 1989) than those containing just two paternal genomes (diploid androgenetics) (Surani et al., 1986b) suggests that the maternal genome can supplement the functional deficiencies of the imprinted paternal genome. The results also show that imprinted differences last well into postimplantation development (a point confirmed by the molecular analysis of *H19* by Bartolomei et al. [1991]). The results shown in Figure 8.5 are consistent with the general observation, mentioned in chapter 5, that many of the differences in imprinting that have been identified seem to involve aspects of differential growth (Cattanach and Beechey, 1990; Surani et al., 1990). Consistent with the parental "tug-of-war" hypothesis (Moore and Haig, 1991) (see chapter 5), AG cells seem to promote general growth of chimeric embryos while PG cells seem to retard it (Surani et al., 1991).

CLONAL ANALYSIS OF POSTIMPLANATATION DEVELOPMENT

One important element in the characterization of postimplantation development has been clonal analysis. It has been utilized to estimate the numbers and types of progenitor cells of the different tissues and organs and also their patterns of proliferation and migration. Indeed, much of murine developmental genetics in the 1970s and 1980s centered around the delineation of the various component cell lineages of the tissues and organs of the postimplantation embryo, by means of clonal analysis.

In contrast to the *Drosophila* work, however, the use of mitotic recombination has played no part in this work. Mitotic recombination events are either rare or, more probably, simply too difficult to detect in mammalian embryos, for most of the markers available (Panthier and Condamine, 1991). Instead, principal reliance has been placed on the construction and analysis of chimeras, in particular in the earlier studies. In addition, other studies have employed both natural and artificial forms of mosaicism to trace lineages. The first of the mosaic analyses employed the phenomenon of single X inactivation in 2X (female) embryos, while more recent ones have utilized the production of mosaics carrying retroviral constructs with marker genes; the latter approach will probably prove to be the most useful of all the methods.

In most of the chimera experiments, the chimeras have been made either by injecting marked cells, of particular stages or types, into blastocysts or, in some of the earlier studies, by aggregation of cells to preblastocyst embryos. The donor cell(s) become integrated into the recipient embryos and the latter are then placed in pseudopregnant foster mothers. Following the requisite period of development, the animals are dissected and the tissues or structures of interest analyzed for donor cell genotype contribution (Fig. 5.8). From the pattern of distribution of the marked cells, inferences are made about the capacity of the donor cells to contribute to the tissues/structures of interest.

As in all chimera studies, the success of the approach depends, in large part, on the choice of the marker used to distinguish the two cell populations from one another. The ideal marker is one that shows perfect hereditability, ubiquitous expression, and unambiguous detectability and does not perturb development significantly (Rossant, 1987). Because injected biochemical labels ("extrinsic markers") are rapidly diluted during growth, they are poor markers for long-term postimplantation development, in which much growth takes place. Such markers have been employed, however, both in studies of preimplantation development, where there are just a few cell divisions, and for short-term periods in postimplantation development, as in the fate mapping experiments of Lawson et al. (1991), described earlier.

In contrast, genetic markers, which, by definition, show perfect heritability, have been the ones of choice. As in preimplantation development, some experiments have employed isozyme variants as indirect markers, but the most useful have been direct markers, which allow cell typing *in situ* at different stages of development. Three kinds of natural direct marker and one class of artificial genetic marker have proved particularly useful.

The first group of natural markers includes the H2 antigens; chimeras are made from embryos of different H2 type and subsequently immunostained for one or other of the H2 isotypes to allow tracing of lineages (Ponder et al., 1983). However, H2 markers only begin to be expressed around day 10 of development and therefore their use is limited to relatively late stages. A second useful marker has been a null allele for cytoplasmic malic dehydrogenase (*Mod-1*), which maps to chromosome 9. Under the appropriate staining conditions, the activity difference between the mutant and wild-type cells can be clearly seen in tissue sections of those tissues which express it. This marker has been particularly useful in the tracking of cell lineage relationships among the extraembryonic membranes (Gardener, 1984). The third useful natural genetic marker is the difference in major satellite DNA species between *M. musculus* and *M. caroli.* Interspecific chimeras can be formed and the difference in satellite composition between cells scored by *in situ* hybridization to one or the other species (Siracusa et al., 1983; Rossant, 1987). This marker satisfies the criteria of ubiquity and unambigous detectability, but the two species differ in embryonic growth rates, and it is not impossible that inferences drawn about normal development from the interspecific chimeras may not be fully applicable to normal development.

The most versatile and perhaps the most satisfactory genetic markers are artificial ones, genetically engineered retroviruses that are injected into early embryos, infect a limited number of cells, where they are copied into DNA, and give rise to single inserts in the genome, becoming heritable cell markers (Price, 1987). The retroviral markers have been used to produce genetic mosaics for clonal analysis.

Chimera analysis has proved particularly effective in analyzing the initial events of postimplantation development and, in particular, the origins of the main histotypes during this phase. For development following the emergence of the primitive streak, the method has been applied principally to estimating the number of founder cells for different organs or tissues. Similarly, the use of X-inactivation to create mosaics has also been applied particularly to the estimation of founder cell numbers for particular tissues and organs in the fetus. Both chimera and X-inactivation mosaic analyses, however, when used to estimate founder cell numbers, suffer from certain ambiguities in the interpretation of the results, as will be explained.

Tracing Postimplantation Lineages

At implantation, the blastocyst consists of two types of trophectodermal (TE) cell, the cells of the ICM and a layer of primitive endodermal cells that enclose the primitive ectoderm. As discussed in chapter 5, the developmental fates of the two TE lineages are relatively clear. The mural TE cells, which occupy the sides of the blastocyst, have begun the transformation to primary giant cells (GCs), which eventually reach ploidy levels of several hundred; their function is to achieve the initial implantation of the blastocyst in the uterine wall. The other TE cell group, the polar TE, produces the ectoplacental cone (EPC) (which can give rise to secondary GCs of the mural TE) and the extraembryonic ectodermal (EE) cells. The EPC and EE give rise to the ectodermal component of the placenta. The only remaining uncertainty about the TE lineages concerns the precise stage when ICM cells cease to be able to contribute to polar TE, though it seems certain that by the expanded blastocyst stage, most or all of the ICM cells have lost this capability.

In contrast to the TE lineages, the origins of two fetal tissues, derived from ICM, have been the subject of much debate. The first of these is the fetal endoderm. From the classic embryological description of the formation of the gut, it had been assumed that the proximal primitive endoderm gave rise to the lining of the gut; without question, endodermal cells from this layer are carried along with the invaginations that give rise to the fore- and hindguts. As discussed in chapter 5, however, Gardner and Rossant (1979), using the indirect marker GPI-1, showed that primitive endodermal cells of 4.5 day embryos contribute to the yolk sac endoderm but not that of the fetus itself. Single cell injection experiments have confirmed and extended this finding: When single primitive endodermal *Mod-1*[+] cells from 5.0 day pc blastocysts are injected into *Mod-1⁻* 4.0 day blastocysts, and tissues stained for *Mod* activity at subsequent stages of development, the chimeric patches are found not in the fetus but in the extraembryonic endoderm, namely the parietal and visceral endoderm layers (Gardner, 1985). (There may also be some contribution in a small number of embryos to the visceral mesoderm of the yolk sac.) Thus, fetal endoderm derives from the "embryonic ectoderm" of the blastocyst, as do fetal ectoderm and fetal mesoderm.

The origins of the germ line, the gamete-producing cells, were, for a long time, the second contentious issue. It was initially believed that the primordial germ cells (PGCs), like the gut, originated in the proximal endoderm. A number of observations have forced a revision of this view; it is now apparent that the germ line, like all other components of the fetus, originates from the epiblast, in the region just posterior to the primitive streak.

The first significant feature of the mammalian germ line is that, unlike that of *Caenorhabditis* and *Drosophila*, it does not arise from specialized, germ line blastomeres but from early cells that serve also as somatic cell precursors. This was initially shown by Kelly (1975), in the experiments described in chapter 5. She found that in each of two four-cell embryos dissociated into individual blastomeres, at least three of the donor cells gave rise to germ line chimeras. In all cases, the germ line chimeras also displayed mosaicism in their somatic tissues, indicating the absence of segregation for germ line-forming ability by as late as the second cleavage division. Even more conclusive are the experiments of Gardner et al. (1985), who injected single, marked primitive ectoderm cells from 4.5 day blastocysts and scored for subsequent somatic as well as germ line mosaicism. In 21 of 37 chimeras whose sex was the same as that of the donor cells, mosaicism was found in both the soma and germ line. Furthermore, for two sets of chimeras, two individual cells from the same donor were found to give chimerism for both soma and germ line (out of an estimated 20–40 primitive ectoderm cells at this stage).

These experiments show that the mammalian oocyte does not contain any specialized germ line "determinant" and that the primary ectoderm is the source of the germ line. The results are also compatible with the possibility that all primitive ectoderm cells at this stage have the potential to give rise to both soma and germ line, though they do not prove it. The certain conclusion is that in the mouse, and, by extension, in mammals, the continuity of the germ line is maintained through the soma rather than independently of it in a specialized cell lineage, as occurs, for instance, in *Drosophila*.

For the two days that elapse between the 4.5 day stage and the emergence of the primitive streak (7.0 days), little is known about the germ line precursor cells. However, a biochemical marker for the germ line precursor cells, or PGCs, is available in the form of high alkaline phosphatase (AP) activity. There has, however, been some controversy about where these AP-rich cells first make their appearance. Ozdzenski (1967) located presumptive PGCs in the 8.0 day pc embryo near but posterior to the caudal end of the primitive streak, within the base of the allantois, while Tam and Snow (1981) reported the location of PGCs in 8.0 day embryos to be within the caudal end of the primitive streak. Ginsburg et al. (1990), employing a more sensitive staining technique, and examining whole mounts, were able to detect AP-rich cells in 7 day embryos in the extraembryonic mesoderm just posterior to the primitive streak, as did Ozdzenski. Stained PGCs, in day 7.5 embryos, are shown in Figure 8.6. Although there is a large increase in AP-staining cells from day 7.0 to day 8.0, it is not known

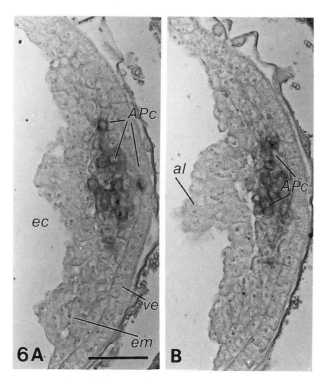

Fig. 8.6. PGCs in a primitive streak, 7.25–7.5 day pc embryo. The figure shows alkaline phosphatase positive stained cells (APc), at the base of the allantois, in two consecutive transverse sections. al, allantois; ec, exocoelom; em, extraembryonic mesoderm; ve, visceral endoderm. (Reproduced from Ginsburg et al., 1990, with permission of the Company of Biologists, Ltd.)

whether this increase reflects cell proliferation or recruitment from the surrounding cells.

Starting at about day 9, the PGCs move by an active chemotactic process, along the dorsal mesentery within the developing body cavity to the genital ridges, which they then colonize. As in *Drosophila,* this movement of PGCs occurs in embryos of both sexes; the differences in germ cell maturation that distinguish the sexes begin to be apparent after colonization of the genital ridges. The process of migration is complete by about 11 days pc and by day 13.5, the number of histochemically detectable PGCs in the genital ridge is about 25,000 (Tam and Snow, 1981).

While the germ line has thus been traced to the extraembryonic mesoderm, by use of a biochemical marker, the origins of the mesodermal and endodermal components of the extraembryonic membranes that surround the postimplantation embryo have been determined by means of chimera experiments. The mesodermal components, which contribute to both the allantois and the yolk sac, derive, like the fetus, from the primitive ectoderm, while the extraembryonic endodermal layers (including the visceral and parietal endoderm) derive from the primitive endodermal layer, visible in the implanting blastocyst (Gardner, 1978). The complete sequence of major tissue derivations in the mouse embryo, reconstructed primarily from chimera studies, is summarized in Figure 8.7. The chief conclusions are that all the cells of the fetus are derived from a portion of the embryonic ectoderm and that the placenta is comprised

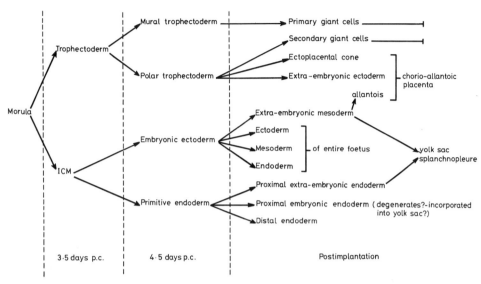

Fig. 8.7. A summary of the principal cell lineages in the pre- and early postimplantation mouse embryo. (Reproduced from Gardner, 1978, with permission of the publisher.)

of several discrete lineages derived from both the TE and the ICM. The pluripotency of the embryonic ectoderm revealed by these experiments is but another illustration of the limited usefulness of conventional germ layer designations (ectoderm, endoderm, mesoderm) in predicting developmental outcomes.

Estimating Founder Cell Numbers in the Mouse

The period of major tissue determination in the mouse embryo occurs within 24 hr of the formation of the primitive streak. However, many of the details of both these events and subsequent organogenesis are obscure. In particular, it is not known when the first determinative restrictions occur within the primitive streak, nor are the number of founder cells, for particular tissues or structures, known. Presumably, the first restrictions of developmental potential within the primitive streak occur between 6.5 days pc and the beginnings of somitogenesis. In avian embryos, and presumably also in mammalian embryos, cells in the embryonic ectoderm prior to formation of the primitive streak stage are pluripotential, while by the late somite stages, restrictions have been imposed.

A useful preliminary approach to ascertaining times of determination, however, would be a fuller description of cell allocation in the early embryo. Allocation is defined as the restriction of interchange of cells between different primordia, and it must precede or accompany determination. A characterization of the timing and number of founder cells allocated to various tissues or organ primordia should provide a basis for the subsequent analysis of determination. If one can determine the time of allocation, then, if one can introduce genetic heterogeneity at that point, one can, from the sizes and patterns of the clones, make estimates of the founder cell numbers. In *Drosophila,* for instance, one first determines the approximate time of allocation by finding that point at which mitotically generated recombinant clones cease to mark neighboring primordia (whose proximity can be judged from the gynandromorph fate map). Mitotic recombination is then induced at that stage, and from the fraction of tissue or organ occupied by the marked cells, the number of founder cells can be calculated (under the assumption that the greater the marked fraction, the smaller the number of founder cells).

Unfortunately, a comparable approach in the mouse has not yet proved possible and, indeed, in chimera experiments, the genetic heterogeneity is introduced during preimplantation development, well before the time of allocation of organ or tissue primordia in the postimplantation embryo. The effect of late cell allocation relative to the time of cell marking is illustrated in Figure 8.8. Imagine, for instance, that in a certain region eight cells are allocated out of a possible 32 to form a particular primordium and that, in a particular set of chimeras, the proportion of minority, marked donor cells was one-eighth. Suppose further that during the period of growth, all cells in the chimera grow equally rapidly and that there is free intermingling in this group. Then, at the time of allocation, the overall proportion of marked cells will also be one-eighth and any marked cell may contribute one, two, three, or four cells to the primordium, making proportionate contributions to the fully developed tissue (or region). Some of the corresponding estimates of primordial cell number are shown in the figure. In most cases, the observed contribution to the tissue or structure among those that show any such contribution will be *disproportionately* large, although if one could score *all* of the descendant cells of the original four, the marked cells would always constitute four and hence one-eighth of the total. Since the chimeras showing no marked cell contribution to the tissue or region under study are discarded, the data from such experiments yield a systematic overrepresentation of marked cells within the area of interest and, hence, systematic underestimates of founder cell number(s).

Another statistical approach to estimating founder cell numbers involves inferences made from the size of clonal patches relative to that of the area of the tissue or

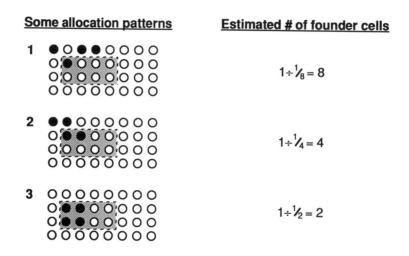

Fig. 8.8. The consequences of clonal marking prior to allocation. Some patterns of inclusion of marked cells within the allocated region/structure (indicated by the dotted line and shaded box), with marked cells indicated in black (left) and corresponding estimates of founder cell number, calculated from proportion of allocated area that is marked.

organ as a whole. These estimates, however, have often been based on the assumption that one patch = one clone. In *Drosophila,* this assumption is generally valid; the descendants of single cells tend to remain contiguous and, hence, most visibly coherent patches are true clones. In the mouse, this is much less true. Extensive cell mingling and cell migration take place, particularly in later development (McLaren, 1976a). The result is that a given early clone will usually fragment into separate daughter subclones. If one mistakenly equates patches of marked cells with individual descendant clones, the estimate of primordial cell number will be inflated to a degree corresponding to the fragmentation of descendant clones. Furthermore, the estimate can be affected depending on whether cell migration occurs before or after tissue allocation. Some of the effects of different patterns of migration and coherent clonal growth on final patch sizes and discreteness are shown in Figure 8.9. When mingling is very extensive, individual patches may consist of two or more subclones that happen to be contiguous. It follows that estimates from minimal patch size can be misleading.

However, the opposite assumption, namely that there is complete and random cell mixing until the moment of tissue foundation, which we used above in assessing the problem of time of allocation relative to marking, can be equally misleading. The consequences of this assumption were first described by McLaren (1972). If there is a period of regional coherent clonal growth preceding the moment of tissue allocation, then the estimate of founder cell number will in reality be an estimate of the the number of marked coherent *clones* from which the primordium originates. The greater the extent of coherent clonal growth preceding tissue allocation, the lower the numerical estimate of cells (= clones). Indeed, all such estimates for fetal tissues probably reduce to the number of *original fetal progenitor cells present at the introduction of genetic heterogeneity* (McLaren, 1972). For chimeras made by aggregating four to eight cell embryos, the procedure of morula aggregation, which was the initial

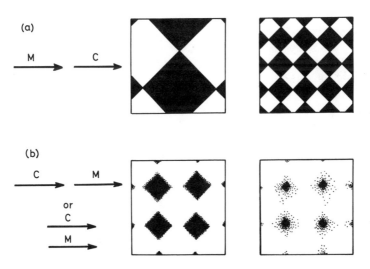

Fig. 8.9. Patch pattern as a function of the timing of coherent clonal growth and cell migration. **a:** Change in a pattern of initial large patches produced by a period of migration (M) and subsequent coherent daughter clone growth (C). **b:** Patterns produced by clonal growth followed by migration or by a balance of clonal growth and migration. (Adapted from McLaren, 1976a.)

way of making chimeras, the estimates of presumptive founder cell number for a given fetal structure therefore in all probability refer to the number of cells present at aggregation which contribute descendants to the structure in question, or about five to eight cells.

The estimate of five to eight cells derives from the following considerations. From single cell injections into blastocysts, it is apparent that single ICM cells can contribute to most or all fetal tissues and hence organs (Gardner and Rossant, 1979). Therefore, most or all ICM cells in the late blastocyst probably contribute to all fetal tissues. (Such widespread descendancy of single ICM cells presumably reflects a thorough mixing of cells prior to or during primitive streak formation.) Tracking of prelabeled cells in morula aggregates, however, shows little cell mixing until blastocyst formation (Garner and McLaren, 1974), with the consequence that the ICM primarily derives from the five to eight cells that are present within the aggregate. This consideration bears on some early estimates of founder cell number of the fetus based on the assumption that its formation can be likened to a lottery drawing in which cells are picked at random, with binomial probability reflecting the initial proportions of the two marked cell types in the chimeras. Such estimates cannot be reliable. Indeed, most binomially derived estimates of fetal tissue founder cell number are in the range of two to five or slightly more cells (West, 1978).

There is a second problem with the statistical approach: the influence of selection during cell growth within chimeras. When the cells of the two input strains differ in background genotype, it often happens that they contribute differentially to different tissues or organs (McLaren, 1976a). A consistent underrepresentation of one genotype in a tissue relative to its frequency in other tissues is a certain indication of relative tissue-specific differences in growth or competitive ability between cells of the two genotypes. Such selective differences can of course influence relative contributions to several tissues, undermining the presumptive significance of correlative frequencies for inferences about similar or disjoint cellular origins.

Another genetic procedure that has been used for estimating founder cell numbers employs X-inactivation mosaics (Nesbitt, 1971; McMahon et al., 1983). As noted earlier, in all X/X (female) embryos, one X is transcriptionally inactivated, eventually giving rise to the Barr body, a characteristic heterochromatic clump in the nuclei of 2X-containing cells. The genetic consequence of X-inactivation is that, if the two Xs differ at one or more loci, the result will be genetically mosaic embryos (Lyon, 1961). While in the principal extraembryonic membranes the paternal X is invariably inactivated, the pattern of inactivation in the embryo proper is nearly random, with either the maternal or paternal allele being subject to inactivation, though there is a slight bias toward maternal X-inactivation. The inactivation of the X takes place during postimplantation development, but the precise time has been the subject of controversy (Gardner, 1985).

The use of X-inactivation to estimate founder cell numbers is based on the use of variances in the different mosaics; the greater the number of founder cells, the smaller the variances. Indeed, the X-inactivation mosaics yield larger estimated numbers of precursor cells than analyses based on chimeras. Thus, Nesbitt (1971) estimated that there are 13–20 fetal precursor cells and 20–50 precursor cells for tissues, while McMahon et al. (1983) estimated the numbers at 47 and 193, respectively. In contrast, from aggregation chimera studies, the estimated number of fetal precursor cells and tissue precursor cells was on the order of three to eight (reviewed in West, 1978).

Although the estimates from the X-inactivation mosaics are undoubtedly more accurate than those from chimera studies based on morula aggregation, because the time of allocation is closer to the time of tissue determination, many of the same kinds of uncertainties pertain and the best estimates are still hedged about with uncertainty. Only when new techniques for ascertaining times of allocation have been developed will it be possible to make better estimates of founder cell numbers.

In Situ Clonal Analyses of Later Postimplantation Development Events

While statistical estimates of founder cell numbers from clonal analyses have proved unsatisfactory to date, studies of both chimeric and mosaic tissue *in situ* have permitted some interesting characterizations of later developmental events, in particular of cell proliferation patterns and of developmental restrictions. In a few instances, the data also permit some inferences about founder cell number. When the screening is only carried out on adult or newborn mice, such inferences are indirect and will be affected by assumptions as to clonal growth patterns. When, however, the marker or markers of choice are expressed over a long period of time, and screening is carried out throughout this period, the results can be informative.

One of the first *in situ* analyses using chimeras concerned the origins of the melanocytes, the cells responsible for pigmentation of the mouse coat. The melanocytes are the daughter cells of the cells termed melanoblasts that derive from migratory cells of the neural crest, the region of epithelium that marks the site of dorsal closure of the neural tube. The neural crest gives rise to two major populations of migratory cells: those which migrate deep into the body between the neural tube and somites to give rise to neurons of the autonomic nervous system and those which migrate just underneath the surface of the ectoderm (Fig. 8.10). It is the latter group that includes the melanoblasts.

By making chimeras between strains that differ in melanocyte pigment-forming capacities, it is possible to determine something about the clonal histories of the

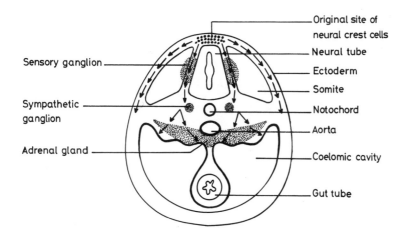

Fig. 8.10. Two pathways of neural crest migration. The outer arrows indicate cells that give rise to melanocytes; the internal arrows denote neural crest cells that give rise to parts of the neural system and adrenal gland; stippled areas indicate neural crest-derived areas. (Adapted from Alberts et al., 1983.)

melanoblast cell populations. Such experiments were first performed by Mintz (1967). Chimeras were made between albino (*c/c*) and wild-type (*c/c*) strains and between strains differing in melanocyte genotype for color (e.g., black [*B/B*] and brown [*b/b*]). In all the experiments, the visually impressive outcome was striped chimeric mice; an example is shown in Figure 8.11. Initially, Mintz reported that the patterns show a regular alternation of the different colors, but larger samples have shown that there can be considerable variability in the width of the stripes.

Despite this variability, an underlying or "archetypal" striping pattern can be discerned, when all the chimeras are compared. The wider bands observed in the animals simply consist of multiple unit widths, where the unit width corresponds to the narrowest single stripe seen. The striping pattern itself is explicable if coloration in the coat is produced by small numbers of melanoblast clones arranged in an initial linear series along the longitudinal axis, in which the clones that neighbor any particular clone may either be of the same genotype (giving a broader band) or of the alternative genotype (giving a central band of unit width). Comparisons of the banding patterns in many such mice suggest that the head region consists of three unit stripe widths, the body of six, and the tail of eight. Because left and right sides at any position along the longitudinal axis can be of different color, the clonal origins of the left and right sides must be independent.

Equating stripe widths with founder clones, one obtains an estimate of 17 founder clones per side or a total of 34 melanoblast founder clones. The independence of left and right sides suggests that the cells that give rise to melanocytes are committed to do so before dorsal closure of the neural tube is complete and also that the founder cells/clones begin their lateral migration before closure takes place (Mintz, 1974). Were it otherwise, cross-contamination would produce a high frequency of mixed unit stripes. With dorsal closure beginning at 8.5–9.0 days pc, the migration of melanoblasts must take place before this period.

Does the estimate of 34 melanoblast founder clones mean that there are only 34 founder cells of the melanoblast lineages? Not necessarily. From the fact that the clarity of the striping is sharper in chimeras composed of cells from more distantly related strains than from more closely related ones, it appears that there is a partial sorting out according to genotype (McLaren, 1976a). If a certain amount of self-segregation is occurring early in the melanoblast lineage, the founder clones might consist

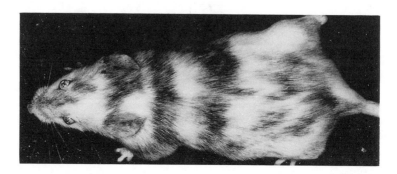

Fig. 8.11. Adult chimeric mouse showing black (*C/C*) and albino (*c/c*) coat color striping pattern. (Photograph courtesy of Dr. B. Mintz.)

of two or more cells of like genotype. However, the number is unlikely to be more than two or three since the higher the number, the more likely would be the occurrence of "unit" width stripes containing small distinct substripes; such patterns are not seen.

Another approach to studying developing cell lineages involves mosaics rather than chimeras and utilizes retroviral constructs containing markers readily scorable *in situ* (Price, 1987). The most useful of these approaches involves defective retroviruses that express an easily detectable, cell autonomous activity, which is not seen in nonmarked animals and which is not harmful to the expressing cells. For one set of experiments, such a defective retrovirus, which expresses the *Escherichia coli lacZ* gene under the control of an internal promoter, was constructed. Control transformation experiments showed that neither insertion of the construct nor expression of *lacZ* was harmful to recipient cells. The plasmid was then packaged into viral particles, through transfection into a specially designed, so-called "packaging" line of cells that can synthesize retroviral coats and package retroviral genomes without making active retrovirus itself. The packaged virons were then injected, a few hundred at a time, through the uterine wall into developing mouse embryos at various stages and the embryos were removed several days later, dissected, and stained for *lacZ* β-galactosidase activity (Sanes et al., 1986).

Clusters of stained cells were found in a minority of the injected embryos (Fig. 8.12) and the proliferation dynamics suggested that each cluster corresponded to a single clone. By generating clones at successively later times, patterns of restriction of cell fate, with respect to cell type, were observed in both the visceral yolk sac mesodermal tissues and in the skin; in both sites, the restriction corresponded with the time of visible cell layer delamination. The deduced lineage in the visceral yolk sac is shown in Figure 8.13.

These experiments illustrate a principal advantage of retroviral marking over chimera studies: By marking at successive time intervals, one can map, with reasonable certainty, times of developmental restriction. In this respect, the method is comparable, if more difficult technically, to that of clonal analysis by mitotic recombination in *Drosophila*. And, as in *Drosophila,* it is possible to use the technique not only for the study of lineage relationships and divergences of cell fate at early and mid-developmental stages but at late times as well.

A *lacZ*-marked retroviral vector has been used, for instance, to explore cell lineage relationships in the rat retina, where mitotic divisions continue for one week after birth and give rise to photoreceptor cells, neuronal cells, and glial cells (Price et al., 1987). Viral particles were injected between the pigment epithelium and the retina in newborn rats, at various times from day 0 to 7 days after birth, and the retinas were then removed and stained for β-galactosidase activity 4–6 weeks later (Turner and Cepko, 1987). Clusters of *lacZ*⁺ cells were found and the indications were that each cluster corresponds to a clone: Cluster numbers are directly proportional to virus titer, consist of more cells when produced at early times, and are the same average size at a given time whether produced following injection of high or low viral particle numbers. The particularly significant finding is that all possible combinations of cell type (photoreceptor cells, bipolar [interneuron] cells, amacrine [interneuron] cells, and Muller glial cells) were found in the clones. However, earlier clones (having experienced more cell divisions) show greater diversity; some two-cell clones produce very different sister cells, such as photoreceptor and Muller glia. The simplest inter-

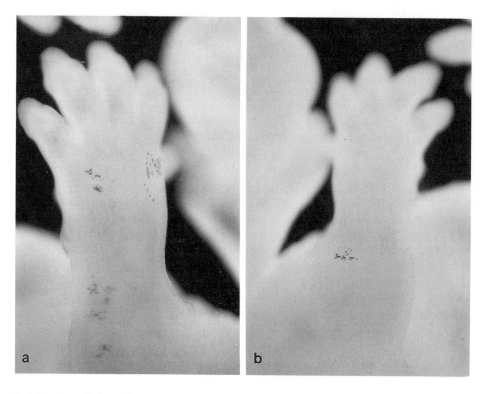

Fig. 8.12. Retroviral marking of a mouse embryo with *lacZ* construct, injected at 8 days pc and assayed at 14 days pc. **a:** Clonal pattern following relatively high titer injection—many clones are seen. **b:** Pattern seen following lower titer injection—only a single clone is marked. (Photographs courtesy of Dr. J.F. Nicholas.)

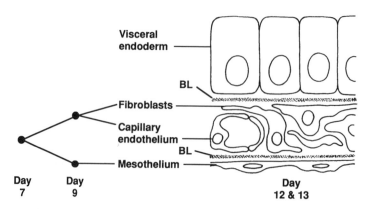

Fig. 8.13. Lineage restrictions in the visceral yolk sac, inferred from clonal marking patterns obtained from marking at successive times with a *lacZ* retroviral construct. Restrictions are inferred from absence of marking of both the mesothelium and the capillary endothelium/fibroblast at day 9 where earlier exposure (day 7) to construct led to joint marking. (Adapted from Sanes et al., 1986.)

pretation is that the four kinds of marked cell found in the retina probably have a common precursor; in the possibility that cells as different as photoreceptors and glia can derive from the same mother cells, the findings are highly reminiscent of the lack of lineage restrictions on ommatidial cell type in the *Drosophila* eye (see chapter 7).

DEVELOPING GENETIC RESOURCES IN THE MOUSE: CLONING "OLD" GENES, FINDING NEW ONES

Cell lineage analyses of the kind described above are inherently of a descriptive nature. For a deeper understanding of the cellular events of postimplantation development, one needs to understand their genetic underpinnings. To do so, one needs the appropriate genetic resources: A large and diverse set of mutants, in which the mutations themselves have been or can be characterized with respect to their expressional properties. With these resources, one can then proceed to molecular characterization of the genes—cloning them, studying their expression patterns and gene products—and, then, reconstructing their functions in wild-type development. Although a century of mouse genetics has produced many mutants of potential interest (Gruneberg, 1952; Lyon and Searle, 1989), the majority have not been fully characterized genetically and only a small number of the "classic" genes have yet been cloned.

The consequence of this comparative paucity of genetic material has been to limit the pace of analysis of murine development. Not surprisingly, therefore, a major goal of many laboratories has been to create new genetic resources. This effort has taken two principal forms. The first has been the cloning of previously discovered genes, whose mutants produce interesting phenotypes, and the second has consisted of the search for and identification of new genes of developmental interest. As an example of new approaches to "old" genes, we will, in this section, look at recent work on two genes of the classical mouse genetics canon, the *T* and *W* loci. Then, to illustrate the second avenue, we will take a look at some of the techniques being used to identify "new" genes.

Two Classic Loci, *W* and *T*: Cloning the Genes and Studying Their Properties

One of the first mutants of the mouse to be phenotypically characterized in any depth was the dominant short-tailed *T* or *Brachyury* mutant, reported by N. Dobrovolskaia-Zavovskaia in 1927. The mutant was obtained from a cross in which one of the parents was X-rayed and is now known to be a deletion within the *T* gene (see below). It is therefore almost certainly an X-ray–induced mutant, though Dobrovolskaia-Zavovskaia, finding normal littermates and evidently regarding the known effects of X-ray treatment as teratogenic rather than mutagenic, classified the mutant as a spontaneous one, a misclassification that went uncorrected for decades, as discussed by Silver (1990).

The developmental interest of the mutant became apparent when short-tailed (*T*/+) mice were crossed to one another: A 2:1 ratio of short-tailed to normal mice was obtained instead of the expected 3:1 dominant to recessive ratio (1 T/T: 2$T/+$: 1 $+/+$). This observation suggests that, like *yellow* (A^y), T is lethal when homozygous. Confirmation of such lethality was provided by Chesley (1935), who performed the first detailed examination of the homozygous embryos. Using the standard light microscopical techniques available at the time, Chesley observed that the first abnormalities in T/T embryos appeared in early somite, 8.5 day pc stage embryos. In the posterior third of the embryo, the notochord is absent, the somites are disorganized, and the neural tube shows aberrant morphology. These early defects are succeeded by progressive disorganization of the entire notochordal and neural tube regions and a failure of posterior trunk and hindlimb development. In contrast, the heterozygous ($T/+$) littermates do not show any defects until about 11 days pc. At this time, they exhibit a diminished neural tube and notochord in the tail region and minor irregularities of the notochord anteriorly. The heterozygous syndrome is thus a milder version, with a later onset, of the homozygous developmental defect and is consistent with the idea that the dominant effect involves haplo-insufficiency.

The cause–effect relationships between the various abnormalities could not be determined at the time, but Chesley speculated that the notochordal defect was primary and the neural tube defects secondary. The simplest interpretation of the basic defect is that the gene acts and is required in the embryonic mesoderm and that the disruption of posterior axial development in the absence of T gene activity reflects a subsequent failure of mesoderm-notochord inductive effects on the overlying ectoderm (Gluecksohn-Waelsch and Erickson, 1970).

The T locus thus provides a valuable genetic handle on a key process in mouse postimplantation development, the elaboration of fetal mesodermal tissue and its role in axial development. Yet, for more than 60 years after its discovery, the precise function of the gene and the causal chain of defects in the mutant remained obscure. Other T mutants were isolated, some being deletions for T and surrounding material (we have discussed one of these deletions, T^{hp}, in chapter 5), and the gene was early assigned a linkage group and subsequently mapped to the proximal region of chromosome 17. However, beyond this mapping and the categorization of the dominant mutant phenotype as reflecting haplo-insufficiency, there was no further direct characterization.

Instead, the principal initial significance of the T mutations was the discovery that they produce a tailless phenotype when the homologous chromosome carries certain mutant regions, originally referred to as "t mutants," now as "t haplotypes," that are obtained from certain wild populations of mice. Many, though not all, of these t haplotypes are lethal as homozygotes, but even the lethals differ among themselves in their developmental effects. For the lethal t mutants, since the t regions include the T locus but cannot undergo recombination with wild-type chromosomes, T/t strains are balanced lethals. An additional surprise about the t haplotypes is that some show preferential meiotic transmission in males when present in $+/t$ heterozygotes. The entire chromosomal region, embracing T and all the t haplotypes, covers approximately 15 cM (or approximately 30 million bp) and is known as the "t complex." The search for the basis of the developmental effects, and the nature of the t mutants, constituted one of the longest-running sagas in developmental genetics and will not be gone into here. Suffice it to say that the developmental and genetic effects associated with the t haplotype interactions involve multiple linked mutations within a rearranged part of the proximal part of chromosome 17 and that, as befits a set of polygenic syndromes, they are complex (reviewed in Silver, 1985).

Progress on the T gene itself, however, had to await the development of a microdissection technique for isolating small regions of mouse metaphase chromosomes. Physical microdissection of small regions of chromosomes was first developed on *Drosophila* late larval polytene chromosomes (Scalenghe et al., 1981). Since a single excised region from a mature polytene chromosome contains thousands of copies, cloning of DNA from such microdissections is comparatively easy. To do a comparable cloning from a given mammalian metaphase chromosome requires doing the microdissection on many copies of the same chromosome. Given that all *M. musculus* chromosomes are acrocentric and of approximately the same size, a technique for distinguishing chromosome 17 was needed, and this was provided by the use of a Robertsonian translocation that links 17 to the slightly longer chromosome 8. Metaphase cells from an embryo carrying this translocation were isolated, and the proximal region, containing T, was then isolated by direct mechanical microdissection from a group of 270 chromosomes, digested with *Eco*RI, and cloned into phage lambda (Rohme et al., 1984). Small clones and clones containing repetitive DNA were eliminated and the rest were screened for hybridization specifically to mouse chromosome 17. These clones were then screened for linkage to RFLPs in the region of T, and several were identified as being within a few centimorgans of the region of interest (Rohme et al., 1984). Subsequent molecular analysis identified a clone very close to the T locus (Rohme et al., 1984) and further work provided a second (Herrman et al., 1986).

These clones, linked to T, permitted the cloning of the gene (Hermann et al., 1990). The definitive identification of T involved analysis of T^{wis}, a spontaneous mutant of T (Shedlovsky et al., 1988). This mutation consists of a small insertion of 5.5 kb DNA, which changes a splicing site in the reading frame of the gene to produce a shorter protein. The sequence of the wild-type gene reveals a 426 amino acid residue protein that does not have the characteristics of either a transmembrane protein or of any of the known classes of transcriptional regulator (Herrman et al., 1990).

In the wild-type, T expression is first discernible in 7.0 day embryos in mesoderm next to the primitive streak and in the embryonic ectoderm that will form embryonic mesoderm, but not in extraembryonic mesoderm; transcripts are restricted to notochordal tissue by 9.5 days, and progressively diminish by 10.5–12.5 days, subsequently becoming undetectable (Wilkinson et al., 1990). In T^{wis} embryos, whose transcript can be detected by *in situ* hybridization, expression is normal until 8 days pc, but then ceases in the primitive streak and the head process (the anterior end of the notochordal region) between 8 and 8.5 days (the eight-somite stage) (Herrman, 1991). It is, in fact, near the head process that the first deviation from wild-type development occurs in T homozygotes (Herrman, 1991).

From the sum total of the results, it appears that the initial development of mesoderm is normal in the absence of T activity but that the later stages are abnormal, with consequent disruption of cell interactions involving the mesoderm. The ensuing defects may then feed back upon and modulate subsequent T expression within mesodermal/notochordal tissue. From chimera studies involving the construction of chimeras between T/T and $+/+$ embryonic cells, it appears that the T defect is cell autonomous, as the mutant cells are not rescued by the wild-type component (Rassbash et al., 1991). This finding, in turn, suggests that the T gene acts intracellularly, providing support for the idea that the derangements of axial development are consequential upon an intracellular effect within the mesodermal cells. The finding of cell autonomy within chimeras also implies that there is a threshold requirement for T gene action (Rassbash et al., 1991).

The molecular work has not yet solved the mystery of how the T product participates in mesodermal development, but it has brought the prospect of a solution considerably closer. Furthermore, the recent work possesses a twofold significance that goes beyond the matter of the T gene itself. First, the results have confirmed the early hypothesis that the T protein is required for embryonic mesodermal (and subsequently, notochordal) development. Second, and more generally, the strategy for cloning the T gene illustrates the potentiality of molecular techniques, when combined with an appropriate set of mutants (those whose mutations cause detectable alterations in restriction enzyme fragments), for the cloning of any mammalian gene.

The identification of a second classic gene, *White-spotting* (*W*), situated on chromosome 5, was achieved by a very different route, and the analysis, to date, of its gene product has proceeded somewhat further than that of T. This case history illustrates the fact that molecular characterization of a gene product can help to explain not only the normal function of a gene product but the previously obscure genetic properties of mutants for that gene. W is named for its dominant coat color phenotype in heterozygotes and was first described in 1916 by one of the pioneer mouse geneticists, C.C. Little. Many W mutations have been identified since the first allele, and the mutants are found to be pleiotropic, showing dominant white-spotting (often extensive, the degree depending upon the allele), anemia, and infertility (in both sexes). These three phenes reflect defects, respectively, in the melanocytes, the hematopoietic stem cells, and the germ line (oocytes and spermatogonia). Although the mutants are pleiotropic, there is a great range of severity in the defects, and the defects in the three tissues are not always correlated in severity (Russell, 1979). As homozygotes, the mutants range from recessive lethals, whose homozygotes die around the time of birth (perinatal lethals), to those that are viable, fertile, and exhibit only mild anemia.

The essential clue to the identification of the W gene product was the fortuitous discovery of a linkage relationship of W to the human proto-oncogene c-*kit*. c-*kit* encodes a transmembrane receptor protein with tyrosine kinase activity in its cytoplasmic domain. This structure signifies that the protein undoubtedly functions in signal transduction across the membrane.

In humans, c-*kit* is found to map on chromosome 4, near the loci that encode phosphoglucomutase 2, α-fetoprotein, and albumin. These loci are also linked in the mouse genome—indeed, there is, in general, a high level of parallel linkages, or synteny, between man and mouse—on murine chromosome 5, and, as it happens, they are closely linked to W. To determine whether c-*kit* is linked to W, use was made of a system that generates many RFLPs and, hence, many molecular markers. It consists of the use of *M. musculus*–*M. spretus* hybrids; as in the use in *C. elegans* of Bergerac-Bristol strain hybrids, the existence of many nucleotide differences between these two murine species produces many restriction fragment differences, which can be used for linkage studies. Using this approach, it was found that c-*kit* and W are, in fact, extremely tightly linked, proving inseparable by recombination in the experiments performed (Chabot et al., 1988).

Earlier work on c-*kit* expression had shown that it is expressed in hematopoietic tissues, one of the sites of W action, raising the possibility that c-*kit* is encoded within W. This proved to be the case. Analysis of two spontaneous strong W alleles showed the mutations to be associated with breakpoints within the coding region of c-*kit* (Geissler et al., 1988). Subsequently, three other spontaneous alleles were found to be associated with particular amino acid substitutions in conserved regions of c-*kit* and reduced specific activities of the kinase; in the absence of other detectable genetic changes in the sequence, these amino acid substitutions are almost certainly the basis of the mutant defects (Reith et al., 1990).

The discovery that *W* is identical to c-*kit* is significant in several respects. In the first place, it is a striking instance of the developmental roles of proto-oncogenes. These genes are the cellular homologs and prototypes of variant, cancer-promoting genes carried by retroviruses, the oncogenes. Initially, the focus of interest in proto-oncogenes concerned their roles in normal cellular growth controls. Yet, as more and more proto-oncogenes were found to play specific roles in developmental events, this emphasis changed. Not least, the conceptual division between gene products required for cellular growth control and those required for development is increasingly seen to be an artificial one. Nor is the developmental role of the cellular homologs of oncogenes limited to mammalian development. For instance, we have already noted one, *Drosophila 1(1)ph*, which shares homology with the oncogene D-*raf* and which is important in the terminal system of the fruit fly (Fig. 7.7).

The molecular properties of receptor tyrosine kinases, the class of molecules to which c-*kit* belongs, helps to explain three aspects of the genetics of *W*. The first is the dominance of the mutant alleles. Although, in principle, such dominant effects might reflect haplo-insufficiency, as is the case for *T*, the discovery of several milder but still dominant alleles that reduce but do not abolish activity (Reith et al., 1990) suggests that something other than simple haplo-insufficiency is involved. This additional factor may involve the aggregation state of the receptor molecules in the membrane, and the consequent equilibrium between monomers and aggregates, which in turn affects activity of tyrosine kinase receptors (Carpenter, 1987). If, for instance, dimers of c-*kit* have enhanced activity, relative to monomers, then either reductions in total amount of protein, produced by regulatory mutants, or novel monomers, generated by point mutations, affecting dimerization or activity of dimers, could affect this equilibrium and hence total activity (Fig. 8.14).

A second genetic property of *W* that becomes easier to explain is the occasional discordance in severity of the three mutant phenes (white-spotting, anemia, infertility) seen with certain *W* mutants. This could reflect either the existence of qualitatively different substrates for the c-*kit* tyrosine kinase activity in the different cell types affected (melanocytes, hematopoietic stem cells, germ line cells) or the existence of different quantitative requirements for c-*kit* activity in these cell types. It should be possible, by investigating c-*kit* activities in the affected cell types from different mutants, to determine which explanation is the correct one.

The third genetic property of *W* that the new findings explain is the apparent cell autonomy of the mutant effects. Bone marrow transplants from wild-type into *W* mice can completely rescue the anemia defect, indicating that wild-type hematopoietic stem cells can function normally in an otherwise *W* mutant environment. Comparably, transplantation of marrow or spleen cells from viable but anemic *W/W*ᵛ animals into lethally irradiated +/+ animals (such irradiation destroying all wild-type hematopoietic stem cells) can rescue some of the recipients, who, nevertheless, show the low red cell counts that are typical of the donor (Russell, 1979). Chimera experiments reveal that the melanogenesis and fertility defects are also cell autonomous (Kuroda et al.,

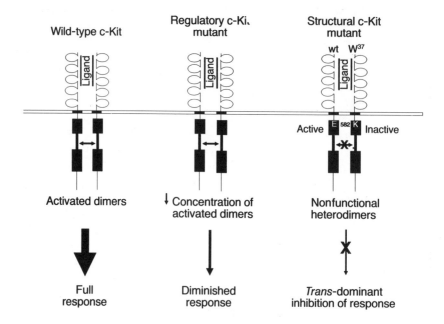

Fig. 8.14. Hypothesized c-*kit* dimerization and responses produced by regulatory mutations (center) that diminish c-*kit* expression and dominant structural gene mutations (right) that produce inactive dimers. (Reproduced from Pawson and Bernstein, 1990, with permission from the publisher.)

1989). The cell autonomony of the *W* gene product is fully consistent with that product being a cell-intrinsic membrane component.

There is an interesting contrast, in this respect, to another gene whose mutants also show white-spotting and anemia. This is the *Steel* (*Sl*) gene, of chromosome 10, which does *not* show cell autonomy for these defects. Thus, bone marrow from viable heterozygous *Sl* animals is just as effective in rescuing the *W* defect as wild-type bone marrow, indicating that its stem cells are fully functional, while wild-type marrow cannot rescue the *Sl* anemia (McCulloch et al., 1964; Bernstein, 1970). Evidently, *Sl* in some way affects the cellular environment in which the hematopoietic stem cells normally function.

Indeed, the similarity of the *W* and *Sl* phenotypes, when contrasted with their different cell autonomy properties, suggests an interesting hypothesis, namely that while *W* is a receptor for one or more signals, *Sl* may be involved in production of the ligand(s) that constitute the signal (Chabot et al., 1988). This suggested relationship would be similar to that between the *sev*-encoded receptor tyrosine kinase and the *boss* gene product in the *Drosophila* ommatidium. The putative *W*–*Sl* interaction has been confirmed: Molecular studies show that *Sl* encodes a secreted product, mast cell growth factor (MGF), which had been previously implicated in hematopoiesis (Copeland et al., 1990), and MGF has been shown, by biochemical means, to be a ligand for the c-*kit* gene product (Flanagan and Leder, 1990; Williams et al., 1990).

This would, however, still leave a puzzle about melanogenesis: Why do heterozygotes for both genes usually show white-spotting rather than pure white coats? The explanation would seem to be that the particular cellular "microenvironment," in which the cells expressing *W* and *Sl* act, is crucial (Russell, 1979). Evidently, when melanoblast numbers are relatively small, the local environment in which melanoblasts become activated for pigment production can either influence the degree of *W* or *Sl* activity or call forth supplementary activities that partially substitute for them. The absence of systemic *Sl* effects suggests that its gene product, MGF, is not freely diffusible but probably, like *boss* in the *Drosophila* eye epithelium, presented on the surface of certain cells.

Obtaining New Mutants: Several Approaches

Though increasing numbers of classical mouse genes, identified on the basis of viable mutants with distinctive phenotypes, are being cloned and characterized, it is clear that any full developmental genetic analysis of the mouse cannot be based on such genes alone. The several hundred genes identified in this fashion comprise just a small proportion of the estimated 10,000–40,000 genes in the mouse genome. Clearly, a major expansion in the numbers and kinds of genes, identified by their genetic properties and developmental roles, is needed. Three kinds of approaches will be mentioned here.

The first of these can be broadly described as "insertion mutagenesis." In one version, cloned DNA is injected into the male pronucleus of the fertilized egg, a typical injection carrying several hundred copies of the DNA molecules. Injection into the pronucleus is possible in the mouse embryo because of the slow pace of the first cleavage division. Following injection, one or more of these molecules may integrate into the haploid genome, disrupting a gene at the integration site. The injected one-cell embryos are given a brief culture period *in vitro* and then implanted into foster mothers. In a second approach, early embryos are infected with retroviruses, whose DNA copies insert into the genome at many different locations.

Insertion is sufficiently frequent to generate mutants at frequencies higher than those obtained by many of the older mutagenesis methods (though not as high as produced by a mutagen, to be described below). Several such insertion mutants have been described. One recessive lethal, an insertion into a structural gene for collagen, was produced by a late insertion, into a postimplantation embryo, by a Moloney leukemia virus (Jaenisch, 1983). In another set of experiments, two out of six one-cell embryos injected with plasmids carrying the gene for human growth hormone (HUGH)-containing plasmids were found to contain recessive lethals (Wagner et al., 1983). Both kinds of experiments generate mutations at random and the mutants obtained then have to be characterized phenotypically and mapped. In principle, however, one can obtain mutants in particular genes, from the distinctive phenotypes generated. A transgenic mouse carrying the bacterial CAT gene on its insert was recognized by its white-spotting effect, which resembled that of *W* or *Sl*. Mapping and complementation experiments revealed it to be a mutation in the *Sl* locus (Keller et al., 1990).

In general, however, the mutations produced by insertion mutagenesis are a fairly random sample. The other two methods that have been employed to generate new mutations focus on obtaining mutations in particular chromosome regions. The first of these more targeted methods makes use of an old genetic strategy, the "uncovering" of a recessive, viable autosomal mutant allele when the point mutant is over a deletion for that locus; deletions are induced by X-rays or chemical treatment and the treated males are mated to females homozygous for the marker mutation. The only offspring that will show the trait are those with deletions (or new point mutations) that have removed the wild-type allele. These deletions are often missing other genes, on either or both sides of the marker gene, and these newly deleted loci can subsequently be identified in a multistep procedure.

The procedure is illustrated by the search for mutants of the *albino* (c) locus on chromosome 7. This gene was long suspected, and is now known to be, the structural gene for tyrosinase, the enzyme that synthesizes the major melanic pigments (Kwon et al., 1987). To collect new albino mutants, females homozygous for the known albino allele (c/c) (the C allele being the dominant, wild-type allele) were mated to irradiated wild-type males, and all albino progeny were outcrossed to isolate the new c alleles (Gluecksohn-Waelsch, 1979; Russell et al., 1982). Most of the mutants isolated are viable as homozygotes and these mutations are either point mutations in C or small deletions. A proportion, however, are lethal when homozygous, these chromosomes being deleted for C and adjacent loci.

The next step in the analysis is the identification and crude mapping of "functional units" by making heterozygotes between the different deletions and scoring the heterozygotes for survival and any unusual phenotypes. Such functional units are regions of the wild-type chromosome whose presence prevents the appearance of a particular developmental abnormality. A functional unit may contain only one gene or several; further analysis is needed to reveal the genetic basis of each functional unit (as discussed further, below).

The principle is essentially the same as in any complementation test and is illustrated in Fig. 8.15. If two deletions do not overlap any vital genes, the heterozygotes should be fully viable; if they do overlap one or more essential genes, they will not be fully viable. Failure to complement one or more of the "anchor" markers in the region (the visible mutant *taupe* [tp], the biochemical marker *Mod-2* [mitochon-

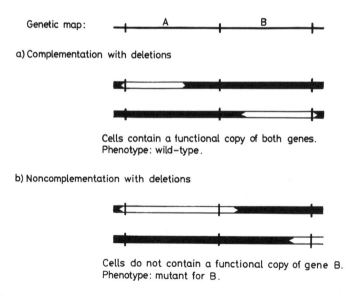

Genetic map:

a) Complementation with deletions

Cells contain a functional copy of both genes.
Phenotype: wild–type.

b) Noncomplementation with deletions

Cells do not contain a functional copy of gene B.
Phenotype: mutant for B.

Fig. 8.15. Deletion mapping of functions, employing tests for complementation.

drial malic dehydrogenase], or the behavioral mutant *shaker-1* [*sh-1*], helps to place the limits of particular deletions. Comparison of the end points of different deletions in a set allows, in principle, the construction of a linear map of known genes and functional units.

In these tests, the heterozygote class should either be distinctively marked, so that its presence or absence can be readily detected, or, when this is not possible, a reduction in litter size can serve as an indication of heterozygote death. When heterozygotes show clear noncomplementation, one examines the embryos *in utero* to determine the time, stage, and characteristics of embryo or fetal death; if the heterozygotes are viable, one examines them for characteristic phenes. From the aggregate set of complementation tests, one can make a self-consistent deletion map that places particular functional units on the chromosome (Fig. 8.16).

One of the first functional units to be characterized is one required to prevent perinatal mortality, described by Glucksohn-Waelsch (1979). The defect in the four perinatal mutants has special interest with respect to genetic regulation. In these mutants, the cause of death appears to be a combination of defects in the liver, kidney, and thymus, with the liver defects being the most serious. The latter involves the failure of six hormonally inducible liver proteins, all involved in energy metabolism,

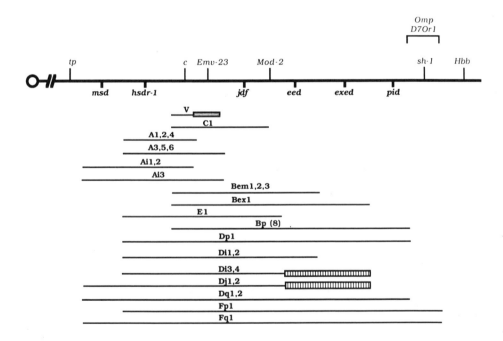

Fig. 8.16. Deletion map of the albino (*c*) region of chromosome 7. The extents of the deletions are indicated below the map. Loci defined by point mutations are as follows: *taupe* (*tp*), *c*, *Emv-23* (integration site 23 of ecotropic murine leukemia provirus), *Mod-2* (mitochomdrial form of malic dehydrogenase enzyme), *shaker-1*, (*sh-1*), *Omp* (olfactory marker protein), *Hbb* (B-hemoglobin). (*D770r1* is a DNA clone, mapped at OAk Ridge.) Functional loci defined by the deletions include *msd* (mesoderm deficient; formerly IS on earlier maps); *hsdr-1* (*hepatocyte-specific developmental regulator-1*; see text for discussion); *jdf* (juvenile development and fertility; formerly *JS*); *eed* (embryonic ectoderm development; formerly *EE*); *exed* (extraembryonic ectoderm development; formerly *ExEE*); *pid* (preimplantation development; fromerly *PS*). (Figure courtesy of Dr. E.M. Rinchik.)

to respond to hormonal induction and to increase after birth, as occurs in wild-type, although other liver-specific enzymes show normal activities.

This alteration involves a specific regulatory defect. When mouse/rat hybrid cells were made by fusing mouse cells homozygous for one of the deletions with rat cells, the hybrid cells were found to express the mouse glucose-6-phosphatase activity, one of the deficient activities of the mutant mouse lines; evidently, the mouse genome is not lacking the structural gene but merely part of the regulatory apparatus, which the rat genome can supply (Cori et al., 1983). Nor is the defect a deficiency or abnormality for the receptors for the glucocorticoid hormones (Gluecksohn-Waelsch, 1983; De Franco et al., 1991). Instead, the failure of increase of the affected enzymes may be due to the failure of activation of a transcriptional activator, the C/EBP protein. The concentration of this protein increases sharply in liver prior to birth but fails to increase in the deletion mutants that experience perinatal mortality (McKnight et al., 1989).

Although the gene encoding C/EBP is also on chromosome 7, it is not included in the deletions but is 30 cM distal from the functional unit, defined by the lethal albino deletions, that is required for perinatal survival (McKnight et al., 1989). The latter gene (or genes) probably maps proximal to c (Fig. 8.16) and includes sequences termed *hepatocyte-specific developmental regulator* (*hsdr*) (De Franco et al., 1991). It is ultimately *hsdr* that directly or indirectly affects the expression of several structural genes, in response to glucocorticoids. One simple hypothesis is that *hsdr* acts as a transcriptional regulator of *C/ebp,* whose product then transcriptionally activates the various structural genes in the liver. Another possibility is that, since *hsdr* mediates the ability of the liver to respond to glucocorticoids, it, in addition to an effect on *C/ebp,* also influences states of chromatin that make the structural genes competent to respond to the hormonal stimulus (Gluecksohn-Waelsch and De Franco, 1991).

Other functional units in the vicinity of c have been identified through the analysis of the defects in noncomplementing deletions. In particular, several centimorgans distal to c are functional units required, respectively, for formation of the embryonic ectoderm, formation of the extraembryonic ectoderm, and for preimplantation development (Fig. 8.16). Initially, the first two were grouped as a single functional unit required for postimplantation survival (Russell et al., 1982). However, careful examination of the affected embryos, for deletions with different end points in this region, resolved two different developmental syndromes and, hence, two different functional units.

Such findings provide the promise of a genetic dissection of tissue and organ development in postimplantation development. One must take the analysis a step further, however, before one can conduct such an analysis. While the functional units certainly contain genes that are necessary for completion of the developmental processes highlighted by their absence in the deletions, it is not clear how many genes are involved, whether one, two, or many. The additional step required is the induction of point mutations within the regions demarcated as functional units and the analysis of the defects found in homozygotes for such mutations.

The mutagen of choice is ethylnitrosurea (ENU), a highly powerful mutagen (reviewed in Rinchik, 1991). Unlike X-rays, which create both deletions and point mutants, ENU is believed to induce solely, or predominantly, single base pair changes. Male mice are injected intraperitonally, with either one dose or several administered at weekly intervals, and mated after they have recovered their fertility (the treatment causing sterility for a number of weeks, due to spermatogonial killing).

To probe the nature of functional units uncovered in the deletion analysis, one combines ENU mutagenesis with use of the requisite deletions. One scheme, designed to detect new lethals in the region covered by one of the longest chromosome 7 deletions, *Fp1*, is shown in Figure 8.17. In this approach, the absence of an albino class in the second generation (G_2) indicates the existence of a new lethal mutation within the region under scrutiny. The method will also identify mutations other than lethals; any consistent abnormality in the G_2 albino class for a tested chromosome signals a potential new mutation. In one screen of 972 gametes from ENU-mutagenized males, this approach produced 13 new mutations in total (Rinchik et al., 1990). Complementation tests revealed new *sh-1* mutations, the existence of a locus (*fit*) whose deficiency produces runting, and lethal mutations defining four new complementation groups (Rinchik et al., 1990). In this experiment, two independent mutations were recovered at both the *sh-1* and *fit* loci, suggesting that saturation mutagenesis for all functions in this 6–11 cM interval is a possibility. It will be of interest to see if further searches reveal an *hsdr* gene, as predicted (see above) from the functional unit analysis; mutations in *hsdr* would be indicated by perinatal lethal albinos for the mutation-bearing chromosome. Efficient as ENU mutagenesis is, however, true saturation mutagenesis with it may not be possible; analysis of ENU-induced mutation rates at different visible loci suggests that this mutagen is more effective for some genes than others (which is, in fact, generally the case for mutagens) (Russell et al., 1979).

The search for new mutations, affecting developmental processes of interest, is, of course, not limited to the region around the *c* locus or, indeed, to those regions for which deletions are available. In principle, any region which includes an easily scorable mutation can be screened for new mutations, given the effectiveness of ENU. The strategy is to mutagenize males homozygous for the marker, *m*, outcross them to

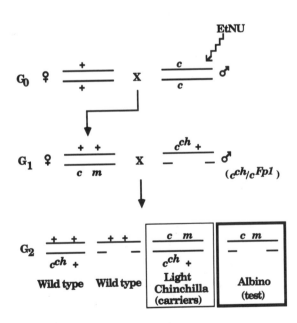

Fig. 8.17. Procedure for detecting new ENU-induced mutations in *c* region; the absence of the albino class in the G_2 signifies the presence of a new lethal mutation, *m*, which will be carried by the light chinchilla sibs in that line. (Figure courtesy of Dr. E.M. Rinchik.)

produce G_1 individuals, cross these to produce G_2 brother and sister heterozygotes (where the homolog contains a visible dominant mutation), then mate these and score in the G_3 generation for the absence of the homozygous marker class.

Although this strategy involves one extra generation, compared to that depicted in Figure 8.17, it has proved successful in a search for new mutations in the proximal region of chromosome 17. This is the wild-type region corresponding to the *t* haplotypes and covers approximately 15 cM (Shedlovsky et al., 1986, 1988). In the first screenings of over 350 tested gametes, 11 lethal and two perinatal complementation groups were identified, and the measured frequency, when compared to that for visible mutations, suggests that there may be over 70 lethal complementation groups in the region. If this figure is correct, it indicates that there is a considerably greater density of such genes per centimorgan than for the *c* region, on chromosome 7.

As regions become progressively "filled in," with the discovery of new genes, it will be necessary to map these genes precisely. The existence of RFLPs, either within the *M. musculus* genome or between *M. musculus–M. spretus* hybrids, in combination with visible markers and biochemical markers, should enable mapping to a precision of millimorgans (1 mM = approximately 200 kb) (Shedlovsky et al., 1988). Strategies for obtaining and mapping DNA clones in desired regions will provide the necessary molecular complement to the genetics.

One approach involves the physical isolation of particular chromosomes, using microdissection or flow sorting of translocation chromosomes of distinctive size, followed by library construction and screening for clones specific to that chromosome. The DNA clones can then be mapped within particular regions, either against RFLPs or with respect to deletions, where these are available (Johnson et al., 1989). New molecular mapping methods, involving use of the polymerase chain reaction (PCR), which can selectively and geometrically amplify specific sequences from genomic DNA, are also being developed and hold the promise of producing considerably more detailed maps (Cox and Lehrach, 1991).

HOMEOBOX GENES, PATTERN FORMATION, AND THE *DROSOPHILA* CONNECTION

Besides "classic" murine genes and new genes identified by viral insertions or ENU mutations, there is a third major source of genetic material that has been tapped for mouse developmental studies: the key pattern formation genes from *Drosophila*. These genes have been used to "fish out" their murine homolog, by using the appropriate hybridization regimes and screening procedures on genomic libraries. The mouse cognate loci have subsequently been mapped by a combination of cytological and molecular techniques and their roles in mouse development are beginning to be analyzed.

The most significant findings to date have involved the mouse genes of the so-called *Antp* homeobox group. As noted in chapter 7, homeobox-containing genes can be grouped in various families, depending upon their degree of relatedness. The first such family to be characterized was the so-called *Antennapedia* family, named for the *Antp* homeobox and found in the homeotic genes of both the Antennapedia and bithorax complexes. The *Antp* homeobox was the first to be identified in *Drosophila* (through its cross-hybridization with the 3' end of *Ubx*) and it was also the first to be used in the search for homologs in other organisms; these experiments revealed the existence of such homologs in a wide variety of animals, both invertebrates and vertebrates, including mammals (McGinnis et al., 1984b). A second homeobox gene family is the *engrailed* family, named after the *en* gene in which this homeobox group was first identified. Yet a third is located within the so-called POU domain, of which the *unc-86* gene in *C. elegans* is a member.

Extensive analyses of the *Antp* family members have illuminated some fundamental developmental and evolutionary relationships. The discovery that there are homologous sequences to *Drosophila Antp*-type homeodomains, detectable on Southern blots (McGinnis et al., 1984b), was quickly followed by the cloning of several of these genes (McGinnis et al., 1984c; Hauser et al., 1985). From the analysis of chromosome walks, starting with the cloned genes, it became apparent that these genes are organized in four separate clusters on different chromosomes. These gene clusters, termed Hox groups, are each about 40–50 kb in length, and have been named Hox-1 (on chromosome 6), Hox-2 (on chromosome 11), Hox-3 (on chromosome 15), and Hox-4 (on chromosome 2). (Other putative clusters, Hox-5 and Hox-6, were named for freshly isolated homeobox genes that seemed, initially, not to belong to the other clusters but that were subsequently found to be part of Hox-4 and Hox-3, respectively.) Within each cluster, all genes are transcribed in the same direction, and, hence, from the same DNA strand.

Within each mouse Hox cluster, the genes have been named in the order in which they were discovered; thus, in the Hox-2 cluster, the first to be identified was *Hox-2.1* and the most recently identified, *Hox-2.9*. Though all Hox clusters are approximately the same range, they are not identical in the numbers of genes they contain. Hox-2, with nine members, is the largest group; Hox-3, with five members known to date, is the smallest. (As the regions surrounding each Hox cluster are further explored, new members may, of course, be discovered.) Despite such differences, certain striking similarities between Hox clusters have been found. These relationships have come to light from comparisons of the expression patterns of these genes (reviewed in Gaunt, 1991). Transcripts for the earliest-expressed Hox genes are first detectable in the early primitive streak stage (about 7.5–8.0 days pc) (Gaunt et al., 1986, 1988; Gaunt, 1987) and then evolve to fairly stable patterns in the central nervous system (CNS) and axial mesoderm by 10.5–12.5 days pc. Though these patterns are complex in the aggregate, with numerous regional, histospecific, and temporal differences exhibited among the *Hox* genes, there are certain general aspects observed for expression during embryogenesis.

The first general rule is that for all Hox groups, there is a correspondence between the chromosomal order of the genes and their relative expression along the a-p axis of the embryo. Thus, genes that are progressively more 5' within the cluster show progressively more posterior domains of expression during embryonic development. This a-p progression is seen in both mesodermal (prevertebral) and CNS axial elements (Gaunt et al., 1988, 1989; Duboule and Dollé, 1989; Graham et al., 1989). In the CNS, the progression is most clearly seen at the anterior boundary, with progressively more 3' genes showing progressively more anterior boundaries. In the mesodermal axial system, the prevertebral column, the a-p progression is particularly clear and is shown for several genes in the Hox-1 and Hox-3 clusters (see Figs. 8.18 and 8.19).

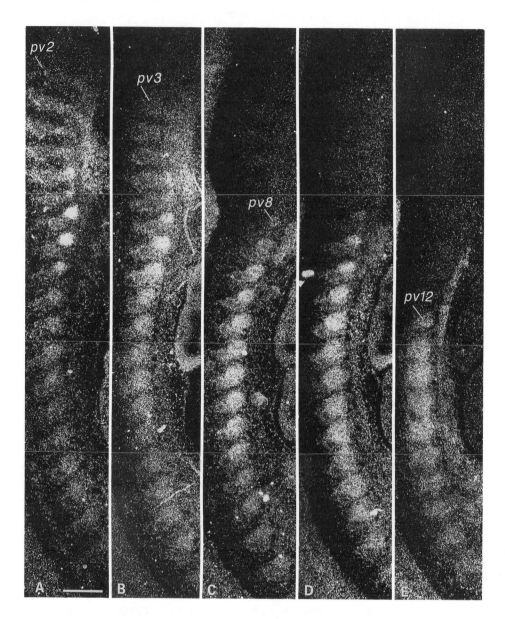

Fig. 8.18. Patterns of expression of *Hox* genes in the prevertebral column, as determined by *in situ* hybridization. **A:** *Hox-1.4.* **B:** *Hox-1.3.* **C:** *Hox-3.3.* **D:** *Hox-1.2.* **E:** *Hox-3.1.* The sections are parasagittal and all are from the same embryo. The more 3′ the gene is in relative position (see Fig. 8.19), the further anterior is its anterior boundary. (Reproduced from Gaunt et al., 1988, with permission of the Company of Biologists, Ltd.)

In itself, this parallel between gene chromosomal positions of genes and their a-p expression domains is highly reminiscent of the *Drosophila* organization of clustered homeotic genes. In the BX-C, it will be recalled, the most 5' located member is *Abd-B*, expressed in the most posterior abdominal region, and the most 3' located member is *Ubx*, expressed in the thorax and anterior abdominal region (Fig. 7.12). The parallel between mouse and *Drosophila* gene organization, however, goes considerably further than this correspondence. When detailed sequence comparisons are made between the homeoboxes of the *Drosophila* homeotic genes and the *Hox* genes of the mouse, it is found that each *Hox* gene has a *Drosophila* homolog. All murine *Hox* genes can be grouped in subfamilies based upon these relationships—in some instances, the homology relationships have been shown to extend beyond the homeodomains—and members of the same subfamily are found to occupy the same positions with respect to one another in their respective Hox groups.

The prospective significance of these relationships becomes most apparent, however, when one compares the relative order of the Hox gene members with their *Drosophila* counterparts. If one brings the BX-C and ANT-C clusters together as one group, the HOM genes, with the former on the left, one finds a distinct correspondence between the order of the HOM genes and their murine cognate loci (Duboule and Dollé, 1989; Graham et al., 1989). Furthermore, members of the same subfamily usually show the same, or nearly the same anterior limits of expression in the CNS and prevertebral column, though these correspondences are generally closer for the more anteriorly expressed subfamilies.

The gene order relationships between the *Hox* clusters and the *Drosophila* homologs of the different subfamilies (the HOM genes) are depicted in Figure 8.19.

Conservation Between Homeobox Complexes

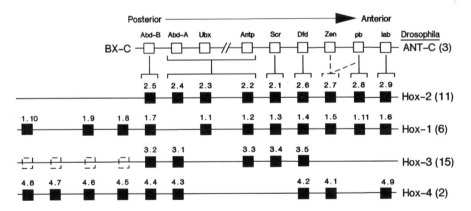

Fig. 8.19. *Hox* gene clusters in the mouse, as compared to the HOM genes of *Drosophila*. The *Hox* genes are named in the order in which they were discovered and their placement with respect to the *Drosophila* genes is based on the extent of similarity in their homeobox regions. (The homeobox regions of *Antp*, *abd-A*, and *Ubx* are too similar to each other to permit unambiguous placement in relationship to the subset of *Hox* genes that they most resemble.) Each *Hox* cluster is depicted in terms of direction of transcription, from 5' on the left to 3' on the right. The numbers in parentheses at the right are the designated chromosome numbers on which the HOM and *Hox* gene clusters are found. (Figure courtesy of Dr. R. Krumlauf.)

Although the HOM genes are found in two clusters in *Drosophila,* it will be recalled that the corresponding genes of a more primitive insect, *Tribolium,* are grouped in one, like the Hox groups, and this gene arrangement is probably the ancestral one (Beeman, 1987). Some of the other differences between the HOM and Hox genes, however, may have some functional significance. One is that in the *Drosophila* gene arrangement, the clusters occupy considerably longer stretches of DNA, measuring in hundreds of kilobase (kb) pairs rather than the 40–50 kb that characterize the Hox clusters. Furthermore, as discussed above, the Hox clusters differ somewhat among themselves in terms of their genetic content; *Hox-5,* for instance, clearly seems to lack several of the cognate genes found in the other *Hox* groups. Finally, while ANT-C contains nonhomeotic loci such as *ftz* and *bcd,* the *Hox* gene clusters seem to lack such additional genes entirely.

Since the last common ancestor of the fruit fly and the mouse lived between 550 and 900 million years ago (Holland, 1990), the existence of differences in genetic organization between the insect HOM and the vertebrate Hox genes is hardly surprising. In particular, the multiple-cluster organization in mammals in contrast to the single-cluster (split or unsplit) organization in insects presumably reflects the occurrence of several duplication events in the line of descent leading to vertebrates, which multiplied the number of clusters. It is the correspondences that are remarkable. The simplest explanation for the observed similarities is that *in both sets of organisms, these genes perform essential and perhaps similar functions* and that these commonalities have been preserved by natural selection.

Beyond individual gene functions, the conservation of gene order indicates that it too has been selected, presumably because it facilitates the appropriate sequence of transcription of these genes. Gaunt and Singh (1990) have proposed that in anterior body positions, essentially all members of each cluster may be "open" for transcription, while as one moves posteriorly in the embryo, 3′ members of the cluster (those expressed anteriorly) are progressively "closed" for transcription by some form of spreading heterochromatinization process. Whether or not this explanation accounts for the retention of the clustered arrangement in evolution, there is some evidence from transgenic mouse lines that clustering *per se* is not essential for the normal initial expression of each *Hox* gene (Wolgemuth et al., 1989; Püschel et al., 1990). One set of experiments that demonstrate that Hox genes can show normal initial expression outside of a cluster involves tests of expression of a *lacZ* coding sequence hooked to a *Hox-1.1* promoter in transgenic mice. In the embryos of three different lines of transgenic mice carrying this reporter gene at each of three different sites, expression was found to be initially identical to that of the endogenous *Hox-1.1* gene and to show the same anterior boundary at day 12 (Püschel et al., 1990). Interestingly, however, differences in expression emerge in later development, when many mesodermal cells continue to express the reporter gene after the endogenous gene has been silenced in these lines.

Comparably, expression of *Hox-1.4* in two multicopy single-site insertion transgenic lines is essentially normal in early embryos but is abnormally elevated in certain tissues later in development (Wolgemuth et al., 1989). The results indicate that, as predicted in the Gaunt and Singh model, the termination of normal expression, in some cells at least, may require or be facilitated by placement within a cluster.

The crucial question, of course, concerns the developmental roles of the *Hox* genes. Sequence similarities between HOM and *Hox* genes can be highly suggestive of, but cannot directly establish, functional relationships. Expression domains can also provide clues to function but, similarly, cannot definitively establish it. In particular, the fact that most of the Hox genes begin to be expressed during primitive streak formation suggests a set of important early functions in the establishment of embryonic pattern, while the similar domains of expression of Hox genes of the same subfamily implies that the members of a given subfamily perform similar functions. The differences in expression between related *Hox* genes that do exist might reflect either slightly different roles for members of the same subfamily (Gaunt et al., 1989) or the need for a degree of regulatory flexibility that cannot be achieved in a single cluster. Possibly, of course, both explanations might apply. Furthermore, late expression patterns for individual *Hox* genes may indicate specialized late roles that have no obvious phenotypic connection to earlier functions. Thus, for instance, *Hox-1.4* is expressed in the adult but only in the testes of male mice and, specifically, only in those spermatogenic cells that have entered meiosis (Wolgemuth et al., 1989).

Ultimately, the only tests of function are genetic ones: the alteration of particular gene activities by mutational inactivation or by other means and assessment of the resulting developmental effects. While conventional searches for mutations in particular genes, even aided by ENU mutagenesis, are still impracticable, the use of gene targeting techniques makes it possible to eliminate particular *Hox* gene activities and assess the developmental consequences, as described in the next section. On the other hand, if, as seems likely, there is a large measure of functional redundancy between members of the same subfamily, as suggested by several of the experiments (discussed below), single *Hox* gene inactivations may, in general, show only slight effects and definitive tests might require inactivation of all the members of a given subfamily. For the *Hox-2.9* subfamily (the homologs of the *Drosophila lab* gene), which contains only two members, this may prove feasible; for the larger subfamilies, such multiple eliminations may prove to be too difficult.

It is simpler, at present, to engineer ectopic expression of particular genes and to assay for the effects of such expression. Transgenic mice are constructed in which a particular *Hox* gene is overexpressed, either from a promoter that gives ubiquitous expression or by virtue of its elevated copy number or position in the recipient genome and, subsequently, the developmental phenotype is scored in heterozygotes or homozygotes. Such tests have been carried out with *Hox-1.4* (*Drosophila* homeodomain homolog *Dfd*) and *Hox-1.1* (*Drosophila* homeodomain homologs *abd-A, Ubx,* and *Antp*). For the particular *Hox-1.4* construct used, which carries 10 kb of normal sequence upstream of the homeobox, the chief result was elevated expression in the embryonic gut and an overdevelopment of the colon itself (Wolgemuth et al., 1989). (In embryos, the principal domain of expression was similar to that of the endogenous gene, as noted above.) For *Hox-1.1,* ubiquitous expression during development resulted in primarily craniofacial abnormalities and the development of an extra vertebra at the craniocervical transition region, the latter result possibly reflecting a homeotic transformation (Kessel et al., 1990). Although these results show that inappropriate expression of *Hox* genes has developmental consequences, they have not yet defined the normal function of these genes. Indeed, it appears probable that the identity of a particular region is determined by the combination of Hox genes expressed, its *Hox code* (Hunt and Krumlauf, 1991; Kessel and Gruss, 1991) and isolated ectopic expression of a single *Hox* gene might produce either a nonstandard coding, of unclear phenotypic effect, or only a slight transformation effect.

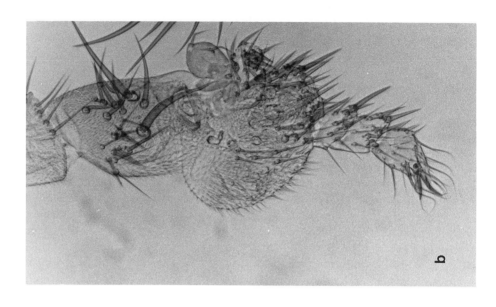

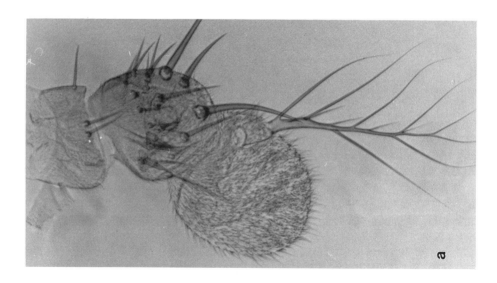

Other ectopic expression experiments, however, have confirmed the importance of *Hox* genes for establishing regional phenotypic identity and, simultaneously, established a point of evolutionary significance. These have involved the placement and expression of mammalian *Hox* genes in *Drosophila* and show that the *Hox* genes can carry out HOM gene-like specification of regional identity along the a-p axis. In this approach, the cDNAs of individual *Hox* genes have been put under the control of the *hsp70* promoter and then used to construct transgenic flies; the experimental animals, carrying one, two, or four doses of the insert, are then given single or short, multiple heat shocks during embryogenesis or late third larval instar and their expression patterns and subsequent phenotypes are scored.

Some particularly informative results are seen when this experiment is performed with *Hox-2.2*, one of the *Antp* homologs. Within the entire set of *Hox* genes, the homeodomain of *Hox-2.2* is one of the closest in sequence to that of *Antp* itself and, in addition, contains two other short sequences characteristic of *Antp* which are 5′ to the homeodomain. When several 20 min heat shocks are given in early embryogenesis to embryos carrying two copies of the insert, approximately 50% of the unhatched embryos display a homeotic transformation, with the appearance of thoracic-type ventral denticle belts on the lateral and dorsolateral sides of their heads (Malicki et al., 1990). Control experiments establish that these transformations are a direct result of *Hox-2.2* expression. The transformations are not seen in control embryos (which do not receive heat shocks; the heat shocks do not induce transcription of the *Antp* gene itself; and the effects are seen in an *Antp⁻* genetic background (under otherwise comparable conditions).

An even more impressive result is seen when a multiple heat shock regimen is given to late third instar larvae; antennal-to-leg transformations are observed for distal antennal structures in many of the unhatched, but fully developed flies (Fig. 8.20). These transformations are thus comparable, qualitatively and quantitatively, to those seen when *Antp* is itself put under the control of the *hsp70* promoter and late third instar larvae carrying this construct are given heat shocks (Schneuwly et al., 1986, 1987) (Fig. 7.27). In particular, in a small proportion of the *Hox-2.2* transformants, where the second antennal segment is transformed, the finding of an apical spur on the transformed appendage (in 3 out of 35 heads) identifies these *Hox-2.2*-induced leg structures as, specifically, T2-like (Malicki et al., 1990). Evidently, *Hox-2.2* not only possesses sequence homology to *Antp* but distinctly *Antp*-like function, when expressed in *Drosophila*. Whether the specificity for these transformations lies in the homeodomain of *Hox-2.2* or in the other *Antp*-like regions of homology has yet to be ascertained.

The second gene for which this kind of test has been performed is the human *Hox* gene, *Hox-4.2*, one of the *Dfd* homologs (McGinnis et al., 1990). In this instance, the capacity of an induced *Hox* gene to mimic a *Drosophila* HOM gene, as judged from phenotypic transformations, is less clear, though induction in late third larval instar produces some head and eye defects resembling that of the original, dominant *Dfd* mutant. Most strikingly, however, *Hox-4.2* can induce the expression of the endoge-

---

Fig. 8.20. Effect of ectopic expression of mouse *Hox-2.2* in *Drosophila*—the transformation of antenna to leg. **a**: Control heat-shocked: result, wild-type antenna. **b**: Experimental heat-shocked transformant with *Hox-2.2* construct: result, partial transformation of antenna to leg. (Photographs courtesy of Dr. W. McGinnis.)

nous *Dfd* gene, as shown by *in situ* hybridization. Such induction resembles the autogenous self-induction of *Dfd* itself, described in chapter 7, but is somewhat weaker in extent (McGinnis et al., 1990). Thus, in the case of this *Hox* gene, too, some of the properties of the *Drosophila* cognate gene have been preserved in the line leading to mammals.

These results focus attention on the precise nature of the spatial domains in which *Hox* gene activities are expressed. In *Drosophila,* the HOM genes are expressed in either parasegmental or segmental units and one would like to know whether there are comparable delimitations in *Hox* gene expression. In the mouse, at least in the 12 day embryo, the domains in the prevertebral column are long, with sharply defined boundaries at particular prevertebral segment borders, showing some relationship to segmental boundaries. However, in the CNS, little can be said about the morphological nature of the domains since there is no distinct segmentation. Nevertheless, in the earlier embryo, transient segmental subdivisions appear in the hindbrain. For these rhombomeres, as they are termed, there is a clear double-segment periodicity for several of the *Hox-2* genes, while the most anteriorly expressed *Hox-2* gene, *Hox-2.9* (the homolog of *lab*), is found exclusively in rhombomere 4 (Wilkinson et al., 1990).

These results, while confirming the importance of *some* segmental domains as expression domains for the *Hox* genes, also raise questions about the evolutionary basis of this connection. In particular, there is general agreement that segmentation has arisen independently as a process several times in evolution and, specifically, in the animal lines leading to insects and vertebrates. To the extent that there are parallel relationships between HOM/*hox* gene expression patterns and segmental domains in these two distantly diverged animal groups, the independent origins of segmentation would suggest that the ancestral HOM/hox domains defined morphological regions prior to the emergence of segmentation. One possibility is that the ancestral genetic structure served to specify a capacity for serial repetition of structures, encoded in a HOM/*hox*-type gene; this gene would have been expressed at repeated intervals along the body axis. Duplication of such a gene could have been followed by diversification of its structure and expression, ultimately providing the basis for differences of morphology along the body axis. Different lines of descent (to insects and vertebrates) might then have produced morphological segmentation independently. The possibility of serial repetition as a forerunner to true (morphological) segmentation is not a new idea. It was, in fact, first proposed nearly a century ago by William Bateson, in his classic text on homeosis (1894), and has been discussed recently in connection with the HOM and *Hox* genes by Holland (1990).

Apart from gene targeting experiments, described below, two further lines of investigation should help to clarify the developmental roles of the *Hox* genes. The first involves careful analysis of the kinetics of appearance of particular *Hox* gene products in particular regions and cell types, for clues to the sequence of developmental changes. Two such studies have been carried out for the hindbrain region of the head and have produced somewhat conflicting results. Frohman et al. (1990) report that expression of *Hox-2.9* occurs first in the head mesoderm and subsequently in the neural plate. These workers interpret their results as indicating an inductive influence of mesoderm upon ectodermal pattern, as occurs in early amphibian development.

In contrast, Hunt et al. (1991) do not find such early mesodermal expression and report that, in the hindbrain region, migrating neural crest cells (which are derived from the neuroepithelium) first express particular *Hox-2* genes, followed by expression of the same genes in the branchial arch regions into which these cells migrate. These workers interpret their results as signifying an early instructive role of the neural crest cells on the ectoderm in the head. Further studies of the *Hox-2* genes and the other *Hox* gene clusters should help to resolve this controversy and elucidate the inductive relationships between *Hox* gene-expressing tissues.

The second area, of course, is the identification of the target genes whose transcription is controlled by the various *Hox* genes. As in the comparable case in *Drosophila*, for a definitive understanding of the phenotypic changes produced by the expression of *Hox* genes, the downstream genes need to be identified and characterized and their activities related to the phenotypes of the regions in which the *Hox* genes are expressed. Such studies have only just begun.

GENE TARGETING: STRATEGY AND PROSPECTS

The analysis of gene function in the mouse inevitably requires mutations in the genes of interest, whether they be early pre- or postimplantation lethals or *Hox* and other genes. The saturation mutagenesis strategies discussed earlier provide one set of approaches but they are labor intensive, involving large numbers of crosses and extensive screening of progeny (adults or embryos). Where particular genes of interest are "uncovered" by existing chromosomal deficiencies, hunts for mutations are feasible, but only just barely, with current technology. Where the deficiencies are not available, conventional mutant hunting, for mutations in particular genes, is not practical.

In recent years, an alternative set of approaches has been devised, under the label of "gene targeting" or, sometimes, "reverse genetics." The starting point of all such analyses is a gene or gene product of potential interest, whose biological function is unknown and to be ascertained, and the end point is the creation of heritable ("germ line transmissible") mutations, whose effects in homozygotes can then be studied. As in classical mutant hunts, these methods also involve intensive screening, but the latter is done on clones of early embryo-derived cells, totipotential embryonic stem cells (ES cells), and thus is more akin to searches for microbial mutants than to conventional mouse mutant hunts, which involve screening of litters of mice after two or three generations of breeding.

There are now numerous methodologies for gene targeting (see reviews by Capecchi, 1989, and Frohman and Martin, 1989) but the essence of the technique can be described simply: the replacement of one copy of the wild-type gene in ES cells by a mutant version engineered *in vitro*, followed by the screening of the cell population for the comparatively rare recombination event (approximately 1 in 1000 cells under optimal conditions). These cells, heterozygous for the mutation (unless the insertion is on an X chromosome in an XY cell) are then used to make chimeras and the latter are allowed to develop to term. Viable chimeras containing the heterozygous cells are then mated and the litters examined for transmission of marker genes. Those chimeras that do transmit the marker mutations are the desired type, being chimeric in the germ line, and are then crossed to one another, to produce homozygotes. Where the mutation is suspected of being a recessive lethal, the litters are dissected from the mothers of such crosses at various embryonic stages and analyzed for any developmental defects. The latter, if found, become the basis for further examination at the cellular and molecular levels. The procedure is summarized in the diagram in Figure 8.21.

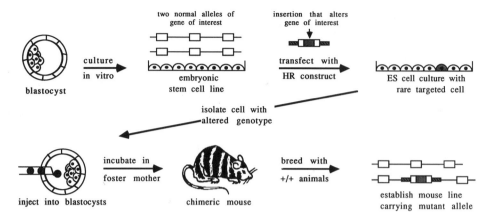

Fig. 8.21. Procedure for gene targeting. See text for description. (Reproduced from Frohman and Martin, 1989, with permission of the publisher. © Cell Press.)

The development of gene targeting techniques was made possible by two prior discoveries. The first was the finding that it is possible to obtain totipotential embryonic cells (ES cells), which can be grown and manipulated in culture without loss of totipotency, at least for moderate periods of time, and then placed in chimeras where they contribute to the development of the embryo (Evans and Kaufman, 1981). As will be apparent from the above description of the gene targeting technique, a crucial fact about ES cells is that they can contribute to the germ line in these chimeras. However, different ES cell lines, established from different strains, have different intrinsic capabilities for doing so.

The second discovery was that such cells have the capacity to undergo homologous recombination with introduced genes (Thomas et al., 1986). This permits the replacement, during *in vitro* culture of the ES cells, of wild-type genes by mutant alleles. Such genetically engineered cells, when transmitted through chimeras, have the ability to found new lines of mice, specifically altered in particular genes.

In principle, the general procedure seems to offer the promise of producing transmissible mutations in any gene of interest, providing that the gene has been, or can be, cloned. In practice, there are several difficulties that have impeded the widespread application of the method. While gene cloning and introduction of the DNA to ES cells (the latter either by microinjection or, more usually, by electroporation) are usually not rate-limiting steps, the relative rarity of homologous recombination (HR) events, ranging from approximately 10^{-2} to 10^{-5} of all insertion events, depending upon the gene employed, has proved to be one. This has required the development of fairly ingenious screening or selection techniques (using linked, selectable markers) for detecting the HR events (Capecchi, 1989). A second possible complication involves the transcriptional state of the gene to be replaced; if the gene is not transcribed in ES cells and is silent, in part at least, because of the associated chromatin state of its region, then selecting for HR events by means of a linked, expressed marker gene might prove impossible, and this would restrict searches to nonselective screens (Frohman and Martin, 1989).

The second hurdle, which slowed the application of gene targeting for a number of years, has been the fact that different ES cell lines differ in their intrinsic capacity to give rise to germ line chimeras. Certain strains have been identified, however, that reproducibly yield ES cells with these traits and with fairly high frequency.

A few successful germ line transmissions of engineered genes have now been reported. One of these is of one of the two *engrailed* homologs found in the mouse genome, *En-2* (Joyner and Martin, 1987). As in all vertebrates that have been examined, the principal region of expression of *en* homologs is in the CNS. In the mouse, the two *En* genes are expressed in overlapping domains within the CNS, with *En-2* showing a seemingly limited domain in the metencephalon (the cerebellum) of 12.5 days pc mice and *En-1* a much broader region of expression. A loss-of-function *En-2* mutation has now been targeted and transmitted through the germ line and the homozygotes analyzed. The result is a set of comparatively subtle defects in the cerebellum, including an alteration in foliation pattern and a reduction in cerebellar size (Joyner et al., 1991). It appears likely that these defects stem from lack of *En-2* function in those specific cells in the cerebellum that express *En-2* but not *En-1*. In the other cells, where both *en* homologs are expressed, *En-1* activity may make up for the deficiency of *En-2,* through partial functional redundancy.

A second instance of targeted inactivation of a gene involves *Hox-1.5,* one of the murine homologs of the *Drosophila zen* and *pb* genes. Though *pb* mutants in *Drosophila* are associated with homeotic changes, no such transformations are observed in *Hox-1.5*-deficient mice (Chisaka and Capecchi, 1991). Instead, there is a range of defects in various anteriorly derived structures and tissues: the thymus, parathyoid and thyroid glands, the heart and some of its connecting blood vessels, and the anterior skeleton, among others (Chisaka and Capecchi, 1991). Significantly, there were no defects in organs and tissues arising from much of the posterior region of the domain of *Hox-1.5* expression. These unaffected regions also show overlapping expression with other members of the same *Hox* gene subfamily, and, as in the case of *En-2,* the relative absence of effects may reflect the existence of some measure of functional redundancy for this gene.

Targeted gene inactivation is, clearly, just beginning to show its potential. After a number of years in which technical problems limited its success, it now appears to be a technique that can be applied to virtually any cloned gene. Where homologous genes are also expressed, however, interpretation of the results may not always be easy. Multiple knockouts may prove to be part of the answer, but, additionally, other strategies are certain to make their contribution to analyzing the developmental roles of various genes in the mouse. Such strategies may include further ectopic expression experiments, involving fusion genes with various kinds of hybrid promoters for genes implicated in pattern formation, and analysis of the resulting developmental effects.

SEX DETERMINATION

Identifying the Key Sex Determination Gene

Although the mouse was the first experimental mammal in developmental genetics, and the nature of sex determination one of the first developmental problems to be addressed by geneticists, the chromosomal basis of sex determination in the mouse, and in mammals generally, was not known until 1959. It was clear, of course, that the difference in sex chromosome composition between the sexes was crucial, with females possessing two X chromosomes and males an X and a Y, and it was suspected that the sex determining "signal" might, as in *Drosophila,* be a consequence of the X:A ratio. However, the demonstration of rare phenotypic XO females in both humans and mice (Jacobs and Strong, 1959; Welshons and Russell, 1959) revealed that there is an essential difference in sex determining mechanisms between fruit flies and mammals. In mammals, it is the presence or absence of the Y chromosome, not the X:A ratio, that determines sex. The subsequent discovery of XXY humans who are phenotypically male confirmed that sex is determined by the presence or absence of a Y and that maleness in mammals is the "dominant" sexual phenotype, with femaleness the "neutral" or "default" state (McCarrey and Abbott, 1979).

There is a further difference in sex determination mechanisms between the fruit fly and the mouse. In *Drosophila,* sexual phenotype is determined autonomously, cell by cell on the basis of each cell's X:A ratio. In mammals, in contrast, gonadal sexual determination is primary, with all other aspects of sexual phenotype following from gonadal state determination (reviewed in Jost et al., 1973). Initial determination of the previously "indifferent" (that is, undifferentiated) gonad to become testes, in the presence of a Y chromosome in a certain critical cell population, leads to the secretion of testosterone by the gonad. This secretion, in turn, triggers further male genital development. In the absence of testosterone, the gonad develops sex characteristics that are female. In mammals, in effect, sex determination is a function of the presence or absence of testes development. And the genetic basis of testes determination must lie in one or more genetic factors present on the Y chromosome.

The first differences in gonadal differentiation begin to become apparent at about 12.5 days pc, two days after the arrival of the first PGCs. The male gonad begins to organize itself into testis cords, consisting of solid strings of spermatogonial cells encased in mesodermal somatic cells. In the absence of determination to become testis, the gonad retains a compartmented appearance, and the PGCs become oogonial cells, clustered in groups surrounded by a matrix of mesodermal cells. The urogenital system experiences parallel differentiation. Initially both sexes possess both kinds of urogenital structures, the female Mullerian ducts and the male Wolffian ducts, and in each sex, one duct system develops while the other degenerates. In Y chromosome-bearing animals, the regression of the Mullerian ducts is caused by the production of a substance, belonging to the TGF-β family (the same family as that of the *dpp* product in *Drosophila*), known as Mullerian inhibiting substance (MIS) or anti-Müllerian hormone (AMH). If ovarian grafts are placed in fetuses that have begun to secrete MIS, there is a general regression of ovarian tissue and the elimination of premeiotic oocytes from the graft. The developing male gonad subsequently secretes testosterone, which further directs sexual development along the male pathway.

Of the three main somatic cell components in the male gonad (Sertoli cells, Leydig cells, and tunica cells), it is almost certainly the Sertoli cells that play the critical part in initiating male development. The Sertoli cells are derived from the so-called supporting cell lineage and it is these cells which come to surround the developing spermatogonia. During early male development, it is the supporting cells that become the first recognizable somatic component to differentiate, preceding the steroid-secreting Leydig cells and the tunica cells (the latter forming the outer casing of the testes). Furthermore, in studies of XY ↔ XX chimeras that developed into phenotypic males, the Sertoli cell population was found to be overwhelmingly XY, in contrast to the other two components to or to the chimeric tissues as a whole, which can have variable percentages of XY cells (Burgoyne et al., 1988). Differentiation of the Sertoli cells (in normal males and in chimeras) then prompts the Leydig cells to secrete testosterone, promoting further male development. In contrast, in either XX females or in XY↔XX chimeras that develop into females, the supporting cells become the follicle cells of the ovary instead, and female gonadal development ensues.

Thus, the central conclusion, from a variety of studies, is that the sex determining signal is produced by the Y chromosome and expressed in a key somatic cell lineage, that of the supporting cells, which, in males, gives rise to the Sertoli cells (reviewed in McLaren, 1991). This leads, in turn, to male development of the gonad and suppression of female gonadal cells and structures. The sex determining gene (or genes) encoded by the Y has been termed *TDF* (for testis determining factor) in man and *Tdy* (for testis determining Y element) in the mouse.

In mice, the timing of the *Tdy* signal is crucial; it must take place early in the supporting cells. If expression is delayed (as occurs in certain hybrids where the Y is from a more slowly developing strain), these precursor cells begin to differentiate into follicle cells, permitting the germ line cells to enter meiosis. Once this has occurred, the germ line cells start developing as oocytes, which, in turn, promote further follicle formation. The sequence of events triggered by early *Tdy* action or its failure to take place are diagramed in Figure 8.22.

In these pathways, the germ line cells are passive at first. Their initial developmental course is set by their somatic environment and not by their chromosomal composition. (However, later development of female gonads requires the presence of

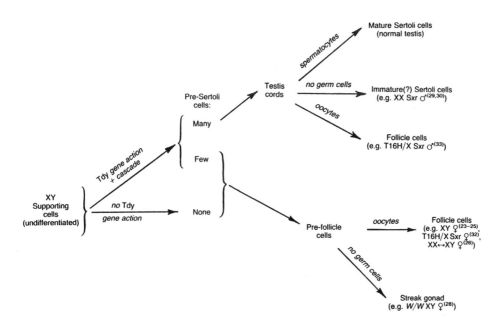

Fig. 8.22. Effect of early expression of *Tdy* in supporting cell lineage in the mouse. (Reproduced from McLaren, 1991, with permission of the publisher. © ICSU Press.)

oocytes, as indicated by the underdeveloped or "streak" gonads found in W or Sl females, which develop as such because of defective germ line development [Fig. 8.22].) Thus, in male XX↔XY chimeric fetuses, XX germ line cells can develop as prospermatogonia (though they later degenerate) while in female XX↔XY chimeras, XY germ line cells enter meiosis and will form oocytes (though these are lost before ovulation) (McLaren, 1991). The ability of XX cells to begin spermatogenic development, in a male somatic cell background, is reminiscent of the nonautonomy of XX germ line cells in the presence of an XY soma in $Drosophila$ but the developmental plasticity of XY mouse germ cells is not seen in the fruit fly (Steinmann-Zwicky et al., 1989).

The critical genetic questions about sex determination concern the identity and mode of action of Tdy. One early hypothesis proposed that Tdy encodes a major transplantation antigen, the H-Y antigen (for the Y histocompatibility antigen) which is both highly conserved and male specific throughout the vertebrates (Ohno, 1976). However, there are genetic conditions known that give both H-Y positive females and H-Y negative males, the latter possessing normal testes, showing that H-Y expression is neither determinative of nor even necessary for male development (Silvers et al., 1982).

In a sense, the hypothesis that H-Y is the Tdy substance was formulated $faute$ de $mieux:$ At the time the idea was suggested, there were no other Y-encoded genes known. Fifteen years of genetic and molecular characterization of the Y have changed that picture. The key genetic resource has been the development of a bank of X and Y chromosomes clones and the key approach has involved what is, in effect, a deletion analysis of the Y. It entails the analysis of those rare human males that have two X chromosomes but which, given their phenotype, must carry a functional TDF (with the ascertainment of those Y chromosome sequences they possess) and the complementary analysis of rare human XY females, which must, given their nonmale phenotype, lack TDF despite possession of a Y chromosome (and ascertaining those Y sequences that are missing, which must include TDF). As pointed out by Ferguson-Smith (1966), the aberrant X and Y chromosomes in these "sex-reversed" individuals could result from rare, so-called illegitimate crossovers during male meiosis that transpose TDF between the Y and the X (Fig. 8.23). With the advent of efficient DNA sequence cloning techniques, the identification of many X and Y clones as carrying particular molecular markers, and the application of such cloning to the aberrant chromosomes in the sex-reversed individuals, the way was open to identifying the particular (small) region of the Y that when missing allows XY individuals to develop as females and, which when present in an X, forces XX individuals to develop as male.

The first systematic analysis of this kind produced an exciting result and a seemingly strong candidate gene for TDF. A small, presumptive sex determining region on the Y was localized on the basis that it was present in a small set of XX males and absent from one Y-autosome translocation XY female. This 140 kb region was then screened for conserved protein-coding sequences and a gene encoding a conserved zinc-finger protein was found (Page et al., 1987). The gene was named ZFY (for zinc-finger Y) and, by the mapping procedure, seemed to be the only gene consistent with the mapping that could be TDY. Its identity as a zinc-finger protein, and hence, as a putative transcriptional regulator, seemed to strengthen the case for its being Tdy.

Various problems with the interpretation, however, soon became apparent. Most critically, exceptional ZFY^- XX males were found (Palmer et al., 1990) and the

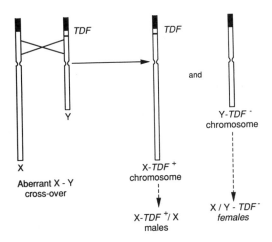

Fig. 8.23. Production of sex chromosomes with altered *TDY* content through rare, unequal crossing-over events (Adapted from Affara, 1991.)

homologous genes in mice (there are two on the mouse Y) were found to be expressed in the *germ cells* of the testes and not in the Sertoli cells, known to be the site of *Tdy* action (Koopman et al., 1990).

While the evidence thus militated conclusively against *ZFY* as the Y determining gene, the mapping data seemed to eliminate all but *ZFY* as *TDF*. The solution became apparent when it was discovered that the exceptional Y-autosome translocation XY phenotypic female in the study of Page et al. (1987) was, in fact, missing a small additional piece of the Y (which was present in the exceptional XX males) (Page et al., 1990).

This additional region, which lies near to *ZFY,* has now been analyzed and cloned, using further exceptional male XX individuals, with the resulting identification and isolation of a gene, *SRY,* that seems to fit all the criteria for the Y sex determining gene (Sinclair et al., 1990). *SRY* shows homology, over an 80 amino acid residue stretch, to several DNA binding proteins and eukaryotic transcriptional regulators and is thus, probably, a transcriptional regulator itself. Just as significantly, it is highly conserved in mammals and localizes only to the Y (with some diverged homologs present on the autosomes but not on the X), as expected for *TDF*. Furthermore, its mouse homolog, designated *Sry* (present in just one copy on the Y), has been cloned and found to be expressed only in XY embryos, beginning just before testes differentiation and specifically in the urogenital ridge, not in the germ cells (Gubbay et al., 1990). Using a sensitive technique to detect transcripts, namely reverse transcription followed by PCR-mediated amplification of the resulting DNA sequences, Koopman et al. (1990) have detected transcripts and shown the period of fetal transcription to be from 10.5 to 17.5 days pc. (*Sry* is also expressed later, in adult testes, and more abundantly than in the embryo, perhaps signifying a second role, one in testis differentiation.)

All of these facts are highly suggestive, but fall short of definitive proof that *SRY/Sry* is *TDF/Tdy.* Two strong proofs have now been obtained. The first is that three sex-reversed XY individuals, phenotypically female, have been found to have mutations within *SRY.* Two of these mutations consist of single base changes, and the third a four-base, frame-shifting deletion (Berta et al., 1990; Jager et al., 1990). The second proof is even stronger. If *Sry* is *Tdy,* then, in principle, adding it to the genome of XX embryos should convert them to full phenotypic males. The test has been performed, by injecting a 14 kb fragment containing *Sry* (and no other genes, as judged by the absence of any other open reading frames into the pronuclei of fertilized eggs and then examining the phenotype of fetuses or mice that develop to term and scored karyotypically (Koopman et al., 1991). Out of 158 embryos derived from injected eggs that developed to 14 days pc, when gonadal development can be scored, eight XX embryos (identified by their sex chromatin and absence of *Zfy* genes) were found to have integrated *Sry.* Of these, two had developed testes. Of 93 experimental embryos allowed to develop to term, one out of the three transgenic XX mice proved to a phenotypically and behaviorally normal male (though sterile, the 2X condition being incompatible with sperm development). These results show that *Sry* addition to a 2X genome *can be* sufficient to produce male development and thus constitute the best evidence to date that *Sry* is *Tdy.* Yet they also raise some questions for future investigation, in particular the failure of several *Sry* XX animals to develop as males. The most probable explanation of these cases is "position effects," the fact that the site of insertion may, in many instances, sufficiently inhibit *Sry* expression to prevent sufficient supporting cells from becoming Sertoli cells.

Yet, leaving such complications aside for the moment, it might seem that on the face of things, the genetic basis of sex determination in mammals has been solved. Further reflection, however, suggests that the curtain has come down only on act I of this story. If *SRY/Sry* encodes a transcriptional regulator, then one would like to know which genes in the supporting cell lineage are activated or repressed to produce differentiation of Sertoli cells. Is there, in fact, a cascade of gene regulatory events, comparable to that seen in *C. elegans* or *Drosophila,* or is it a direct single-step process? To put it in slightly different terms, what are the downstream genes? In comparison to *Drosophila,* the situation is comparable to that which would exist if only *Sxl* had been identified and none of the genes involved in the sex determination or dosage compensation pathways. Furthermore, while *Sry* can be sufficient to produce full male development in the mouse, the finding of several partially sex-reversed human males that carry *SRY* but lack *Zfy* (Palmer et al., 1990) indicates that sex determination, at least in man, may be a multifactorial condition. Indeed, it may also be so in mice, if the factors are to some extent interchangeable, with there being a threshold for *total* activity of these factors to give male development. This might help to explain the failure of several *Sry* transgenic XX mice to develop as males, if the "successful" ones are expressing the gene at higher than average levels in the Sertoli cells. It is, perhaps of interest, in this connection that *SRY* apparently cannot substitute for *Sry* in mice to promote male development (Koopman et al., 1991). Ideas about multifactorial effects and thresholds were proposed by Burgoyne (1989), before the discovery of *SRY.*

In addition, one would like to know what event(s) trigger *Sry* expression in the supporting cells at the crucial time. In particular, one might ask: Why does delayed *Tdy/Sry* expression *not* trigger Sertoli cell differentiation? Several genetic conditions of sex reversal in XY mice are known in which the Y chromosome appears fully normal but in which there is some "mismatch" between the Y and one or more autosomal genes. The "mismatch" seems to be one of timing, in which the initial Y "decision" does not commit the gonad to testis development sufficiently quickly, with resulting development of ovotestes (Eicher and Washburn, 1986).

Post Sex Determinative Steps

The sex determinative "decision" produced by *Tdy* is, of course, just the first step in setting the course of male sexual development, and is followed by the actual differentiation of the testes. This differentiation is dependent on the continued presence of the secreted sex hormones: If androgen production is stopped by removal of the testes in male fetuses, the further course of somatic sexual development becomes shifted toward the production of female characteristics. Similarly, mutations that interfere with either gonadal sex hormone production or hormone binding would affect the later stages of sexual development.

Although mutants deficient in hormone production have not been reported, a sex differentiation mutant, characterized by insensitivity to testosterone, has long been known. It is the X chromosomal testicular feminization (*Tfm*) mutant, and its phenotype is that of partially feminized males. (The mutation has no effect in females.) XY mice that carry *Tfm* develop testes, but these are small and underdeveloped; externally, the animals resemble females, though not perfectly (Lyon and Hawkes, 1970). The animals have normal testosterone levels and administration of either testosterone or dihydrotestosterone does not rescue the male phenotype. Evidently, the defect is not in testosterone production but in the response to testosterone. Similar genetic syndromes have also been reported in rats and man. In all cases, the specific defect appears to be a deficiency of the androgen receptor protein, which is distributed in all or most cell types in nonmutant mice (Ohno, 1976).

The most dramatic effects of *Tfm* are on secondary sexual characteristics, but there is also a block to spermatogenesis. In *Tfm*/Y individuals, spermatogonia are prevented from giving rise to sperm although primary spermatocytes are formed. To determine whether the effect is a direct one within the germ line or a secondary effect of an inappropriate gonadal soma, Lyon et al. (1975) constructed chimeras of *Tfm*/Y and +/Y embryos and then mated the mature chimeric males to *Tfm*/+ females. Some of the progeny obtained were found to be *Tfm/Tfm* embryos, showing transmission of some *Tfm*-carrying sperm from the male chimeras. The result indicates that the presence of some *Tfm*[+] in the gonadal soma can rescue spermatogenesis in some of the *Tfm* spermatogonia, and that the effect of the mutation on spermatogenesis is in the soma rather than the germ line.

Here, too, as in the sex determination decision itself, one sees the importance of the somatic cell in setting nonautonomous developmental "decisions." Yet much remains to be elucidated here also, in particular, the details of the cell biology involved in the *Tfm*[+]-mediated signal and the characterization of the downstream genes involved in both male gonadal differentiation and in the default state of female gonadal differentiation.

PROSPECTS

The history of the developmental genetics of the mouse can be divided into three phases. In the first, which lasted approximately 60 years, the primary emphasis was on studying the developmental effects of particular spontaneous or induced mutations—the discipline of classical developmental genetics. The second phase occupied much of the 1960s and 1970s, and emphasized chimera studies to explore cellular contributions to particular structures and the points of cell fate divergence. The third phase began in the 1980s, with the accelerated cloning of mouse genes, the development of new techniques for saturation mutagenesis and gene targeting, and the isolation of genes that almost certainly play major roles in pattern formation in this animal. This present phase promises to provide answers to many of the fundamental questions of mammalian development. In these new investigations, both conventional mutant analysis (and earlier-existing mutants) and clonal analysis will continue to play important roles.

Nevertheless, the magnitude of the task that lies ahead in understanding mouse, and mammalian, development should not be underestimated. There is, in particular, still much fundamental information lacking about the cellular basis of most developmental changes in the mouse. Thus, compared to *Caenorhabditis* or even *Drosophila*, the numbers of precursor cells for particular tissues, organs, and other structures, such as somites, and the times at which cells are allocated to these fates are either still unknown or determined only approximately. For instance, while the origins of the germ line can be traced to the embryonic ectoderm of the 5.5 day embryo, the events between this point and the visible appearance of PGCs at 7.25 days pc remain mysterious. Similarly, the nature of gastrulation in the primitive streak and the first cellular commitments during this ingression of cells are largely unknown.

In a further contrast to the knowledge we have of fruit fly and nematode developmental genetics, where certain hierarchies of gene action are known in some detail, if not completely—for instance, vulval development in *Caenorhabditis*, segmentation in *Drosophila*, sex determination in both organisms—comparably complete sequences of genetic "control" in the mouse are virtually largely unknown.

Both in sex determination and in axial development, however, there is beginning to be somewhat comparable information, although, to date, the information pertains primarily to the first stages of the hierarchy. Apart from the *Hox* genes, other regulator genes involved in axial development have been identified, such as the *undulated* (*un*) gene, which is encoded by a transcriptional regulator of the *Pax* gene family (identified in the mouse by virtue of its homology to the so-called "paired domain" of the *Drosophila prd* gene) (Balling et al., 1988; Chalepakis et al., 1991). Whether the *Hox* genes and the *Pax* genes are part of a regulatory hierarchy, or act independently and additively in axial development, is a question for resolution in the future.

Yet, while we continue to lack definite knowledge, and even cogent hypotheses, about most of the causal events in mouse embryonic development, the prospects are infinitely more favorable than they appeared at the start of the 1980s. It seems probable that by the end of the 1990s, our knowledge of the fundamental genetic events in mammalian development, not least of pattern formation events, will be considerably advanced relative to the present. One portent of success is the increasing number of genes known to be involved in coat color development (and other developmental events) that have been isolated and characterized at the molecular level. The cases of *W* and *Sl* have been noted in this chapter. In the next chapter, we will look at several others, where the developmental defect associated with alleles of particular genes, identified initially on the basis of their coat color phenotype, can be related to information derived from their sequences and their expression patterns.

NINE

SURFACE PATTERNS
Developmental Genetics of the Integument

What can the bristles of *Drosophila* teach us? Many things, undoubtedly. The structure of bristles, their development, their function—all are worthy of study. In this discussion, however, we shall consider them from still another aspect and shall ask two closely related questions: (1) Why is the surface of the fly not a homogeneous sheet but differentiated into regions without bristles and regions with bristles? (2) Why are the bristles formed at fixed, specific places, and not simply at random?

C. Stern (1954)

INTRODUCTION

The outer surface of the developing animal is typically one of the last places where definitive form and pattern take shape, these events occurring long after the basic body plan has been laid down and the formation of the internal tissues and organs has taken place. Taking development chronologically, therefore, it is fitting to conclude with a review of the elaboration of these surface structures. There are two additional reasons, however, for devoting a separate, and penultimate, chapter to them. First, these processes exemplify all of the key phenomena of pattern formation and thus provide a microcosm of this central biological problem, Second, this branch of developmental genetics has historical importance. Two of the subjects to be discussed here, coat color patterns in the mouse and bristle patterns in the fruit fly, were among the first to receive the attention of geneticists interested in development, in the first decades of this century. The ideas ventured to explain these patterns, ranging from concepts about gene–enzyme relationships to those about spatial patterning mechanisms in general, had a major impact on the subsequent development of ideas throughout genetics and developmental biology.

The integumentary ectodermal patterns examined in this chapter differ dramatically in their outward aspect, though all reflect the activities of the outer ectodermal cells. In the nematode, the external surface patterns are those of the cuticle, the exoskeleton of the animal. The components of the cuticle are secreted by the underlying hypodermal cells and consist of a large set of collagen-like molecules, which assemble in characteristic fashion. In consequence, the development of cuticular pattern in this animal reflects both the regulated synthetic patterns of the hypodermal cells and the dynamics of molecular self-assembly of the secreted molecules.

In the fruit fly, in contrast, the features that comprise the external surface patterns in the imago are primarily bristles and hairs, though additional features, such as nonbristle sensilla and wing veins, are also present in characteristic arrays. (In the larva, as we have seen, the surface features that make up the pattern are the denticle hook bands of the ventral segmental regions, hairs, and various sensilla.) The questions of surface patterning in *Drosophila* are not primarily ones of self-assembly (though self-assembly phenomena are relevant in the formation of individual hairs and bristles) but of placement: What factors govern the spatial relationships between hairs and bristles? What genetic and molecular events determine the time of appearance of these structures?

Finally, in the mouse, the outer ectodermal patterns that are of primary interest are those of coat color pattern and the questions concern the underlying events that govern the production of these pigmentary patterns and their regional variations. This subject was the center of attention for the mouse "fancy" of the 1890s and thus antedates the rediscovery of Mendel's work at the start of the 20th century. In the last few years, more and more of the genes whose products affect coat color, or coat color pattern, have been cloned; in consequence, there is an increasingly precise description, in both cellular and molecular terms, of coat color pattern development. In chapter 8, for instance, we saw how the analysis of two of the "classic" genes of coat color, *W* and *Sl,* has revealed the involvement of their gene products with an essential cell signal transduction pathway.

Despite the biological differences in the surface patterns of these three animals, there are certain underlying themes that connect them. These are the roles of cell lineage in setting certain capacities (particularly apparent in the nematode and the fruit fly), the involvement of short-range and long-range cellular interactions in setting the patterns (the fruit fly and the mouse), the general phenomenon of the regulation of gene expression in response to inter- and intracellular signals (all three systems), and the existence of genetically based functional redundancy. We will begin, as before, with *Caenorhabditis* and then proceed to discuss the work on *Drosophila* and *Mus* integumentary patterns.

THE *CAENORHABDITIS* CUTICLE: A COLLAGENOUS CONCOCTION

The intricate structure of the adult cuticle of *Caenorhabditis* consists of three major layers, as shown in Figure 9.1. The basal layer, immediately adjacent to the hypodermis and the last to be secreted during the L4-adult transition, consists of two sublayers of fibers which run in opposite, helical directions around the body of the animal. Above these basal layers is the median or "clear" layer, which is fluid filled and contains many struts that support the topmost layer, the cortical layer. The latter is traversed longitudinally on both sides of the animal by the long triridged structures termed the alae, which are secreted by the underlying lateral seam cells of the hypodermis. Circumferentially, the cuticle exhibits a sequence of rings, or annuli, along the main axis.

In contrast to certain other nematodes, however, which produce essentially identical (and simpler) cuticles at each molt, *C. elegans* produces visibly different cuticles

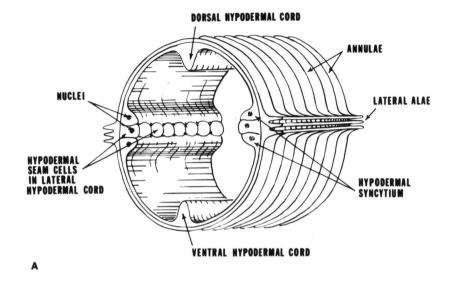

A

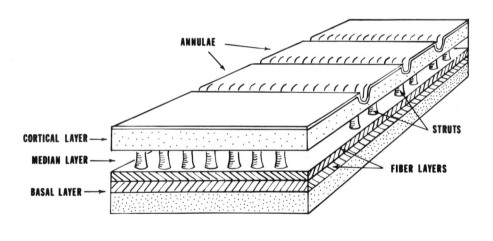

B

Fig. 9.1. Cuticle structure in *C. elegans*. **A:** A transverse section illustrating the cuticle and underlying hypodermis of the adult worm. **B:** Schematic of the ultrastructure of the adult nematode cuticle. (Reproduced from Cox et al., 1980, with permission of the publisher.)

at several developmental stages, thus showing stage specificity. One such difference is the median layer, which only appears in the adult cuticle. A second involves the alae of the adult cuticle; though the L1 and dauer larva produce alae, they lack the typical three ridges seen in the adult form (Cox et al., 1981b). Furthermore, while the cortical layer appears much the same at all stages, the basal layers clearly differ among several. Of the stages that have been most thoroughly examined, distinctively different cuticles are found in the L1 (produced by the embryo shortly before hatching), the dauer, the L4, and the adult (Cox et al., 1981b). The L2 and L3 cuticles have been studied less extensively (because of the difficulties of preparing synchronous batches of L2 and L3 animals), but appear fairly similar to the L4 cuticle. The L1 cuticle is the thinnest and the dauer is the thickest (with the adult cuticle the next most so). Other differences involve specific ultrastructural features, such as the alae (Cox et al., 1981a). Indeed, it was the visible differences in cuticular structure, which typify each stage, that first revealed the existence of heterochronic mutants in C. elegans.

The differences in ultrastructure reflect, ultimately, the different patterns of synthesis by the hypodermis of the component molecules of the cuticle (Cox et al., 1981b; Cox and Hirsh, 1985). Nevertheless, all stages have cuticles that consist predominantly of collagen-like molecules, as first indicated by the collagen-like amino acid composition of the cuticles (being high in glycine, proline, and hydroxyproline) and their susceptibility to collagenases (Cox et al., 1981b). In contrast to vertebrate collagens, those of C. elegans are small, though cross-linked into products of higher molecular weight. As determined from in vitro translation experiments, the polypeptides are one-third to one-half that of their vertebrate homologs, ranging from 30 to 50 kDa, and are encoded by small mRNAs, about 1.2 kb in length (Politz and Edgar, 1984). The genes themselves are also small, being about 3 kb in length, and lack the numerous introns seen in vertebrate collagen genes (Cox et al., 1984).

Estimates of the number of collagen genes, from cloning experiments, suggest that, altogether, the genome contains from 50 to 150, considerably more than the 10 or so large collagen genes of vertebrates, with the majority of these genes not in clustered arrangements (Cox et al., 1984). Though all are members of the same gene family, they are, for the most part, not identical, as shown by both renaturation studies (Cox et al., 1984; Cox and Hirsh, 1985) and direct sequence analysis (Kramer et al., 1988; von Mende et al., 1988).

This large genetic repertoire provides the basis for the stage-specific alterations in visible structure seen in the different larval structures, as first inferred from the changing patterns of bands seen on one-dimensional protein gels of the isolated cuticles (Cox et al., 1981b). Studies with a battery of different cloned collagen genes have confirmed that there are diverse and, in some instances, large transcriptional changes in different members of the family (Cox and Hirsh, 1985). In addition, there seems to be an increase in the number of collagen genes expressed, as the animal progresses from the L1 to the adult, paralleling the increase in ultrastructural complexity (Cox and Hirsh, 1985). These general, stage-specific differences in cuticle composition provide one element in the development of cuticular patterns.

As discussed in chapter 6, the heterochronic mutants, such as lin-4 and lin-14, provide the necessary genetic "handle" for investigating these major regulatory shifts, which take place within the hypodermal cell lineages during the lineages. Thus, for lin-14 mutants, the occurrence of either accelerated or delayed formation of the adult alae correlates with the acceleration or retardation of the lineage behavior of the lateral seam cells. These seam cells, which normally cease division and secrete alae at the L4-adult molt, will, in "precocious" (recessive, loss-of-function) mutants, acquire the late L4 seam cell characteristics at earlier stages, while the "retarded" (dominant, gain-of-function) mutants, behave like younger seam cells at the L4 molt (Ambros and Horvitz, 1984). Understanding the stage-specific aspects of cuticle formation, will, eventually, require understanding how the heterochronic genes determine the partic- ular stages of the hypodermal cell lineages and how these lineage properties, in turn, determine the regulation of the collagen genes.

There is a second aspect to the developmental genetics of the cuticle, and it concerns the genes that either directly encode or modify the individual components of the cuticle. The genetic analysis of these genes began long before it was known that the cuticle was composed of collagen-like genes, with the discovery of a set of mutants, the Roller or Rol mutants, distinguished by a behavioral abnormality rather than an immediately apparent structural defect (Brenner, 1974). Instead of the nearly linear path segments that wild-type worms trace while moving through the bacterial lawn, Rol mutants transcribe circular or near circular craters in the lawn. This feature makes the isolation of recessive Rol mutants easy: One simply inspects the F_2 generation of a mutagenized population for those individuals which make circular craters. (Dominant Rol mutants have also been found and these reveal themselves in the F_1.)

The Rol phenotype can be understood in terms of the normal mechanism of movement. Wild-type worms move on their sides in a sinusoidal motion in the dorsoventral (d-v) plane; once a worm has chosen a side, it stays on that side for the rest of its life, unless flipped to the other side by external forces. Traction against the medium is provided by the alae, which, as noted above, possess a three-ridged tread (Fig. 9.1). Body movement results from rhythmic contractions of the longitudinal muscles working against the cuticle, which acts as a flexible external skeleton.

Rol mutants differ from wild-type both in the shape of their paths and in their slow rotation about the long axis of the body as they move. The relationship of this movement to cuticle structure was first explored by Higgins and Hirsh (1977). In isolating Rol mutants, they found two distinguishable types: left rollers (LR), who make counterclockwise circles, and right rollers (RR), who make clockwise circles. For both classes of mutants, their cuticles and inner longitudinal organs show helical twists of the same handedness. The LR mutants display left-handed helically twisted alae and neural cords instead of linear alae and neural cords; RR mutants show right-handed helices in their alae and internal organs. The helical path of the alae explains the rolling motion of the animal as it moves forward: Gliding along the twisted alae automatically produces a rotation around the long axis of the body.

In a more extensive collection and classification of Rol mutants, 88 in all, Cox et al. (1980) identified five distinct classes, consisting of mutants in 14 distinct and nonclustered genes. LR mutants were all recessive and mapped to four genes. Left dumpy rollers (LDR), also recessive, were produced by mutation in any of six genes; the mutants were both Dpy and Rol in phenotype (but some alleles produced just the Dpy phenotype and others just the Rol phenotype). RR mutants, recessive mutations, all mapped to a single gene (*rol-6*). "Left squat" (LS) mutants, produced by mutation in a single gene (*sqt-3*), were found to be semidominant, displaying left rolling in heterozygotes and a mild Dpy phenotype, when homozygous, in juveniles. Finally, the "right squat" (RS) mutants, produced by mutation in either of two genes (*sqt-1* and *sqt-2*), exhibit right rolling in heterozygotes and a mild Dpy phenotype as homozygotes.

The mutants are all zygotic rather than maternal in character and the behavioral phenotype is expressed late in development, not in the early larval stages. For most mutants in the LR, LDR, and LS classes, it is only the adult stage that is affected, while for RR and RS mutants, expression is first seen at the L3 stage (Cox et al., 1980). The only major developmental events that occur late and which therefore might be the source of the defects are the completion of the gonad and the final larval molts. However, the gonad is unlikely to be the source of an imposed helicity since it effectively floats within the body. Rather, the cuticle itself must be the site of the initial developmental defects, leading directly to the abnormality of movement and imposing helicity on the internal organs. If the cuticle is produced with a helical twist in it, then the muscles and neural cords, which appear to be anchored to the cuticle, will assume a corresponding helical twist.

All of the mutant classes show visible abnormalities in one or more aspects of cuticle structure. Thus, in the LR and RR classes, the primary visible alteration is in the helicity of their alae, with little visible disruption of the internal cuticular architecture. The LDR, LS, and RS mutants, however, show various abnormalities of cuticle ultrastructure, for instance in the struts and the alae, and various degrees of altered helicity.

A probable site for the origin of the defects, in at least the LR and RR mutants, is the basal layers, which display counterhelical paths in the wild-type cuticle. A change in the orientation of one layer might produce an overall helical twist in one direction; a change in the construction of the second basal layer could produce the opposite change in helical twist. Aberrant placement of the topmost basal layer, to which the struts attach in the adult cuticle, might then transmit the imposed helicity "upward" resulting in the observable twist of the alae (Cox et al., 1980). A similar twisting could also be produced by an alteration in the deformability of either of the basal layers. Such an alteration could result from altered thickness or in the degree of cross-linking or from the absence or alteration of a key cuticular constituent.

The late development of the Rol phenotype in all mutants presumably reflects the differences, compositional and ultrastructural, between the early larval cuticles and those developing later. One obvious difference, as mentioned above, is the presence of alae on the adult cuticle and its absence of alae on the adult cuticle and its absence in the cuticles of the L2, L3, and L4 stages (Cox et al., 1981b). The absence of alae in these larval forms would help to explain the prevalence of the Rol phenotypes in adults, since without alae, the animals would lack their helical glide tracks. However, the display of the Rol phenotype in the L3 and L4 forms in unusual mutants of the LR and LDR classes and in most of the RR and RS mutants indicates that other cuticular structural features can also be relevant. (Although the L1 wild-type possesses alae, albeit not of the three-ridge structure, they do not show the Rol phenotype, presumably because of compositional differences between the L1 and adult.)

The central genetic question about the mutants is how different point mutations in many genes lead to such dramatic and global changes in cuticular structure. Mutations which completely abolish synthesis of particular constituents of the cuticle are one conceivable category. However, another possibility is that comparatively subtle alterations in the structure of individual components, created by point mutations, are sufficient to alter assembly of the cuticle, with consequent global effects on its pattern.

Several genetic observations support this notion of complex interactive effects in assembly (Cox et al., 1980; Kusch and Edgar, 1986). These can be illustrated by the genetics of the *sqt-1* and *sqt-3* loci. Thus, *sqt-1* can mutate to give four different phenotypes, namely LR, RR, Dpy, and Lon, while *sqt-3* can mutate to give three, specifically Dpy, severe Dpy, and LR. That the multiplicity of phenotypes does not reflect, in some manner, different quantitative amounts of product is indicated by the fact that putative amorphic *sqt-1* alleles (identified as such by the similarity of their complementation behavior to deficiencies of the locus) exhibit a wild-type phenotype. Therefore, simple elimination of the gene product has no phenotypic effect, indicating that the various mutant phenotypes reflect altered states of the gene product.

Furthermore, while most alleles at all three *sqt* loci are dominant, a fact suggestive of neomorphic products (since it cannot reflect reduced gene dosage, at least for *sqt-1*), a few are recessive unless present in certain *dpy* gene backgrounds. Thus, *sqt-1* (*sc13*) is a recessive LR, unless in a homozygous *dpy-12* background, where heterozygosity is now sufficient to produce the LR phenotype. Such genetic behavior, involving intergenic noncomplementation, has been termed "cryptic dominance" (Kusch and Edgar, 1986). The null alleles for *sqt-1* do not show such cryptic dominance, as indeed is not surprising if it reflects interactions between different gene products. Other unusual interactions involve noncomplementation between particular *recessive* alleles of different genes; for instance, both *sqt-1* (*sc13*) and *rol-8* (*sc15*) individually show LR behavior only when homozygous, yet the double heterozygote shows a distinct LR phenotype (Kusch and Edgar, 1986).

Such genetic effects are explicable if the genes that give the Rol and Sqt phenotypes in fact encode cuticle components that interact in complex ways during cuticle assembly (Kusch and Edgar, 1986). If the members of this gene family are indeed distinctive, as shown by the molecular characterization, then their interactions during cuticle assembly at the hypodermal cell surface (from which they are secreted) might also be distinctive, as is indicated by the genetic observations. On the other hand, the fact that certain genes, such as *sqt-1*, are dispensable (judging from the wild-type phenotype of the homozygous nulls) indicates that some cuticle gene products can replace others. Nevertheless, while the simplest explanation of the results, especially for genes like *sqt-1*, which can mutate to give several markedly different phenotypes, is that they directly encode collagen components of the cuticle, it is possible that some, or many, encode enzymes that modify particular cuticle products.

This issue can, in principle, be settled by cloning of the genes; the genes characterized thus far have been found to encode collagens rather than collagen-modifying activities. Using Tc1 insertion mutagenesis, Kramer et al. (1988) sought and found, through a multistep procedure, a Tc1 insertion in *sqt-1*, within the putative signal sequence (required for secretion). By sequencing, the wild-type gene was found to encode a 32 kDa collagen and, in common with many of the collagen genes, a 1.2 kb transcript; the homozygote, however, makes no transcript. In contrast, three mutants whose homozygotes do show aberrancies are associated with single base pair changes within the gene.

The results support the notion of complex multicomponent assembly interactions, suggested by the genetic findings. Furthermore, the fact that the null mutants of *sqt-1* are wild-type is explicable if the wild-type gene can be effectively substituted by another collagen gene. The paradox that remains, however, as noted by Kramer et al. (1988), concerns the fact that *sqt-1* is the only gene in the family that mutates to give such a diverse array of phenotypes. If its gene product can be substituted by others, then one might have expected the substitutable gene(s) to be able to mutate in similar ways.

Two other genes, associated with visible abnormalities, have been cloned and shown to encode collagens. One of these is *rol-6,* the gene associated with the RR phenotype and which interacts with *sqt-1* (see above) (R. French and J. Kramer, cited in Kramer et al. [1988]). The third is *dpy-13,* whose mutants give a classical Dpy phenotype (von Mende et al., 1988). Kramer et al. (1988) speculate that the Dpy phenotype results from greater extensibility of the cuticle, in response to the internal hydrostatic pressure of the animal. Conversely, the Lon phenotype may reflect reduced radial extensibility. As more of the genes that produce the various cuticular phenotypes are cloned and sequenced, in particular the Dpy and Lon phenotypes, it will be interesting to see how many either directly encode cuticle components or enzymes that modify them or, alternatively, regulatory switches that affect the synthesis of either individual collagen molecules or small sets of them. Coupled with the studies on how heterochronic genes affect the hypodermal lineages, these analyses should transform this subject into one of the best-understood examples of morphogenetic processes in all of biology.

DROSOPHILA: CELL–CELL INTERACTIONS AND THE PATTERNING OF BRISTLES

The integuments of both the larva and adult of *Drosophila* are richly patterned with a variety of different cuticular structures, each in a characteristic array. These kinds of processes, however, can be broadly divided into sensilla, those elements that are innervated, and those that lack innervation, such as the denticle belt hooks of the larva and the single-cell-derived hairs found in both larva and adult. Of the two kinds of structures, the sensillae are inherently more complicated, each being a multicellular unit or "organule"; collectively, they are designated SOs (for "sensory organs"). We will concentrate on one class of SOs, the bristles, the first to be investigated and the best understood of the various SOs.

Bristle Developmental Biology

The bristles, or chaetae, of the adult are conventionally divided into the macrochaetes, those that are particularly large, and the microchaetes, which are much more abundant and widely distributed. Despite the size difference, the microchaetes are not, in fact, a homogeneous grouping but consist of several distinct subtypes. The majority of them are, like the macrochaetes, mechanoreceptors, while a minority, having a hollow shaft, are chemoreceptors. (The latter are also sometimes referred to as "recurved" bristles.)

The typical mechanoreceptor bristle consists of four cells: the trichogen cell (which secretes the bristle shaft), the tormogen (which secretes the socket), a neuron (which makes contact via its dendrite with the bristle shaft and connects via its axon with the

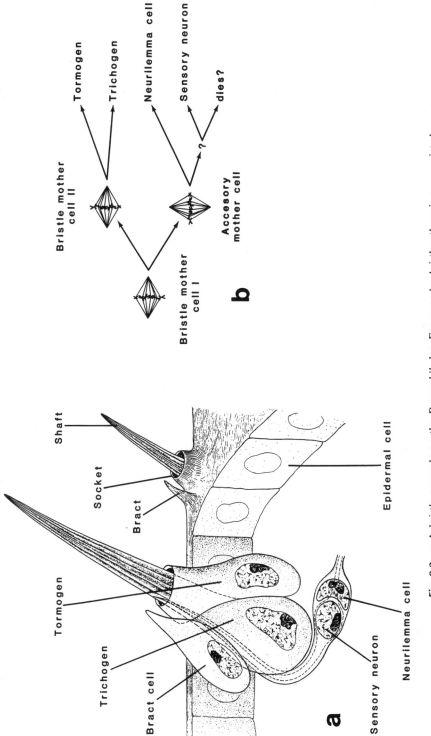

Fig. 9.2. **a:** A bristle organule on the *Drosophila* leg. For some leg bristles, there is an associated structure termed the bract, as diagramed here. **b:** Typical cell lineage relationships in the bristle organule. (Reproduced from Held and Bryant, 1984, with permission of the publisher.)

nervous system), and a neural accessory cell, the neurilemma (which secretes the sheath for the axon of the neuron) (Fig. 9.2a). The traditional view of the relationship of these cells to one another is that they always comprise a clone, descended from an individual mother cell, termed either a bristle mother cell (BMC) or, more generally, a sensory mother cell (SMC). One of the daughters of the SMC divides to produce the trichogen and tormogen, while its sister gives rise to the neuron and neurilemma (Fig. 9.2b). In *Drosophila,* the SMCs experience the first of these two divisions after puparium formation, the differentiation of the full bristle apparatus taking place late in the development of the pupa.

While most bristle organules in *Drosophila* show this pattern of clonal relationships, not all do (Hartenstein and Posakony, 1989). The tracing of cell relationships, by BUdR pulse-labeling during S phases, indicates that certain mechanoreceptor bristles of the wing margin are not always clonal in composition. For some of these bristles, trichogen and tormogen cells are not always sisters, as shown by the presence of label in one but not in the other; furthermore, in some cases, apparent sister trichogen and tormogen cells can be donated to adjacent bristles (Hartenstein and Posakony, 1989).

There is some older genetic evidence that the SMCs that give rise to bristles are singled out for their fates long before they undergo their differentiation divisions. Using mitotic recombination and marked clones, Garcia-Bellido and Merriam (1971b) showed that only up to 40 hr before puparium formation (BPF) could both adjacent bristles and hairs be marked; after this point in development, X-irradiation produces only marked bristles or hairs in single clones. Yet, as noted above, the actual cell divisions that give rise to the microchaetes do not begin until about the time of puparium formation, and these divisions take place in two waves, with some SMCs dividing just after puparium formation (APF) and another large group dividing between 9 and 20 hr APF. The dividing SMCs are within islands of dividing cells, most of which give rise to epidermal cells. These patterns of SMC cell division have been detected in several investigations, but the clearest demonstration involves immunostaining with a monoclonal antibody, MAb 22C10, which stains cells of both neural character and the accessory nonneural cells, associated with bristles; the staining is cytoplasmic and the antigen detected may be a cytoskeletal component (Hartenstein and Posakony, 1989).

If the SMCs that give rise to bristles are "committed" to their fates by about 40 hr BPF, but do not commence their divisions until 0–20 hr APF, there are only two possible explanations for their behavior during this interval. The first is that they might proliferate as SMCs, each committed to forming a bristle; however, this has not been seen for nonbristle SMCs, it doesn't accord with the fact that bristles arise separated from other bristles, and, finally, there is some direct evidence (see below) against the formation of SMC clusters. The second, and more probable explanation is that the SMCs undergo a long period of mitotic quiescence, from the time of their singling out, ranging from about 40 to 60 hr. During this period, the majority of the cells in the imaginal discs are actively proliferating, and during the period of SMC mitotic quiescence, their dividing neighbors may experience four to five rounds of cell division. Since the SMCs, when they begin to divide, are embedded in clusters of similarly dividing cells, it would seem probable that one of the first steps in SMC commitment takes place in islands or strips of cells, which become mitotically inactive. The definitive SMCs are then "chosen" from within these populations of mitotically quiescent cells.

To verify or refute this hypothesis, one would like to have an independent method of detecting, and following, SMCs. Such a method exists and utilizes an "enhancer trap" strain, which detects SMC states specifically. The principle of the "enhancer trap" is simultaneously simple and ingenious (O'Kane and Gehring, 1987; for a review see Bellen et al., 1990). One fuses a *lacZ* reporter gene to a weak promoter and inserts the construct into a P element. Germ line transformation is then carried out with this engineered P element and survivors are then bred to produce independent single-site insertion lines, which are then scored during development for their spatial patterns of β-galactosidase expression (detected with Xgal, a blue staining reagent). Because the promoter is weak, only those elements that have inserted near an enhancer will be able to express the reporter gene, and with the conventional staining produce blue cells. The majority of these strains show a particular pattern of tissue- or cell-specific expression. One line, A101, specifically detects pre-SMC and SMC cells (Huang et al., 1991).

When the pattern of *lacZ* expression is scored in the wing discs of A101, it is found that within that part of the disc that gives rise to the notum, single or sometimes pairs of cells expressing β-galactosidase appear in the locations of the future large bristles, the macrochaetes, beginning around 30 hr BPF (and, thus, not long after the period of commitment detected in the mitotic recombination experiments). Where two cells show staining, one "wins" subsequently, the other losing its ability to express β-galactosidase. Evidently, the enhancer trap detects both the SMC state and a short-lived pre-SMC state (Huang et al., 1991). Furthermore, these cells continue to express *lacZ* throughout the rest of the pupal period but do not divide until after puparium formation, the sequence of subsequent divisions mirroring that of the pattern of onset of staining (Huang et al., 1991). These observations confirm the inference, from the earlier work, that the process of commitment to bristle organule formation is accompanied by a long period of mitotic quiescence of the SMC and its immediate cell neighbors.

Patterns and Prepatterns: Some Hypotheses

When one turns from the biology of the individual bristle to the overall spatial arrays of these organules on the surface of the fly, the immediately apparent features of these patterns are their diversity and regional distinctiveness. Of the approximately 5000 bristles present on the adult's cuticular surface, each is present within an array characteristic of its neighborhood. Thus, on the surface of the thorax, the microchaetes are arranged in fairly regular rows, while on the dorsal surface of the abdomen, the bristles are spaced quasi-randomly but with approximately uniform spacing ("isotropic spacing"). On the basitarsus of the second leg (SL), the majority of bristles are arranged in highly regular rows (Fig. 9.3). Yet another pattern is that of the dense rows of bristles (including both chemosensory and mechanosensory bristle organules) along the anterior margin of the wing (Fig. 9.4A). However, the most unusual, and visually the most distinctive, of the bristle patterns is the fixed but nonregular arrangement of the macrochaetes of the mesonotum, shown in Figure 2.7 and schematically in Figure 9.4B, which has been termed a "constellation" pattern (Held, 1991).

The numbers of bristles are similarly a characteristic of the region, with each region showing a characteristic variance. At one extreme, there are the macrochaetes of the

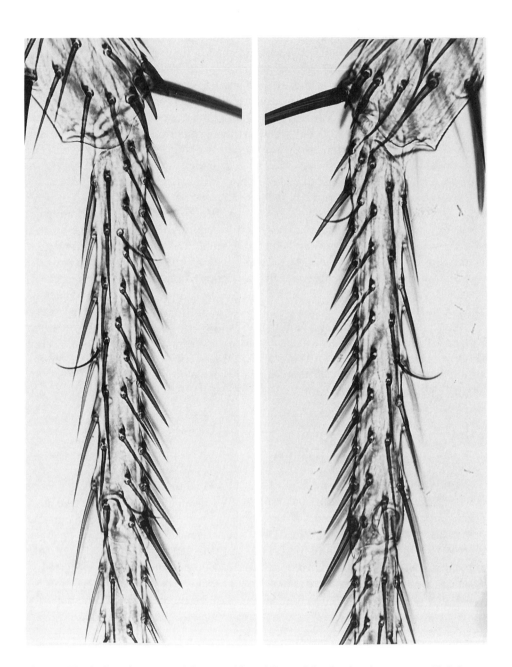

Fig. 9.3. The basitarsal segment of the second leg of *Drosophila,* showing the patterns of bristle rows. (Reproduced from Held, 1990a, with permission of the publisher.)

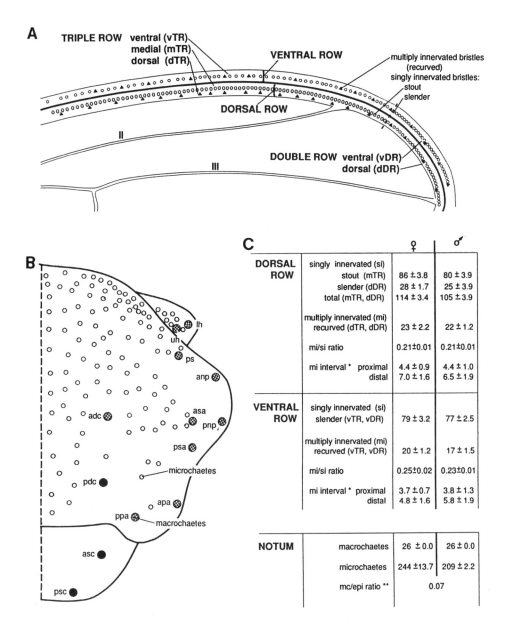

A TRIPLE ROW ventral (vTR)
medial (mTR)
dorsal (dTR)

VENTRAL ROW

multiply innervated bristles
(recurved)
singly innervated bristles:
stout
slender

DORSAL ROW

II

III

DOUBLE ROW ventral (vDR)
dorsal (dDR)

B

lh
uh
ps
anp
asa
adc
pnp
psa
microchaetes
pdc
apa
ppa
macrochaetes
asc
psc

C

		♀	♂
DORSAL ROW	singly innervated (si)		
	stout (mTR)	86 ± 3.8	80 ± 3.9
	slender (dDR)	28 ± 1.7	25 ± 3.9
	total (mTR, dDR)	114 ± 3.4	105 ± 3.9
	multiply innervated (mi) recurved (dTR, dDR)	23 ± 2.2	22 ± 1.2
	mi/si ratio	0.21±0.01	0.21±0.01
	mi interval * proximal	4.4 ± 0.9	4.4 ± 1.0
	distal	7.0 ± 1.6	6.5 ± 1.9
VENTRAL ROW	singly innervated (si) slender (vTR, vDR)	79 ± 3.2	77 ± 2.5
	multiply innervated (mi) recurved (vTR, vDR)	20 ± 1.2	17 ± 1.5
	mi/si ratio	0.25±0.02	0.23±0.01
	mi interval * proximal	3.7 ± 0.7	3.8 ± 1.3
	distal	4.8 ± 1.6	5.8 ± 1.9
NOTUM	macrochaetes	26 ± 0.0	26 ± 0.0
	microchaetes	244 ±13.7	209 ± 2.2
	mc/epi ratio **	0.07	

Fig. 9.4. Some bristle patterns in *Drosophila*. **A:** Bristle row patterns on the anterior edge of the wing. The chemosensory, or recurved, bristles (represented by triangles) are multiply innervated, in contrast to the mechanosensory bristles (circles and ovals), which are innervated by single neurons. **B:** Bristle patterns on the notum, with the positions of the macrochaetes indicated by the large circles, microchaetes by the smaller circles. Differential shading within the macrochaetes signifies different times of appearance. (See legend to Fig. 9.5 for key to notal macrochaete bristles.) **C:** Table of bristle numbers, with standard deviations, for the different regions. (Reproduced from Hartenstein and Posakony, 1989, with permission of the Company of Biologists, Ltd.)

mesonotum, which are invariant in position and numbers, while in contrast, the microchaetes of the anterior wing margin show some variability in numbers, for both the mechanosensory and chemosensory bristles (Fig. 9.4C).

These differences in both spatial array and in the variances associated with them suggest that there may not be a single, unitary mechanism of bristle pattern formation, but perhaps several contributory processes. Indeed, several analyses suggest that even within an ostensibly single bristle field, several mechanisms may be at work for the different kinds of bristles. This seems particularly clear for the basitarsus (the most proximal segment) of the SL (Fig. 9.3). As discussed in chapter 2, most models of pattern formation tend to accentuate either "global" field properties or sequences of local interactions. The SL basitarsus seems to have, rather nondogmatically, employed mechanisms of both classes.

The SL basitarsal leg segment possesses a characteristic arrangment of bristles of two kinds: eight proximodistal (p-d) rows of, approximately 75 in total, "bracted" bristles (the bract being a thick, darkened cuticular process) (Fig. 9.2) and five bractless bristles, the latter each at a particular location outside of the rows. (The bracted bristles are all mechanosensory elements while the bractless bristles on this segment are chemosensory.) To clarify the mechanisms that produce the characteristic bristle patterns, investigators have used various mutant and environmental conditions that alter either the number of cells or cell size during development and then scored the final spacing of the bristles on the SL basitarsus. The results show that two kinds of mechanisms account for the respective patterns of bracted and bractless bristles. For the bracted bristles, situated within the rows, the spacing within each row is a function of the number of cells within that row rather than the absolute distance (Tokunaga and Gerhart, 1976; Held, 1979; Held and Bryant, 1984). Thus, bristle spacing within the rows may involve some kind of cell-counting mechanism. On the other hand, the relative positions of the five bractless bristles are maintained irrespectively of the absolute numbers of cells between them; their spacing, therefore, seems to involve some general field property (Held and Bryant, 1984). Thus, the two sets of bristles appear to be assigned their spacing arrangements by two different mechanisms that operate within the same group of epidermal cells.

The mechanism of spacing of the bristles within the rows of the SL basitarsus has been investigated further. One hypothetical possibility is that there is a wave of specification of some kind proceeding along the p-d axis of the segment, much as there is a wave of specification of ommatidia along the anterposterior (a-p) axis in the developing eye. In this model, the wave would count intervals in terms of cell numbers, assigning a bristle mother cell at the end point of each interval. If such is the case, then treatments which interfere with bristle development, such as X-rays following pupariation, might be expected to reveal a sequence of developing insensitivity along each bristle row during the sensitive period. In contrast to the eye, where such a sequence along the a-p axis is detected, following the morphogenetic furrow, when defects are induced by heat pulses to temperature-sensitive mutants during the third larval instar (Cagan and Ready, 1989), no comparable p-d wave is found for the row bristles of the SL basitarsus (Held and Bryant, 1984; Held, 1990a). The spacing mechanism appears to involve, instead, an approximate specification within inhibitory fields, and then a process of "fine-tuning." These fine-tuning processes evidently include adjusted cell movements within each row to produce the correct spacing. Thus, heat shocks delivered late in development (22–24 hr APF) produce bristle rows in which the spacing is uneven (Held, 1990a). The simplest explanation is that, in normal development, the bristle cells adjust their spacing with respect to each other, perhaps by some mutual repulsive process, and that the process is sensitive to heat shock at this point in development (Held, 1990a,b).

Despite the evident complexity and diversity of bristle patterns, and of the processes that contribute to these patterns, there are, however, certain common factors to be kept in mind. One important commonality is that the number of epidermal cells that are *potentially* capable of generating bristles is much greater than the number of cells which do so (Lawrence, 1973). This has been shown, in particular, by various wounding and regeneration experiments; following elimination of bristle SMCs, new bristles are produced by cells that normally do not form them. Furthermore, manipulation of the genetic background of certain mutants that affect bristle patterns can substantially alter the number of bristles formed. Evidently, the "decision" of a particular epidermal mother cell whether or not to form a bristle involves some form of lateral inhibition, involving interactions with that cell's neighbors (reviewed in Simpson, 1990). The situation is analogous to the singling out of a fraction of the neuroectodermal cells to become neuroblasts, each such cell then inhibiting its neighbors from following suit, with the nonneuroblasts giving rise to epidermis. (The two processes, bristle and neuroblast patterning, also involve some of the same gene products, as will be discussed below.)

Such lateral inhibition may be sufficient to account for the simplest pattern, that of quasi-random but isotropic spacing, as first suggested by V. Wigglesworth in 1940. He suggested that bristle formation is induced by a specific morphogen, a "chaetogen," produced initially by every epidermal cell. Induction first occurs randomly and is immediately followed either by reduction in chaetogen in the immediate vicinity or by production of an inhibitor by the induced cell. The consequence of induction of a bristle, therefore, is an inhibition of further bristle formation in its immediate vicinity. The pattern will indeed show isotropic spacing, with some "noise" in the particular positionings of bristles with respect to one another.

Such a relatively simple mechanism, however, cannot in itself account for the pattern at the other extreme, the "constellation" pattern of nonequidistant but fixed spacing between particular bristles seen in the array of macrochaetes in the mesothorax (Fig. 9.4B). Within each half mesothorax, there is a characteristic placement of 11 macrochaetes on the main portion (the scutum) and of two on the posterior (the scutellum). The positioning of these bristles demands a more precise mechanism than the Wigglesworth model can provide, and, it would seem, necessarily, some form of global spatial patterning, at least in part.

One such explanation was the "prepattern" hypothesis, proposed by Stern (1954), who suggested that there is an initial underlying "prepattern" of chaetogen and that the concentration high points of the prepattern prefigure the positions of the bristles (Fig. 2.6b). Although Stern's hypothesis was a formative concept for the field of pattern formation as a whole, it has proved less useful as an explanation, at least in its original guise, than it first seemed to be. First, it demands a precise shaping of the prepattern itself, which only pushes the question of origins of pattern one step back without solving it. Second, if prepatterns exist there must be genes responsible for the synthesis of the morphogen that comprises the prepattern, whose viable mutants (should such exist) would show globally transformed patterns. However, no convincing candidate genes for such morphogens have been identified, despite many searches.

Nevertheless, Stern's hypothesis has served as the basis for later models. An example is that of Richelle and Ghysen (1979), who proposed that there is a probability distribution within each bristle-producing field for chaetogen synthesis by each epidermal cell, this probability distribution being the prepattern. Induction of bristle formation takes place when the local chaetogen concentration exceeds a certain threshold value and each induced cell then secretes a diffusible inhibitor, whose action is to prevent neighboring cells from making bristles. With the appropriate adjustment of the various parameters, the model can account for both the distinctive placement of the macrochaetes and the production of bristle rows. A strength of the model is that the overall distribution is a summed function of all the bristle-producing regions rather than a global property and becomes subject to the local factors that influence bristle formation. However, the model does not address a key question: the nature of the probability distribution, which, in this case, constitutes the prepattern.

In effect, both modeling and various forms of experimental manipulation of the developing primordia can take one only so far in exploring the underlying phenomena. In recent years, however, a major breakthrough has taken place, and it has involved the molecular genetic analysis of a key gene complex involved in bristle specification, the *achaete-scute* complex (AS-C) of the X chromosome (1-0.0), which we have previously encountered (chapter 7), first as one of the "proneural" sets of genes required for formation of the neuroblasts and hence of the central nervous system (CNS) and, then, as the unexpected source of one of the "numerator elements" involved in sex determination. Long before those roles had been discovered, however, the AS-C had been implicated as essential for development of the bristles, elements of the peripheral nervous system (PNS), from the effects of certain of its mutants on particular sets of macrochaetes of the mesothorax. Indeed, the first mutant, one of the so-called *scute* series, was first isolated in T.H. Morgan's laboratory in 1916. Three-quarters of a century later, the nature of the AS-C and its component genes have begun to be explicable. We will first look at the molecular genetics of the AS-C, and its role in bristle patterning, and then return to some of the questions that remain.

The AS-C and Bristle Patterns

The first two decades of work on the bristle mutants that mapped to the tip of the X revealed a surprising genetic complexity. Interpretation was further complicated by the fact that nearly all the mutant conditions were associated with visible chromosome breaks or deletions. The central finding was that the majority of the mutants, termed generically *scute* (*sc*), involved the elimination of one or more of the macrochaetes on the notum except for two, the dorsocentral macrochaetes, the anterior dorsocentral (adc) and the posterior dorsocentral (pdc) (see Fig. 9.4B). For the *sc* mutants, the particular subsets removed and the frequency with which particular ones are deleted (the expressivity) were found to be a function of the particular *sc* allele used. However, the two macrochaetes of the notum never removed by *sc* mutants can be eliminated by mutations in the closely linked gene *achaete* (*ac*), which can also eliminate many of the microchaetes on the thorax, the two macrochaetes of the region just posterior to the notum (the scutellum), and also one of the *sc* set of bristles (the posterior supra-alar [psa]). The severity of the *ac* effects are also a function of the particular allele examined.

In addition, a third, closely linked gene was later identified to be essential for viability, and found to map between two regions that can mutate to the *sc* phenotype (Muller and Prokofyeva, 1935). This gene was termed *lethal of scute* (*l'sc*) and found to be required for development of the embryonic CNS, unlike the *sc* and *ac* functions, but not needed for bristle development (Garcia-Bellido and Santamaria, 1978). Nevertheless, despite the distinctive phenotypes of *ac, sc,* and *l'sc,* they are clearly part of some form of gene complex, as shown by their linkage, the partial overlap of requirements for *ac* and *sc* (for the psa bristle), and the position of *l'sc* between two noncomplementing *sc* regions (subsequently termed *sc alpha* and *sc beta*).

The essential puzzle of the findings was why particular alleles of *sc* would affect only certain bristle subsets, the members of each such subset not being obviously related by position or any other discernible property. Two schools of thought came into being on this issue (reviewed in Ghysen and Dambly-Chaudière, 1988). On one side was the view that the entire locus consists of a whole set of subgenes within the *sc* region, each of which is responsible for particular bristles; on the other side, it was contended that *sc* was not divisible into many separate genes but that the differences in pattern produced by different mutant conditions reflected quantitative regional differences in sensitivity by the epidermis, with the different alleles affecting that sensitivity to bristle induction differently. A third possibility, that there is just one, or a small number, of gene product(s) and that the mutant alleles differentially affect the synthesis of the product(s), as a function of the region, could not be clearly formulated in the 1930s and 1940s.

The issue remained unresolved until the cloning of the AS-C in the 1980s, but two important sets of genetic findings, prior to the molecular characterization, provided some essential information. The first involved mosaic studies, using *ac* gynandromorphs, and showed that genetically *ac* cells could produce one of the three macrochaetes normally deleted by *ac* mutations, if the mutant tissue was closely surrounded by *ac+* cells (Stern, 1956; Roberts, 1961). Thus, *ac* shows some degree of nonautonomy and evidently is not required directly for the differentiation of bristles in the cells that produce them. Rather, *ac* (and presumably, *sc*) must be required for the initial induction steps that lead, eventually, to bristle induction and which take place within small groups of cells, not all of whom need to express the wild-type *ac* function.

This interpretation is supported by the well-known dominant phenotype of *Hairy wing* (*Hw*) mutants, which, as their name indicates, produce extra bristles (both on the wing and on the notum). *Hw* mutants also map to the AS-C and are hypermorphs (Richelle and Ghysen, 1979; Campuzano et al., 1986; Balcells et al., 1988). The excess numbers of bristles in *Hw* mutants are more readily explained by the idea that overexpression of AS-C gene products recruits additional cells to form bristles than on the basis that it provides extra differentiation products to cells already singled out.

The second set of studies involved the analysis of the effects of removing specific parts of the AS-C, through the use of translocations that split the region and the segregation of the two halves of each such translocation, and the generation of deficiencies by recombination between overlapping inversions. These results indicated that there is some duplication and partial redundancy for both *sc* and *ac* functions within the AS-C (Garcia-Bellido, 1979). Thus, the *sc alpha* and *sc beta* regions, though separated by the presence of *l'sc,* affect overlapping sets of bristles. The dichotomous interpretation of the classic genetic results can, in fact, be resolved if the AS-C contains several genes of related function, which can be separated by rearrangements and regulated quasi-independently. In the phrase of Garcia-Bellido (1979), the region might consist of "reiterative signals," namely, a certain degree of internal functional redundancy. This suggestion was a prescient one and, as it turns out, has helped to explain much of the otherwise puzzling genetic data.

The cloning of the AS-C was first reported in 1985 and established that it is indeed a large region, greater than 100 kb in length, and encodes several transcripts (Campuzano et al., 1985). The key transcripts and their locations are shown in Figure 9.5. Of the nine transcripts eventually identified within and on either side of the AS-C, those designated T5 and T3 are associated with *ac* and *l'sc*, respectively, while a third transcript, T4, is associated with *sc alpha*. A fourth transcript, T8, has been identified with the gene *asense*, which is required for the development of certain larval sensory organs but not for bristle development (Dambly-Chaudière and Ghysen, 1987; Alonso and Cabrera, 1988).

The molecular mapping of the AS-C and the placement of particular transcripts was informative in itself. First, it served to identify particular transcripts with particular gene functions, as given above (Campuzano et al., 1985). Second, because the coding regions occupy a relatively small portion of the AS-C, it is necessary to conclude that much of the region affected by the *sc* rearrangements is involved in complex, long-range *cis* effects on some of these transcripts, in particular T4. (The *sc*

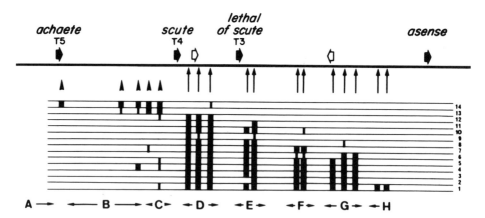

Fig. 9.5. Molecular map of the *achaete-scute* complex (AS-C). The line at the top indicates the chromosomal segment, of approximately 100 kb, that contains the complex; the genes and directions of their transcripts are as shown. Arrows below the line indicate the breakpoints of various translocations and inversions; arrowheads, the mapped end points of several deletions (see Ruiz-Gomez and Modolell, 1987). The lower portion indicates the *sc* phenotype associated with these mutants, with thick lines representing the suppression of more than 50% of the indicated bristles (numbered on the right), and thinner lines, more than 10% suppression. The letters at the bottom represent eight phenotypic groups, based on the bristle sets deleted for breakpoints within those regions. The identities of macrochaetes affected are indicated by numerals in the right-hand column (see Fig. 9.4 for positions of these) 1: asc and psc, anterior and posterior scutellars. 6: apa and ppa, anterior and posterior postalars. 7: anp, anterior notopleural. 9: ps, presutural. 11: asa, anterior supra-alar. 13: pnp, posterior notopleural. 14: psa, posterior supra-alar. The other bristle numbers refer to macrochaetes located on other regions of the thorax or on the head. The dorsocentrals (adc and pdc) shown in Figure 9.4 are affected by *ac* mutations, which can also delete the psa (no. 14), as indicated in the panel. (Reproduced from Ghysen and Dambly-Chaudière, 1988, with permission of the publisher.)

beta region, in fact, is one of these *cis*-controlling elements, rather than a separate gene function [Ghysen and Dambly-Chaudière, 1988].) In general, the closer a breakpoint is to T4 on the proximal side, the more severe are the *sc* effects, in terms of numbers of bristles affected (Campuzano et al., 1985); the chromosomal region, in which these effects are seen, stretches for more than 50 kb.

The situation is highly similar to that of the BX-C, which also has long regions of *cis*-acting control sequences that govern the spatial expression of a small number of protein-coding sequences (Fig. 7.12). Evidently, much of the AS-C similarly consists of enhancer-like elements that regulate where and when the gene products will appear. In this interpretation, the different *sc* chromosomal breakpoints affect both the number and identity of the enhancers that the *sc* gene responds to; the spatial array of bristles seen in a particular *sc* condition would be a reflection of the resulting spatial profile of expression modulated by the retained enhancer elements.

Furthermore, with the cloned regions corresponding to particular transcripts and gene functions, it became possible to determine, using *in situ* hybridization, the spatial distribution and temporal patterns of expression of these transcripts. Thus, as expected from the identification of T4 with *sc,* there is a strong correlation between the sites of T4 transcript abundance in the imaginal discs of third instar larvae and the sites at which various sensory organs, including bristles, will develop (Romani et al., 1989). *ac* shows a broadly similar, though somewhat more restricted, pattern of expression while T3 (*l'sc*) and T8 (*asense*), which is required for larval but not imaginal sensory organ development, are expressed only slightly in the wing disc (Romani et al., 1989).

Significantly, the initial regions of *sc-ac* expression are clusters of cells, of about 20–30 cells per cluster, even for those regions that will give rise to a single macrochaete (Cubas et al., 1991; Skeath and Carroll, 1991). Some hours after each cluster of *sc-ac* expression becomes detectable, by immunostaining of the protein products, the SMCs become visible in the enhancer trap strain A101, by their *lacZ* expression. Strikingly, for the macrochaete *sc*-expressing clusters, only a single SMC is seen and this occupies a characteristic position within each cluster (Cubas et al., 1991). Evidently, the clusters represent small regions of *potentiality,* with a further singling out of the definitive bristle-producing cells taking place subsequently. That the *sc* expression patterns are not just a concomitant of the bristle development process but are required for it is proven by the abolition of *sc* expression in particular regions by mutants and rearrangements that delete the bristles that would arise in those regions (Romani et al., 1989).

The kinetics of transcript appearance are also informative. Thus, both *sc* and *ac* are first detected around 48–40 hr BPF, the stage at which specific commitment to bristle formation had been earlier identified from the clonal analytic experiments. About the time of puparium formation, expression begins to diminish, as scored both by *in situ* hybridization (Romani et al., 1989) and by immunostaining of the *ac* and *sc* proteins (Skeath and Carroll, 1991). Indeed, the proteins tend to disappear just before the first SMC cell division and well before actual differentiation of the bristle organules (Skeath and Carroll, 1991). The kinetics of expression, therefore, fully support the earlier inferences that the AS-C is required for the initial developmental decisions that commit cells to bristle formation rather than for the later, visible differentiative events.

The initial sequencing of the transcripts of the AS-C revealed a further crucial fact: T4 (*sc*) and T5 (*ac*) share homology in two regions and both of these regions are homologous to signature sequences of a mammalian proto-oncogene, c-*myc* (Villares and Cabrera, 1987). Indeed, there are four *myc*-related sequences in the AS-C: In addition to T4 (*sc*) and T5 (*ac*), these are T3 (*l'sc*) and T8 (Alonso and Cabrera, 1988). The existence of a cluster of genes with sequence relatedness, even apart from the specific identity of the shared sequence, is noteworthy in itself. It indicates that the gene complex probably arose through gene duplication events. It also provides a molecular basis for the earlier suggestion of Garcia-Bellido's (1979), derived solely from genetic analysis, of reiterative elements within the complex possessing a degree of functional overlap.

Thus, while *ac* and *sc* appear to be independent genes, in terms of their bristle pattern effects and complementation behavior, they can fully substitute for one another in promoting the development of certain larval sensilla (Dambly-Chaudière and Ghysen, 1987). Furthermore, when *sc* is linked to a heat shock promoter and induced at appropriate times in development, it can rescue the production of those bristles normally dependent on *ac* expression (Rodriquez et al., 1990). Evidently, the seemingly different requirements for *ac* and *sc* among different notal macrochaete SMCs reflects primarily their different spatial regulation rather than intrinsic qualitative differences in their activities.

The relationship of the AS-C genes to the vertebrate gene c-*myc* is significant in at least two respects. The first, and more general, is that the proteins encoded by these genes are now known to be part of a large group of transcriptional factors possessing the helix-loop-helix (HLH) structure and capable of forming heterodimers as well as homodimers (Murre et al., 1989; reviewed in Garrell and Campuzano, 1991). This capacity for heterodimer formation probably accounts for certain aspects of *sc* regulation. It has been known for a long time that loss-of-function mutations in a number of other unlinked genes, such as *hairy* (*h*) (at 3-26.5) and *extramacrochaete* (*emc*) (at 3-0.0), cause the appearance of extra bristles, as indicated by their names. (For both *h* and *emc*, these effects are seen with hypomorphs, not amorphs, which are lethal.) The wild-type products of these genes form heterodimers with the *sc alpha* protein and reduce its effectiveness as a transcriptional activator; partial loss-of-function mutations prevent such dimerization and allow maximal activity for the *sc* protein, with a corresponding increase in bristle formation (Moscoso del Prado and Garcia-Bellido, 1984; Ellis et al., 1990). The ubiquitously expressed *emc* protein probably produces a general reduction in *sc* activity, lowering its background activity (Cubas et al., 1991).

The second aspect of interest in connection with the c-*myc* homology is that several of these proteins seem to be directly involved in cell division control. In mammalian cells, c-*myc* expression is associated with active proliferative states, while another member of the family, *MyoD*, a protein that promotes myogenesis, has a seemingly opposite effect, being able to inhibit cell division independently of its role in muscle development (Sorrentino et al., 1990). Perhaps, the *sc-ac* proteins have a similar inhibitory effect, with the prolonged state of mitotic quiescence of the SMCs due, in part, to their accumulation of these proteins. Nevertheless, homozygous clones for *sc* and *ac* can be induced 10–25 hr before prospective SMCs show enhanced accumulation of *sc* and *ac* (Cubas et al., 1991). This result is explicable if the regions of mitotic quiescence precede the AS-C expression and if the recombinogenic effects of X-rays can persist in the irradiated cells for long periods until the time of first division and clone production. Some of the aspects of the singling out of SMCs are depicted in Figure 9.6.

Lateral Inhibition Aspects

Though the genes of the AS-C are essential for the formation of bristle-generating cells in the appropriate places, there must be other genes whose products ensure that not too many bristles appear at particular locations, through the phenomenon of lateral inhibition (Simpson, 1990). Viable mutants of such genes would exhibit extra bristles, either ubiquitously throughout the epidermis or clustered around the normal sites. While not all of the genes whose mutants produce extra bristles are involved in lateral inhibition—as we have seen, *h* and *emc* appear to be involved in the transcriptional regulation of the AS-C—several specify products that act at the cell surface and are candidates for lateral inhibition functions.

One of these genes we have encountered previously, namely *Notch* (*N*), one of the neurogenic genes. Since homozygotes for amorphs of the neurogenic genes are lethal, their effects have had to be explored by other means. These have been primarily clonal analysis and the use of temperature-sensitive mutants. In the first thorough study of the effect of neurogenic mutations on bristle formation, Dietrich and Campos-Ortega (1984) examined the effect of homozygosity for mutations in *N* and the other neurogenics on bristle development and found that clones for *N, neu,* and *mam* could give either "nude" (bristle-deficient) or "bushy" (extra bristle) clones, depending upon the allele and the region of the fly. The allele dependence is particularly interesting: All clones for an amorphic mutation were found to be "nude" and apparently lacked all neural structures, whereas homozygous clones for a temperature-sensitive allele (in flies grown at 29°C) could be either "bushy" or "nude," depending upon their location and the time of the heat pulse. "Bushy" clones, produced by homozygosity for N^{ts1} had also been found by Shellenbarger and Mohler (1978).

These seemingly paradoxical dichotomous effects associated with loss of N^+ activity—either complete loss of bristles or production of supernumerary bristles—were resolved by a study of the effect of temperature upshift with N^{ts1} as a function of time with respect to bristle development (Hartenstein and Posakony, 1990). Hartenstein and Posakony found that heat pulses following puparium formation but prior to the time of bristle precursor cell division (about 14 hr APF) cause supernumerary bristles to form, at the expense of epidermal cells, while heat pulses during and immediately after these cell divisions cause a loss of accessory cells (the trichogen and tormogen cells) and production of supernumerary neurons (apparently through conversion of cells that would have formed the accessory cells).

Evidently, the N^+ gene product performs two functions: Prior to BMC division, it serves to inhibit recruitment of additional bristle precursor cells from the population, while subsequently it promotes formation of the accessory cells, preventing them from turning into sensory cells. Its second role is at least superficially similar to that in embryonic epidermal development, where it limits neuronal cell formation and promotes epidermal development. Its function in bristle formation is also presumably connected both to its promotion of cell adhesion and its involvement in cell signaling (Hoppe and Greenspan, 1986; Hartenstein and Posakony, 1989).

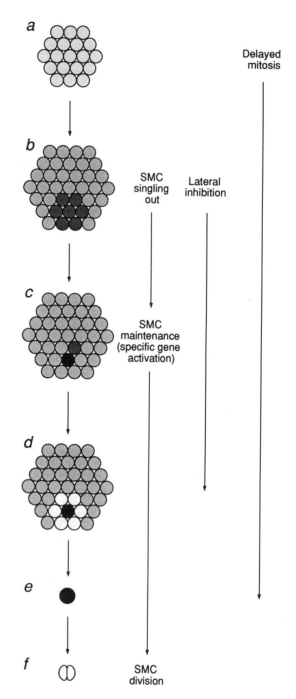

Fig. 9.6. Development of the proneural cluster and the emergence of the SMC. The singling out of the SMC occurs within the group of cells expressing AS-C, within a mitotically quiescent group, and involves lateral inhibition; the particular cell that becomes the SMC is always within a characteristic position within the cluster. (Reproduced from Cubas et al., 1991, with permission of the publisher.)

At least one gene has been identified whose product seems to be involved directly in signaling and, specifically, in the inhibitory signal generated by BMC's that prevents other neighboring cells from embarking on bristle development. This gene is *scabrous* (*sca*) (located on the second chromosome, at 49D1-3) and was first identified by its rough eye mutant phenotype, displaying disordered arrangements of ommatidia. Mutant flies also display supernumerary bristles. *sca* has been cloned and, from its sequence, shown to encode a secreted product related to the fibrinogen gene family of vertebrates (Baker et al., 1990). This property would make the Sca protein a good candidate for the lateral inhibition substance, or one such substance (there may of course be more than one). From observations of certain genetic interactions between *sca* and *N* mutations, involving suppression or enhancement, depending upon the particular mutations, it has been proposed that *sca* may produce a ligand for *N* (Mlodzik et al., 1990). This hypothesis predicts that in the absence of the *sca* gene product, *N* would fail in its initial action to limit BMC recruitment, yielding supernumerary bristle cells. This hypothesis should prove testable.

Some Conclusions and Some Remaining Conundrums

In the last decade, in particular, there has been great progress in understanding the mechanisms that underlie bristle development (reviewed in Held, 1991). The process clearly involves several steps: 1) a rough "blocking out" of the areas in which bristles will appear, characteristic of each imaginal disc, possibly concomitant with the imposition of mitotic quiescence in those areas; 2) the appearance of AS-C gene products within (most of) these regions (a few bristle-producing regions require neither *sc* nor *ac* but presumably other proneural genes); 3) the selection of specific cells within the AS-C expressing groups to become SMCs; 4) inhibition of neighboring cells, by some form of lateral inhibition, from becoming SMCs, with the pattern of inhibition being characteristic of the region; 5) division of the SMC and its daughters to produce the differentiated bristle organules. The first steps of the process are illustrated in Figure 9.6.

Yet there are still some unresolved, and not insignificant, questions. The most obvious of these concerns the process in step 1, that by which regions of AS-C expression are delimited. If, as discussed in chapter 7, there is a common system of positional coordinates—perhaps linked to some of the segment polarity genes— among the different discs, with the ultimate cellular patterns, including those of the sensory organs, determined by these coordinates, then one needs to explain how either the "interpretation" or spacing of these coordinates can yield the dramatically different kinds of bristle patterns shown. It has been suggested that the greatly reduced expression of *sc* and *ac*, and the relative paucity of bristles, in the posterior compartment of the wing may reflect repression by the *engrailed* (*en*) gene in that compartment (Skeath and Carroll, 1991). If that is the case, then one needs to explain how bristles arise in the posterior compartment of the leg despite *en* expression there (Brower, 1986). Further work on segment polarity gene expression patterns and requirements in the discs should help elucidate the relationships of bristle patterns to the overall patterning system.

A second question concerns how the SMC within each macrochaete-forming cluster on the notum is selected from the 10–30 cells within that cluster expressing

AS-C proteins. The crucial finding to be accommodated is that the definitive SMC within each such cluster, detected by *lacZ* staining in the A101 strain, occupies a characteristic place within that cluster (Cubas et al., 1991). Lateral inhibition among equals is, clearly, not a sufficient explanation; particular cells within each cluster are, evidently, "chosen" by some more precise mechanism, whose nature we do not yet know.

The third general question concerns the phenomenon of "competence," the ability of particular cells to generate bristles. A particular set of experiments shows that while expression of the genes of the AS-C is necessary for bristle formation, it is not sufficient to induce them in all epidermal cells. Rodriquez et al. (1990) demonstrated this by placing *sc* under a heat shock promoter, in wild-type flies and in flies lacking both *sc* and *ac,* and inducing generalized *sc* expression for various periods both before and after PF. As noted earlier, this treatment can restore normally *ac*-dependent bristles. The singular finding, however, is that in the mutant background, while most bristles and sensillae are restored, there aren't large numbers of ectopic SOs. Further-more, the SOs that appear are those that are "right" for that particular location. Apparently, local factors determine what kind of SO should appear in each spot. When the experiment is carried out in a wild-type background, one obtains more ectopic SOs, especially additional microchaetes, but these tend to cluster near the normally placed ones and, again, are always of the kind normally found in that location. (Thus, for instance, only ectopic campaniform sensillae, but not bristles, appear along the wing veins with this treatment.) What are the additional factors that confer competence on the "right" cells to form the "right" structures, when all are expressing AS-C? Are they additional particular transcriptional factors or special chromatin states that only form in those cells? Or is something else involved? And how does the spatial allocation of states of competence relate to the general pattern-ing that assigns particular sites of AS-C expression?

There is one final feature of bristle pattern that will not be discussed here, but should be noted. This is topographical polarity—the direction in which cuticular processes point with respect to body axes—and it is shown by both bristles and hairs on the surface of the fruit fly body. These orientations are regular and characteristic for the different regions of the body surface, and a number of genes can mutate to disturb these characteristic polarities (Gubb and Garcia-Bellido, 1982; Adler et al., 1990). The investigation of the cell polarity phenomenon is still at an early stage but the results should eventually be informative with respect to general mechanisms of cell morphogenesis.

MOUSE COAT COLOR: DEVELOPMENTAL BIOLOGY AND GENETICS

The hereditary basis of coat color began as the breeder's hobby known as the "mouse fancy" in the 19th century, but the modern analysis of coat color genetics began with the work of Cuénot, Castle, and Wright in the first two decades of this century. The work of Wright on the genetics of guinea pig coat color was particularly important; it was the first attempt to use genetic differences to probe the biochemistry of development. The reason for this early attention to the coat color of rodents is its ready accessibility to genetic analysis. Although the mammalian coat still conceals many of its secrets, the broad outlines of its biology are now visible. Excellent, detailed treatments of the classic genetics can be found in Searle (1968) and Silvers (1979).

The mammalian coat serves two functions. The primary one is insulation. By constituting an efficient air-trapping layer next to the skin, the packed hairs of the coat provide substantial protection against thermal loss. The coat typically consists of two classes of hair: the shorter and thinner underhairs—about 80% of the total hair on a mouse—and the longer, thicker overhairs. Within each category, several different hair types exist. The underhairs collectively provide most of the direct insulating capacity of the coat, while the chief function of the overhair layer is to protect the underhair coat. The overhairs also contribute predominantly to the overall color of the coat. In addition, some overhairs have a specialized sensory role, such as the vibrissae ("whiskers") of the mouse.

The second major function of the coat is as a visual signaling device; in this role, color becomes crucial. The color pattern serves as a recognition device for other members of the species, a function that is particularly important in mate selection. In some species, such as the skunk, it serves additionally as a warning signal to other animals. In some mammalian species, color serves the function of concealment through camouflage, as in the snowshoe rabbit.

All hair shafts develop from multicellular structures termed follicles, which arise late in fetal development as inpocketings of the epidermis into the dermis. The process of hair follicle development is illustrated in Figure 9.7. As the epidermal cells push down to form an inverted blunt-ended cylinder, mesenchymal cells of the dermis aggregate at the base or hair bulb. These mesenchymal cells form a thickened papilla at the end of the follicle which is partially enclosed by the epidermal cells of the hair bulb. It is from the hair bulb that the shaft originates. Each hair consists of an outer cylinder of thin, compressed cells, the cortex, and an internal set of cells, the medullary cells, separated by spaces within the shaft of the hair (Fig. 9.8). Both cortex and medullary cells, but especially the cortical cells, are rich in the structural protein keratin. Both cell types are also repositories of pigment. These melanic pigments are secreted directly into the cells of the hair shaft by melanocytes, located within the region of the hair bulb. Pigment is transferred in the form of small pigmentary

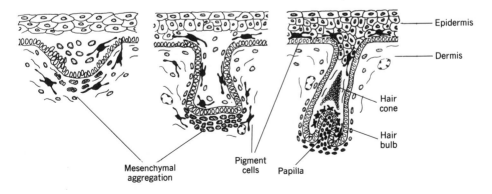

Fig. 9.7. Sequence of hair follicle development in the mouse coat. (Reproduced from Silvers, 1979, with permission of the publisher.)

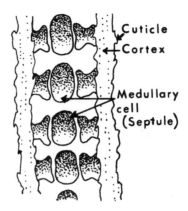

Fig. 9.8. Internal structure of a hair shaft in the mouse coat. (Reproduced from Silvers, 1979, with permission of the publisher.)

granules termed melanosomes to the basal cells of the shaft throughout the period of hair growth.

The developmental biology of the melanocytes that populate the hair follicles is complex. As discussed earlier, the melanocytes of the skin are derived from the migratory melanoblasts of the neural crest. They arrive at the dermis-epidermis interface about day 12 of development and form a network of interconnected cells throughout the epidermis. Each active melanocyte eventually transfers its melanosomes either into cells of a developing hair shaft or to neighboring epidermal cells that are not part of a follicle. The transferred melanosomes are complex entities in their own right; each consists of a fibrous protein granule attached to a melanic pigment. These pigments are of either of two kinds: eumelanin and phaeomelanin. Eumelanin is the prototypical black melanic pigment, while phaeomelanin is yellow. Both are synthesized from tyrosine in a series of sequential oxidations, carried out by tyrosinase. Phaeomelanin differs chiefly from eumelanin in having attached cysteinyl groups. Maturation of the protein moiety of the melanosome takes place gradually and only mature melanosomes are transferred to epidermal cells. The final color of a hair or hair subregion is a function of which pigment is deposited, the number and spatial arrangement of the melanosomes within the cells of the hair shaft, and the shape and size of the melanosomes.

Given the developmental and biochemical complexity of hair construction, it is not surprising that many genes directly or indirectly exert effects on the color, color pattern, or structure of the coat. To date, more than 50 loci have been identified on the basis of their mutant effects on coat color; a large proportion of these have pleiotropic effects, and for most of these, the precise relationship between the coat color and other mutant phenes is unknown.

As ever greater numbers of mouse genes are cloned and characterized, the particular functions of the genes affecting coat color development are increasingly better understood (reviewed in Jackson, 1991). The various steps in the development of the coat are diagramed schematically in Figure 9.9, and several of the better-known coat color genes can be placed within this scheme, in terms of their sites and modes of action.

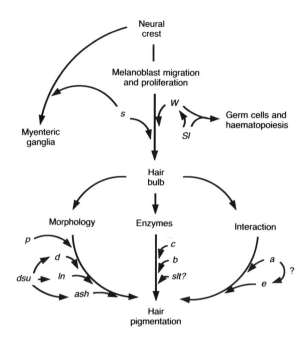

Fig. 9.9. A depiction of the pathway of coat color development. See text for details. (Reproduced from Jackson, 1991, with permission of the publisher. © ICSU Press.)

The first step is that of melanocyte formation from migratory melanoblasts derived from the neural crest. Two coat color genes identified by dominant alleles, *W* (on chromosome 5) and *Sl* (on chromosome 10), affecting melanocyte formation have been described earlier (chapter 8). Both are part of an essential cell signaling system, with *W* encoding a membrane receptor molecule, the c-Kit protein, and *Sl* encoding the growth factor ligand for this receptor; both gene products are required for melanoblast migration or proliferation. Failure of normal melanoblast development in *W* or *Sl* mice leads directly to a deficit of melanocytes and consequent failure of pigmentation. The third locus that can mutate to give a white-spotting phenotype (though a recessive one), *piebald* (*s*) (on chromosome 14), has not yet been cloned or characterized in molecular terms. Loss of *s* activity is associated with inviability, the cause of death being the condition of megacolon, produced by paralysis of the gastrointestinal tract. This paralysis reflects a defect in the myenteric ganglia, also of neural crest origin, and the *s* defect in pigmentation, like that of *W* and *Sl* reflects some developmental abnormality in the melanoblasts (Market and Silvers, 1956).

Once melanoblasts have populated the hair bulb, they undergo differentiation into melanocytes, a process involving various interactions with the other cell types of the hair follicle. Several genes are known to be required for attainment of proper melanocyte morphology, in whose absence pigment transfer is aberrant. These genes include *dilute* (*d*), *leaden* (*ln*), and *ashen* (*ash*). Homozygotes exhibit an apparent diminution of pigment intensity, the particular reduction depending upon the particular allelic composition for the other pigment genes. Thus, homozygosis for any of these mutants causes mice that would otherwise be black to be Maltese blue and those that would be dark brown to be light brown. The dilution effects for mutants of these genes stem from a broadly similar change in melanocyte morphology, though the precise biochemical basis of the change probably differs between them as evidenced by their genetic interactions with other loci (see, for instance, Silvers, 1979, pp. 84–86). Normal melanocytes have a highly branched, or dendritic, morphology that facilitates transfer of melanosomes. In the mutants, the melanocytes are much more compact, failing to show the normal dendritic extensions. In consequence, melanosome transfer takes place in an irregular, bottleneck pattern of release. The individual hairs of d and ln mutants show large aggregations of pigment that are unevenly spaced within the hair shafts. Though the macroscopic impression is of a reduction in pigment, there is as much pigment present as in wild-type.

Of these three genes, only d has been cloned and sequenced to date. The cloning was made possible by the discovery of tight linkage between a certain d mutant and a retroviral insertion. Since reversion of the mutant was found to be accompanied by loss of the retrovirus, it was apparent that the latter was within the gene (Copeland et al., 1983). The sequence of d reveals that the gene encodes an unusual heavy myosin chain; it would thus appear to be involved in construction of the melanocyte cytoskeleton (Mercer et al., 1991). Nearly all d mutants also show neurological defects, and it seems probable that both the melanocyte and neural defects reflect a similar failure of dendritic extension produced by a cytoskeletal abnormality. The molecular basis of action of ln and ash is unknown, as is that of *dilute suppressor* (*dsu*), which, as its name indicates, ameliorates the d phenotype (and that of ln and ash) when homozygous. Mutants of another gene that affect melanocyte differentiation, *pink eyed dilution* (*p*), produce melanocytes that have dendritic extensions but the pigment granules themselves, the melanosomes, are smaller than normal melanosomes and are abnormally shaped. The effects of all of these mutations are probably intrinsic (autonomous) to the melanocytes, though this has not yet been shown definitively.

The precise patterns of pigment formation, however, are strongly influenced by interactions with the other principal cellular components of the hair follicle, in particular, the epidermal and dermal cells, and these interactions are genetically conditioned. The most thoroughly studied locus in this category is the *agouti* (*A*) gene (on chromosome 2), which is characterized by an allelic series that runs from dominants producing a yellow coat to recessive producing full black pigmentation. Thus, in an otherwise wild-type background, A^y (the classic yellow lethal of Cuénot) and A^{vy} (a viable dominant) produce a yellow coat; the standard allele, A, produces a gray coat; and a^e, extreme nonagouti, produces a completely black coat in homozygotes. The *agouti* locus evidently governs, in some fashion, the balance between phaeomelanin and eumelanin synthesis. A diagram of several of the *agouti* Allele hair color phenotypes is shown in Figure 9.10. A general feature of these phenotypes, except for the extreme black-pigmented form, is that hairs on the ventral side are always lighter than on the dorsal side.

The complexity of the action of the *agouti* locus is revealed by considering the color pattern of individual hairs in the standard A genotype. These hair shafts have black tips, yellow shafts, and black bases. The apparent grayness of the coat in the A strain results from this internal distribution of the two pigments. The transitions between the black and yellow within the hair are not abrupt but take place over the length of

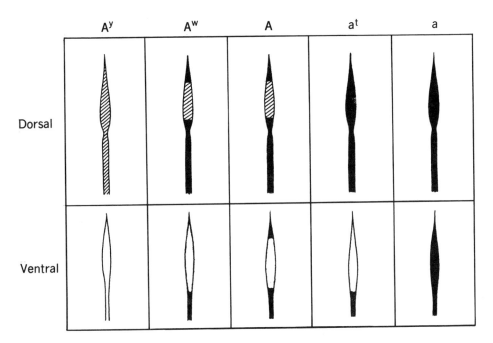

Fig. 9.10. Hair color phenotypes corresponding to representative alleles of the *agouti* locus. Note the consistent d-v differences. Black represents eumelanin, and hatching, phaeomelanin; unmarked areas are nonpigmented. (Reproduced from Searle, 1968, with permission of the publisher.)

three to four medullary cells. Indeed, some medullarly cells in the transition zone contain both phaeomelanin and eumelanin granules. This suggests that the gene somehow affects the synthetic behavior of the melanocyte rather than its intrinsic synthetic capacity for the two melanic pigments. In fact, the *agouti* locus acts within the nonmelanocyte cells of the follicle rather than in the melanocytes, as shown by the creation of skin chimeras in grafting experiments (Silvers and Russell, 1955).

The pigmentation pattern is evidently the product of a subtle interaction between two very different cell types, dermal cells and melanocytes. A second locus, the *extension* (*e*) locus (on chromosome 8), would also seem to be involved in this interaction; it too produces a phenotypic series, ranging from yellow to black, but with the opposite dominance relationships. Thus, the dominant *E* alleles are black while the recessive *e* alleles produce a yellow coat. In contrast to *agouti*, however, grafting experiments of skin to strains with a different melanocyte genotype for *e* show that the grafts follow the host melanocyte genotype, rather than that of the donor graft, and that, therefore, the *e* locus acts within the melanocytes (Lamoreux and Mayer, 1975). The reciprocal nature of the allelic series for the *A* and *e* loci and their different cellular sites of action is reminiscent of the *W-Sl* signaling system; indeed, cell signaling hypotheses accounting for the inverse allelic series can be formulated (Jackson, 1991). Definitive tests of these hypotheses, and definitive resolution of the action(s) of the *agouti* and *extension* gene products, must await the cloning and sequencing of these genes.

Chimera experiments have also yielded information about the initial cellular origins and patterns of the contributing cell types. As discussed earlier, the melanocytes derive from the neural crest, a tissue that is classified as ectodermal; in contrast, the major part of the body dermis is mesodermal and derives from the somites. Because the melanocytes and somites have different tissue origins, genes expressed in one of these hair follicle components but not the other show very different spatial patterns of expression in chimeras. All the melanocytes of skin, as we have seen, may ultimately trace their origins to 34 melanoblast clones in the body; what seems certain is that the coat patterns of melanoblast chimeras can be reduced to a basic archetypal pattern of 17 stripes on each side. In contrast, *agouti-nonagouti* (gray-black) chimeras show a different and much more finely striped pattern (McLaren, 1976a). Examination of large numbers of these animals suggests that there may be as many as 85 basic stripe widths per side, including about 18 in the head region, altogether a much larger number than seen in the comparable melanoblast-generated patterns.

Two interpretations of the *agouti* chimera pattern have been offered. The first (Mintz, 1974) is that each unit stripe derives from a hair follicle progenitor clone. Apart from the 18 or so in the head, there are about 65 per side for the body and tail, a number that matches the somite number for these regions. In effect, each somite may contain one hair follicle progenitor clone, the 18 head clones perhaps revealing the existence of small transient head somites. (The question of "invisible" head somites has been a long-running controversy in mammalian embryology.)

The other interpretation of the fine-grained pattern is based on the observation that there is much greater regularity of striping in these hair follicle chimeras than in melanocyte chimeras. McLaren (1976a) suggested that this regularity signifies a systemic patterning of some kind that is superimposed on the chimeric skin rather than an alternation of genetically determined clones. On the basis of random clonal placement, such regularity is unexpected. That there are *some* systemic influences that regulate *agouti* locus expression is apparent from the consistent differences in pigmentation within single genotypes (Fig. 9.10). Indeed, many of the mutants show distinct and regional modulations of color, indicating the existence of regional influences on pigmentation pattern.

Finally, among the genes affecting coat color are those directly involved in the synthesis of the melanic pigments. The *albino* (*c*) gene, discussed in chapter 8 in connection with mutant hunting, is the most essential in this respect. Its mutants form a phenotypic series whose effects range from only a mild diminution of the pigments to a complete absence of melanin. This last is the original *c* mutant, characterized by a completely white coat and pink eyes. In standard genetic backgrounds, the wild-type (*C*) allele is dominant to all other members of the series. However, in the presence of certain nonallelic genetic differences that affect coat color, *C/c* animals can be observed to make less pigment than *C/c* homozygotes.

The *albino* locus, long suspected to be the structural gene for tyrosinase, the key enzyme in melanin synthesis, has in recent years been proved to be so. Cloned tyrosinase cDNAs map to the *c* locus (Kwon et al., 1987), and the original mutant has been shown to have two amino acid substitutions in the coding region, of which apparently only one is the inactivating mutation (Shibahara et al., 1991). Perhaps the strongest proof is that tyrosinase cDNA can rescue the albino phenotype in cell culture and in transgenic mice (Larue and Mintz, 1990; Tanaka et al., 1990).

A second locus, *brown* (*b*) (on chromosome 4), is also involved in the biosynthesis of melanosomes. Mutants of this gene cause the replacement of the black granules with brown melanosomes. When examined microscopically, the melanosomes of *b* mutants are seen to be smaller than those of the wild-type; the visible alteration in color is a direct consequence of this change in melanosome size. *brown* has been cloned and shown to encode a tyrosinase-related protein (TRP-1) (Jackson, 1988). It appears that the conversion of brown to black eumelanin requires this gene activity and that failure of the conversion to take place prevents melanosomes from achieving their normal size.

Although the analysis of the mouse color mutants has served to explain much about coat color genetics, these studies have left largely untouched the basis of some of the more interesting coat color patterns found in various other mammals. Thus, many mammals show a species-specific coat pattern of spots or stripes or sometimes both spots and stripes. Some common examples are zebras, giraffes, and many members of the cat and squirrel families. In contrast to the color patterns of genetic chimeras or those of X-inactivation mosaics, each of these species-specific coat patterns is produced in animals whose cells are essentially identical in genotype. The question of the origins of such patterns is therefore, at present, one of developmental physiology rather than developmental genetics.

In formal terms, these patterning processes can be thought of as morphogen systems acting throughout the presumptive field of pigmentation within the embryo or fetus. The morphogens may either suppress or enhance pigment formation. A two-morphogen model that produces a number of the typical spotting and striping patterns, and a general discussion of the problem, can be found in Bard (1981). Alternatively, these patterns may involve repetitive series of relatively localized interactions, which sum to give the global patterns. The work on the molecular basis of pigmentation control in the mouse may, it is to be hoped, ultimately contribute to an understanding of the more dramatic coat color patterns found in the mouse's distant mammalian cousins.

TEN

TOWARD THE FUTURE
Goals, Questions, and Prospects

...development is not a necessary consequence of life per se, but rather something added to the secret basis of life, something which therefore perhaps can be analyzed successfully for a certain stretch of the way without having gained previous insight into the basis itself..

W. Roux (1885); as translated by K. Sander (1991)

INTRODUCTION

The critical realization of Wilhelm Roux—that development could be analyzed as a problem in itself, one distinct from the evolutionary and cell biological concerns of 19th-century biologists—made possible the foundation of experimental embryology as a separate discipline within biology. Indeed, for more than 70 years, the research program outlined by Roux took investigators a considerable "stretch of the way." Yet by the late 1950s and early 1960s, it was becoming apparent that something more was needed; it would be necessary to grapple with the "secret basis of life," in particular the way in which the genetic material works, in order to go the full distance. It is because this step has been taken that we are at last capable of analyzing biological development in depth and with precision.

The 1980s brought a further change to the field. Evolutionary questions, banished by Roux and his colleagues and disciples from developmental studies, again began to be insistent in certain areas. For instance, it is impossible to think about why different animals can have such different sex determination mechanisms (Hodgkin, 1990), why the *Drosophila* segmentation hierarchy is the way it is (particularly in the future abdominal region of the embryo) (Hülskamp and Tautz, 1991), or why mammals have genomic imprinting (chapter 5) without inquiring into the evolutionary history of these processes.

This book has looked at animal development from the particular standpoint of genetics, taking as its focus the three animal systems most susceptible to the combination of genetic and molecular forms of analysis. In closing this account, however, it may be useful to step back from the particulars of the nematode, the fruit fly, and the mouse and attempt a more general assessment of the field today. In the past decade, developmental genetics has passed from being but one of several subdisciplines of developmental biology (and one that was considered rather esoteric by many biologists) to a position of centrality. Its ideas, and increasingly its techniques, conjoined with those of molecular biology, are permeating the entire field. A recent but important element in this evolution has been the development of alternative modes of functional analysis, which are being applied to organisms that either have little conventional genetics or where genetic analysis is difficult. These methods, which may be termed "quasi-genetic," include both means of selectively "knocking out" particular gene functions (by inactivating the gene itself, or inhibiting the synthesis or action of its RNA or protein), thus mimicking amorphic or hypomorphic conditions for that gene, and methods for provoking the synthesis of particular gene products in unusual amounts, places or times, thus simulating a hypermorphic or neomorphic condition. As a result, a *de facto* developmental genetics of frogs and sea urchins may well be achieved in the near future. In addition, new animal systems directly susceptible to traditional mutant hunts and genetic analysis, such as the zebra fish, are being explored explicitly for their potentialities of developmental genetic analysis (Kimmel, 1989).

What follows is one perspective on the field today, ranging over its implicit, yet central postulates to the specific questions and areas that are likely to be the principal foci of attention in the near future. In part, this discussion involves a return to some of the major, traditional questions about biological development, which were initially raised by experimental embryology (chapter 2). In addition, however, some of the questions about the genetic foundations of development that have arisen from genetic analysis itself, as touched upon in the preceding seven chapters, will be briefly reviewed. We will begin with the most fundamental fact about biological development, namely its stereotypic character within species.

DEVELOPMENTAL CONSTANCY AND GENETIC "PROGRAMS"

The cardinal axiom of developmental genetics is that the characteristic biological development of each species is based on the genetic constitution of that species. This idea seems so obvious as to be hardly worth remarking upon, yet it has ramifications that are worth considering.

It is, in fact, the observed general constancy of the developmental process and result that produces species-specific outcomes and allows one to classify individual animals quickly as members of particular species or species groups. Such constancy of development has, of course, been recognized implicitly for millenia. Noah, for instance, clearly relied on it, in recruiting just a single breeding pair of each kind of animal for his voyage to ensure propagation of that species.

Yet to speak of the genetic determinacy of development does not imply that such determinacy is absolute. The development of each individual animal takes place within a particular physical environment, whether that environment be a pond or the sea, the confines of a large, calcified eggshell, or the mother's womb, and whatever the environment, it will exert effects on development. At the least significant level, this takes the form of "developmental noise," the minor random perturbations that can influence, for instance, the number of ommatidia in the eye of a fly or which produces slight left–right asymmetries between otherwise mirror-symmetric features in bilateral animals. At the other extreme, particular large environmental perturbations experienced at specific times in development can produce characteristic developmental abnormalities in many or all wild-type embryos or larvae of that species, aberrancies which will often resemble particular known mutant phenotypes (for those species where a large stock of mutants exists).

Such mutant-mimicking aberrations, produced by environmental shocks, are termed "phenocopies," and the basis of their resemblance to certain mutants is no longer mysterious. Just as a mutant phenotype resulting from a loss-of-function pheno-type reflects the absence of a key, rate-limiting gene product, phenocopy production almost certainly reflects a similar, preferential loss of synthesis or activity of a required, rate-limiting gene product at a key moment in development (Mitchell and Lipps, 1978; Mitchell and Petersen, 1985). Since, in general, the environmental shock will affect the synthesis or activity of more than one gene product, the resemblance between pheno-copy and mutant is often not precise but, nevertheless, can be very close.

Despite environmental influences, however, the general aspect of animal develop-ment is its high degree of constancy and resilience. In effect, development possesses internal buffering mechanisms, which must ultimately be traceable to the genome and the "fail-safe" devices built into the genome. In classical genetic parlance, the function of the genome in development is to "set the norm of reaction"; yet the singular fact, as noted above, is that this norm has a surprisingly small variance. The existence of such buffering of developmental processes was also recognized many years ago by, among others, C.H. Waddington, who demonstrated that the genetic background could be manipulated either to increase or decrease this buffering and to maintain or alter the direction of development accordingly; his key finding, in this respect, was that the genetic background could be manipulated to exaggerate and "lock in" certain mutant phenotypes. The term he coined to describe these genetically based stabilizing processes is "canalization" (Waddington, 1941).

Unfortunately, Waddington was probably ahead of his time in describing this phenomenon; concepts of genetic regulation and determination of development were too imprecise in the 1940s and 1950s to permit a significant analysis of the basis of canalization and the nature of genetic background effects. Not surprisingly, therefore, this whole subject fell into neglect, and relatively little attention has been paid to it in the last three decades. Today, however, one can begin to grapple with the phenom-enon of genetically based stabilization and channeling of development, in terms of contemporary ideas of genetic regulation. These ideas include both self-reinforcing regulatory circuits and the existence of various kinds of functional redundancy, matters that we will return to later.

While the Waddingtonian approach and outlook fell into abeyance, however, the recognition of the constancy of development by molecular biologists gave rise to another term, which came into popular currency in the 1960s, that of the "genetic program." This expression is sometimes used simply as a shorthand phrase for denoting the sequential events of development, the visible progression of develop-mental change which is underlain by the sequence of gene expression changes. However, as usually employed, the idea of the genetic program is a metaphor that has deeper implications, reflecting the conjunction of two scientific revolutions in our time, those of molecular biology and cybernetics.

The word "program" means different things in different contexts but, in biology, the reference is to computers: A program in this sense is a set of coded instructions for a digital computer. In this context, the genetic program is a set of instructions for development embedded within the genome. In this view of things, development is simply the automatic computation and "readout" of these genetic instructions.

The idea of a genetic program of development evolved from the first successes of molecular biology in the 1950s and 1960s. It was, in part, the conscious application of the principles of information theory, borrowed from the then new science of cybernetics, that propelled the conceptual revolution in biology. The crucial insight, implicit in the Watson-Crick model of DNA structure, was that each gene is a mini-program, a set of coded instructions, for the construction of a protein. For instance, if one takes a bacterial gene for the enzyme β-galactosidase, copies it into mRNA, and then places the transcripts into an appropriate cytoplasmic extract from another organism (which contains the essential protein-synthesizing components), β-galactosidase polypeptides are synthesized and are identical to those encoded by the bacterial genome. Because the DNA code and the decoding and processing instructions are essentially universal (some mitochondria and protozoans have variant codes), the gene program can be inserted into virtually any cell computer and "read" to give a fixed result.

At first sight, the extrapolation from the gene-as-program (for polypeptide synthesis) to the genome-as-program (for development) seems reasonable because the genome contains all the necessary information for development. Furthermore, the genomes of different animals are ultimately responsible for the differences in development between these animals. Nevertheless, despite the seductiveness of the extrapolation, it is probably a misleading one (Brenner, 1981; Jacob, 1982; Nijhout, 1990).

The key differences between the putative developmental program and the gene program concern the *completeness* and *temporal sequence* of information utilization. When the gene program of a prokaryote is decoded, *all* of that information is utilized and processed in a strict linear sequence from a fixed initiation point. In contrast, each stage of biological development consists of the *selective* utilization of sets of information that are scattered around the genome. Furthermore, the selected information indirectly determines what information will be utilized at the next step. If the genome were a program in the cybernetic sense, it would contain *direct* instructions on where the information selection process is to begin and it would specify the rules for the successive selection steps. There is, at present, no evidence for such information in any coded form within the genome. On the contrary, everything we know about information retrieval from the genome indicates that it occurs by means of individual gene products that act in highly (molecular) context-dependent operations. The "go to" instructions emerge indirectly from the three-dimensional molecular structures of the gene regulatory proteins and particular chromatin states and each (or many of the) "go to" instructions generates some more of the molecular "hardware" for eliciting more of the information from the (DNA) "software." If there are additional "go to" rules written into the genome, they have proved singularly hard to detect, or, even to hypothesize about, on the basis of what we currently know about gene expression.

"UNIVERSALS," EVOLUTION, AND DEVELOPMENT

Yet, while organisms may lack a program, in the sense of *a priori* and complete instructions, there is an implicit set of programmatic goals for the field of developmental genetics itself. These goals are 1) to understand the genetic foundations of development in particular animal systems and 2) if possible, to elucidate the general underlying "rules" of the usage of genetic information in the development of complex animal systems, in effect to extract the "universal" features of animal development to the extent that such aspects exist.

The latter search, that for general principles, is, of course, typical of all scientific disciplines. Unlike physicists, however, who openly search for such unifying concepts, developmental biologists are, as a rule, considerably more reticent about stating such goals, and justifiably so. In contract to chemistry and physics, whose principles, as far as we know, are valid throughout the universe, the biologist is restricted to earthbound "universals," indeed, "terrestrials." Generalizations about biological organization and processes are necessarily tempered by the accidents and contingencies of biological evolution. Evolution does not create new organisms through a program of minimalist engineering principles but through lengthy repeated processes of tinkering, or "bricolage" (Jacob, 1977). Each stage builds on the existing material created earlier, with the "materials" at hand, rather than designing new ones from scratch. This does not mean that "optimal" design solutions are never achieved in evolution, only that the evolutionary process itself is not a continual exercise in optimization. Given this character of evolutionary history, one might well believe that it makes as much sense to look for widely shared genetic principles in animal development as to find the shared principles of construction in a hundred Rube Goldberg or Heath Robinson devices.

Indeed, one may well ask whether, in the animal world, which encompasses forms as different as *Hydra,* annelids, fruit flies, and humans, there are any significant general features in development beyond the obvious and clichéd ones of diffusible signals, cell contact-mediated interactions, sequences of transcriptional switches, and the like. The answer is unknown but there are at least two broad reasons for believing that there might be some significant general rules.

The first is the fact of evolutionary relatedness itself, which guarantees that the closer the relationship between organisms, the less changed will be their fundamental biological properties. The whiteness of swans in the Northern Hemisphere is (as we know) not a reliable indicator of swan color in the Southern Hemisphere, but it would be very surprising if both white and black swans experienced different modes of development. Indeed, not only do all birds show obvious similarities to each other in the middle stages of embryogenesis, when the body plan has been laid down, but strong similarities can be seen with all vertebrate embryos, a key fact for 19th-century biology.

Furthermore, even the obvious differences in very early development may obscure more fundamental connections. Thus, for instance, in contrast to the standard *Xenopus* pattern of relatively small eggs and tadpole development as an intermediary stage, several frog genera (which live in trees, away from water) are known that produce large, yolky eggs which give rise to embryos that develop on the surface, somewhat similarly to the blastodisc embryos of birds and reptiles, and which develop directly into tiny frogs, without the tadpole stage (Elinson, 1987). Other cases of "direct development," in which an intermediate larval stage is eliminated by species closely related to others that retain it, are also known in ascidians and sea urchins (reviewed in Jeffrey and Swalla, 1992; Raff, 1992). In effect, while the overt differences in early development between different animal groups are undeniable (see review by Davidson, 1990), such instances of direct development suggest that only a relatively small *number* of genetic differences can produce large-scale phenotypic differences in early development.

Molecular findings are, increasingly, reinforcing this sense of hidden connections. Mesoderm formation provides one example. Though the processes of gastrulation and mesoderm formation in *Xenopus* and the mouse embryo are outwardly very different, they may utilize much of the same molecular machinery. Thus, the *Brachyury* (*T*) gene, long known to be required for mesodermal development in mice (chapter 7), has recently been cloned from *Xenopus* (using the mouse homolog for the isolation) and its properties and expression pattern studied. The *Xenopus* homolog, *Xbra,* not only shows a large degree of sequence conservation, relative to the mouse gene but, just as significantly, a comparable expression pattern in the mesoderm and notochord (Smith et al., 1991). The evidence, though still incomplete, strongly suggests that *Xbra* plays a similar role in the frog embryo to that of *T* in mesodermal development in the mouse embryo.

The difficult question is whether there are any significant underlying genetic principles in development not only within a phylum, such as vertebrates, but between highly disparate phyla. In such comparisons, the visible differences in form and development would seem to argue against commonalities, while molecular relatedness between gene products need not imply relatedness of developmental function. An instance of disparities in use of homologous gene products is that of *glp-1* of *Caenorhabditis elegans* and *N* of *Drosophila* (chapters 6 and 7, respectively) which show clear similarity of sequence to vertebrate epidermal growth factor (EGF) but which, unlike EGF, have little to do with promoting cellular growth. Rather they are cell membrane molecules required to mediate certain fundamental binary cell fate decisions. Or, to take another set of examples, comparable differences in biological function are seen between many of the vertebrate proto-oncogenes and their *Drosophila* homologs. Evidently, evolution will often preserve a fundamental biochemical catalytic activity but put it to very different biological/developmental uses in highly different animals.

Hence, for such distantly related animal lines, where morphology and developmental pattern speak of difference rather than similarity and where the molecular homologies may reveal little about developmental relatedness, the supposition of any significant connecting links might seem unwarranted. Nevertheless, in recent years, one set of findings stands out as indicating connections between at least one class of invertebrates and mammals that may be significant. These are the relationships, in terms of chromosomal gene sequence and of relative regions of spatial expression between the HOM genes of fruit flies and the *Hox* genes of mice (chapter 8). Despite the obvious differences in developmental biology between insects and vertebrates, the genetic relationships between these sets of genes seems too strong to be purely coincidental. We cannot yet say what the precise significance of these relationships is, but further combined genetic and molecular analyses of these genes and their products in the two sets of organisms is certain to provide some clarification of the connections.

In contrast, some other comparative analyses of developmental processes that appear outwardly similar in very different organisms have revealed some surprising differences. An excellent example is sex determination in nematodes and fruit flies, as reviewed by Hodgkin (1990). In both organisms, it is the X:A ratio that provides the initial sex determination signal. Yet the regulatory cascades initiated by the alternative X:A ratios in the two species seem to operate by highly different rules (Table 10.1a). Thus, dosage compensation in fruit flies operates by differential activation of the X in males, while in nematodes, it involves damping down of X activity

Table 10.1. Comparison of Sex Determination Systems

	D. melanogaster	*C. elegans*
A. Differences		
Sexes	Female, male	Hermaphrodite, male
Compensation mechanism	Positive	Negative?
Soma and germ line	Different genes	Same genes
Cascade interactions	Positive	Negative
Interaction mechanisms	Alternative spiicing	Transcriptional and post-transcriptional
B. Similarities		
X: A ratio as basic signal	XX and X(Y) sexes	XX and XO sexes
Multiple numerator elements on X	*sis-a, -b...*	Octamers?
Master regulators of sex and compensation	*Sxl*	*xol, sdc*
Dosage compensation genes	*MSL* set	*DCD* set
Regulatory cascade for somatic sex	*tra, ix, dsx*	*her, fem, tra*
Germline specific controls	*ovo?*	*fog*

Reprinted by permission from *Nature*, vol. 344, p.724. Copyright 1990 Macmillan Magazines, Ltd.

in the hermaphrodites. In fruit flies, sexual phenotype of the germ line involves a different set of genes from the soma, while in nematodes, it is largely the same set; in fruit flies, the cascade involves a series of activation events, while in nematodes, it is largely the same set; in fruit flies, the cascade involves a series of activation events, while in nematodes, it entails a series of inhibitions or repressions. Lastly, in fruit flies, the cascade operates primarily via differential splicing events, but in nematodes, transcriptional control events seem to comprise the main mechanism.

Yet, while the molecular processes are very different, there is, at the same time, a large degree of formal similarity in the way the hierarchy operates in the two organisms: In both, the X:A ratio sets one or a small number of master regulatory genes that regulate both dosage compensation and phenotypic sex determination through independent subordinate subsets of regulator genes (Table 10.1b). Since it is possible to conceive of alternative regulatory schemes, this general similarity in the character of the cascades is almost as striking as the manifest differences.

Whether other animal sex determination systems will ultimately show such similarities—for instance, the temperature-dependent sex determination systems of certain fish and reptiles and the *Sry*-dependent system of mammals—remains to be determined. On the surface, of course, the dosage compensation mechanism of mammals, namely X-inactivation, looks obviously different from that of the comparative fine-tuning mechanisms of fruit flies and nematodes. It should be kept in mind, however, that the regulatory machinery that produces X-inactivation in females is still largely unknown and may involve initial steps bearing some relationship to those of other organisms that do not show X-inactivation.

OLD QUESTIONS: NEW APPROACHES AND PERSPECTIVES

Ten years ago, it was generally agreed that there were only two general strategies for analyzing developmental change, in terms of its underlying genetic basis. The first was that of classical developmental genetics itself, which has been termed the "top down" mode. One would start with a mutant phenotype, which reflects a derangement of development as scored at the end point, the phenotype, and then one would try to work down to the initial set of changes at the molecular level. In contrast, there was the molecular biologist's procedure, that of "bottom up" analysis. One began with specific molecules and would try to determine how their properties and expression patterns ultimately produced effects "upward" on the phenotype. In principle, the two directions of analysis would meet in the middle and so produce a complete picture of the sequence of events. In practice, the difficulty was that often the genes of interest, defined by their mutant phenotype, could not be related by cause-and-effect chains to particular molecular events in the cell, while the molecules accessible to the molecular biologists would usually not suffice to explain the phenotypes of interest. Quite simply, the two modes of analysis, all too frequently, never converged on the same problem. A second difficulty was that in both camps there was a tendency to settle for what came to hand, whether it was a mutant of an interesting phenotype or a molecule that could be readily obtained. In large degree, the priorities were determined by the material available.

In the early 1990s, the picture has completely changed. One can isolate virtually any gene of interest, starting either with a related gene or the gene product; one can then mutate the gene and alter its activity, and, in several organisms, insert these engineered genes back into the genome and examine the developing organism for the gene's expression and effects. Indeed, in general, one can analyze in depth, by means of an exquisite battery of techniques, the details of change in any developmental sequence, ranging from the changes in protein and mRNA sets synthesized to the conformations of cytoskeletal components to nuclear structure alterations and changes in ionic composition. Coupled with various forms of cell lineage analysis, it is possible, in principle, to assemble as complete a profile of a developmental change as time, manpower, technical ingenuity, and financial re-sources allow. The limitations are no longer intrinsic insufficiencies in the techniques available.

In consequence, the major questions of developmental biology, those outlined in chapter 2, can now be addressed directly and with precision. The mechanisms of morphogenesis provide an example. Thus, one can ask about the behavior of and requirements for any of a large class of cell adhesion molecules, now known to be required for many instances of morphogenetic change (reviewed in Takeichi, 1989). Furthermore, one can, in many instances, test specific general hypotheses about the triggering and coordination of such changes. For instance, an early and influential hypothesis about morphogenetic movements postulated the existence of a cortical net connecting the apices of invaginating cells, that could, upon an appropriate local stimulus, yield a propagated contraction throughout an epidermal sheet, resulting in the infolding of the sheet (Odell et al., 1981). However, from careful analysis of the movements of individual cells (Leptin and Grunewald, 1990; Kam et al., 1991) and of mutants blocked in different stages of gastrulation (Sweeton et al., 1991), it appears rather that gastrulation is a two-stage process involving, initially, a random rather than coordinated intracellular nuclear movement (Bard, 1991). As other instances of gastrulation and morphogenesis are further analyzed, the common features of mech-anism will almost certainly emerge. Genetics will continue to play a role in this analysis. Mutants in *Drosophila,* and possibly *C. elegans,* should help to define crucial genes and gene products; the activities of their homologs can then be studied in other organisms.

Pattern formation is also a general and major problem—perhaps the central problem in developmental biology—and one where the whole panoply of techniques can be applied to the analysis of fundamental mechanisms. It is now fairly certain that a number of critically important genes in various aspects of pattern formation have been identified in *Drosophila* and other organisms. With respect to the segmentation genes, it should now be possible to see if, for instance, some of the segment polarity genes are involved in the epiphenomena described by the polar coordinate or Meinhardt hypotheses.

Critical questions about differences between regions should also be amenable to investigation. If, for example, different imaginal discs obey certain common rules of pattern formation (Bryant et al., 1981; Meinhardt, 1986), we would like to know how the differences that produce different patterns of sensory organs and shapes are produced. And, between the imaginal discs and the embryo, if a common system of positional values is employed, why does its deployment seem to require cell division in the former but not be dependent on such in the latter? To take a comparative phylogenetic approach, if the regeneration behavior of insect and salamander legs seem to obey the rules described by the polar coordinate model, does this behavior involve homologous genes or, alternatively, a convergence of processes that employ different molecules and mechanisms? Furthermore, the identification of the HOM/*Hox* genes, and their similarities of organization and deployment in insects and mammals, provokes comparable questions about similarity and difference in their employment.

Despite the wealth of technical approaches available, however, it would be a mistake to believe that answers to fundamental questions will emerge from sheer brute force application of the methods. The difficulties and the intellectual challenges ahead should not be underestimated. Identifying the "downstream" genes controlled by these regulators is an essential approach but will not, in itself, answer the question of why, for instance, the mesothorax of the fruit fly looks the way it does. We will also need to know what factors produce the mosaic patterns of expression of the homeotic selector genes themselves (Figs. 7.15 and 7.17) and how these fine-scale modulations produce the visible morphological characteristics of segments and parasegments. Imaginative hypotheses will continue to be essential if such questions are to be answered.

It should also be possible to address some of the phenomena described both by earlier observations and by more recent genetic data. Chief among these are the nature of effects of "genetic background," a preoccupation of the the classical period of developmental genetics and, in particular, as noted earlier, of the Waddington school. In principle, there are two ways in which such background differences might modify the expression of mutant genes (and, by extension, that of wild-type genes). The first is by direct regulation of the mutationally altered genes themselves; for hypomorphs or neomorphs, any such regulatory influences can, in principle, modify the "strength" of the mutant effect. The second is by supplementation or compensation through partially redundant functions or pathways. This simple distinction was generally not made in the classical literature but it should now be possible, given the range of molecular techniques available, both to classify individual effects of background as to which category they belong to, and, with the increasing precision of molecular mapping techniques, to map the modifier genes involved, as a step toward further characterization.

Another crucial set of questions concerns the nature of "combinatorial" controls in regulation. It is increasingly clear that single regulator genes do not "control" individual steps but rather act in concert with other regulatory genes. Some of these effects are purely additive ones, and involve different regulatory genes, acting on different sets of downstream genes, while many involve direct physical interactions between the proteins in complex transcriptional complexes, which mediate repression or activation. Such complexes have been termed "reglomerates" (Brenner et al., 1990) or "aggregulons" (F. Jacob). Whatever the ultimate accepted term for such complexes, it is of importance to determine whether they share certain general features of significance. To take but one of many possible specific questions about such complexes, do those containing self-activating regulator genes, like *Ubx* or *Dfd,* exhibit any shared characteristics that are not seen in the complexes of these or other genes under conditions where autoregulation is not taking place?

It is apparent that much of a fundamental nature about the genetic basis of development remains to be learned. Nevertheless, while our ideas about many aspects of development remain imprecise, the analytic techniques that are now available, in conjunction with the relevant gene libraries, will allow fairly precise and incisive tests of particular hypotheses to be made. This ability to subject models to precise molecular tests is the central difference between the early 1990s and the early 1980s. In the work that is yet to come, however, it seems fairly certain that the traditional and central strategy of developmental genetics, namely "knocking out" particular genes of interest and analyzing the effects, will remain a potent one.

FUNCTIONAL REDUNDANCY:
THE LIMITS OF DEVELOPMENTAL GENETICS?

There is, however, one genetic phenomenon, referred to both above and often in the preceding chapters, which might be thought to pose a limit to conventional genetic analysis of developmental processes, based on "subtracting" particular genetic functions. This is the existence of possibly widespread functional redundancy (Brenner et al., 1990; Tautz, 1992). In the succinct sentence of Brenner et al. (1990): "Redundancy strikes fear in the heart of geneticists." And it is not hard to see why. Saturation mutagenesis for a pathway or process demands that individual genes play unique roles in that pathway, such that any loss-of-function mutations will show up as unique phenotypes. The technique has been obviously successful in *Drosophila* for analyzing the five maternally specified patterning systems and the zygotic segmentation genes. Yet, if there are many processes where each or many genes are "covered" by genes that carry out similar functions, any analysis of a pathway by the saturation mutagenesis approach is likely to leave those steps undetected.

Functional redundancy, indeed, can come in several forms, involving either similar genes, similar processes (carried out by nonhomologous genes), or cells that can substitute for one another (e.g., equivalence groups). Experimental embryologists long ago learned how to deal with groups of cells that perform the same function, at least when they are physically close and can be jointly removed, and that is by simple physical extirpation. It is the other two kinds of redundancy—gene and process based—that pose a new challenge. Yet while in principle they may pose a problem, there are, in reality, an increasing number of cases where the existence of functional redundancy has not deterred or prevented analysis. Thus, for instance, while the partial interchangeability of genes of the AS-C for one another has produced a complex set of genetic and functional relationships, it has been possible to produce a thorough analysis of the region and their actions because 1) they are differentially regulated; 2) they are linked; and 3) despite some interchangeability, there are probably some functional differences between them (chapter 9).

The partial interchangeability of members of a gene family has also been seen for the collagen-like genes of the *C. elegans* cuticle (chapter 9). Indeed, one feature of this analysis is that while loss-of-function mutations may have no detectable phenotype, other mutations can create dominant phenotypes through neomorphic properties (Kusch and Edgar, 1986). Other instances of this phenomena have been described by Park and Horvitz (1986). Thus, while the traditional attitude of developmental geneticists to neomorphic mutations is that, in general, they are to be shunned, given the complexity of possible interpretations one can make, these mutations may yet prove an invaluable instrument for analyzing instances of functional redundancy based on similar, interacting gene products (Brenner et al., 1990). Where the gene family is small, consisting of only two to three members, and where the latter can be mapped, one can then proceed, in principle, to analyze the loss-of-function phenotype by inactivating all the members of the group. This has been done, for instance, with the three CLN genes of yeast, involved in the G1–S phase transition (Richardson et al., 1989), and a comparable approach may, ultimately, be employed for some of the smaller *Hox* gene families in the mouse.

The analysis of functional redundancy where similar processes but nonhomologous gene products are involved may prove a more difficult task but, here also, studies on *C. elegans* may provide a precedent. This was the work that flowed from the discovery of the "synthetic" Muv mutation, involving loss-of-function mutations in both *lin-8* and *lin-9*. By finding mutations that create a phenotype with either of those genes, Ferguson and Horvitz (1988) were able to identify whole sets of genes whose activities can affect vulval development, including genes that were not detected in conventional screens. For readily genetically manipulable organisms, this kind of approach provides a hopeful augury.

In sum, genetically based functional redundancy may present a challenge to developmental geneticists but it need not be an insuperable one; there are ways of dissecting its components in particular situations. There is, however, another way to view it, which is not as a problem to be overcome but as an intrinsic aspect of the developmental process, which has been repeatedly built in by evolution, ensuring the reliability of those developmental processes in which it participates (Tautz, 1992). The redundant function may be initially selected for some other property but then retained, in addition, because of the selective value it confers in stabilizing another developmental process. Indeed, a correlate of functional redundancy seems to be the utilization of many gene products, or their homologs, in many different developmental situations. In such circumstances, the consequences of the employment of the gene product (or its homolog) are highly "context dependent," that is, a function of the other gene products present, namely the other members of the reglomerate.

As a result, the products of genes named for their original mutant effect often participate in, and are required for, very different developmental purposes as well. Thus, to take several of an innumerable set of possible examples, *ftz* and *eve,* discovered and named for their zygotic roles in embryonic segmentation in *Drosophila,* are also required for neural development (Goodman et al., 1988); most of the so-called *male abnormal* or *mab* genes of *C. elegans* are not devoted to male development *per se* but are basic cell lineage genes whose mutants produce a readily detectable effect on male nematode morphology (Hodgkin, 1983a); and the *White-spotting (W)* gene of the mouse has effects not only on pigmentation, for which it was named, but on hematopoeisis and germ cell development.

In the context of functional redundancy, the example of *ftz* is particularly instructive because it is not strictly essential for the development of certain neuronal patterns but raises the probability that these connections will be properly made (Goodman et al., 1988). In effect, *ftz* in neural development acts as a "fine-tuning" gene (Williams and Newell, 1976) or an "optimization" one (Brenner, 1981), while in segmentation it plays an essential and indispensable role. To put this in more general terms, much of the genetic determination of development may consist of "locking in" developmental pathways by building in the numbers and particular elements of functional redundancy, utilizing gene products that play key roles in other processes.

Indeed, the fact that the winnowing of animal phyla that occurred at the end of the Precambrian era was not followed by any later burst of new phyla (Gould et al., 1990) may reflect, at least in part, the self-fortification of developmental processes by functional redundancy. To the extent that new gene products can both play new roles and be employed as backup functions for old ones, there should be a distinct selection pressure for adding on elements of functional redundancy. This speculation is untestable at present but it provides a framework for thinking about the genetic nature of developmental "constraints" in evolution, a topic of much interest in evolutionary biology in recent years (Smith et al., 1984). For developmental geneticists, the topic of functional redundancy may graduate from the status of nuisance to worthy subject in its own right.

THE FUTURE

As noted at the beginning of this book, the field of developmental genetics has experienced in the last decade what amounts to an almost complete transformation. Its methods and strategies of analysis, which gave it its distinctive identity within developmental biology until the late 1970s, have been incorporated and subsumed within a much larger methodological framework, a framework that embraces virtually all the techniques of molecular biology and contemporary cell biology. Yet, in losing its identity, it has also, in its central precepts and concerns, moved, to a very large extent, to center stage, not only within developmental biology but within biology as a whole. From evolutionary studies, focused on changes in the distant past, to human genome research efforts, involved with mapping the genes involved in particular hereditary defects, questions about development and the ways particular gene products are deployed in development are, increasingly, matters of general interest.

In the work to come, one growing challenge will be the integration of data and observations from different biological systems and approaches; a certain amount of this can be done by computer. Ultimately, however, acts of synthesis must take place in the minds of individual human beings. The elaboration of significant connections between processes and systems will remain a task for human intelligence. Furthermore, the fundamental questions of development—which are all, essentially, variants of one question: Why do particular changes of cell type and form happen at particular times in particular places?—are still in need of answers. In recent years, some of the long-standing general concepts of differentiation and pattern formation have been framed in considerably more precise terms. That task, of continually refining the questions, also continues as a central and significant one. Though "theory" is, perhaps, too grand a word for what is needed in organizing ideas about development—given the role of evolutionary contingency in shaping developmental systems—there will always be a need for imaginative, general, testable hypotheses. It will, in part, be the clarity and testability of such ideas that will determine our eventual success in finding, for some of the oldest questions in biology, the significant answers.

APPENDIX

Certain conventions in genetic nomenclature are general. In particular, gene names are always italicized (while allele designations may or may not be). Wild-type alleles are designated by a +, either directly (when the gene reference is unambiguous) or superscripted; for example, w^+ signifies the wild-type allele of the *Drosophila* white gene. In contrast to gene names, the corresponding phenotypes are not italicized; for instance, Cy indicates the curly phenotype produced by the dominant *Cy* mutation. There are also certain general conventions for particular kinds of chromosome aberrations: In (inversion), T (translocation), Tp (transposition—a gene or small chromosomal region moved into an atypical chromosomal site), Dp (duplication), and Df (deficiency). When particular aberrations are to be indicated, the symbol becomes italicized; for example, *Df(3)P9* refers to the P9 deletion on chromosome 3. The heterozygous condition is indicated by a slash (e.g., w/w^+), while mutations or genes on different linkage groups are separated by semicolons (e.g., $+/a$; $+/b$ symbolizes a dihybrid in which loci *a* and *b* are on different chromosomes). In general, recessive mutations are indicated by lowercase designations and dominants by uppercase designations of one to four letters. Gene products are usually indicated by a three-letter, nonitalicized symbol, in which either the first letter or all three are capitalized. Enzymes are always indicated by a three-letter, fully capitalized symbol; for example, ADH stands for the alcohol dehydrogenase enzyme, encoded by the *Adh* gene, while the product of the bicoid *bcd* can be symbolized by either Bcd or BCD.

The nomenclatural rules for *Caenorhabditis elegans* have been briefly described in chapter 3. For this organism, alleles are usually indicated by italicized one or two letters (representing the laboratory of origin, by country or city, e.g., alleles starting with *e* represent mutants isolated in England) followed by a number. Where several genes on the same chromosome are being referred to, they are listed from left to right, as on the standard genetic map. The full set of nomenclatural rules is described in Horvitz et al. (1979). One set of fairly complete genetic maps, along with a listing of many of the genes, can be found in Appendix 4 of Wood (1988).

The genetic maps are based on recombination frequencies, as are all genetic maps, but as discussed in chapter 3, measuring recombination in a hermaphroditic organism is less simple than in a standard outcrossing system. The basic principle is illustrated in the accompanying table and the principles are discussed at greater length in Brenner (1974). The principles for constructing the combined genetic and molecular maps of *C. elegans* are discussed in Coulson et al. (1991).

The nomenclatural rules for *Drosophila* are somewhat more complex than those for the nematode, reflecting the longer history of the fruit fly as an experimental genetic organism. As in other cases, wild-type alleles are indicated by a + and gene

Table App. 1. Measuring the Map Distance Between Two Genes in *Caenorhabditis elegans*[a]

		Egg chromosome frequencies			
		Parental ($1 - p$, total)		Recombinant (p, total)	
Sperm chromosome		$++$	ab	$+a$	$+b$
frequencies		$\frac{1}{2}(1 - p)$	$\frac{1}{2}(1 - p)$	$\frac{1}{2}p$	$\frac{1}{2}p$
Parental	$++$	W	W	W	W
($1 - p$, total)	$\frac{1}{2}(1 - p)$	$\frac{1}{4}(1 - p)^2$	$\frac{1}{2}(1 - p)^2$	$\frac{1}{2}(1 - p)p$	$\frac{1}{4}(1 - p)p$
	ab	W	AB	A	B
	$\frac{1}{2}(1 - p)$	$\frac{1}{4}(1 - p)^2$	$\frac{1}{4}(1 - p)^2$	$\frac{1}{4}(1 - p)p$	$\frac{1}{4}(1 - p)p$
Recombinant	$a+$	W	A	A	W
(p, total)	$\frac{1}{2}p$	$\frac{1}{4}(1 - p)p$	$\frac{1}{4}(1 - p)p\ \frac{1}{2}p^2$	$\frac{1}{4}p^2$	$\frac{1}{4}p^2$
	$+b$	W	B	W	B
	$\frac{1}{2}p$	$\frac{1}{4}(1 - p)p$	$\frac{1}{4}(1 - p)p$	$\frac{1}{4}p^2$	$\frac{1}{4}p^2$

[a]The parental genotype: the *cis* heterozygote $+; +/ab$.

Notes: a and b are the recessive mutant alleles; A and B are their corresponding mutant phenotypes, and W is the wild-type phenotype; p is the map distance (percentage of recombinant chromosomes).

Let R stand for the total recombinants, A and B. Adding the A and B frequencies, one obtains $R = p(1 - p) + \frac{1}{2}p^2$. Expanding and solving the quadratic for p yields $p = 1 - \sqrt{1 - 2R}$.

One can arrive at the same conclusion in a slightly different way. It can be seen from the table that half of the recombinant chromosomes end up in zygotes that give a wild-type phenotype. It follows that the frequency of zygotes produced by matings involving only *nonrecombinant* chromosomes is $1 - 2R$, and since this frequency is the square of the nonrecombinant chromosome frequency itself, it follows that the recombinant chromosome frequency, which equals 1 minus the frequency of the nonrecombinant chromosomes, is $p = 1 - \sqrt{1 - 2R}$. For semidominant mutations, in which every zygote containing a recombinant is distinguishable, the relationship becomes $p = 1 - \sqrt{1 - R}$.

An example will illustrate the procedure: The X-linked recessive mutants *dpy-6* and *unc-3* were put into a *cis* double heterozygote: *dpy-6 unc-3/++* (wild-type in phenotype) and the segregants scored. The results were: 826 wild-type; unc, dpy, 213; unc, 110; dpy, 97. The last two classes are the recombinant classes, and comprise $R = 207/1246 = 15.8\%$. Solving for p, $p = 18.3\%$ (example cited from Brenner, 1974).

When scoring both recombinant types is difficult, one can calculate R simply as the scorable recombinant class divided by either parental class (when both parental classes are equal).

names range from one to four letters. Lethals are designated by an *l* followed by the chromosome number in parentheses and an allele designation; for instance, the hypothetical lethal mutation m^5 on the second chromosome would be given as $l(2)m^5$. Because of the complexity and confusion regarding lethal allele designations, a new system was introduced in 1986 for lethals that have been mapped to a particular cytogenetic location but not (necessarily) otherwise characterized. Under the new system, the gene name incorporates the cytogenetic interval. Thus for instance, a lethal in complementation group a of the 62A region of the third chromosome would be designated *l(3)62Aa* (Lindsley and Zimm, 1986).

Deficiencies are indicated as noted above, while duplications and transpositions incorporate the original chromosomal origin and the present site of the transposed material. For example, *Dp (1;2)rb^{71g}* denotes a particular piece of the X chromosome (chromosome 1) carrying *rb⁺* that has been moved to the second chromosome, in a background containing at least one nonrearranged X chromosome. Inversions in either of the large autosomes (chromosomes 2 and 3) are pericentric if they involve material on both chromosome arms, for example, *In (2LR),* or paracentric if they involve just one arm, for example, *In (2L)*. Information on the synthesis of the *Drosophila* molecular and genetic maps can be found in Merriam et al. (1991).

The nomenclatural rules for mouse genetics were briefly described in chapter 5 and are given in full in Lyon (1981). Recent progress in the filling in of the mouse genetic map, by molecular and genetic means, has been described in Copeland and Jenkins (1991).

GLOSSARY

allele: One of several distinguishable forms of a gene.

amorph: A mutant allele of a gene having no phenotypically detectable activity.

aneuploidy: The state of having one, two, or a small number of chromosomes more or less than the standard number of chromosomes for the species.

antimorph: An allele of a gene whose phenotypic effect is opposite or antagonistic to the wild-type gene activity.

autonomy (cellular autonomy): The confinement of a mutant gene's effect to the cell or clone expressing that mutation.

autosome: A chromosome that is not one of the sex chromosomes.

backcross: A cross between a strain, usually bearing recessive alleles, and its offspring by a previous cross.

balanced lethal system: The condition of having two nonallelic lethal genes on opposite members of a pair of homologous chromosomes, where recombination frequencies are either low because of proximity of the two genes or where recombination has been suppressed by the presence of inversions.

centimorgan: A map unit; a distance equivalent to 1% recombination.

chimera: An individual containing cells derived from two or more different zygotes, usually of different and distinguishable genotypes.

clone: A set of genetically identical members, either cells, organisms, or genes, all descended directly from a single progenitor.

complementary DNA (cDNA): Complementary DNA sequences synthesized from mRNA, usually by the enzyme reverse transcriptase.

complementation: The restoration in whole or part of a wild-type phenotype in cells or organisms containing two mutations on opposite homologs (in *trans*).

deficiency (deletion): The loss of a DNA sequence from a chromosome; the portion that is missing.

diploid: The condition of having two complete homologous sets of chromosomes.

dominant: An allele of a gene that produces the same phenotypic effect in heterozygotes as in homozygotes.

enhancer: A site within a chromosome that serves to boost transcription from the promoters of one or more nearby genes; often, enhancers act at distances of several kilobases (kb) and can function in either orientation.

enhancer trap: A gene construct possessing a weak promoter, which when inserted in a genome near an enhancer is transcribed efficiently, to express the coding portion of the gene. Such constructs are used to find genes with new spatial or temporal expression patterns.

epistasis: The condition in which one gene mutation masks the expression of a nonallelic gene mutation.

exon: A portion of a eukaryotic gene that is found in the mature transcript, usually one that codes for a polypeptide sequence, located between noncoding portions (introns).

expressivity: The strength of mutant gene expression, as compared to the extreme mutant phenotype.

F_1: The first-generation offspring of a cross between two strains.

F_2: The second generation of offspring following a cross between two strains; produced by intercrossing the F_1.

flanking sequence: The DNA sequences on either side (both 5' and 3') of a polypeptide-coding sequence.

gamete: A mature germ line cell; in animals, either a spermatozoan or an ovum.

gene: A stretch of DNA (or of RNA in RNA viruses) possessing a definable function; usually but not always a DNA sequence specifying a polypeptide chain.

genome: The complete haploid set of chromosomes of an organism.

genotype: The genetic constitution of an organism or a cell within that organism.

germ line: The reproductive cells within a complex organism that give rise to the sex cells or gametes involved directly in reproduction.

haploid: The condition of having a single complete set of chromosomes, the chromosomal constitution of the gametes.

hemizygote: A diploid individual that carries only one copy of a gene, usually either a deficiency heterozygote for that gene or the heterogametic sex (e.g., XY individuals in an organism with X-Y sex determination).

heterochromatin: The highly compacted and densely staining regions of eukaryotic chromosomes that are transcriptionally inactive.

heterogametic sex: The sex that produces two different kinds of gamete, distinguished by their sex chromosome constitution (e.g., X- and Y-bearing sperm).

heterozygote: A diploid individual carrying two different allelic forms of a gene.

homeobox: A distinctive DNA sequence motif, of 180–183 bp, specifying a polypeptide region, the homeodomain, with DNA binding activity. Found in many eukaryotic transcriptional regulators and similar to the DNA binding regions of certain prokaryotic transcriptional regulator molecules.

homeotic mutation: A mutation causing the replacement of one part of an individual by a recognizable structure or region that normally develops in a different location.

homozygote: A diploid individual carrying two identical alleles of a gene.

hypomorph: A mutant gene producing reduced, but not nil, gene activity.

intron: A transcribed portion of a eukaryotic gene that is excised by the RNA processing machinery and which is, therefore, not found in the mature mRNA.

inversion: A chromosome rearrangement in which a section has been turned through 180°, to give a reversal of gene order in that section.

linkage: An association in gene inheritance such that parental allele combinations tend to be inherited.

locus: The position of a gene on a chromosome; the gene itself.

maternal effect gene: A gene expressed in oogenesis, capable of mutating to give a maternal effect in the progeny.

maternal effect mutation: A mutation whose expression in the female gamete-producing parent creates an effect, usually deleterious, in the progeny regardless of the progeny's genotype.

meiosis: The two successive nuclear divisions that accompany gamete formation in diploid organisms and which reduce the diploid state in gamete precursor cells to the haploid state in gametes.

messenger RNA (mRNA): The complementary RNA copy of a protein-coding gene.

mid-repetitive sequence: A sequence present in multiple copies, often slightly different from one another, in the genome of a eukaryote. A typical mid-repetitive sequence is present in several hundred copies.

mitotic recombination: The form of genetic recombination that takes place in diploid somatic cells, which can yield homozygous daughter cell progeny.

mosaic: 1. (noun) An individual composed of cells of different genotypes, though all derived initially from a single genotype zygote. 2. (adj.) In classical embryological literature, a type of egg whose territories possess specific capacities for specifying particular parts of the embryo. 3. (adj.) In more recent literature, may refer to an embryo whose different cellular regions develop independently of one another.

mutant: An individual manifesting the expression of an altered gene. Also (adj.) the state of bearing a mutation, referring either to a gene or to an individual.

mutation: 1. The process by which a gene undergoes a transmissible alteration. 2. The hereditary alteration in the gene itself.

neomorph: A dominant mutant gene producing a qualitatively novel phenotypic effect.

nonautonomy (cellular nonautonomy): The expression of either the mutant or wild-type character in a mosaic or chimera within cells of the other genotype.

null allele: A mutant allele producing no wild-type gene product or activity. Sometimes used synonymously for amorph, but, if classification involves a biochemical test, categorization as a null allele is a more definitive test of nil activity.

operator: A repressor binding site adjacent to one or more structural genes.

penetrance: The percentage of mutant individuals which exhibit the mutant phenotype.

perdurance: The persisting activity of a wild-type gene product, and phenotype, in cells lacking the wild-type allele but whose progenitor(s) possessed it.

phene: A specific characteristic or trait associated with a mutant condition.

phenocopy: An individual whose phenotype has been altered by external factors during development, mimicking a known mutant condition.

phenotype: The appearance of an organism with respect to one or more specific gene-affected characters; the phenotype is produced by the interaction of genotype and environment.

plasmid: A self-replicating circular DNA molecular found in bacteria; used in cloning of exogenous DNA sequences.

pleiotropy: The multiple phenotypic effects produced by the expression of a single mutant gene.

polyploidy: The condition of possessing an integral number of gene sets greater than two.

polytene: The condition of multiple aligned chromatids in certain "giant" chromosomes; found in certain insects and plants.

posttranscriptional control: The regulated, delayed translation of preformed mRNAs.

posttranslational modification: The addition or removal of certain metabolite groups (such as phosphates) to preformed polypeptides.

promoter: The sequence that is adjacent 5′ to a gene required for binding of RNA polymerase and, hence, initiation of transcription.

recessive: A mutant allele that only produces a phenotypic effect in homozygotes.

reciprocal crosses: A pair of crosses of the form X × Y and Y × X, where X and Y are two strains of differing genotypes.

recombination: The process of genetic exchange, usually of reciprocal homologous chromosome sections between paired chromosomes.

regulative: (adj.) The property of those eggs whose different regions can take on the developmental capacities displayed in normal development only by other regions. Contrasted with mosaic eggs.

regulator gene: A gene whose product directly controls the expression of other genes.

regulator site: A DNA sequence that binds the product of a regulator gene and controls the expression in *cis* of neighboring genes.

reporter gene: A gene construct consisting of the promoter of a gene whose expressional characteristics are under study and a coding sequence whose activity is readily scorable (e.g., *Escherichia coli* β-galactosidase).

repressor: The gene product of a regulator gene whose primary and direct action is the inhibition of transcription or translation of one or more structural genes.

restriction enzyme: An endonuclease that cleaves at a particular short DNA sequence; each such enzyme is found in and isolated from a particular bacterial species.

restriction fragment length polymorphism (RFLP): A characteristic DNA sequence generated by the possession of a novel restriction enzyme-susceptible site, revealed by digestion of the DNA with that enzyme. RFLPs are often used as molecular markers in linkage studies.

reverse transcriptase: An enzyme encoded by certain RNA tumor viruses that copies RNA into single-stranded complementary DNA.

satellite sequence: A simple DNA sequence reiterated thousands to millions of times in the genome, and usually located in or around the centromere.

sex chromosomes: The two chromosomes which share some homology that are dissimilar in the heterogametic sex and whose activities provide the initial signal for sex determination.

sex linkage: The coinheritance of a gene and a sex chromosome, usually reflecting gene location on the X chromosome in animals with X-Y sex determination.

single-copy (unique) sequence: A DNA sequence present only once, or a very few times, per haploid genome; defined operationally by renaturation kinetics.

structural gene: A gene that codes for a polypeptide chain involved in some aspect of metabolism or cell structure; a gene whose expression is controlled by a regulator gene.

suppressor mutation: A mutation that abolishes the phenotypic expression of a mutation in one or more other genes.

temperature-sensitive mutant: A mutant whose phenotypic manifestation is temperature dependent; usually refers to a heat-sensitive mutant.

transgenes: Heritable gene sequences inserted into genomes by artificial means, usually at novel locations, with respect to the endogenous homologs.

translocation: A chromosome rearrangement involving the interchange of segments between nonhomologous chromosomes.

transposable element: A short DNA sequence capable of independent movement and relocation within the genome. There are many different types, each species containing characteristic kinds, and they move at frequencies that are high compared to standard mutation rates.

transposon tagging: The generation of new mutations by the experimenter using transposable element insertion; the new inserts can then be used to clone the mutated gene.

wild-type: The character or strain taken as the representative genetic standard or nonmutant type.

zygotic genome: The genome of the embryo or the fertilized cell from which the embryo derives; as distinguished from the genome of the mother from which the embryo derives (the maternal genome).

REFERENCES

Adler, P.N. (1979). Position-specific interactions between cells of the imaginal wing and haltere discs of *Drosophila melanogaster. Dev. Biol.,* **70,** 262–267.

Adler, P.N., Vinson, C., Park, W.J., Conaver, S., and Klein, L. (1990). Molecular structure of *frizzled,* a Drosophila tissue polarity gene. *Genetics,* **126,** 401–416.

Affara, N.A. (1991). Sex and the single Y. *BioEssays,* **13,** 475–478.

Affara, N., Jacquet, M., Jakob, H., Jacob, F., and Gros, F. (1977). Comparison of polysomal polyadenylated RNA from embryonal carcinoma and committed myogenic and erythropoietic lines. *Cell,* **12,** 509–520.

Agnel, A., Kerridge, S., Vola, C., and Griffin-Shea, R. (1989). Two transcripts from the rotund region of *Drosophila* show similar positional specificities in imaginal disc tissues. *Genes Dev.,* **3,** 85–95.

Akam, M.E. (1983). The location of *Ultrabithorax* transcripts in *Drosophila* tissue sections. *EMBO J.,* **2,** 2075–2084.

Akam, M. (1989). Hox and HOM: Homologous gene clusters in insects and vertebrates. *Cell,* **57,** 347–349.

Akam, M., and Martinez-Arias, A. (1985). The distribution of *Ultrabithorax* transcripts in *Drosophila* embryos. *EMBO J.,* **4,** 1689–1700.

Alberga, A., Boulay, J.L., Kempe, E., Dennefel, C., and Haenlin, M. (1991). The *snail* gene required for mesoderm formation in Drosophila is expressed dynamically in derivatives of all three germ layers. *Development,* **111,** 983–992.

Albert, P.S., Brown, S.J., and Riddle, D.L. (1981). Sensory control of dauer larva formation in *Caenorhabditis elegans. J. Comp. Neurol.,* **198,** 435–451.

Alberts, B., Bray, D., Lewis, J., Raff, M., Roberts, K., and Watson, J.D. (1983). *The Molecular Biology of the Cell.* New York: Garland Publishing.

Albertson, D.G. (1984). Formation of the first cleavage spindle in nematode embryos. *Dev. Biol.,* **101,** 61–72.

Albertson, D.G., and Thomson, J.N. (1982). The kinetochores of *Caenorhabditis elegans. Chromosoma,* **86,** 409–428.

Albertson, D.G., Sulston, J.E., and White, J.G. (1978). Cell cycling and DNA replication in a mutant blocked in cell division in the nematode *Caenorhabditis elegans. Dev. Biol.,* **63,** 165–178.

Allen, G.E. (1986). Origin of the embryological tradition in the United States. In T.J. Horder, J.A. Witkowski, and C.C. Wylie (Eds.), *A History of Embryology.* Cambridge: Cambridge University Press, pp. 113–146.

Alonso, M.C., and Cabrera, C.V. (1988). The *achaete-scute* gene complex of *Drosophila melanogaster* comprises four homologous genes. *EMBO J.,* **7,** 2585–2591.

Ambros, V., and Horvitz, H.R. (1984). Heterochronic mutants of the nematode *Caenorhabditis elegans. Science,* **226,** 409–416.

Ambros, V., and Horvitz, H.R. (1987). The *lin-14* locus of *Caenorhabditis elegans* controls the time of expression of specific postembryonic developmental events. *Genes Dev.,* **1,** 398–414.

Ambrosio, L., Mahowald, A.P., and Perrimon, N. (1989). *1(1) pole hole* is required maternally for pattern formation in the terminal regions of the embryo. *Development,* **106,** 145–158.

Amrein, H., Gorman, M., and Nöthiger, R. (1988). The sex-determining gene *tra-2* of Drosophila encodes a putative RNA binding protein. *Cell,* **55,** 1025–1035.

Ananiev, E.V., and Gvozdev, V.R. (1974). Changed pattern of transcription and replication in polytene chromosomes of *Drosophila melanogaster* resulting from eu-heterochromatin rearrangement. *Chromosoma*, **45**, 173–191.

Anderson, K.V., and Lengyel, J.A. (1979). Rates of synthesis of major classes of RNA in *Drosophila* embryos. *Dev. Biol.*, **70**, 217–231.

Anderson, K.V., and Lengyel, J.A. (1980). Changing rates of histone mRNA synthesis and turnover in Drosophila embryos. *Cell*, **21**, 717–727.

Anderson, K.V., and Nüsslein-Volhard, C. (1984). Genetic analysis of dorsal-ventral embryonic pattern in *Drosophila*. In G.M. Malacinski and S.V. Bryant (Eds.), *Pattern Formation*. New York: Macmillan, pp. 269–289.

Anderson, K.V., and Nüsslein-Volhard, C. (1984). Information for the dorsal-ventral pattern of the *Drosophila* embryo is stored as maternal mRNA. *Nature*, **311**, 223–227.

Anderson, K.V., Jurgens, G., and Nüsslein-Volhard, C. (1985). Establishment of dorso-ventral polarity in the *Drosophila* embryo—Genetic studies on the role of the *Toll* gene product. *Cell*, **42**, 779–783.

Aristotle (1963). *Generation of Animals*. Cambridge, MA: Harvard University Press.

Aroian, R.V., Koga, M., Mendel, J.E., Ohshima, Y., and Sternberg, P.W. (1990). The *let-23* gene necessary for *Caenorhabditis elegans* vulval induction encodes a tyrosine kinase of the EGF receptor subfamily. *Nature*, **348**, 693–699.

Arthur, C.G., Weide, C.M., Vincent, W.S., and Goldstein, E.S. (1979). mRNA sequence diversity during early embryogenesis in *Drosophila melanogaster*. *Exp. Cell Res.*, **121**, 87–94.

Ashburner, M. (1989). Drosophila: *A Laboratory Handbook*. Cold Spring Harbor, NY: Cold Spring Harbor Laboratory Press.

Austin, J., and Kimble, J. (1987). *glp-1* is required in the germ line for regulation of the decision between mitosis and meiosis in *C. elegans*. *Cell*, **51**, 589–599.

Austin, J., and Kimble, J. (1989). Transcript analysis of *glp-1* and *lin-12*, homologous genes required for cell interactions during development of *C. elegans*. *Cell*, **58**, 565–571.

Avner, P., Amar, L., Dandolo, L., and Guenet, J.L. (1988). Genetic analysis of the mouse using interspecific crosses. *Trends Genet.*, **4**, 18–23.

Babu, P. (1974). Biochemical genetics of *Caenorhabditis elegans*. *Mol. Gen. Genet.*, **135**, 39–44.

Bachvarova, R., and De Leon, V. (1980). Polyadenylated RNA of mouse ova and loss of maternal RNA in early development. *Dev. Biol.*, **74**, 1–8.

Bachvarova, R., De Leon, V., Johnson, A., Kaplan, G., and Paynton, B.V. (1985). Changes in total RNA, polyadenylated RNA and actin mRNA during meiotic maturation of mouse oocytes. *Dev. Biol.*, **108**, 325–331.

Baker, B.S. (1989). Sex in flies: The splice of life. *Nature*, **340**, 521–524.

Baker, B.S., and Ridge, K.A. (1980). Sex and the single cell. I. On the action of major loci affecting sex determination in *Drosophila melanogaster*. *Genetics*, **94**, 383–423.

Baker, B.S., Smith, D.A., and Gatti, M. (1982). Region-specific effects on chromosome integrity of mutations at essential loci in *Drosophila melanogaster*. *Proc. Natl. Acad. Sci., U.S.A.*, **79**, 1205–1209.

Baker, B.S., and Wolfner, M.F. (1988). A molecular analysis *double-sex*, a bifunctional gene that controls both male and female sexual differentiation in *Drosophila melanogaster*. *Genes Dev.*, **2**, 477–489.

Baker, N.E. (1987). Molecular cloning of sequences from *wingless*, a segment polarity gene in *Drosophila*: The spatial distribution of a transcript in embryos. *EMBO J.*, **6**, 1765–1773.

Baker, N.E. (1988a). Transcription of the segment-polarity gene *wingless* in the imaginal discs of *Drosophila*, and the phenotype of a pupal-lethal *wg* mutation. *Development*, **102**, 489–497.

Baker, N.E. (1988b). Localization of transcripts from the *wingless* gene in whole Drosophila embryos. *Development*, **103**, 289–298.

Baker, N.E., and Rubin, G.M. (1989). Effect on eye development of dominant mutations in *Drosophila* homologue of the EGF receptor. *Nature*, **340**, 150–153.

Baker, N.E., Mlodzik, M., and Rubin, G.M. (1990). Spacing differentiation in the developing *Drosophila* eye: A fibrinogen-related lateral inhibitor encoded by *scabrous*. *Science*, **250**, 1370–1377.

Baker, W.K. (1968). Position effect variegation. *Adv. Genet.*, **14**, 133–166.

Bakken, A. (1973). A cytological and genetic study of oogenesis in *Drosophila melanogaster. Dev. Biol.*, **33**, 100–122.

Balcells, L., Modolell, J., and Ruiz-Gomez, M. (1988). A unitary basis for different *Hairy-wing* mutations of *Drosophila melanogaster. EMBO J.*, **7**, 3899–3906.

Balling, R., Deutsch, U., and Gruss, P. (1988). *undulated,* a mutation affecting the development of the mouse skeleton, has a point mutation in the paired box of *Pax 1. Cell*, **55**, 531–535.

Banerjee, U., Renfranz, P.J., Hinton, D.R., Rubin, B.A., and Benzer, S. (1987). The *sevenless* protein is expressed apically in cell membranes of the developing Drosophila retina; it is not restricted to R7. *Cell*, **51**, 151–158.

Bard, J.B.L. (1981). A model for generating aspects of zebra and other mammalian coat patterns. *J. Theor. Biol.*, **93**, 363–385.

Bard, J.B.L. (1991). Epithelial rearrangement and *Drosophila* gastrulation. *BioEssays*, **13**, 409–411.

Barlow, D.P., Stagen, R., Hermann, B.G., Saito, K., and Schweifer, N. (1991). The mouse insulin-like growth factor type-2 receptor is imprinted and closely linked to the *Tme* locus. *Nature*, **349**, 84–87.

Barnett, T., Pachl, C., Gergen, T.P., and Wensink, P.C. (1980). The isolation and characterization of *Drosophila* yolk protein genes. *Cell*, **21**, 729–738.

Bartolomei, M.S., Zemei, S., and Tilghman, S.M. (1991). Parental imprinting of the mouse H9 gene. *Nature*, **351**, 153–155.

Barton, S.C., Surani, M.A.H., and Norris, M.L. (1984). Role of paternal and maternal genomes in mouse development. *Nature*, **311**, 374–376.

Barton, S.C., Adams, C.A., Norris, M.L., and Surani, M.A.H. (1985). Development of gynogenetic and parthenogenetic tissues in reconstituted blastocysts in the mouse. *J. Embryol. Exp. Morphol.*, **90**, 267–285.

Basler, K., and Hafen, E. (1989). Dynamics of *Drosophila* eye development and temporal requirements of *sevenless* expression. *Development*, **107**, 723–731.

Basler, K., and Hafen, E. (1991). Specification of cell fate in the developing eye of *Drosophila. BioEssays*, **13**, 621–631.

Basler, E., Christen, B., and Hafen, E. (1991). Ligand-independent activation of the sevenless receptor tyrosine kinase changes the fate of cells in the developing Drosophila eye. *Cell*, **64**, 1069–1081.

Bate, M., and Martinez-Arias, A. (1991). The embryonic origin of imaginal discs in *Drosophila. Development*, **112**, 755–761.

Bate, M., Rushton, E., and Currie, D.A. (1991). Cells with persistent *twist* expression are the embryonic precursors of adult muscle in *Drosophila. Development*, **113**, 79–89.

Bateson, W. (1894). *Materials for the Study of Variation Treated with Especial Research to Discontinuity in the Origin of Species.* London and New York: Macmillan.

Baumgartner, S., Bopp, D., Burri, M., and Noll, M. (1987). Structure of two genes at the *gooseberry* locus related to the *paired* gene and their spatial expression during *Drosophila* embryogenesis. *Genes Dev.*, **1**, 1247–1267.

Beachy, P.A., Helfand, S.L., and Hogness, D.S. (1985). Segmental distribution of bithorax complex proteins during Drosophila development. *Nature*, **313**, 545–551.

Beachy, P.A., Krasnow, M.A., Gavis, E.R., and Hogness, D.S. (1988). An *Ultrabithorax* protein binds sequences near its own and the *Antennapedia P1* promoters. *Cell*, **55**, 1069–1081.

Beadle, G.W., and Ephrussi, B. (1937). Development of eye colors in *Drosophila:* Diffusible substances and their interrelations. *Genetics*, **22**, 76–86.

Beadle, G., and Tatum, E.L. (1941). Genetic control of biochemical reactions in *Neurospora. Proc. Natl. Acad. Sci., U.S.A.*, **27**, 499–506.

Beeman, R. (1987). A homeotic gene cluster in the red flour beetle, *Tribolium. Nature*, **327**, 247–249.

Beer, G., Technau, G.M., and Campus-Ortega, J.A. (1987). Lineage analysis of transplanted individual cells in embryos of *Drosophila melanogaster.* IV. Commitment and proliferative capabilities of mesodermal cells. *Wilhelm Roux Arch. Dev. Biol.*, **196**, 222–230.

Bell, L.R., Maine, E.M., Schedl, P., and Cline, T.W. (1988). *Sex-lethal,* a Drosophila sex determination switch gene, exhibits sex-specific RNA splicing and sequence similarity to RNA binding proteins. *Cell,* **55,** 1037–1046.

Bellen, H.J., Wilson, C., and Gehring, W.J. (1990). Dissecting the complexity of the nervous system by enhancer detection. *BioEssays,* **12,** 199–204.

Belote, J.M., and Lucchesi, J.C. (1980a). Male-specific lethal mutations of *Drosophila melanogaster. Genetics,* **96,** 165–186.

Belote, J.M., and Lucchesi, J.C. (1980b). Control of X-chromosome transcription by the maleless gene in *Drosophila. Nature,* **285,** 573–575.

Belote, J.M., and Baker, B.S. (1982). Sex determination in *Drosophila melanogaster:* Analysis of *transformer-2,* a sex-transforming locus. *Proc. Natl. Acad. Sci., U.S.A.,* **79,** 1568–1572.

Belote, J.M., McKeown, M.B., Andrew, D.J., Scott, T.N., Wolfner, M.F., and Baker, B.S. (1985). Control of sexual differentiation in *Drosophila melanogaster. Cold Spring Harbor Symp. Quant. Biol.,* **50,** 605–614.

Bender, W., Spierer, P., and Hogness, D.S. (1983a). Chromosomal walking and jumping to isolate DNA from the *Ace* and *rosy* loci and the bithorax complex in *Drosophila melangoaster. J. Mol. Biol.,* **168,** 17–33.

Bender, W., Akam, M., Karch, E., Beachy, P.A., Peifer, M., Spierer, P., Lewis, E.B., and Hogness, D.S. (1983b). Molecular genetics of the bithorax complex in *Drosophila melanogaster. Science,* **221,** 23–29.

Bennett, D., Dunn, L.C., Spiegelman, M., Artzt, K., Cookingham, J., and Schermerhorn, E. (1975). Observations on a set of radiation-induced dominant T-like mutations in the mouse. *Genet. Res.,* **26,** 95–108.

Bensaude, O., Babinet, C., Morange, M., and Jacob, F. (1983). Heat shock proteins, first major products of zygotic gene activity in mouse embryos. *Nature,* **305,** 331–333.

Benzer, S. (1967). Behavioral mutants of Drosophila isolated by countercurrent distribution. *Proc. Natl. Acad. Sci., U.S.A.,* **58,** 1112–1116.

Berleth, T., Burri, M., Thoma, G., Richstein, S., Frijerio, G., Noll, M., and Nüsslein-Volhard, C. (1988). The role of localization of *bicoid* RNA in organizing the anterior pattern of the *Drosophila* embryo. *EMBO J.,* **7,** 1749–1756.

Bernstein, S.E. (1970). Tissue transplantation as an analytic and therapeutic tool in hereditary anemias. *Am. J. Surg.,* **119,** 448–451.

Berta, P., Hawkins, J.R., Sinclair, A.H., Taylor, A., Griffith, B.L., Goodfellow, P.N., and Fellous, M. (1990). Genetic evidence equating *Sry* and the testis-determining factor. *Nature,* **348,** 448–450.

Bienz, M., and Tremml, G. (1988). Domain of *Ultrabithorax* expression in *Drosophila* visceral mesoderm from autoregulation and exclusion. *Nature,* **333,** 576–578.

Bier, E., Jan, L.-Y., and Jan, Y.N. (1990). *rhomboid,* a gene required for dorsoventral axis establishment and peripheral nervous system development in *Drosophila melanogaster. Genes Dev.,* **4,** 190–203.

Bischoff, W.L., and Lucchesi, J.C. (1971). Genetic organization in *Drosophila melanogaster:* Complementation and fine structure analysis of the deep orange locus. *Genetics,* **69,** 453–466.

Bodenstein, D. (1950). The postembryonic development of *Drosophila.* In M. Demerec (Ed.), *The Biology of Drosophila.* New York: John Wiley & Sons, pp. 275–367.

Boggs, R.T., Gregor, P., Idrias, S., Belote, J.M., and McKeown, M. (1987). Regulation of sexual differentiation in *D. melanogaster* via alternative splicing of RNA from the *transformer* gene. *Cell,* **50,** 739–747.

Boswell, R.E., and Mahowald, A.P. (1985). *Tudor,* a gene required for assembly of the germ plasm in *Drosophila. Cell,* **43,** 97–104.

Botas, J., Cabrera, C.V., and Garcia-Bellido, A. (1988). The reinforcement-extinction process of selector gene activity: A positive feed-back loop and cell-cell interactions in *Ultrabithorax* patterning. *Wilhelm Roux Arch. Dev. Biol.,* **197,** 424–434.

Boulay, J.L., Dennefeld, C., and Alberga, A. (1987). The *Drosophila* developmental gene *snail* encodes a protein with nucleic acid binding fingers. *Nature,* **330,** 395–398.

Bourois, M., Moore, P., Ruel, L., Grau, Y., Heitzler, P., and Simpson, P. (1990). An early embryonic product of the gene *shaggy* encodes a serine/threonine protein kinase related to the CDC28/cdc2^{+} subfamily. *EMBO J.,* **9,** 2877–2884.

Boveri, T. (1910). Die potenzen der *Ascaris*—Blastomeren bei abgeänderter furchang. Zugleich ein beitrag zur frage qualitativiv-ungleicher chromosomen-teilung. In *Festschrift f. R. Hertwig's.* Jena: Fischer, vol. III, p. 133.

Bownes, M., and Kalthoff, K. (1974). Embryonic defects in *Drosophila* eggs after partial UV-irradiation at different wavelengths. *J. Embyrol. Exp. Morphol., 31,* 1–17.

Bownes, M., and Sang, J.H. (1974). Experimental manipulations of early *Drosophila* embryos. I. Adult and embryonic defects resulting from microcautery at nuclear multiplication and blastoderm stages. *J. Embryol. Exp. Morphol., 32,* 253–272.

Bownes, M., Steinmann-Zwicky, M., and Nöthiger, R. (1990). Differential control of yolk protein gene expression in fat bodies and gonads by the sex-determining gene of *tra-2* of *Drosophila. EMBO J., 9,* 3975–3980.

Bowtell, D.D.L., Simon, M.A., and Rubin, G.M. (1989). Ommatidia in the developing Drosophila eye require and can respond to *sevenless* for only a restricted period. *Cell, 56,* 931–936.

Boycott, R.E., and Diver, C. (1923). On the inheritance of sinistrality in *Limnaea peregra. Proc. R. Soc. Lond., 95B,* 207–213.

Bradley, A., Evans, M., Kaufman, M.H., and Robertson, E. (1984). Formation of germ-line chimaeras from embryo-derived teratocarcinoma cell lines. *Nature, 309,* 255–256.

Brand, M.A. and Campos-Ortega, J.A., (1988). Two groups of interrelated genes regulate early neurogenesis in *Drosophila melanogaster. Wilhelm Roux Arch. Dev. Biol., 197,* 457–470.

Braude, P.R. (1979). Time-dependent effects of α-amanitin on blastocyst formation in the mouse. *J. Embryol. Exp. Morphol., 52,* 193–202.

Braude, P., Pelham, H., Flach, G., and Lobatto, R. (1979). Post-transcriptional control in the early mouse embryo. *Nature, 282,* 102–105.

Breen, T.R., and Lucchesi, J.C. (1986). Analysis of the dosage compensation of a specific transcript in *Drosophila melanogaster. Genetics, 112,* 483–491.

Brennan, M.D., Weiner, A.J., Goralski, T.J., and Mahowald, A.P. (1982). The follicle cells are a major site of vitellogenin synthesis in *Drosophila melanogaster. Dev. Biol., 89,* 225–236.

Brenner, S. (1974). The genetics of *Caenorhabditis elegans. Genetics, 77,* 71–94.

Brenner, S. (1981). Genes and development. In C.W. Lloyd and D.A. Rees (Eds.), *Cellular Controls in Differentiatiation.* New York: Academic Press.

Brenner, S., Dove, W., Herskowitz, I., and Thomas, R. (1990). Genes and development: Molecular and logical themes. *Genetics, 126,* 479–486.

Bridges, C.B. (1916). Non-disjunction as proof of the chromosome theory of heredity. *Genetics, 1,* 1–52.

Bridges, C.B. (1921). Triploid intersexes in *Drosophila melanogaster. Science, 54,* 252–254.

Bridges, C.B. (1925). Sex in relation to chromosomes and genes. *Am. Nat., 59,* 127–137.

Bridges, C.B. (1935). Salivary chromosome maps. *J. Hered., 26,* 60–64.

Broadie, K.S., and Bate, M. (1991). The development of adult muscles in *Drosophila:* Ablation of identified muscle precursor cells. *Development, 113,* 103–118.

Brosseau, G.E. (1960). Genetic analysis of the male fertility factors on the X chromosome of *Drosophila melanogaster. Genetics, 45,* 257–274.

Brower, D.L. (1986). *engrailed* gene expression in *Drosophila* imaginal discs. *EMBO J., 5,* 2649–2656.

Brown, D.D. (1984). The role of stable complexes that repress and activate eukaryotic genes. *Cell, 37,* 359–365.

Brown, E.H., and King, R.C. (1972). Studies on the events resulting in the formation of an egg chamber in *Drosophila melanogaster.* In *Invertebrate Oogenesis I.* New York: MSS Information, pp. 7–47.

Bryant, P.J. (1970). Cell lineage relationships in the imaginal wing disc of *Drosophila melanogaster. Dev. Biol., 22,* 389–411.

Bryant, P.J. (1988). Localized cell death caused by mutations in a *Drosophila* gene coding for a transforming growth factor-‰ homolog. *Dev. Biol., 128,* 386–395.

Bryant, P., and Schneiderman, H.A. (1969). Cell lineage, growth, and determination in the imaginal leg discs of *Drosophila melanogaster. Dev. Biol., 20,* 263–290.

Bryant, P.J., and Simpson, P. (1984). Intrinsic and extrinsic control of growth in developing organs. *Q. Rev. Biol.,* **59,** 387–415.

Bryant, P.J., and Fraser, S.E. (1988). Wound healing, cell communication, and DNA synthesis during imaginal disc regeneration in *Drosophila. Dev. Biol.,* **127,** 197–208.

Bryant, S.V., French, V., and Bryant, P.J. (1981). Distal regeneration and symmetry. *Science,* **212,** 993–1002.

Bull, A.L. (1966). *Bicaudal,* a genetic factor which affects the polarity of the embryo in *Drosophila melanogaster. J. Exp. Zool.,* **161,** 221–242.

Burglin, T.R., Finney, M., Coulson, A., and Ruvkun, G. (1989). *Caenorhabditis elegans* has scores of homeobox-containing genes. *Nature,* **341,** 239–242.

Burgoyne, P.S. (1989). Thumbs down for zinc finger? *Nature,* **342,** 860–862.

Burgoyne, P.S., Buehr, M., Koopman, P., Rossant, J., and McLaren, A. (1988). Cell autonomous action of the testis-determining gene: Sertoli cells are exclusively XY in XX ↔ XY chimaeric testes. *Development,* **102,** 443–450.

Burtis, K.C., and Baker, B.S. (1989). Drosophila *doublesex* gene controls somatic sexual differentiation by producing alternatively spliced mRNAs encoding related sex-specific polypeptides. *Cell,* **56,** 997–1010.

Burtis, K.C., Coschigano, K.T., Baker, B.S., and Wensink, P.C. (1991). The Doublesex proteins of *Drosophila melanogaster* bind directly to a sex-specific yolk protein gene enhancer. *EMBO J.,* **10,** 2577–2582.

Byerly, L., Cassada, R.C., and Russell, R.L. (1976). The life cycle of the nematode *Caenorhabditis elegans* I. Wild-type growth and reproduction. *Dev. Biol.,* **51,** 23–33.

Cabrera, C.V. (1990). Lateral inhibition and cell fate during neurogenesis in *Drosophila:* The interactions between *scute, Notch* and *Delta. Development,* **109,** 733–742.

Cabrera, C.V., Martinez-Arias, A., and Bate, M. (1987). The expression of three members of the *achaete-scute* gene complex correlates with neuroblast segregation in Drosophila. *Cell,* **50,** 425–433.

Cagan, R.L., and Ready, D.F. (1989). *Notch* is required for successive cell divisions in the developing *Drosophila* retina. *Genes Dev.,* **3,** 1099–1112.

Campos-Ortega, J.A., and Hofbauer, A. (1977). Cell clones and pattern formation: On the lineage of photoreceptor cells in the compound eye of *Drosophila. Wilhelm Roux Arch. Dev. Biol.,* **181,** 227–245.

Campos-Ortega, J.A., and Hartenstein, V. (1985). *The Embryonic Development of* Drosophila melanogaster. Berlin: Springer-Verlag.

Campuzano, S., Carrmolino, L., Cabrera, C.V., Ruiz-Gomez, M., Villares, R., Boronat, A., and Modolell, J. (1985). Molecular genetics of the *achaete-scute* gene complex of D. melanogaster. *Cell,* **40,** 327–338.

Campuzano, S., Balcells, L., Villares, R., Carramolino, L., Garcia-Alonso, C., and Modolell, J. (1986). Excess of function *Hairy-wing* mutations caused by gypsy and copia transposable elements inserted within structural genes of the *achaete-scute* locus of Drosophila. *Cell,* **44,** 303–312.

Capecchi, M. (1989). The new mouse genetics: Altering the genome by gene targeting. *TIG,* **5,** 70–76.

Carpenter, A.T.C. (1975). Electron microscopy of meiosis in *Drosophila melanogaster* females. I. Structure, arrangement, and temporal change of the synaptinemal complex in wild-type. *Chromosoma,* **51,** 157–182.

Carpenter, G. (1987). Receptors for epidermal growth factor and other polypeptide mitogens. *Annu. Rev. Biochem.,* **56,** 881–914.

Carroll, S.B., and Scott, M.P. (1986). Zygotically active genes that affect the spatial expression of the *fushi tarazu* segmentation gene during early *Drosophila* embryogenesis. *Cell,* **45,** 113–126.

Casanova, J. (1990). Pattern formation under the control of the terminal system in the Drosophila embryo. *Development,* **110,** 621–628.

Casanova, J., Sanchez-Herrero, E., and Morata, G. (1986). Identification and characterization of a parasegmental specific regulatory element of the *Abdominal-B* gene of *Drosophila. Cell,* **47,** 627–636.

Casanova, J., and Struhl, G. (1989). Localized surface activity of *torso,* a receptor tyrosine kinase, specifies terminal body pattern in *Drosophila. Genes Dev.,* **3,** 2025–2038.

Cassada, R.C., and Russell, R.L. (1975). The dauerlarva, or post-embryonic developmental variant of the nematode *Caenorhabditis elegans. Dev. Biol.,* **46,** 326–342.

Cassada, R., Issenghi, E., Culotti, M., and von Ehrenstein, G. (1981). Genetic analysis of temperature-sensitive embryogenesis mutants in *Caenorhabditis elegans. Dev. Biol.,* **84,** 193–205.

Castle, W.E., and Little, C.C. (1910). On a modified Mendelian ratio among yellow mice. *Science, 32,* 868–870.

Catalano, G., Eilbeck, C., Monroy, A., and Parisi, E. (1979). A model for early segregation of territories in the ascidian egg. In N. LeDouarin (Ed.), *Cell Lineage, Stem Cells, and Cell Determination.* Amsterdam: Elsevier/North-Holland, pp. 15–29.

Cattanach, B.M. (1975). Control of chromosome inactivation. *Annu. Rev. Genet., 9,* 1–18.

Cattanach, B.M. (1986). Parental origin effects in mice. *J. Embryol. Exp. Morphol., 97* (suppl.), 137–150.

Cattanach, B.M., and Kirk, M. (1985). Differential activity of maternally and paternally derived chromosome regions in mice. *Nature, 315,* 496–498.

Cattanach, B.M., and Beechey, C.M. (1990). Autosomal and X-chromosome imprinting. *Development,* (Suppl.), pp. 63–73.

Caudy, M., Vassin, H., Brand, M., Tuma, R., Jan, L.Y., and Jan, Y.N. (1988). *daughterless,* a *Drosophila* gene essential for both neurogenesis and sex determination, has sequence similarities to *myc* and the *achaete-scute* complex. *Cell, 55,* 1061–1067.

Cavener, D., Corbett, G., Cox, D. and Whetten, R. (1986). Isolation of the eclosion gene cluster and the developmental expression of the *Gld* gene in *Drosophila melanogaster. EMBO J., 5,* 2939–2948.

Celniker, S.E., Sharma, S., Keelan, D.J., and Lewis, E.B. (1990). The molecular genetics of the bithorax complex of *Drosophila: Cis*-regulation in the *Abdominal-B* domain. *EMBO J., 9,* 4277–4286.

Chabot, B., Stephenson, D.A., Chapman, V.M., Besmer, P., and Bernstein, A. (1988). The proto-oncogene c-*kit* encoding a transmembrane tyrosine kinase receptor maps to the mouse *W* locus. *Nature, 335,* 88–89.

Chalepakis, G., Fritsch, R., Fickensher, H., Deutsch, U., Goulding, M., and Gruss, P. (1991). The molecular basis of the *undulated/Pax-1* mutation. *Cell, 66,* 873–884.

Chalfie, M., and Au, M. (1989). Genetic control of differentiation of the *Caenorhabditis elegans* touch receptor neurons. *Science, 243,* 1027–1033.

Chalfie, M., Horvitz, H.R., and Sulston, J.E. (1981). Mutations that lead to reiterations in the cell lineages of *C. elegans. Cell, 24,* 59–69.

Chan, L-N., and Gehring, W. (1971). Determination of blastoderm cells in *Drosophila melanogaster. Proc. Natl. Acad. Sci., U.S.A., 68,* 2217–2221.

Chapman, K.B., and Wolfner, M.F. (1988). Determination of male-specific gene expression in *Drosophila* accessory glands. *Dev. Biol., 126,* 195–202.

Chapman, V.M., Adler, D., Labarca, C., and Wudl, L. (1976). Genetic variation of β-glucuronidase during early embryogenesis. In M.H. Johnson (Ed.), *Early Development of Mammals.* Amsterdam: Elsevier/North-Holland, pp. 115–124.

Chesley, P. (1935). Development of the short-tailed mutant in the house mouse. *J. Exp. Zool., 70,* 429–459.

Chisaka, O., and Capecchi, M. (1991). Regionally restricted developmental defects resulting from targeted disruption of the mouse homeobox gene *hox-1.5. Nature, 350,* 473–479.

Church, R.B., and Schultz, G.A. (1974). Differential gene activity in the pre- and postimplantation mammalian embryo. *Curr. Top. Dev. Biol., 8,* 179–202.

Churchill, F.B. (1991). The rise of classical descriptive embryology In S.F. Gilbert (Ed.), *Developmental Biology: A Comprehensive Synthesis.* New York: Plenum Press, vol. 7, pp. 1–29.

Cleavinger, P.J., McDowell, J.W., and Bennett, K.L. (1989). Transcription in nematodes: Early *Ascaris* embryos are transcriptionally active. *Dev. Biol., 133,* 600–604.

Cline, T.W. (1978). Two closely linked mutations in *Drosophila melanogaster* that are lethal in opposite sexes and interact with *daughterless. Genetics, 90,* 683–698.

Cline, T.W. (1979). A male-specific lethal mutation in *Drosophila melanogaster* that transforms sex. *Dev. Biol., 72,* 266–275.

Cline, T.W. (1984). Autoregulatory functioning of a Drosophila gene product that establishes and maintains the sexually-determined state. *Genetics, 107,* 231–277.

Cline, T.W. (1986). A female-specific lethal lesion in an X-linked positive regulator of the Drosophila sex determination gene, *Sex-Lethal. Genetics, 113,* 641–663.

Cline, T.W. (1988). Evidence that *sisterless-a* and *sisterless-b* are two of several discrete "numerator elements" of the X/A sex determination signal in Drosophila that switch *Sxl* between two alternative stable expression states. *Genetics,* **119,** 829–862.

Cohen, S.M., and Jurgens, G. (1989). Proximal-distal pattern formation in *Drosophila:* Cell autonomous requirement for *Distal-less* gene activity in limb development. *EMBO J.,* **8,** 2045–2055.

Cohen, S.M., Bronner, G., Kuttner, F., Jurgens, G., and Jackle, H. (1989). *distal-less* encodes a homeodomain protein required for limb development. *Nature,* **338,** 432–434.

Colberg-Poley, A.M., Vass, S.D., Chaudbury, K., Stewart, C.L., Wagner, E.F., and Gruss, P. (1985). Clustered homeoboxes are differentially expressed during murine development. *Cell,* **43,** 39–45.

Cole, E.S., and Palka, J. (1982). The patterns of campaniform sensillae on the wing and haltere of *Drosophila melanogaster* and several of its homoeotic mutants. *J. Embryol. Exp. Morphol.,* **71,** 41–61.

Collins, J., Saari, B., and Anderson, P. (1987). Activation of a transposable element in the germ line but not the soma of *Caenorhabditis elegans. Nature,* **328,** 726–728.

Conway, K., Feiock, K., and Hunt, R.K. (1980). Polyclones and patterns in growing *Xenopus* eye. *Curr. Top. Dev. Biol.,* **15,** 217–317.

Copeland, N.G., and Jenkins, N.A. (1991). Development and application of a molecular genetic linkage map of the mouse genome. *TIG,* **7,** 113–118.

Copeland, N.G., Jenkins, N.A., and Lee, B.K. (1983). Association of the lethal Yellow (*A y*) coat color mutation with an ecotropic murine leukemia virus genome. *Proc. Natl Acad. Sci. USA,* **80,** 247–249.

Copeland, N.G., Gilbert, D.J., Cho, B.C., Donovan, P.J., Jenkins, N.A., Cosman, D., Anderson, D., Lyman, S.D., and Williams, D.E. (1990). Mast cell growth factor maps near the *Steel* locus on mouse chromosome 10 and is deleted in a number of *Steel* alleles. *Cell,* **63,** 175–183.

Cori, C.F., Glueksohn-Waelsch, S., Shaw, P.A., and Robinson, C. (1983). Correction of a genetically caused enzyme defect by somatic cell hybridization. *Proc. Natl. Acad. Sci., U.S.A.,* **80,** 6611–6614.

Costa, M., Weir, M., Coulson, A., Sulston, J., and Kenyon, C. (1988). Posterior pattern formation in C. elegans involves position-specific expression of a gene containing a homeobox. *Cell,* **55,** 747–756.

Coulson, A., Kozono, Y., Lutterbach, B., Shownkeen, A., Sulston, J., and Waterston, R. (1991). YACs and the *C. elegans* genome. *BioEssays,* **13,** 413–417.

Coulter, D.E., and Wieschaus, E. (1988). Gene activities and segmental patterning in Drosophila—Analysis of odd-skipped and pair-rule double mutants. *Genes Dev.,* **2,** 1812–1823.

Counce, S.J., and Ede, D.A. (1957). The effect on embryogenesis of a sex-lethal female-sterility factor in *Drosophila melanogaster. J. Emb. Exp. Morph.,* **5,** 404–421.

Cowan, A.E., and McIntosh, J.R. (1985). Mapping the distribution of differentiation potential for intestine, muscle and hypodermis during early development in *Caenorhabditis elegans. Cell,* **41,** 923–932.

Cox, G.N., and Hirsh, D. (1985). Stage-specific patterns of collagen gene expression during development of *Caenorhabditis elegans. Mol. Cell Biol.,* **5,** 363–372.

Cox, G.N., Laufer, J.S., Kusch, M., and Edgar, R.S. (1980). Genetic and phenotypic characterization of roller mutants of *Caenorhabditis elegans. Genetics,* **95,** 317–339.

Cox, G.N., Kusch, M., DeNevi, K., and Edgar, R.S. (1981a). Temporal regulation of cuticle synthesis during development of *Caenorhabditis elegans. Dev. Biol.,* **84,** 277–285.

Cox, G.N., Staprons, S., and Edgar, R.S. (1981b). The cuticle of *Caenorhabditis elegans.* II. Stage-specific changes in ultrastructure and protein composition during postembryonic development. *Dev. Biol.,* **86,** 456–470.

Cox, G.N., Kramer, J.M., and Hirsh, D. (1984). Number and organization of collagen genes in *Caenorhabditis elegans. Mol. Cell Biol.,* **4,** 2389–2395.

Cox, R.D., and Lehrach, H. (1991). Genome mapping: PCR based meiotic and somatic cell hybrid analysis. *BioEssays,* **13,** 193–198.

Crick, F.H.C., and Lawrence, P.A. (1975). Compartments and polyclones in insect development. *Science,* **189,** 340–347.

Cronmiller, C., and Cline, T.W. (1987). The Drosophila sex determination gene *daughterless* has different functions in the germ line versus the soma. *Cell,* **48,** 479–487.

Crouse, H.V. (1960). The controlling element in sex chromosome behavior in *Sciara*. *Genetics,* **45,** 1429–1443.

Cubas, P., de Celis, J.-F., Campuzano, S., and Modolell, J. (1991). Proneural clusters of *achaete-scute* expression and the generation of sensory organs in the *Drosophila* imaginal wing disc. *Genes Dev.,* **5,** 996–1008.

Cuénot, L. (1908). Sur quelques anomalies apparentes des proportions Mendeliennes. *Notes Renne,* **Ib, 9,** 7–15.

Dalcq, A.M. (1957). *Introduction to General Embryology.* London: Oxford University Press.

Dambly-Chaudière, C., and Ghysen, A. (1987). Independent subpatterns of sense organs require independent genes of the *achaete-scute* complex in *Drosophila* larvae. *Genes Dev.,* **1,** 297–306.

Dambly-Chaudière, C., Ghysen, A., Jan, L.Y., and Jan, Y.N. (1988). The determination of sense organs in *Drosophila:* Interaction of *scute* with *daughterless.* *Rouxs Arch. Dev. Biol.,* **197,** 409–423.

Davidson, E.H. (1986). *Gene Activity in Early Development,* 3rd ed. New York: Academic Press.

Davidson, E.H. (1990). How embryos work: A comparative view of diverse modes of cell fate specification. *Development,* **108,** 365–389.

Davidson, E.H., and Britten, R.J. (1979). Regulation of gene expression: possible role of repetitive sequences. *Science,* **204,** 1052–1059.

Davidson, E.H., Jacobs, H.T., Thomas, T.L., Hough-Evans, B.R., and Britten, R.J. (1983). Poly (A) RNA of the egg cytoplasm: Structural resemblance to the nuclear RNA of somatic cells. In R. Proter and J. Whelan (Eds.), *Molecular Biology of Egg Maturation.* London: Pitman, pp. 6–24.

De Chiara, T.M., Robertson, E.J., and Efstradiatis, A. (1991). Parental imprinting of the mouse insulin-like growth factor II gene. *Cell,* **64,** 849–859.

De Franco, D., Bali, D., Torres, R., De Pinho, R.A., Erickson, R.A., and Glueksohn-Waelsch, S. (1991). The glucocorticoid hormone signal transduction pathway in mice homozygous for chromosomal deletions causing failure of cell type specific inducible gene expression. *Proc. Natl. Acad. Sci., U.S.A.,* **88,** 5607–5610.

Degelmann, A., Hardy, P.A., Perrimon, N., and Mahowald, A.P. (1986). Developmental analysis of the torso-like phenotype in *Drosophila* produced by a maternal-effect locus. *Dev. Biol.,* **115,** 479–489.

de La Concha, A., Dietrich, U., Weigel, D., and Campos-Ortega, J.A. (1988). Functional interactions of neurogenic genes of *Drosophila melanogaster. Genetics,* **118,** 499–508.

De Leon, V., Johnson, A., and Bachvarova, R. (1983). Half lives and relative amounts of stored and polyribosomal ribosomes and poly A+ RNA in mouse oocytes. *Dev. Biol.,* **98,** 400–411.

DeLong, L., Casson, L.P., and Meyer, B.J. (1987). Assessment of X chromosome dosage compensation in *Caenorhabditis elegans* by phenotypic analysis of *lin-14. Genetics,* **117,** 657–670.

DeLorenzi, M., and Bienz, M. (1990). Expression of *Abdominal-B* homeoproteins in *Drosophila* embryos. *Development,* **108,** 323–329.

De Lotto, R., and Spierer, P. (1986). A gene required for the specification of dorsal-ventral pattern in *Drosophila* appears to encode a serine protease. *Nature,* **323,** 688–692.

del Pino, E.M., and Elinson, R.P. (1983). A novel developmental pattern for frogs: Gastrulation produces an embryonic disk. *Nature,* **306,** 589–591.

Denell, R.E. (1978). Homeosis in Drosophila: II. A genetic analysis of Polycomb. *Genetics,* **90,** 277–289.

Denell, R., Hummels, R.K., Wakimoto, B.T., and Kauffman, T.C. (1981). Developmental studies of lethality associated with the Antennapedia gene complex in *Drosophila melanogaster. Dev. Biol.,* **81,** 43–50.

Denich, T.R., Schierenberg, E., Isnenghi, E., and Cassada, R. (1984). Cell-lineage and developmental defects of temperature-sensitive embryonic arrest mutants of the nematode *Caenorhabditis elegans. Wilhelm Roux Arch. Dev. Biol.,* **193,** 164–179.

Deppe, U., Schierenberg, E., Cole, T., Krieg, C., Schmitt, D, Yoder, B., and von Ehrenstein, G. (1978). Cell lineages of the embryo of the nematode *Caenorhabditis elegans. Proc. Natl. Acad. Sci., U.S.A.,* **75,** 376–380.

Dickinson, W.J. (1988). On the architecture of regulatory systems: Evolutionary insights and implications. *BioEssays,* **8,** 204–208.

Dietrich, U., and Campos-Ortega, J.A. (1984). The expression of neurogenic loci in imaginal epidermal cells of *Drosophila melanogaster. J. Neurogen.*, **1**, 315–332.

DiNardo, S., and O'Farrell, P.H. (1987). Establishment and refinement of segmental pattern in the *Drosophila* embryo: Spatial control of *engrailed* expression by pair-rule genes. *Genes Dev.*, **1**, 1212–1225.

DiNardo, S., Sher, E., Heemskerk-Jurgens, J., Kassis, J.A., and O'Farrell, P.H. (1988). Two-tiered regulation of spatially patterned *engrailed* gene expression during *Drosophila* embryogenesis. *Nature*, **332**, 604–609.

Dobrovolskaia-Zavovskaia, N. (1927). Sur la modification spontanée de la queue chez la souris nouveau-née et sur l'éxistence d'un charactère (facteur) héréditaire "non-viable". *C. R. Soc. Biol. (Paris)*, **97**, 114–116.

Doe, C.Q., and Goodman, C.S. (1985). Early events in insect neurogenesis. II. The role of cell interactions and cell lineage in the determination of neuronal precursor cells. *Dev. Biol.*, **111**, 206–219.

Doe, C.Q., Hiromi, Y., Gehring, W.J., and Goodman, C.S. (1988). Expression and function of the segmentation gene *fushi tarazu* during *Drosophila* neurogenesis. *Science*, **239**, 170–175.

Doniach, T. (1986). Activity of the sex-determining gene *tra-2* is modulated to allow spermatogenesis in the *C. elegans* hermaphrodite. *Genetics*, **114**, 53–76.

Doniach, T., and Hodgkin, J. (1984). A sex-determining gene, *fem-1*, required for both male and hermaphrodite development in *Caenorhabditis elegans. Dev. Biol.*, **106**, 223–235.

Dove, W.F. (1987). Molecular genetics of *Mus musculus:* Point mutagenesis and millimorgans. *Genetics*, **116**, 5–8.

Doyle, H.J., Harding, K., Hoey, T., and Levine, M. (1986). Transcripts encoded by a homeobox gene are restricted to dorsal tissues of *Drosophila* embryos. *Nature*, **323**, 76–79.

Driever, W., and Nüsslein-Volhard, C. (1988a). A gradient of *bicoid* protein in *Drosophila* embryos. *Cell*, **54**, 83–93.

Driever, W., and Nüsslein-Volhard, C. (1988b). The *bicoid* protein determines position in the *Drosophila* embryo in a concentration-dependent manner. *Cell*, **54**, 95–104.

Driever, W., and Nüsslein-Volhard, C. (1989). The bicoid protein is a positive regulator of *hunchback* transcription in the early *Drosophila* embryo. *Nature*, **337**, 138–143.

Driever, W., Thoma, G., and Nüsslein-Volhard, C. (1989). Determination of spatial domains of zygotic gene expression in the *Drosophila* embryo by the affinity of binding sites for the bicoid morphogen. *Nature*, **340**, 363–367.

Driscoll, M., and Chalfie, M. (1992). Developmental and abnormal cell death in *C. elegans. TINS*, **15**, 15–19.

Dubendorfer, K., and Nöthiger, R. (1982). A clonal analysis of cell lineage and growth in the male and female genital disc of *Drosophila melanogaster. Wilhelm Roux Arch.* **191**, 42–55.

Dubinin, N.P. (1929). Allelomorphentreppen bei *Drosophila melanogaster. Biol. Zentralbl.*, **49**, 328–339.

Duboule, D., and Dollé, P. (1989). The structural and functional organization of the murine HOX gene family resembles that of *Drosophila* homeotic genes. *EMBO J.*, **8**, 1497–1505.

Ducibella, T. (1977). Surface changes of the developing trophoblast cell. In M.H. Johnson (Ed.), *Development in Mammals*. Amsterdam: Elsevier/North/Holland, vol. 1, pp. 5–30.

Duncan, I.M. (1986). Control of bithorax complex functions by the segmentation gene *fushi tarazu* of D. melanogaster. *Cell*, **47**, 297–309.

Duncan, I., and Lewis, E.B. (1982). Genetic control of body segment differentiation in *Drosophila*. In S. Subtelny and P.B. Green (Eds.), *Developmental Order: Its Origin and Regulation*. New York: Alan R. Liss, pp. 533–554.

Eaton, S.,and Kornberg, T.B. (1990). Repression of *ci-D* in posterior compartments of *Drosophila* by *engrailed. Genes Dev.*, **4**, 1068–1077.

Eberlein, S., and Russell, M.A. (1983). Effects of deficiencies in the *engrailed* region of *Drosophila melanogaster. Dev. Biol.*, **100**, 227–237.

Edgar, B.A., and Schubiger, G. (1986). Parameters controlling transcriptional activation during early Drosophila development. *Cell*, **44**, 871–877.

Edgar, B.A., Kiehle P., and Schubiger, G. (1986). Cell cycle control by the nucleo-cytoplasmic ratio in early Drosophila development. *Cell,* **44,** 365–372.

Edgar, B.A., Odell, G.M., and Schubiger, G. (1987). Cytoarchitecture and the patterning of *fushi tarazu* expression in the *Drosophila* blastoderm. *Genes Dev.,* **1,** 1226–1237.

Edgar, B.A., Odell, G.M., and Schubiger, G. (1989). A genetic switch, based on negative regulation, sharpens stripes in *Drosophila* embryos. *Dev. Genet.,* **10,** 124–142.

Edgar, B.A., and O'Farrell, P.H. (1989). Genetic control of cell-division patterns in the *Drosophila* embryo. *Cell,* **57,** 177–187.

Edgar, L.G., and McGhee, J.D. (1986). Embryonic expression of a gut-specific esterase in *Caenorhabditis elegans. Dev. Biol.,* **114,** 109–118.

Edgar, L.G., and McGhee, J.D. (1988). DNA synthesis and the control of embryonic gene expression in *C. elegans. Cell,* **53,** 589–599.

Eicher, E. (1983). Primary sex determining genes in mice. In R.P. Ammaru and G.E. Seidel (Eds.), *Prospects for Sexing Mammalian Sperm.* Boulder: Colorado University Press, pp. 121–135.

Eicher, E.M., and Washburn, L.L. (1986). Genetic control of primary sex determination in mice. *Annu. Rev. Genet.,* **20,** 327–360.

Eissenberg, J.C. (1989). Position effect variegation in *Drosophila*—Toward a genetics of chromatin assembly. *BioEssays,* **11,** 14–17.

Elinson, R.P. (1987). Change in developmental patterns: embryos of amphibians with large eggs. In R.A. Raff and E.C. Raff (Eds.), *Development as an Evolutionary Process.* New York: Alan R. Liss, pp. 1–21.

Ellis, H.M., and Horvitz, H.R. (1986). Genetic control of programmed cell death in the nematode *C. elegans. Cell,* **44,** 817–829.

Ellis, H.M., Spann, D.R., and Posakony, J.W. (1990). *Extra-macrochaete,* a negative regulator of sensory organ development in *Drosophila,* defines a new class of helix-loop-helix proteins. *Cell,* **61,** 27–38.

Emmons, S.W. (1988). The genome. In W.B. Wood (Ed.), *The Nematode* Caenorhabditis elegans. Cold Spring Harbor, NY: Cold Spring Harbor Laboratory Press.

Emmons, S.W., Klass, M.R., and Hirsh, D. (1979). Analysis of the constancy of DNA sequences during development and evolution of the nematode *Caenorhabditis elegans. Proc. Natl. Acad. Sci., U.S.A.,* **76,** 1333–1337.

Emmons, S., Rosensweig, B., and Hirsh, D. (1980). Arrangement of repeated sequences in the DNA of the nematode *Caenorhabditis elegans. J. Mol. Biol.,* **144,** 481–500.

Englesberg, E., and Wilcox, G. (1974). Regulation: Positive control. *Annu. Rev. Genet.,* **8,** 219–242.

Englesberg, E., Irr, J., Power, J., and Lee, N. (1965). Positive control of enzyme synthesis by gene C in the L-arabinose system. *J. Bacteriol.,* **90,** 946–955.

Ephrussi, A., Dickinson, L.K., and Lehmann, R. (1991). *oskar* organizes the germ plasm and directs localization of the posterior determinant *nanos. Cell,* **66,** 37–50.

Eppig, J.J. (1991). Intercommunication between mammalian oocytes and companion somatic cells. *BioEssays,* **13,** 569–574.

Erickson, R.P. (1990). Post-meiotic gene expression. *Trends Genet.* **6,** 264–269.

Esworthy, S., and Chapman, V.M. (1980). The expression of β-galactosidase during pre-implantation mouse embryogenesis. *Dev. Genet.,* **2,** 1–12.

Evans, M.H., and Kaufman, M.H. (1981). Establishment in culture of pluripotential cells from mouse embryos. *Nature,* **292,** 154–156.

Fehon, R.G., Kooh, P.J., Rebay, I., Regan, C.L., Xu, T., Muscavitch, M.R.T., and Artavanis-Tsakonas, S. (1990). Molecular interactions between the protein products of the neurogenic loci *Notch* and *Delta,* two EGF-homologous genes in Drosophila. *Cell,* **61,** 523–534.

Ferguson, E.L., and Horvitz, H.R. (1985). Identification and characterization of 22 genes that affect the vulval cell lineages of the nematode *Caenorhabditis elegans. Genetics,* **110,** 17–72.

Ferguson, E.L., and Horvitz, H.R. (1989). The multivulva phenotype of certain *Caenorhabditis elegans* mutants results from defects in two functionally redundant pathways. *Genetics,* **123,** 109–121.

Ferguson, E.L., Sternberg, P.W., and Horvitz, H.R. (1987). A genetic pathway for the specification of the vulval cell lineages of *Caenorhabditis elegans. Nature,* **326,** 259–267.

Ferguson-Smith, A.C., Cattanach, B.M., Barton, S.C., Beechey, C.V., and Surani, M.A. (1991). Embryological and molecular investigations of parental imprinting on mouse chromosome 7. *Nature*, **351**, 667–670.

Ferguson-Smith, M.A. (1966). X-Y chromosomal interchange in the aetiology of true hermaphroditism and of XX Klinefelter's syndrome. *Lancet*, **ii**, 475–476.

Ferrus, A., and Kankel, D.R. (1981). Cell lineage relationships in *Drosophila melanogaster*: The relationships of cuticular to internal tissues. *Dev. Biol.*, **85**, 485–504.

Files, J.G., Carr, S., and Hirsh, D. (1983). Actin gene family of *Caenorhabditis elegans*. *J. Mol. Biol.*, **164**, 355–375.

Finnegan, D.J., and Fawcett, D.H. (1986). Transposable elements in eukaryotes. *Int. Rev. Cytol.*, **93**, 281–326.

Finney, M., and Ruvkun, G. (1990). The *unc-86* gene product couples cell lineage and cell identity in C. elegans. *Cell*, **63**, 895–905.

Finney, M., Ruvkun, G., and Horvitz, H.R. (1988). The *C. elegans* cell lineage and differentiation gene *unc-86* encodes a protein with a homeodomain and extended similarity to transcription factors. *Cell*, **55**, 757–769.

Fjose, A., McGinniss, W.J., and Gehring, W.J. (1985). Isolation of a homeobox containing gene from the *engrailed* region of *Drosophila* and the spatial distribution of its transcripts. *Nature*, **313**, 284–289.

Flach, G., Johnson, M.H., Braude, P.R., Taylor, R.A.S., and Bolton, V.N. (1982). The transition from maternal to embryonic control in the 2-cell mouse embryo. *EMBO J.*, **1**, 681–685.

Flanagan, J.G., and Leder, P. (1990). The *kit* ligand—A cell surface molecule altered in *Steel* mutant fibroblasts. *Cell*, **63**, 185–194.

Fleming, T.P. (1987). A quantitative analysis of cell allocation to trophectoderm and inner cell mass in the mouse blastocyst. *Dev. Biol.*, **119**, 520–531.

Fleming, T.P., and Johnson, M.H. (1988). From egg to epithelium. *Annu. Rev. Cell Biol.*, **4**, 459–485.

Foe, V., and Alberts, B.M. (1985). Reversible chromosome condensation induced in *Drosophila* embryos by anoxia—Visualization of interphase nuclear organization. *J. Cell Biol.*, **100**, 1623–1636.

Foe, V.E. (1989). Mitotic domains reveal early commitment of cells in *Drosophila* embryos. *Development*, **107**, 1–22.

Foe, V.E., and Alberts, B.M. (1983). Studies of nuclear and cytoplasmic behavior during the five mitotic cycles that precede gastrulation in *Drosophila* embryogenesis. *J. Cell Sci.*, **61**, 31–70.

Foster, G.G., and Suzuki, D.T. (1970). Temperature-sensitive mutations in *Drosophila melanogaster*. IV. A mutation affecting eye facet arrangement in a polarized manner. *Proc. Natl. Acad. Sci., U.S.A.*, **67**, 738–741.

Freeman, G., and Lundelius, J.W. (1982). The developmental genetics of dextrality and sinistrality in the gastropod *Lymnaea perergra*. *Wilhelm Rouxs Arch. Entwicklungs. Org.*, **191**, 69–83.

French, V., Bryant, P.J., and Bryant, S. (1976). Pattern regulation in epimorphic fields. *Science*, **193**, 969–981.

Freyd, G., Kim, S., and Horvitz, H.R. (1990). Novel cysteine-rich motif and homeodomain in the product of the *Caenorhabditis elegans* cell lineage gene *lin-11*. *Nature*, **344**, 876–879.

Frigerio, G., Burri, M., Bopp, D., Baumgartner, S., and Noll, M. (1987). Structure of the segmentation gene *paired* and the Drosophila PRD gene set as part of a gene network. *Cell*, **47**, 735–746.

Frohman, M.A., and Martin, G.R. (1989). Cut, paste and save: New approaches to altering specific genes in mice. *Cell*, **56**, 145–147.

Frohman, M., Boyle, M., and Martin, G.R. (1990). Isolation of the mouse *Hox-2.9* gene: Analysis of embryonic expression suggests that positional information along the antero-posterior axis is specified by mesoderm. *Development*, **110**, 589–607.

Frohnhöfer, H.G., and Nüsslein-Volhard, C. (1986). Organization of anterior pattern in the *Drosophila* embryo by the maternal gene *bicoid*. *Nature*, **324**, 120–125.

Frohnhöfer, H.G., and Nüsslein-Volhard, C. (1987). Maternal genes required for the anterior localization of *bicoid* activity in the embryo of *Drosophila*. *Genes Dev.*, **1**, 880–890.

Frohnhöfer, H.G., Lehmann, R., and Nüsslein-Volhard, C. (1986). Manipulating the anterior-posterior pattern of the *Drosophila* embryo. *J. Embryol. Exp. Morphol.* **97** (Suppl.), 169–179.

Fryxell, K.J., and Meyerowitz, E.M. (1987). An opsin gene that is expressed only in the R7 photoreceptor of *Drosophila*. *EMBO J.*, **6**, 443–451.

Fullilove, S.L., and Jacobson, A.G. (1971). Nuclear elongation and cytokinesis in *Drosophila montana*. *Dev. Biol.*, **26**, 560–577.

Fullilove, S.L., Jacobson, A.G., and Turner, F.R. (1978). Embryonic development: Descriptive. In M. Ashburner and T.R.F. Wright (Eds.), *The Genetics and Biology of Drosophila*. London: Academic Press.

Galau, G., Klein, W.H., Davis, M.M., Wold, B.J., Britten, B.J., and Davidson, E.H. (1976). Structural gene sets active in embryos and adult tissues of the sea urchin. *Cell*, **7**, 487–506.

Gans, M., Audit, C., and Masson, M. (1975). Isolation and characterization of sex-linked female-sterile mutants in *Drosophila melanogaster*. *Genetics*, **81**, 863–704.

Garcia-Bellido, A. (1975). Genetic control of wing disc development in *Drosophila*. In R. Porter and K. Elliott (Eds.), *Cell Patterning*. Amsterdam: Elsevier/North-Holland, pp. 161–178.

Garcia-Bellido, A. (1979). Genetic analysis of the *achaete-scute* system of *Drosophila melanogaster*. *Genetics*, **91**, 491–520.

Garcia-Bellido, A., and Merriam, J.R. (1969). Cell lineage of the imaginal discs in *Drosophila* gynandro-morphs. *J. Exp. Zool.*, **170**, 61–76.

Garcia-Bellido, A., and Merriam, J.R. (1971a). Genetic analysis of cell heredity in imaginal discs of *Drosophila melanogaster*. *Proc. Natl. Acad. Natl. Sci., U.S.A.*, **68**, 2222–2226.

Garcia-Bellido, A., and Merriam, J.R. (1971b). Parameters of the wing imaginal disc development of *Drosophila melanogaster*. *Dev. Biol.*, **24**, 61–87.

Garcia-Bellido, A., and Santamaria, P. (1972). Developmental analysis of the wing disc in the mutant *engrailed* of *Drosophila melanogaster*. *Dev. Biol.*, **24**, 61–87.

Garcia-Bellido, A., and Santamaria, P. (1978). Developmental analysis of the achaete-scute system of *Drosophila melanogaster*. *Genetics*, **88**, 469–486.

Garcia-Bellido, A., and Moscoso del Prado, J. (1979). Genetic analysis of maternal information in *Drosophila*. *Nature,*, **278**, 346–348.

Garcia-Bellido, A., and Robbins, L.G. (1983). Viability of female germ-line cells homozygous for zygotic lethals in *Drosophila melanogaster*. *Genetics*, **103**, 235–247.

Garcia-Bellido, A., Ripoll, P., and Morata, G. (1973). Developmental compartmentalization of the wing disc of *Drosophila*. *Nature New Biol.* **245**, 251–253.

Garcia-Bellido, A., Ripoll, P., and Morata, G. (1976). Developmental compartmentalization in the dorsal mesothoracic disc of *Drosophila*. *Dev. Biol.*, **48**, 132–147.

Garcia-Bellido, A., Delprado, J.M., and Botas, J. (1983). The effect of aneuploidy on embryonic develop-ment in *Drosophila melanogaster*. *Mol. Gen. Genet.*, **192**, 253–263.

Gardner, R.L. (1978). The relationship between cell lineage and differentiation in the early mouse embryo. In W.J. Gehring (Ed.), *Genetic Mosaics and Differentiation*. Heidelberg: Springer-Verlag, pp. 205–241.

Gardner, R.L. (1984). An *in situ* cell marker for clonal analysis of development of the extraembryonic endoderm in the mouse. *J. Embryol. Exp. Morphol.*, **80**, 251–258.

Gardner, R.L. (1985). Clonal analysis of early mammalian development. *Philos. Trans. R. Soc. Lond. [Biol.]*, **312**, 163–178.

Gardner, R.L., and Rossant, J. (1979). Investigation of the fate of 4.5 day post-coitum mouse inner cell mass cells by blastocyst injection. *J. Embryol. Exp. Morphol.*, **52**, 141–152.

Gardner, R.L., Lyon, M.F., Evans, E.P., and Burtenshaw, M.D. (1985). Clonal analysis of X-chromosome inactivation and the origin of the germ line in the mouse embryo. *J. Embryol. Exp. Morphol.*, **88**, 349–363.

Garner, W., and McLaren, A. (1974). Cell distribution in chimaeric mouse embryos before implantation. *J. Embryol. Exp. Morphol.*, **32**, 495–503.

Garrell, J., and Campuzano, S. (1991). The helix-loop-helix domain: A common motif for bristles, muscles and sex. *BioEssays*, **13**, 493–498.

Gatti, M., and Pimpinelli, S. (1983). Cytological and genetic analysis of the Y chromosome of *Drosophila melanogaster*. I. Organization of the fertility factors. *Chromosoma*, **88**, 349–373.

Gaul, U., and Jäckle, H. (1987). Pole region-dependent repression of the Drosophila gap gene *Krüppel* by maternal gene products. *Cell*, **51**, 549–555.

Gaul, U., and Jäckle, H. (1989). Analysis of maternal effect mutant combinations elucidates regulation and function of the overlap of *hunchback* and *Krüppel* gene expression in the *Drosophila* blastoderm embryo. *Development,* **107,** 651–662.

Gaunt, S.J. (1987). Homoeobox gene *Hox-1.5* expression in mouse embryos: Earliest detection by *in situ* hybridization is during gastrulation. *Development,* **101,** 51–60.

Gaunt, S.J. (1991). Expression patterns of mouse Hox genes: Clues to an understanding of developmental and evolutionary strategies. *BioEssays,* **13,** 505–513.

Gaunt, S.J., Miller, J.R., Powell, D.J., and Duboule, D. (1986). Homeobox gene expression in mouse embryos varies with position by the primitive streak stage. *Nature,* **324,** 662–664.

Gaunt, S.J., Sharpe, P.T., and Duboule, D. (1988). Spatially restricted domains of homeo gene transcripts in mouse embryos: Relation to a segmented body plan. *Development,* **104** (Suppl.), 169–179.

Gaunt, S.J., Krumlauf, R., and Duboule, D. (1989). Mouse homeo-genes within a subfamily, Hox-1.4, -2.6, and -5.1, display similar anteroposterior domains of expression in the embryo but show stage- and tissue-dependent differences in their regulation. *Development,* **107,** 131–141.

Gaunt, S.J., and Singh, P.B. (1990). Homeogene expression patterns and chromosomal imprinting. *Trends Genet.,* **6,** 208–212.

Gehring, W. (1967). Clonal analysis of determination dynamics in cultures of imaginal discs in *Drosophila melanogaster. Dev. Biol.,* **16,** 438–456.

Gehring, W.J. (1976). Determination of primordial disc cells and the hypothesis of stepwise determination. In P.A. Lawrence (Ed.), *Insect Development.* Oxford: Blackwell Scientific Publications, pp. 99–108.

Gehring, W.J. (1978). Imaginal discs: Determination. In M. Ashburner and T.R.F. Wright (Eds.), *The Genetics and Biology of Drosophila.* London: Academic Press, vol. 2c, pp. 511–554.

Gehring, W.J. (1985). Homeotic genes, the homeo box, and the genetic control of development. *Cold Spring Harbor Symp. Quant. Biol.,* **50,** 243–251.

Geissler, E.N., Ryan, M.A., and Housman, D.E. (1988). The dominant-White spotting (*W*) locus of the mouse encodes the c-*kit* proto-oncogene. *Cell,* **55,** 185–192.

Georgi, L.L., Albert, P.S., and Riddle, D.L. (1990). *daf-1,* a C. elegans gene controlling dauer larva development, encodes a novel receptor protein kinase. *Cell,* **61,** 635–645.

Gerasimova, T.I., and Smirnova, S.G. (1979). Maternal effect for genes encoding 6-phospho-gluconate dehydrogenase and glucose-6-phosphate dehydrogenase in *Drosophila melanogaster. Dev. Genet.,* **1,** 97–107.

Gergen, J.P. (1987). Dosage compensation in Drosophila: Evidence that daughterless and *Sex-lethal* control X chromosome activity at the blastoderm stage of embryogenesis. *Genetics,* **117,** 477–485.

Gergen, J.P., and Wieschaus, E.F. (1985). The localized requirements for a gene affecting segmentation in *Drosophila*: Analysis of larvae mosaic for *runt. Dev. Biol.,* **109,** 321–335.

Gergen, J.P., and Wieschaus, E.F. (1986). Localized requirements for gene activity in segmentation of *Drosophila* embryos: analysis of *armadillo, fused, giant* and *unpaired* mutations in mosaic embryos. *Wilhelm Roux Arch. Dev. Biol.,* **195,** 49–62.

Gergen, J.P., Coulter, D.,and Wieschaus, E. (1986). Segmental pattern and blastoderm cell identities. In S. Subtelny (Ed.), *Gametogenesis and the Early Embryo.* New York: Alan R. Liss, pp. 195–200.

Ghysen, A., and Dambly-Chaudière, C. (1988). From DNA to form: The *achaete-scute* complex. *Genes Dev.,* **2,** 495–501.

Gilbert, S.F. (1991). Induction and the origins of developmental genetics. In S.F. Gilbert (Ed): *Developmental Biolog, A Comprehensive Synthesis.* New York and London: Plenum, vol. 7, pp. 181–206.

Ginelli, E., di Lernia, R., and Cornea, G. (1977). The organization of DNA sequences in the mouse genome. *Chromosoma,* **61,** 215–226.

Ginsburg, M., Snow, M.H.L., and McLaren, A. (1990). Primordial germ cells in the mouse embryo during gastrulation. *Development,* **110,** 521–528.

Gluecksohn-Schoenheimer, S. (1938). The development of two tailless mutants in the house mouse. *Genetics,* **23,** 573–584.

Gluecksohn-Waelsch, S. (1979). Genetic control of morphogenetic and biochemical differentiation: Lethal albino deletions in the mouse. *Cell,* **16,** 225–237.

Gluecksohn-Waelsch, S. (1983). Genetic control of differentiation. *Teratocarcinoma Stem Cells*, **10**, 3–13.

Gluecksohn-Waelsch, S., and Erickson, R.P. (1970). The T-locus of the mouse: Implications for mechanisms of development. *Curr. Top. Dev. Biol.*, **5**, 281–316.

Gluecksohn-Waelsch, S., and De Franco, D. (1991). Lethal chromosomal deletions in the mouse, a model system for the study of development and regulation of postnatal gene expression. *BioEssays*, **13**, 557–561.

Goddard, M.J., and Pratt, H.P.M. (1983). Control of events during early cleavage of the mouse embryo: An analysis of the 2-cell block. *J. Embryol. Exp. Morphol.*, **73**, 111–133.

Goldberg, D.A., Posakony, J.W., and Maniatis, T. (1983). Correct developmental expression of a cloned alcohol dehydrogenase gene transduced into the *Drosophila* germ line. *Cell*, **34**, 59–73.

Golden, J.W., and Riddle, D.L. (1982). A pheromone influences larval development in the nematode *Caenorhabditis elegans*. *Science*, **218**, 578–580.

Golden, J.W., and Riddle, D.L. (1984). The *Caenorhabditis elegans* dauer larva: Developmental effects of pheromone, food and temperature. *Dev. Biol.*, **102**, 368–378.

Goldschmidt, R. (1938). *Physiological Genetics*. New York: McGraw-Hill.

Goralski, T.J., Edstrom, J.-E.,and Baker, B.S. (1989). The sex determination locus *transformer-2* of Drosophila encodes a polypeptide with similarity to RNA binding proteins. *Cell*, **56**, 1011–1018.

Gould, A.P., Brookman, J.J., Strutt, D.I., and White, R.A.H. (1990). Targets of homeotic gene control in *Drosophila*. *Nature*, **348**, 308–312.

Gould, S.J. (1977). *Ontogeny and Phylogeny*. Cambridge, MA: Harvard University Press.

Gould, S.J. (1990). *Wonderful Life: The Burgess Shale and the Nature of History*. London: Hutchinson.

Govind, S., and Steward, R. (1991). Dorsoventral pattern formation in *Drosophila*: Signal transduction and nuclear targetting. *Trends Genet.* **7**, 119–125.

Graham, A., Papalopulu, N., and Krumlauf, R. (1989). The murine and *Drosophila* homeobox gene complexes have common features of organization and expression. *Cell*, **57**, 367–368.

Graham, C.F. (1973). The necessary conditions for gene expression during early mammalian development. In F.H. Ruddle (Ed.), *Genetic Mechanisms of Development*. New York: Academic Press.

Graham, C.F. (1974). The production of parthenogenetic mammalian embryos and their use in biological research. *Biol. Rev.*, **49**, 399–422.

Grau, Y., Carteret, C., and Simpson, P. (1984). Mutations and chromosomal rearrangements affecting the expression of *snail*, a gene involved in embryonic patterning in Drosophila melanogaster. *Genetics*, **108**, 347–360.

Green, M.C. (1989). Catalog of mutant genes and polymorphic loci. In M.F. Lyon and A.G. Searle (Eds.), *Genetic Variants and Strains of the Laboratory Mouse*. Oxford: Oxford University Press, pp. 12–403.

Green, S.H. (1981). Segment-specific organization of leg motoneurones is transformed in *bithorax* mutants of *Drosophila*. *Nature*, **292**, 152–154.

Greenwald, I. (1985). *lin-12*, a nematode homeotic gene, is homologous to a set of mammalian proteins that includes epidermal growth factor. *Cell*, **43**, 583–590.

Greenwald, I.S., and Horvitz, H.R. (1980). *unc-93 (1500):* A behavioral mutant of *Caenorhabditis elegans* that defines a gene with a wild-type null phenotype. *Genetics*, **96**, 147–164.

Greenwald, I., and Seydoux, G. (1990). Analysis of gain-of-function mutations in *lin-12* gene of *Caenorhabditis elegans*. *Nature*, **346**, 197–199.

Greenwald, I.S., Sternberg, P.W., and Horvitz, H.R. (1983). The *lin-12* locus specifies cell fates in *Caenorhabditis elegans*. *Cell*, **34**, 435–444.

Gropp, A., Winking, H., and Putz, B. (1981). Critical points in development of trisomic mouse embryos. *Clin. Genet.*, **17**, 170.

Gruneberg, H. (1952). *The Genetics of the Mouse*. The Hague: Martinus Nijhoff.

Gubb, D. (1985a). Domains, compartments and determinative switches in *Drosophila* development. *BioEssays*, **2**, 27–31.

Gubb, D. (1985b). Further studies on *engrailed* mutants in *Drosophila melanogaster*. *Wilhelm Roux Arch. Dev. Biol.*, **194**, 181–195.

Gubb, D. (1986). Intron-delay and the precision of expression of homoeotic gene products in *Drosophila*. *Dev. Genet.*, **7**, 119–131.

Gubb, D., and Garcia-Bellido, A. (1982). A genetic analysis of the determination of cuticular polarity during development in *Drosophila melanogaster. J. Embryol. Exp. Morphol.*, **68**, 37–57.

Gubbay, J., Collignon, J., Koopman, P., Capel, B., Economas, A., Munsterberg, A., Vivian, N., Goodfellow, P., and Lovell-Badge, R. (1990). A gene mapping to the sex-determining region of the mouse Y chromosome is a member of a novel family of embryonically expressed genes. *Nature*, **346**, 245–249.

Guo, X., Johnson, J.J., and Kramer, J.M. (1991). Embryonic lethality caused by mutation in basement membrane collagen of *C. elegans. Nature*, **349**, 707–709.

Gutzheit, H.O. (1979). Expression of the zygotic genome in blastoderm stage embryos of *Drosophila:* Analysis of a specific protein. *Wilhelm Roux Arch.*, **188**, 153–156.

Gutzheit, H.O., and Gehring, W.J. (1979). Localized protein synthesis of specific proteins during oogenesis and early embryogenesis in *Drosophila melanogaster. Wilhelm Roux Arch.*, **187**, 151–165.

Hadorn, E. (1948). Gene action in growth and differentiation of lethal mutants. In *Society of Experimental Biology Symposium*. Cambridge: Cambridge University Press, vol. 2, pp. 177–195.

Hadorn, E. (1961). *Development Genetics and Lethal Factors*. New York: John Wiley & Sons.

Hadorn, E. (1965). Problems of determination and transdetermination. *Brookhaven Symp. Biol.*, **18**, 148–161.

Hadorn, E. (1978). Transdetermination. In M. Ashburner and T.R.F. Wright (Eds.), *The Genetics and Biology of Drosophila*. London: Academic Press, vol. 2c, pp. 556–617.

Haenlin, M., Roos, C., Cassab, A., and Mohier, E. (1987). Oocyte-specific transcription of *fs(1)K10:* A *Drosophila* gene affecting dorsal-ventral developmental polarity. *EMBO J.*, **6**, 801–807.

Hafen, E., Levine, M., and Gehring, W.J. (1984a). Regulation of *Antennapedia* transcript distribution by the bithorax complex in *Drosophila. Nature*, **307**, 287–289.

Hafen, E., Kuroiwa, A., and Gehring, W.J. (1984b). Spatial distribution of transcripts from the segmentation gene *fushi tarazu* during *Drosophila* embryonic development. *Cell*, **37**, 833–841.

Hafen, E., Basler, K., Edstroem, J.-E., and Rubin, G.M. (1987). *sevenless*, a cell-specific homeotic gene of Drosophila encodes a putative transmembrane receptor with a tyrosine kinase domain. *Science*, **236**, 55–63.

Hall, J.G. (1990). Genomic imprinting: Review and relevance to human diseases. *Am. J. Hum. Genet.*, **46**, 857–873.

Han, M., and Sternberg, P.W. (1990). *let-60*, a gene that specifies cell fates during C. elegans vulval induction, encodes a *ras* protein. *Cell*, **63**, 921–931.

Handyside, A.H. (1978). Time of commitment of inside cells isolated from pre-implantation mouse embryos. *J. Embryol. Exp. Morphol.*, **45**, 37–53.

Handyside, A.H., and Johnson, M.H. (1978). Temporal and spatial patterns of the synthesis of tissue-specific polypeptides in the pre-implantation mouse embryo. *J. Embryol. Exp. Morphol.*, **44**, 191–199.

Harlow, G.M., and Quinn, P. (1982). Development of preimplantation mouse embryos *in vivo* and *in vitro. Aust. J. Biol. Sci.*, **35**, 187–193.

Harris, W.A., Stark, W.S., and Walker, J.A. (1976). Genetic dissection of the photoreceptor system in the compound eye of *Drosophila melanogaster. J. Physiol.*, **256**, 415–439.

Hart, C.P., Awgulewitsch, A., Fainsod, A., McGinnis, W., and Ruddle, F.H. (1985). Homeo box gene complex on mouse chromosome 11: Molecular cloning, expression in embryogenesis and homology to a human homeo box locus. *Cell*, **43**, 9–18.

Hartenstein, V., and Posakony, J.W. (1989). Development of adult sensilla on the wing and notum of *Drosophila melanogaster. Development*, **107**, 389–405.

Hartenstein, V., and Posakony, J.W. (1990). A dual function of the *Notch* gene in *Drosophila* sensillum development. *Dev. Biol.*, **142**, 13–30.

Hartley, D.A., Priess, A., and Artavanis-Tsakonas, S. (1988). A deduced gene product from the *Drosophila* neurogenic locus *Enhancer of split* shows homology to mammalian G-protein B subunit. *Cell*, **55**, 785–795.

Hartwell, L.H., Culotti, J., and Reid, B. (1970). Genetic control of the cell-division cycle in yeast. I. Detection of mutants. *Proc. Natl. Acad. Sci., U.S.A.*, **66**, 352–359.

Hashimoto, C., Hudson, K.L., and Anderson, K.V. (1988). The *Toll* gene of Drosophila, required for dorsal-ventral embryonic polarity, appears to encode a transmembrane protein. *Cell*, **52**, 269–279.

Hastie, N.D., and Bishop, J.O. (1976). The expression of three abundance classes of messenger RNA in mouse tissues. *Cell*, **9**, 761–774.

Hauser, C.A., Joyner, A.L., Klein, R.D., Learned, T.K., Martin, G.R., and Tijian, R. (1985). Expression of homologous homeobox-containing genes in differentiated human teratocarcinoma cells and mouse embryos. *Cell*, **43**, 19–28.

Hayes, P. (1982). *Mutant analysis of determination in Drosophila*, Ph.D. thesis. Department of Genetics, University of Alberta, Edmonton, Canada.

Hayes, P.H., Sato, T., and Denell, R.E. (1984). Homeosis in *Drosophila:* The Ultrabithorax larval syndrome. *Proc. Natl. Acad. Sci., U.S.A.*, **81**, 545–549.

Haynie, J.L. (1982). Homologies of positional information in thoracic imaginal discs of *Drosophila melanogaster. Wilhelm Roux Arch.*, **191**, 293–300.

Hecht, N.B. (1986). Regulation of gene expression during mammalian spermatogenesis. In J. Rossant and R.A. Pedersen (Eds.), *Experimental Approaches to Mammalian Embryonic Development.* New York: Cambridge University Press.

Hecht, R.M., Gossett, L.A., and Jeffrey, W.R. (1981). Ontogeny of maternal and newly transcribed mRNA analyzed by *in situ* hybridization during development of *Caenorhabditis elegans. Dev. Biol.*, **83**, 374–379.

Hedgecock, E.M., Sulston, E., and Thomson, J.N. (1983). Mutations affecting programmed cell deaths in the nematode *Caenorhabditis elegans. Science*, **220**, 1277–1279.

Heine, U., and Blumenthal, T. (1986). Characterization of regions of the *Caenorhabditis elegans* X chromosome containing vitellogenin genes. *J. Mol. Biol.*, **188**, 301–312.

Held, L.I., Jr. (1979). Pattern as a function of cell number and cell size on the second leg basitarsus of *Drosophila. Wilhelm Roux Archiv. Dev. Biol.*, **187**, 105–127.

Held, L.I., Jr. (1990a). Sensitive periods for abnormal patterning on a leg segment in *Drosophila melanogaster. Wilhelm Roux Arch. Dev. Biol.*, **199**, 31–47.

Held, L.I., Jr. (1990b). Arrangement of bristles as a function of bristle number on a leg segment in *Drosophila melanogaster. Rouxs Arch. Dev. Biol.*, **199**, 48–62.

Held, L.I., Jr., (1991). Bristle patterning in *Drosophila. BioEssays*, **13**, 633–640.

Held, L.I., Jr., and Bryant, P.J. (1984). Cell interactions controlling the formation of bristle patterns in *Drosophila.* In G.M. Malacinski (Ed.), *Pattern Formation.* New York: Macmillan, pp. 291–322.

Heller, D.T., Cahill, D.M., and Schultz, R.M. (1981). Behavioural studies of mammalian oogenesis: metabolic cooperativity between granulosa cells and growing mouse oocytes. *Dev. Biol.*, **84**, 455–464.

Herman, R.K. (1984). Analysis of mosaics of the nematode *Caenorhabditis elegans. Genetics*, **108**, 165–180.

Herman, R.K. (1988). Genetics. In W.B. Wood (Ed.), *The Nematode* Caenorhabditis elegans. Cold Spring Harbor, NY: Cold Spring Harbor Laboratory Press, pp. 17–45.

Herman, R.K., and Hedgecock, E.M. (1990). Limitation of the size of the vulval primordium of *Caenorhabditis elegans* by *lin-15* expression in surrounding hypodermis. *Nature*, **348**, 169–171.

Herr, W., Sturm, R.A., Clerc, R.G., Corcoran, L.M., Baltimore, D., Sharp, P.A., Ingraham, H.A., Rosenfeld, M.G., Finney, M., Ruvkun, G., and Horvitz, H.R. (1988). The POU domain: A large conserved region in the mammalian *pit-1, oct-1, oct-2,* and *Caenorhabditis elegans unc-86* gene products. *Genes Dev.*, **2**, 1513–1516.

Herrman, B.G. (1991). Expression pattern of the *Brachyury* gene in whole mount T^{Wis}/T^{Wis} mutant embryos. *Development*, **113**, 913–917.

Herrman, B., Bucan, M., Morris, P.E., Frischauf, A.-M., Silver, L.M., and Lehrach, P.M. (1986). Genetic analysis of the proximal portion of the mouse *t* complex: Evidence for a second inversion within *t* haplotypes. *Cell*, **44**, 469–476.

Herrman, B.G., Labeit, S., Paustka, A., King, T.R., and Lehrach, H. (1990). Cloning of the T gene required in mesoderm formation in the mouse. *Nature*, **243**, 617–622.

Heyner, S., Smith, R.M., and Schultz, G.A. (1989). Temporally regulated expression of insulin and insulin-like growth factors and their receptors in early mammalian development. *BioEssays,* **11,** 171–176.

Hidalgo, A., and Ingham, P. (1990). Cell patterning in the Drosophila segment—Spatial regulation of the segment polarity gene *patched. Development* **110,** 291–301.

Higgins, B.J., and Hirsh, D. (1977). Roller mutants of the nematode *Caenorhabditis elegans. Mol. Gen. Genet.,* **150,** 63–72.

Hildreth, P.E. (1965). *Doublesex,* a recessive gene that transforms both males and females of *Drosophila* into intersexes. *Genetics,* **51,** 659–678.

Hillman, N., Sherman, M.I., and Graham, C.F. (1972). The effect of spatial arrangement on cell determination during mouse development. *J. Embryol. Exp. Morphol.,* **28,** 263–278.

Hirsh, D. (1979). Temperature-sensitive maternal effect mutants of early development in *Caenorhabditis elegans.* In S. Subtelny and I.R. Konigsberg (Eds.), *Determinants of Spatial Organization.* New York: Academic Press, pp. 149–165.

Hirsh, D., and Vanderslice, R. (1976). Temperature-sensitive developmental mutants of *Caenorhabditis elegans. Dev. Biol.,* **49,** 220–235.

Hirsh, D., Oppenheim, D., and Klass, M. (1976). Development of the reproductive system of *Caenorhabditis elegans. Dev. Biol.,* **49,** 200–219.

Hirsh, D., Kemphues, K.J., Stinchcomb, D.T., and Jefferson, A. (1985). Genes affecting early development in *Caenorhabditis elegans. Cold Spring Harbor Symp. Quant. Biol.,* **50,** 69–78.

Hodgkin, J. (1980). More sex determination mutants of *Caenorhabditis elegans. Genetics,* **96,** 649–664.

Hodgkin, J. (1983a). Male phenotypes and mating efficiency in *Caenorhabditis elegans. Genetics,* **103,** 43–64.

Hodgkin, J. (1983b). Two types of sex determination in a nematode. *Nature,* **304,** 267–268.

Hodgkin, J. (1985). Novel nematode amber suppressors. *Genetics,* **111,** 287–310.

Hodgkin, J. (1986). Sex determination in the nematode *Caenorhabditis elegans:* Analysis of *tra-3* suppressors and characterization of *fem* genes. *Genetics,* **114,** 15–52.

Hodgkin, J. (1987a). Primary sex determination in the nematode. *C. elegans. Development,* **101,** 5–16.

Hodgkin, J. (1987b). A genetic analysis of the sex-determining gene, *tra-1,* in the nematode *Caenorhabditis elegans. Genes Dev.,* **1,** 731–745.

Hodgkin, J. (1988). Sexual dimorphism and sex determination. In W.B. Wood (Ed.), *The Nematode* Caenorhabditis elegans. Cold Spring Harbor, NY: Cold Spring Harbor Laboratory Press, pp. 243–279.

Hodgkin, J.A. (1990). Sex determination compared in *Drosophila* and *Caenorhabditis. Nature,* **344,** 721–728.

Hodgkin, J.A., and Brenner, S. (1977). Mutations causing transformation of sexual phenotype in the nematode *Caenorhabditis elegans. Genetics,* **86,** 275–281.

Hodgkin, J., Horvitz, H.R., and Brenner, S. (1979). Nondisjunction mutants of the nematode *Caenorhabditis elegans. Genetics,* **91,** 67–94.

Hogan, B., and Tilly, R. (1978). *In vitro* development of inner cell masses isolated immunosurgically from mouse blastocysts. I. Inner cell masses from 3.5 day p.c. blastocysts incubated for 24 hr before immunosurgery. *J. Embryol. Exp. Morphol.,* **45,** 93–105.

Holland, P.W.H. (1990). Homeobox genes and segmentation: Co-option, co-evolution, and convergence. *Semin. Dev. Biol.,* **1,** 135–145.

Holland, P.W.H., and Hogan, B.L.M. (1988). Expression of homeo box genes during mouse development: A review. *Genes Dev.,* **2,** 773–782.

Holliday, R., and Pugh, J.E. (1975). DNA modification mechanisms and gene activity during development. *Science,* **187,** 226–232.

Hooper, J.E. (1986). Homeotic gene function in the muscles of *Drosophila* larvae. *EMBO J.,* **5,** 2321–2329.

Hooper, J.E., and Scott, M.P. (1989). The *Drosophila patched* gene encodes a putative membrane protein required for segmental patterning. *Cell,* **59,** 751–765.

Hoppe, P.E., and Greenspan, R.J. (1986). Local function of the Notch gene for embryonic ectodermal pathway choice in Drosophila. *Cell,* **46,** 773–783.

Horvitz, H.R. (1988). Genetics of cell lineage. In W.B. Wood (Ed.), *The Nematode* Caenorhabditis elegans. Cold Spring Harbor, NY: Cold Spring Harbor Laboratory Press, pp. 157–190.

Horvitz, H.R., and Sulston, J.E. (1980). Isolation and genetic characterization of cell-lineage mutants of the nematode *Caenorhabditis elegans. Genetics,* **96,** 435–454.

Horvitz, H.R., and Sulston, J.E. (1990). Joy of the worm. *Genetics,* **126,** 287–292.

Horvitz, H.R. and Sternberg, P.W. (1991). Multiple intercellular signalling systems control the development of the *Caenorhabditis elegans* vulva. *Nature,* **351,** 535–541.

Horvitz, H.R., Brenner, S., Hodgkin, J., and Herman, R.K. (1979). A uniform genetic nomenclature for the nematode *Caenorhabditis elegans. Mol. Gen. Genet.,* **175,** 129–133.

Horvitz, H.R., Ellis, H.M., and Sternberg, P.W. (1982). Programmed cell death in nematode development. *Neurosci. Comment.,* **1,** 56–65.

Hotta, Y., and Benzer, S. (1973). Mapping of behavior in *Drosophila* mosaics. In F. Ruddle (Ed.), *Genetic Mechanisms of Development.* New York: Academic Press, pp. 129–167.

Hough-Evans, B.R., Wold, B.J., Ernst, S.G., Britten, R.J., and Davidson, E.H. (1977). Appearance and persistence of maternal RNA sequences in sea urchin development. *Dev. Biol.,* **60,** 258–277.

Hough-Evans, B.R., Jacobs-Lorena, M., Cummings, M.R., Britten, R.J., and Davidson, E.H. (1980). Complexity of RNA in eggs of *Drosophila melanogaster* and *Musca domestica. Genetics,* **95,** 81–94.

Howard, K. (1988). The generation of periodic pattern during early *Drosophila* embryogenesis. *Development,* **104** (Suppl.), 35–50.

Howlett, S.K., and Bolton, V.N. (1985). Sequence and regulation of morphological and molecular events during the first cell cycle of mouse embryogenesis. *J. Embryol. Exp. Morphol.,* **87,** 175–206.

Huang, F., Dambly-Chaudière, C., and Ghysen, R. (1991). The emergence of sense organs in the wing disc of *Drosophila. Development,* **111,** 1087–1095.

Hull, D.L. (1974). *Philosophy of Biological Science.* Englewood Cliffs, NJ: Prentice-Hall, p. 142.

Hülskamp, M., and Tautz, D. (1991). Gap genes and gradients—the logic behind the gaps. *BioEssays,* **13,** 261–268.

Hülskamp, M., Schroder, C., Pfeifle, C., Jäckle, H., and Tautz, D. (1989). Posterior segmentation of the *Drosophila* embryo in the absence of a maternal posterior organizer gene. *Nature,* **338,** 629–632.

Hülskamp, M., Pfeifle, C., and Tautz, D. (1990). A morphogenetic gradient of *hunchback* protein organizes the expression of the gap genes *Krüppel* and *knirps* in the early *Drosophila* embryo. *Nature,* **346,** 577–580.

Hunt, P., and Krumlauf, R. (1991). Deciphering the Hox code: Clues to patterning branchial regions of the head. *Cell,* **66,** 1075–1078.

Hunt, P., Whiting, J., Muchamore, L., Marshall, H., and Krumlauf, R. (1991). Homeobox genes and models for patterning the hindbrain and branchial arches. *Development,* (Suppl.), 187–196.

Hunter, C.P., and Wood, W.B. (1990). The *tra-1* gene determines sexual phenotype cell-autonomously in the *C. elegans Cell,* **63,** 1193–1264.

Hunter, C.P., and Wood, W.B. (1992). Evidence from mosaic analysis of the masculinizing gene *her-1* for cell interactions in *C. elegans* sex determination. *Nature,* **355,** 551–555.

Hyafil, F., Babinet, C., and Jacob, F. (1981). Cell-cell interactions in early embryogenesis: A molecular approach to the role of calcium. *Cell,* **26,** 447–454.

Illmensee, K. (1978). *Drosophila* chimaeras and the problem of determination. In W.J. Gehring (Ed.), *Genetic Mosaics and Cell Differentiation,* vol. 9 of *Results and Problems in Differentiation.* Berlin: Springer-Verlag, pp. 51–69.

Illmensee, K., and Mahowald, A.P. (1974). Transplantation of posterior polar plasm in *Drosophila.* Induction of germ cells at the anterior pole of the egg. *Proc. Natl. Acad. Sci., U.S.A.,* **71,** 1016–1020.

Illmensee, K., Mahowald, A.P., and Loomis, M.R. (1976). The ontogeny of germ plasm during oogenesis in *Drosophila. Dev. Biol.,* **49,** 40–65.

Immergluck, K., Lawrence, P.A., and Bienz, M. (1990). Induction across germ layers in Drosophila mediated by a genetic cascade. *Cell,* **62,** 261–268.

Ingham, P.W., and Martinez-Arias, A. (1986). The correct activation of *Antennapedia* and bithorax complex genes requires the *fushi tarazu* gene. *Nature,* **324,** 592–597.

Ingham, P., and Gergen, J.P. (1988). Interactions between the pair-rule genes *runt, hairy, even-skipped* and *fushi-tarazu* and the establishment of periodic pattern in the *Drosophila* embryo. *Development*, **104** (Suppl.), 51–60.

Ingham, P.W., Howard, K.R., and Ish-Horowicz, D. (1985). Transcription pattern of the *Drosophila* segmentation gene *hairy*. *Nature*, **318**, 439–446.

Ip, Y.P., Kraut, R., Levine, M., and Rushlow, C.A. (1991). The *dorsal* morphogen is a sequence-specific DNA-binding protein that interacts with a long-range repression-specific element in Drosophila. *Cell*, **64**, 439–446.

Irish, V.F., and Gelbart, W.M. (1987). The haplo-insufficient region of the *decapentaplegic* sequences required for dorso-ventral patterning of the Drosophila embryo. *Genes Dev.*, **1**, 868–879.

Irish, V.F., Lehmann, R., and Akam, M. (1989). The *Drosophila* posterior-group gene *vasa* functions by repressing *hunchback* activity. *Nature*, **338**, 646–648.

Ish-Horowicz, D., and Pinchin, S.M. (1987). Pattern abnormalities induced by ectopic expression of the Drosophila gene *hairy* are associated with repression of *ftz* transcription. *Cell*, **51**, 405–415.

Isnenghi, E., Cassada, R., Smith, K., Denich, K., Radwa, K., and von Ehrenstein, G. (1983). Maternal effects and temperature-sensitive period of mutations affecting embryogenesis in *Caenorhabditis elegans*. *Dev. Biol.*, **98**, 465–480.

Izquierdo, M., and Bishop, J.O. (1979). An analysis of cytoplasmic RNA populations in *Drosophila melanogaster*, Oregon R. *Biochem. Genet.*, **17**, 473–497.

Jäckle, H., Tautz, D., Schuh, R., Siefert, E., and Lehmann, R. (1986). Cross-regulatory interactions among the gap genes of *Drosophila*. *Nature*, **324**, 668–670.

Jackson, I.J. (1988). A cDNA encoding tyrosinase-related protein maps to the mouse *brown* locus. *Proc. Natl. Acad. Sci., U.S.A.*, **85**, 4392–4296.

Jackson, I.J. (1991). Mouse coat color mutations: A molecular genetic resource which spans the centuries. *BioEssays*, **13**, 439–446.

Jacob, F. (1977). Evolution and tinkering. *Science*, **196**, 1161–1166.

Jacob, F. (1982). *The Possible and the Actual*. New York: Pantheon Books.

Jacob, F., and Monod, J. (1961). Genetic regulatory mechanisms in the synthesis of proteins. *J. Mol. Biol.*, **3**, 318–356.

Jacob, F., and Monod, J. (1963). Genetic repression, allosteric inhibition, and cellular differentiation. In M. Locke (Ed.), *Cytodifferentiation and Macromolecular Synthesis*. New York: Academic Press.

Jacobs, P.S., and Strong, J.A. (1959). A case of human interesexuality having a possible XXY sex determining mechanism. *Nature*, **183**, 302–303.

Jaenisch, R. (1983). Retroviruses and mouse embryos: A model system in which to study gene expression in development and differentiation. In R. Porter and J. Whelan (Eds.), *Molecular Biology of Egg Maturation*. London: Pitman, pp. 44–63.

Jager, R.J. Anvret, M., Hall, K., and Scherer, G. (1990). A human XY female with a frame shift mutation in the candidate testis-determining gene *SRY*. *Nature*, **348**, 452–454.

James, T.C., and Elgin, S.C.R. (1986). Identification of a non-histone chromosomal protein associated with heterochromatin in *Drosophila melanogaster* and its gene. *Mol. Cell Biol.*, **6**, 3862–3872.

Janning, W., Pfreudt, J., and Tiemann, R. (1979). The distribution of anlagen in the early embryo of *Drosophila*. In N. Le Douarin (Ed.), *Cell Lineages, Stem Cells, and Cell Determination*. Amsterdam: Elsevier/North Holland, pp. 83–98.

Janning, W., Lutz, A., and Wissen, D. (1986). Clonal analysis of the blastoderm anlage of the Malpighian tubules in *Drosophila melanogaster*. *Rouxs Arch. Dev. Biol.*, **195**, 22–32.

Jeffrey, W.J., and Swalla, B.J. (1992). Evolution of alternative modes of ascidian development. *BioEssays*, **14**,

Jiminez, F., and Campos-Ortega, J.A. (1981). A cell arrangement system of the *Drosophila* embryo: Its behavior in homoeotic mutants. *Wilhelm Roux Arch.*, **190**. 370–373.

Jiminez, F., and Campos-Ortega, J.A. (1982). Maternal effects of zygotic mutants affecting early neurogenesis in *Drosophila*. *Wilhelm Roux Arch.*, **191**, 191–201.

Johnson, D.K., Hand, R.E., Jr., and Rinchik, E.M. (1989). Molecular mapping within the mouse albino-deletion complex. *Proc. Natl. Acad. Sci., U.S.A.,* **86,** 8862–8866.

Johnson, D.R. (1974). Hairpin-tail: A case of post-reductional gene action in the mouse egg? *Genetics,* **76,** 795–805.

Johnson, D.R. (1975). Further observations on the hairpin-tail (T^{hp}) mutation in the mouse. *Genet. Res.,* **24,** 207–213.

Johnson, K., and Hirsh, D. (1979). Patterns of proteins synthesized during development of *Caenorhabditis elegans. Dev. Biol.,* **70,** 241–248.

Johnson, M.H. (1979). Intrinsic and extrinisic factors in pre-implantation development. *J. Reprod. Fertil.,* **55,** 255–265.

Johnson, M.H., and Ziomek, C.A. (1981). The foundation of two distinct cell lineages within the mouse morula. *Cell,* **24,** 71–80.

Johnson, M.H., and Ziomek, C.A. (1983). Cell interactions influence the fate of mouse blastomeres undergoing the transition from the 16- to the 32-cell stage. *Dev. Biol.,* **95,** 211–218.

Johnson, M.H., Pratt, H.P.M., and Handyside, A.H. (1981). The generation and recognition of positional information in the preimplantation mouse embryo. In S.R. Glasser and D.W. Bullock (Eds), *Cellular and Molecular Aspects of Implantation.* New York: Plenum Publishing, pp. 55–74.

Johnson, M.H., Chisholm, J.C., Fleming, T.P., and Houliston, E. (1986). A role for cytoplasmic determinants in the development of the mouse early embryo? *J. Embryol. Exp. Morphol.,* **97,** (Suppl.), 97–121.

Jorgensen, E.M., and Garber, R.L. (1987). Function and misfunction of the two promoters of the *Drosophila Antennapedia* gene. *Genes Dev.,* **1,** 544–555.

Jost, A., Vigier, B., Prepin, J., and Perchellet, J.P. (1973). Studies on sex differentiation in mammals. *Recent Prog. Horm. Res.,* **29,** 1–41.

Joyner, A.L., and Martin, G.M. (1987). *En-1* and *En-2,* two mouse genes with sequence homology to the Drosophila *engrailed* gene—Expression during embryogenesis. *Genes Dev.,* **1,** 29–38.

Joyner, A.L., Kornberg, T., Coleman, K.G., Cox, D.R., and Martin, C.R. (1985). Expression during embryogenesis of a mouse gene with sequence homology to the Drosophila *engrailed* gene. *Cell,* **43,** 29–37.

Joyner, A.L., Hermys, K., Auerbach, B.A., Davis, C.A., and Rossant, J. (1991). Subtle cerebellar phenotype in mice homozygous for a targeted deletion of the *En-2* homeobox. *Science,* **251,** 1239–1243.

Judd, B.H. (1977). The nature of the module of genetic function in *Drosophila.* In E.M. Bradbury and K. Javaherian (Eds.), *The Organization and Expression of the Eukaryotic Genome.* London: Academic Press, pp. 469–483.

Judd, B.H., Shen, M.W., and Kaufman, T.C. (1972). The anatomy and function of a segment of the X chromosome of *D. melanogaster. Genetics,* **71,** 139–156.

Judson, H.F. (1979). *The Eighth Day of Creation: Makers of the Revolution in Biology.* London: Jonathan Cape.

Jurgens, G. (1985). A group of genes controlling the spatial expression of the bithorax complex in *Drosophila. Nature,* **316,** 153–155.

Jurgens, G., and Weigel, D. (1988). Terminal versus segmental development in the *Drosophila* embryo: The role of the homeotic gene *fork head. Rouxs Arch. Dev. Biol.,* **197,** 345–354.

Jurgens, G., Wieschaus, E., Nüsslein-Volhard, C., and Kluding, H. (1984). Mutations affecting the pattern of the larval cuticle in *Drosophila melanogaster.* II. Zygotic loci on the third chromosome. *Wilhelm Roux Arch. Dev. Biol.,* **193,** 283–295.

Kalthoff, K. (1979). Analysis of a morphogenetic determinant in an insect embryo (*Smittia* Spec., *Chironomidae, Diptera*). In S. Subtelny and I. Konigsberg (Eds.), *Determinants of Spatial Organization.* New York: Academic Press, pp. 97–126.

Kam, Z., Minden, J.S., Agard, D.A., Sedat, J.W., and Leptin, M. (1991). *Drosophila* gastrulation: Analysis of cell shape changes in living embryos by three-dimensional fluorescence microscopy. *Development,* **112,** 365–370.

Karch, F., Weiffenbach, B., Peifer, M., Bender, W., Duncan, I., Celniker, S., Crosby, M., and Lewis, E.B. (1985). The abdominal region of the bithorax complex. *Cell,* **43,** 81–96.

Karlsson, J. (1981a). The distribution of regenerative potential in the wing disc of *Drosophila. J. Embryol. Exp. Morphol.,* **61,** 303–316.

Karlsson, J. (1981b). Sequence of regeneration in the *Drosophila* wing disc. *J. Embryol. Exp. Morphol.,* **65,** 37–47.

Karr, T.L., and Alberts, B.M. (1986). Organization of the cytoskeleton in early *Drosophila* embryos. *J. Cell Biol.,* **102,** 1494–1509.

Karr, T.L., Ali, Z., Drees, B., and Kornberg, T. (1985). The *engrailed* locus of D. melanogaster provides an essential zygotic function in precellular embryos. *Cell,* **43,** 591–601.

Kauffman, S.A. (1973). Control circuits for determination and transdetermination. *Science,* **181,** 310–318.

Kauffman, S.A. (1980). Heterotopic transplantation in the syncytial blastoderm of *Drosophila:* Evidence for anterior and posterior nuclear commitments. *Wilhelm Roux Arch.,* **189,** 135–145.

Kauffman, S.A. (1987). Developmental logic and its evolution. *BioEssays,* **6,** 82–87.

Kauffman, S.A., and Ling, E. (1981). Regeneration by complementary wing disc fragments of *Drosophila melanogaster. Dev. Biol.,* **82,** 238–257.

Kauffman, S.A., Shymko, R.M., and Trabert, K. (1978). Control of segmental compartment formation in *Drosophila. Science,* **199,** 259–270.

Kaufman, M.H., Speirs, S., and Lee, K.K.H. (1989). The sex-chromosome constitution and early postimplantation development of diandric triploid mouse embryos. *Cytogenet. Cell Genet.,* **50,** 98–101.

Kaufman, T.C. (1983). The genetic regulation of segmentation in *Drosophila melanogaster.* In W.R. Jeffery and R.A. Raff (Eds.), *Time, Space, and Pattern in Development.* New York: Alan R. Liss, pp. 365–383.

Kaufman, T.C., Lewis, R., and Wakimoto, B. (1980). Cytogenetic analysis of chromosome 3 in *Drosophila melanogaster:* The homeotic gene complex in polytene chromosome interval 84A-B. *Genetics,* **94,** 115–133.

Keller, S.A., Liptay, S., Hajra, A., and Meisler, A.H. (1990). Transgene-induced mutation of the murine steel locus. *Proc. Natl Acad. Sci. USA,* **87,** 10019–10022.

Kelly, S.J. (1975). Studies of the potency of the early cleavage blastomeres of the mouse. In M. Balls and A.E. Wild (Eds.), *The Early Development of Mammals.* Cambridge: Cambridge University Press, pp. 97–105.

Kemphues, K.J., Wolf, N., Wood, W.B., and Hirsh, D. (1986). Two loci required for cytoplasmic organization in early embryos of *Caenorhabditis elegans. Dev. Biol.,* **113,** 449–460.

Kemphues, K.J., Priess, J.R., Martin, D.G., and Cheng, N. (1988a). Identification of genes required for cytoplasmic localization in early C. elegans embryos. *Cell,* **52,** 311–320.

Kemphues, K.J., Kusch, M., and Wolf, N. (1988b). Maternal-effect lethal mutations in linkage group II of *Caenorhabditis elegans. Genetics,* **120,** 977–986.

Kenyon, C. (1986). A gene involved in the development of the posterior body region of C. elegans. *Cell,* **46,** 477–487.

Kerridge, S., and Thomas-Cavallin, M. (1988). Appendage morphogenesis in *Drosophila:* A developmental study of the *rotund (rn)* gene. *Wilhelm Roux Arch. Dev. Biol.,* **197,** 19–26.

Kessel, M., and Gruss, P. (1991). Homeotic transformations of murine vertebrae and concomitant alteration of *Hox* codes induced by retinoic acid. *Cell,* **67,** 89–104.

Kessel, M., Balling, R., and Gruss, P. (1990). Variations of cervical vertebrae after expression of a Hox-1.1 transgene in mice. *Cell,* **61,** 301–308.

Kimble, J.E. (1981a). Alterations in cell lineage following laser ablation of cells in the somatic gonad of *Caenorhabditis elegans. Dev. Biol.,* **87,** 286–300.

Kimble, J.E. (1981b). Strategies for the control of pattern formation in *Caenorhabditis elegans. Philos. Trans. R. Soc. Lond. [Biol.],* **295,** 539–551.

Kimble, J.E., and Hirsh, D. (1979). The postembryonic cell lineages of the hermaphrodite and male gonads in *Caenorhabditis elegans. Dev. Biol.,* **70,** 396–417.

Kimble, J.E., and White, J.G. (1981). On the control of germ cell development in *Caenorhabditis elegans. Dev. Biol.,* **81,** 208–219.

Kimble, J.E., and Sharrock, W.J. (1983). Tissue-specific synthesis of yolk proteins in *Caenorhabditis elegans. Dev. Biol.,* **96,** 189–196.

Kimble, J.E., and Ward, S. (1988). Germ-line development and fertilization. In W.B. Wood (Ed.), *The Nematode* Caenorhabditis elegans. Cold Spring Harbor, NY: Cold Spring Harbor Laboratory Press.

Kimble, J., Sulston, J., and White, J. (1979). Regulative development in the postembryonic lineages of *Caenorhabditis elegans*. In N. Le Douarin (Ed.), *Cell Lineage, Stem Cells, and Determination*. Amsterdam: Elsevier/North-Holland, pp. 59–68.

Kimble, J.E., Edgar, L., and Hirsh, D. (1984). Specification of male development in *Caenorhabditis elegans:* The *fem* genes. *Dev. Biol., 105*, 234–239.

Kimmel, C.B. (1989). Genetics and early development of the zebrafish. *Trends Genet. 5*, 283–288.

King, R.C. (1970). *Oogenesis in Drosophila*. New York: Academic Press.

King, R.C., and Mohler, J.D. (1975). The genetic analysis of oogenesis in *Drosophila melanogaster*. In R.C. King (Ed.), *Handbook of Genetics*. New York: Plenum Press, vol. 3., pp. 757–791.

King, R.C., Cassida, J.D., and Rousset, A. (1982). The formation of clones of interconnected cells during gametogenesis in insects. In R.C. King and H. Akai (Eds.), *Insect Ultrastructure*. New York: Plenum Press.

Kirby, B.S., Bryant, P.J., and Schneiderman, H.A. (1982). Regeneration following duplication in imaginal wing disc fragments of *Drosophila melanogaster*. *Dev. Biol., 52*, 1–18.

Klass, M., Wolf, N., and Hirsh, D. (1976). Development of the male reproductive system and sexual transformation in the nematode *Caenorhabditis elegans*. *Dev. Biol., 52*, 1–18.

Klingensmith, J., Noll, E., and Perrimon, N. (1989). The segment polarity phenotype of *Drosophila* involves differential tendencies toward transformation and cell death. *Dev. Biol., 134*, 130–145.

Klingler, M., Erdelyi, M., Szabad, J., and Nüsslein-Volhard, C. (1988). Function of *torso* in determining the terminal anlagen of the *Drosophila* embryo. *Nature, 335*, 275–277.

Knipple, D.C., Seifert, E., Rosenberg, U.B., Priess, A., and Jäckle, H. (1985). Spatial and temporal patterns of *Krüppel* gene expression in early *Drosophila* embryos. *Nature, 317*, 40–44.

Knowland, J.S., and Graham, C.F. (1972). RNA synthesis at the two-cell stage of mouse development. *J. Embryol. Exp. Morphol., 27*, 167–176.

Knust, E., and Campos-Ortega, J.A. (1989). The molecular genetics of early neurogenesis in *Drosophila melanogaster*. *BioEssays, 11*, 95–100.

Koopman, P., Gubbay, J., Vivian, N., Goodfellow, P., and Lovell-Badge, R. (1991). Male development of chromosomally female mice transgenic for *Sry*. *Nature, 351*, 117–121.

Kornberg, T. (1981a). *engrailed:* A gene controlling compartment and segment formation in *Drosophila*. *Proc. Natl. Acad. Sci., U.S.A., 78*, 1095–1099.

Kornberg, T. (1981b). Compartments in the abdomen of *Drosophila* and the role of the *engrailed* locus. *Dev. Biol., 86*, 363–372.

Kornberg, T., Siden, I., O'Farrell, P., and Simon, M. (1985). The *engrailed* locus of Drosophila: In situ localization of transcripts reveals compartment-specific expression, *Cell, 40*, 45–53.

Kornfeld, S., and Mehlman, I. (1989). The biogenesis of lysozomes. *Annu. Rev. Cell Biol., 5*, 483–525.

Kornsfield, K., Saint, R.B., Beachy, P.A., Harte, P.J., Peattie, D.A., and Hogness, D.S. (1989). Structure and expression of a family Ultrabithorax mRNAs generated by alternative splicing and polyadenylation in *Drosophila*. *Genes Dev., 3*, 243–258.

Kramer, J.M., Johnson, J.J., Edgar, R.S., Basch, C., and Roberts, S. (1988). The *sqt-1* gene of C. elegans encodes a collagen critical for organismal morphogenesis. *Cell, 55*, 555–565.

Krasnow, M.A., Saffman, E.E., Kornfeld, K., and Hogness, D.S. (1989). Transcriptional activation and repression by Ultrabithorax proteins in cultured Drosophila cells. *Cell, 57*, 1031–1043.

Krieg, C., Cole, T., Deppe, U., Schierenberg, E., Schmitt, D., Yoder, B.,and von Ehrenstein, G. (1978). The cellular anatomy of embryos of the nematode *Caenorhabditis elegans:* Analysis and reconstruction of serial section electron micrographs. *Dev. Biol., 65*, 193–215.

Kuhn, D.T., and Packert, G. (1988). Paternal imprinting of inversion *Uab*[1] causes homeotic transformations in Drosophila. *Genetics, 118*, 103–107.

Kuhn, D.T., Woods, D.F., and Cook, J.L. (1981). Analysis of a new homoeotic mutation (*iab-2*) within the Bithorax complex in *Drosophila melanogaster*. *Mol. Gen. Genet., 181*, 82–86.

Kuroda, H., Nakayama, H., Nawicki, M., Matsumoto, K., Nishimune, Y., and Kitamura, Y. (1989).

Differentiation of germ cells in somniferous tubules transplanted to testes of germ cell–deficient mice of W/W^v and Sl/Sl^o genotypes. J. Cell Physiol., **139,** 329–344.

Kuroda, M.I., Kernan, M.J., Kreber, R., Ganetsky, B., and Baker, B.S. (1991). The *maleless* protein associates with the X chromosome to regulate dosage compensation in Drosophila. *Cell,* **66,** 935–947.

Kusch, M., and Edgar, R.S. (1986). Genetic studies of unusual loci that affect body shape of the nematode *Caenorhabditis elegans* and may code for cuticle structural proteins. *Genetics,* **113,** 621–639.

Kuziora, M.A., and McGinnis, W. (1988). Autoregulation of a *Drosophila* homeotic selector gene. *Cell,* **55,** 477–485.

Kwon, B.S., Haq, A.K., Pomerantz, S.H., and Halaban, R. (1987). Isolation and sequence of a cDNA clone for human tyrosinase that maps at the mouse c-albino locus. *Proc. Natl. Acad. Sci., U.S.A.,* **84,** 7473–7477.

Lambie, E.J., and Kimble, J. (1991). Two homologous regulatory genes, *lin-12* and *glp-1,* have overlapping functions. *Development,* **112,** 231–240.

Lamoreux, M.L., and Mayer, T.C. (1975). Site of gene action in the development of hair pigment in recessive yellow (*e/e*) mice. *Dev. Biol.,* **46,** 160–166.

Larue, L., and Mintz, B. (1990). Pigmented cell lines of mouse albino melanocytes containing a tyrosinase cDNA with an inducible promoter. *Somatic Cell Molec. Genet.,* **16,** 361–368.

Lasko, P.F., and Ashburner, M. (1988). The product of the *Drosophila* gene *vasa* is very similar to eukaryotic initiation factor-4A. *Nature,* **335,** 611–617.

Latham, K.E., Garrels, J.I., Chang, C., and Solter, D. (1991). Quantitative analysis of protein synthesis in mouse embryos. I. Extensive reprogramming at the one- and two-cell stages. *Development,* **112,** 921–932.

Laufer, J.S., and von Ehrenstein, G. (1981). Nematode development after removal of egg cytoplasm: Absence of localized unbound determinants. *Science,* **211,** 402–405.

Laufer, J.S., Bazzicalupo, P., and Wood, W.B. (1980). Segregation of developmental potential in early embryos of *Caenorhabditis elegans. Cell,* **19,** 569–577.

Laugé, E. (1980). Sex determination. In M. Ashburner and T.R.F. Wright (Eds.), *The Genetics and Biology of Drosophila.* London: Academic Press, vol. 2d, pp. 33–106.

Laughan, A., and Scott, M.P. (1984). Sequence of a *Drosophila* segmentation gene: Protein structure homology with DNA-binding proteins. *Nature,* **310,** 25–31.

Lawrence, P.A. (1973). The development of spatial patterns in the integument of insects. In S.J. Counce and C.H. Waddington (Eds.), *Developmental Systems: Insects,* vol. 2. London: Academic Press.

Lawrence, P.A. (1981). The cellular basis of segmentation in insects. *Cell,* **26,** 3–10.

Lawrence, P.A. (1982). Cell lineage of the thoracic muscles of *Drosophila. Cell,* **29,** 493–503.

Lawrence, P.A. (1989). Cell lineage and cell states in the *Drosophila* embryo. In D. Evered and J. Marsh (Eds.), *Cellular Basis of Morphogenesis.* Chichester: John Wiley & Sons, pp. 130–140.

Lawrence, P.A., and Morata, G. (1976a). Compartments in the wing of *Drosophila:* A study of the *engrailed* gene. *Dev. Biol.,* **50,** 321–337.

Lawrence, P.A., and Morata, G. (1976b). The compartment hypothesis. In P.A. Lawrence (Ed.), *Insect Development.* Oxford: Blackwell Scientific.

Lawrence, P.A., and Green, S.M. (1979). Cell lineage in the developing retina of *Drosophila. Dev. Biol.,* **71,** 142–152.

Lawrence, P.A., and Morata, G. (1979). Pattern formation and compartments in the tarsus of *Drosophila.* In S. Subtelny and and I.R. Konigsberg (Eds.), *Determinants of Spatial Organization,* pp. 317–323.

Lawrence, P.A., and Brower, D.L. (1982). Myoblasts from *Drosophila* wing discs can contribute to developing muscles throughout the fly. *Nature,* **295,** 55–57.

Lawrence, P.A., and Struhl, G. (1982). Further studies of the *engrailed* phenotype in *Drosophila. EMBO J.,* **1,** 827–833.

Lawrence, P.A., and Morata, G. (1983). The elements of the bithorax complex. *Cell,* **35,** 595–601.

Lawrence, P.A., and Johnston, P. (1989). Pattern formation in the *Drosophila* embryo: Allocation of cells to parasegments by *even-skipped* and *fushi tarazu. Development,* **105,** 761–769.

Lawrence, P.A., Struhl, G., and Morata, G. (1979). Bristle patterns and compartment boundaries in the tarsi of *Drosophila. J. Embryol. Exp. Morphol.,* **51,** 195–208.

Lawrence, P.A., Johnston, P., MacDonald, P., and Struhl, P. (1987). Borders of parasegments in *Drosophila* embryos are delimited by the *fushi tarazu* and *even-skipped* genes. *Nature*, **328**, 440–442.

Lawson, K.A., Meneses, J.J., and Pedersen, R.A. (1991). Clonal analysis of epiblast fate during germ layer formation in the mouse embryo. *Development*, **113**, 891–911.

Lebowitz, R.M., and Ready, D.F. (1986). Ommatidial development in *Drosophila* eye disc fragments. *Dev. Biol.*, **117**, 663–671.

Lefevre, G., Jr. (1974). The relationship between genes and polytene chromosome bands. *Annu. Rev. Genet.*, **8**, 51–62.

Lehmann, R. (1988). Phenotypic comparison between maternal and zygotic genes controlling the segmental pattern of the *Drosophila* embryo. *Development*, **104** (Suppl.), 17–27.

Lehmann, R., and Nüsslein-Volhard, C. (1986). Abdominal segmentation, pole cell formation, and embryonic polarity require the localized activity of *oskar*, a maternal gene in *Drosophila*. *Cell*, **47**, 141–152.

Lehmann, R., and Nüsslein-Volhard, C. (1987a). Involvement of the *pumilio* gene in the transport of an abdominal signal in the *Drosophila* embryo. *Nature*, **329**, 167–170.

Lehmann, R., and Nüsslein-Volhard, C. (1987b). *hunchback*, a gene required for segmentation of an anterior and posterior region of the *Drosophila* embryo. *Dev. Biol.*, **119**, 402–417.

Lehmann, R., and Nüsslein-Volhard, C. (1991). The maternal gene *nanos* has a central role in posterior pattern formation of the *Drosophila* embryo. *Development*, **112**, 679–691.

Lehmann, R., Dietrich, U., Jiminez, F., and Campos-Ortega, J.A. (1981). Mutations of early neurogenesis in *Drosophila*. *Wilhelm Roux Arch. Dev. Biol.*, **190**, 226–229.

Lehmann, R., Jiminez, F., Dietrich, U., and Campos-Ortega, J.A. (1983). On the phenotype and early development of mutants of early neurogenesis in *Drosophila melanogaster*. *Wilhelm Roux Arch. Dev. Biol.*, **192**, 62–74.

Leptin, M., and Grunewald, B. (1990). Cell shape changes during gastrulation in *Drosophila*. *Development*, **110**, 73–84.

Levey, I.L., Stull, G.B., and Brinster, R.L. (1978). Poly(A) and synthesis of polyadenylated RNA in the preimplantation mouse embryo. *Dev. Biol.*, **64**, 140–148.

Levine, M., Hafen, E., Garber, R.L., and Gehring, W.J. (1983). Spatial distribution of *Antennapedia* transcripts during *Drosophila* development. *EMBO J.*, **2**, 2037–2046.

Levy, L.S., and Manning, J.E. (1981). Messenger RNA sequence complexity and homology in developmental stages of *Drosophila*. *Dev. Biol.*, **85**, 141–149.

Levy, W.B., and McCarthy, B.J. (1975). Messenger RNA complexity in *Drosophila melanogaster*. *Biochemistry*, **14**, 2440–2446.

Lewis, E.B. (1964). Genetic control and regulation of developmental pathways. In M. Locke (Ed.), *The Chromosomes in Development*. New York: Academic Press.

Lewis, E.B. (1978). A gene complex controlling segmentation in *Drosophila*. *Nature*, **276**, 565–570.

Lewis, E.B. (1981). Developmental genetics of the bithorax complex in *Drosophila*. In D.D. Brown and C.F. Fox (Eds.), *Developmental Biology Using Purified Genes*. New York: Academic Press, pp. 1–20.

Lewis, R., Hazelrigg, T., and Rubin, G.M. (1985). Separable *cis*-acting control elements for expression of the *white* gene of *Drosophila*. *EMBO J.*, **4**, 3489–3499.

Lewis, R.A., Kaufman, T.C., Denell, R.E., and Tallerico, P. (1980a). Genetic analysis of the Antennapedia gene complex (ANT-C) and adjacent chromosomal regions of *Drosophila melanogaster*. I. Polytene chromosome segments 84B–D. *Genetics*, **95**, 367–381.

Lewis, R.A., Wakimoto, B.T., Denell, R.E., and Kaufman, T.C. (1980b). Genetic analysis of the Antennapedia gene complex (ANT-C) and adjacent chromosomal regions of *Drosophila melanogaster*. II. Polytene chromosome segments 84A–84B1,2. *Genetics*, **95**, 383–397.

Li, C., and Chalfie, M. (1990). Organogenesis in *C. elegans*: Positioning of neurons and muscles in the egg-laying system. *Neuron*, **4**, 681–695.

Lindsley, D.L., and Grell, E.H. (1968). Genetic variations of *Drosophila melanogaster*. *Carnegie Inst. Wash. Publ. 627*.

Lindsley, D.L., and Zimm., G. (1986). The genome of *Drosophila melanogaster*. Part 2: Lethals, maps. *Drosophila Information Service*, **64**, 1–158.

Lock, L.F., Takagi, N., and Martin, G.R. (1987). Methylation of the *Hprt* gene on the inactive X occurs after chromosome inactivation. *Cell*, **48**, 39–46.

Locke, J., Katarski, M.A., and Tartof, K.D. (1988). Dosage-dependent modifiers of position effect variegation in Drosophila and a mass action model that explains their effect. *Genetics*, **120**, 181–198.

Lohs-Schardin, M. (1982). *Dicephalic*—a *Drosophila* mutant affecting polarity in follicle organization and embryonic patterning. *Wilhelm Rouxs Arch.* **191**, 28–36.

Lohs-Schardin, M., Sander, K., Cremer, C., Cremer, T., and Zorn, C. (1979a). Localized ultraviolet laser microbeam irradiation of early *Drosophila* embryos: Fate maps based on locationand frequency of adult defects. *Dev. Biol.*, **68**, 533–545.

Lohs-Schardin, M., Cremer, C., and Nüsslein-Volhard, C. (1979b). A fate map for the larval epidermis of *Drosophila melanogaster*. Loclaized cuticle defects following irradiation of the blastoderm with an ultraviolet laser microbeam. *Dev. Biol.*, **73**, 239–255.

Lovett, J., and Goldstein, E.S. (1977). The cytoplasmic distribution and characterization of poly(A)+ RNA in oocytes and embryos of *Drosophila*. *Dev. Biol.*, **61**, 70–78.

Loyd, J.E., Raff, E.C., and Raff, R.A. (1981). Site and timing of synthesis of tubulin and other proteins during oogenesis in *Drosophila melanogaster*. *Dev. Biol.*, **86**, 272–284.

Lucchesi, J.C. (1989). On the origin of the mechanism compensating for gene-dosage differences in *Drosophila*. *Am. Nat.*, **134**, 474–485.

Lucchesi, J.C., and Skripsky, T. (1981). The link between dosage compensation and sex differentiation in *Drosophila melanogaster*. *Chromosoma*, **82**, 217–227.

Lucchesi, J.C., and Manning, J.E. (1987). Gene dosage compensation in *Drosophila melanogaster*. *Adv. Genet.*, **24**, 371–429.

Lyon, M.F. (1961). Gene action in the X-chromosome of the mouse (*Mus musculus* L.). *Nature*, **190**, 372–373.

Lyon, M.F. (1989). Rules and guidelines for gene nomenclature. In M.F. Lyon and A.G. Searle (Eds.), *Genetic Variants and Strains of the Laboratory Mouse* Oxford: Oxford University Press, pp. 1–11.

Lyon, M.F., and Hawkes, S.G. (1970). X-linked gene for testicular feminization in the mouse. *Nature*, **227**, 1217–1219.

Lyon, M.F., Glenister, P.H., and Lamoureux, M.L. (1975). Normal spermatozoa from androgen-resistant germ cells of chimaeric mice and the role of androgen in spermatogenesis. *Nature*, **258**, 620–622.

Lyon, M.F., Ward, H.C., and Simpson, G.M. (1976). A genetic method for measuring nondisjunction in mice with Robertsonian translocations. *Genet. Res.*, **26**, 283–295.

MacDonald, P.M., and Struhl, G. (1988). *cis*-acting sequences responsible for anterior localization of *bicoid* mRNA in *Drosophila* embryos. *Nature*, **336**, 595–598.

Madhavan, M., and Schneiderman, H.A. (1977). Histological analysis of the dynamics of growth and imaginal discs and histoblast nests during the larval development of *Drosophila melanogaster*. *Wilhelm Roux Arch.*, **183**, 269–305.

Madl, J., and Herman, R.K. (1979). Polyploids and sex determination in *Caenorhabditis elegans*. *Genetics*, **93**, 393–402.

Magnuson, T. (1986). Mutations and chromosomal abnormalities: How are they useful for studying genetic control of early mammalian development? In J. Rossant and R.A. Pederson (Eds.), *Experimental Approaches to Mammalian Embryonic Development*. New York: Cambridge University Press, pp. 437–474.

Magnuson, T., and Epstein, C.J. (1984). Oligosyndactyly: a lethal mutation in the mouse that results in mitotic arrest very early in development. *Cell*, **38**, 823–833.

Mahaffey, J.W., Diederich, R.J., and Kaufman, T.C. (1989). Novel patterns of homeotic protein accumulation in the head of the *Drosophila* embryo. *Development*, **105**, 167–174.

Mahowald, A.P., and Kambysellis, M.P. (1980). Oogenesis. In M. Ashburner and T.R.F. Wright (Eds.), *The Genetics and Biology of Drosophila*. London: Academic Press, vol. 2c, pp. 141–224.

Mahowald, A.P., Caulton, J.H., and Gehring, W.J. (1979). Ultrastructural studies of oocytes and embryos derived from female flies carrying the *grandchildless* mutation in *Drosophila subobscura. Dev. Biol.,* **69,** 118–132.

Maine, E.M., and Kimble, J. (1990). Genetic control of cell communication in *C. elegans* development. *BioEssays,* **12,** 265–271.

Maine, E.M., Salz, H.K., Schedl, P., and Cline, T.W. (1985a). *Sex-lethal,* a link between sex determination and sexual differentiation in *Drosophila melanogaster. Cold Spring Harbor Symp. Quant. Biol.,* **50,** 595–604.

Maine, E.M., Salz, H.K., Cline, T.W., and Schedl, P. (1985b). The *Sex-lethal* gene of Drosophila: DNA alterations associated with sex-specific lethal mutations. *Cell,* **43,** 521–529.

Malicki, J., Schighart, K., and McGinnis, W. (1990). Mouse *Hox-2.2* specifies thoracic segmental identity in Drosophila embryos and larvae. *Cell,* **63,** 961–967.

Mango, S.E., Maine, E.M., and Kimble, J. (1991). Carboxy-terminal truncation activates *glp-1* protein to specify vulval fates in *Caenorhabditis elegans. Nature,* **352,** 811–815.

Mann, J.R., and Lovell-Badge, R.H. (1984). Inviability of parthenogenesis is determined by pronuclei, not egg cytoplasm. *Nature,* **310,** 66–67.

Mann, R.S., and Hogness, D.S. (1990). Functional dissection of Ultrabithorax proteins in D. melanogaster. *Cell,* **60,** 597–610.

Manseau, L.J., and Schüpbach, T. (1989). *cappucino* and *spire:* Two unique maternal-effect loci required for both the anteroposterior and dorsoventral patterns of the *Drosophila* embryo. *Genes Dev.,* **3,** 1437–1452.

Mariol, M-C. (1981). Genetic and developmental studies of a new grandchildless mutant of *Drosophila melanogaster. Mol. Gen. Genet.* **181,** 505–511.

Mark, G.E., McIntyre, R.J., Digan, M.E., Ambrosio, L., and Perrimon, N. (1987). *Drosophila melanogaster* homologs of the *raf* oncogene. *Molec. Cell Biol.* **7,** 2124–2140.

Markert, C.L. and Silvers, W.K. (1956). The effects of genotype and cell environment on melanoblast differentiation in the house mouse. *Genetics* **41,** 429–450.

Marsh, J.L., and Wieschaus, E. (1978). Is sex determination in germ line and soma controlled by separate genetic mechanisms? *Nature,* **272,** 249–251.

Martin, K.J. (1991). The interactions of transcription factors and their adaptors, coactivators and accessory proteins. *BioEssays,* **13,** 499–503.

Martinez, C., and Modolell, J. (1991). Cross-regulatory interactions between the proneural *achaete* and *scute* genes of *Drosophila. Science,* **251,** 1485–1487.

Martinez-Arias, A. (1989). A cellular basis for pattern formation in the insect epidermis. *Trends Genet.* **5,** 262–267.

Martinez-Arias, A., and Lawrence, P.A. (1985). Parasegments and compartments in the *Drosophila* embryo. *Nature,* **313,** 639–642.

Martinez-Arias, A., Baker, N.E., and Ingham, P.E. (1988). Role of segment polarity genes in the definition and maintenance of cell states in the *Drosophila* embryo. *Development,* **103,** 157–170.

Maynard Smith, J., Burian, M., Kauffman, S., Alberch, P., Campbell, P., Lande, R., Raup, D., and Wolpert, L. (1985). Developmental constraints and evolution. *Q. Rev. Biol.,* **60,** 265–287.

McCarrey, J.R., and Abbott, U.K. (1979). Mechanisms of genetic sex determination, gonadal sex differentiation, and germ-cell development in animals. *Adv. Genet.,* **20,** 217–290.

McClintock, B. (1951). Chromosome organization and genic expression. *Cold Spring Harbor Symp. Quant. Biol.,* **16,** 13–47.

McClintock, B. (1956). Controlling elements and the gene. *Cold Spring Harbor Symp. Quant. Biol.,* **21,** 197–216.

McCoubrey, W.K., Nordstrom, K.D., and Meneely, P.M. (1988). Microinjected DNA from the X chromosome affects sex determination in *Caenorhabditis elegans. Science,* **242,** 1146–1151.

McCulloch, E.A., Siminovitch, L., Till, J.E., Russell, E.S., and Bernstein, S.E. (1964). The cellular basis of the genetically determined hemopoietic defect in anemic mice of genotype Sl/Sl^d. *Blood,* **26,** 399–410.

McGinniss, N., Kuziora, M.A., and McGinnis, W. (1990). Human *Hox-4.2* and Drosophila *Deformed* encode similar regulatory specificities in Drosophila embryos and larvae. *Cell,* **63,** 969–976.

McGinnis, W., Levine, M.S., Hafen, E., Kuroiwa, A., and Gehring, W.J. (1984a). A conserved DNA sequence in homoeotic genes of the *Drosophila* Antennapedia and bithorax complexes. *Nature,* **308,** 428–433.

McGinnis, W., Hart, C.P., Gehring, W.J., and Ruddle, F.H. (1984b). Molecular cloning and chromosome mapping of a mouse DNA sequence homologous to homeotic genes of Drosophila. *Cell,* **38,** 675–680.

McGinnis, W., Hart, C.P., Gehring, W.J., and Ruddle, F.H. (1984). Molecular cloning and chromosome mapping of a mouse DNA sequence homologous to homeotic genes of Drosophila. *Cell,* **38,** 675–680.

McGrath, J., and Solter, D. (1983). Nuclear transplantation in the mouse embryo by microsurgery and cell fusion. *Science,* **220,** 1300–1302.

McGrath, J., and Solter, D. (1984a). Completion of mouse embryogenesis requires both the maternal and paternal genomes. *Cell,* **37,** 179–183.

McGrath, J., and Solter, D. (1984b). Maternal T^{hp} lethality in the mouse is a nuclear, not cytoplasmic defect. *Nature,* **308,** 550–551.

McKnight, G.S., Hammer, R.E., Kuegel, E.A., and Brinster, R.L. (1983). Expression of the chicken transferrin gene in transgenic mice. *Cell,* **34,** 335–341.

McKnight, S.L., and Miller, O.L., Jr. (1976). Ultrastructural patterns of RNA synthesis during early embryogenesis of *Drosophila melanogaster. Cell,* **8,** 305–319.

McKnight, S.L., Lane, M.D., and Gluecksohn-Waelsch, S. (1989). Is CCAAT/enhancer binding protein a central regulator of energy metabolism? *Genes Dev.,* **3,** 2021–2024.

McLaren, A. (1972). The numerology of development. *Nature,* **239,** 274–276.

McLaren, A. (1976a). *Mammalian Chimaeras.* Cambridge: Cambridge University Press.

McLaren, A. (1976b). Genetics of the early mouse embryo. *Annu. Rev. Genet.,* **10,** 361–388.

McLaren, A. (1979). The impact of pre-fertilization events on post-fertilization development in mammals. In D.R. Newth and M. Balls (Eds.), *Maternal Effects in Development.* Cambridge: Cambridge University Press, pp. 287–320.

McLaren, A. (1991). Development of the mammalian gonad: The fate of the supporting cell lineage. *BioEssays,* **13,** 151–156.

McLaren, A., Simpson, E., Tomonari, K., Chandler, P., and Hogg, H. (1984). Male sexual differentiation in mice lacking H-Y antigen. *Nature,* **312,** 552–554.

McMahon, A., Fosten, M., and Monk, M. (1983). X-chromosome inactivation mosaicism in the three germ layers and the germ line of the mouse embryo. *J. Embryol. Exp. Morphol.,* **74,** 207–220.

McNeill, H., Ozawa, M., Kemler, R., and Nelson, W.J. (1990). Novel function of the cell adhesion molecule uvomorulin as an inducer of cell surface polarity. *Cell,* **62,** 309–316.

Meinhart, H. (1977). A model of pattern formation in insect embryogenesis. *J. Cell Sci.,* **23,** 117–139.

Meinhardt, H. (1982). *Models of Biological Pattern Formation.* London: Academic Press.

Meinhardt, H. (1983). Cell determination boundaries are organizing regions for secondary embryonic fileds. *Dev. Biol.,* **96,** 375–385.

Meinhardt, H. (1984). Models for positional signalling, the threefold subdivision of segments and the pigmentation pattern of molluses. J. Embryol. Exp. Morph. 83 (suppl.), 289–311.

Meinhardt, H. (1986). The threefold subdivision of segments and the initiation of legs and wings in insects. *Trends Genet.* **2,** 36–41.

Meneely, P.M. (1990). X-linked expression nad sex determination in *Caenorhabditis elegans. BioEssays,* **12,** 513–518.

Meneely, P.M., and Herman, R.K. (1979). Lethals, steriles, and deficiencies in a region of the X chromosome of *Caenorhabditis elegans. Genetics,* **92,** 99–115.

Meneely, P.M., and Wood, W.B. (1987). Genetic analysis of X-chromosome dosage compensation in *Caenorhabditis elegans. Genetics,* **117,** 25–41.

Mercer, J.A., Seperack, P.K., Strobel, M.C., Copeland, N.G. and Jenkins, N.A. (1991). The murine dilute coat colour locus encodes a novel myosin heavy chain. *Nature,* **348,** 709–713.

Merriam, J.R. (1978). Estimating primordial cell numbers in *Drosophila* imaginal discs and histoblasts. In W. J. Gehring (Ed.), *Genetic Mosaics and Cell Differentiation.* Berlin: Springer-Verlag, pp. 71–94.

Merriam, J., Ashburner, M., Hartl, D.L., and Kafatos, F.C. (1991). Toward cloning and mapping the genome of *Drosophila*. *Science*, **254**, 221–225.

Merrill, V.K.L., Turner, F.R., and Kaufman, T.C. (1987). A genetic and developmental analysis of mutations in the *Deformed* locus in *Drosophila melanogaster*. *Dev. Biol.*, **122**, 379–395.

Meyer, B.J., and Casson, L.P. (1986). Caenorhabditis elegans compensates for the difference in X chromosome dosage between the sexes by regulating transcript levels. *Cell*, **47**, 871–881.

Meyerowitz, E.M., and Kankel, D.R. (1978). A genetic analysis of visual system development in *Drosophila melanogaster*. *Dev. Biol.*, **62**, 112–142.

Miller, L.M., Plenefisch, J.D., Casson, L.P., and Meyer, B.J. (1988). *xol-1*: A gene that controls the male modes of both sex determination and X chromosome dosage compensation in *C. elegans. Cell*, **55**, 167–183.

Miller, O.J., and Miller, D. (1975). Cytogenetics of the mouse. *Annu. Rev. Genet.*, **9**, 285–303.

Mintz, B. (1967). Gene control of mammalian pigmentary differentiation. I. Clonal origin of melanocytes. *Proc. Natl. Acad. Sci., U.S.A.*, **58**, 344–351.

Mintz, B. (1974). Gene control of mammalian differentiation. *Annu. Rev. Genet.*, **8**, 411–470.

Mitchell, H.K., and Lipps, L.S. (1978). Heat shock and phenocopy induction in *Drosophila. Cell*, **15**, 907–919.

Mitchell, H.K., and Petersen, N.S. (1981). Rapid changes in gene expression in differentiating tissues of *Drosophila. Dev. Biol.*, **85**, 233–242.

Mitchell, H.K., and Petersen, N.S. (1985). The recessive phenotype of *forked* can be uncovered by heat shock in Drosophila. *Dev. Genet.*, **6**, 93–100.

Mitchell, P.J., and Tijian, R. (1989). Transcriptional regulation in mammalian cells by sequence-specific DNA binding proteins. *Science*, **245**, 371–378.

Miwa, J., Schierenberg, E., Miwa, S., and von Ehrenstein, G. (1980). Genetics and mode of expression of temperature-sensitive mutations arresting embryonic development in *Caenorhabditis elegans. Dev. Biol.*, **76**, 160–174.

Mlodzik, M., Baker, N.E., and Rubin, G.M. (1990). Isolation and expression of *scabrous*, a gene regulating neurogenesis in Drosophila. *Genes Dev.*, **4**, 1848–1861.

Moerman, D.G., and Waterston, R.H. (1984). Spontaneous unstable *unc-22 IV* mutations in *Caenorhabditis elegans* var Bergerac. *Genetics*, **108**, 859–877.

Mohler, J.D. (1977). Developmental genetics of the *Drosophila* egg. I. Identification of 59 sex-linked cistrons with maternal effects on embryonic development. *Genetics*, **85**, 259–272.

Mohler, J. (1988). Requirements for *hedgehog*, a segmental polarity gene, in patterning larval and adult cuticle of Drosophila. *Genetics*, **120**, 1061–1072.

Mohler, J., and Wieschaus, E.F. (1986). Dominant maternal-effect mutations of *Drosophila melanogaster* causing the production of double-abdomen embryos. *Genetics*, **112**, 803–822.

Monk, M., and Grant, M. (1990). Preferential X-chromosome inactivation, DNA methylation and imprinting. *Development*, (Suppl.), 55–62.

Monk, M., Boubelik, M., and Lehnert, S. (1987). Temporal and regional changes in DNA methylation in the embryonic, extraembryonic and germ cell lineages during mouse embryo development. *Development*, **99**, 371–382.

Monk, M., and Surani, A., eds. (1990). *Genomic Imprinting. Development* Supp. Cambridge: Company of Biologists.

Monod, J., and Jacob, F. (1961). Genetic regulatory mechanisms in the synthesis of proteins. *J. Mol. Biol.*, **3**, 318–356.

Montell, C., Jones, K., Zuker, C.S., and Rubin, G.M. (1987). A second opsin gene expressed in the ultraviolet sensitive R7 photoreceptor of *Drosophila melanogaster. J. Neurosci.*, **7**, 1558–1566.

Moore, T., and Haig, D. (1991). Genomic imprinting in mammalian development: A parental tug of war. *Trends Genet.* **7**, 45–49.

Morata, G., and Lawrence, P.A. (1975). Control of compartment development by the *engrailed* gene in *Drosophila. Nature*, **255**, 614–617.

Morata, G., and Ripoll, P. (1975). Minutes: Mutants of *Drosophila* autonomously affecting cell division rate. *Dev. Biol.,* **42,** 211–221.

Morata, G., and Lawrence, P.A. (1979). Development of the eye-antenna imaginal disc of *Drosophila. Dev. Biol.,* **70,** 355–371.

Morata, G., and Kerridge, S. (1981). Sequential functions of the bithorax complex of *Drosophila. Nature,* **290,** 778–781.

Morgan, T.H. (1934). *Embryology and Genetics.* New York: Columbia University Press.

Morgan, T.H., and Bridges, C.B. (1919). The origin of gynandromorphs. *Carnegie Inst. Wash. Publ. 278.*

Moscoso del Prado, J., and Garcia-Bellido, A. (1984). Genetic regulation of the achaete-scute complex of *Drosophila melanogaster. Wilhelm Roux Archiv. Dev. Biol.,* **193,** 242–245.

Moyzis, R.K., Bonnet, J., Li, D.W., and Ts'o, P.O.P. (1981). An alternative view of mammalian DNA sequence organization. II. Short repetitive sequences are organized into scrambled tandem clusters in Syrian hamster DNA. *J. Mol. Biol.,* **153,** 871–896.

Mukherjee, A.S. (1966). Dosage compensation in *Drosophila*—An autoradiographic study. *Nucleus,* **9,** 83–90.

Mukherjee, A.S., and Beeman, W. (1965). Synthesis of ribonucleic acid by the X chromosomes of *Drosophila melanogaster* and the problem of dosage compensation. *Nature,* **207,** 785–786.

Muller, H.J. (1932). Further studies on the nature and causes of gene mutations. *Proc. 6th Int. Congr. Genet.,* **I,** 213–255.

Muller, H.J. (1950). Evidence for the precision of genetic adaptation. *Harvey Lect.,* **43,** 165–229.

Muller, H.J., and Prokofyeva, A.A. (1935). The individual gene in relation to the chromomere and chromosome. *Proc. Natl. Acad. Sci., U.S.A.,* **21,** 16–26.

Murre, C., McCaw, P.S., and Baltimore, D. (1989). A new DNA binding and dimerization motif in immunoglobulin enhancer binding, *daughterless, MyoD* and *vg* proteins. *Cell,* **56,** 777–783.

Nagoshi, R.N., McKeown, M., Burtis, K.C., Belote, J.M., and Baker, B.S. (1988). The control of alternative splicing at genes regulating sexual differentiation in D. melanogaster. *Cell,* **53,** 229–236.

Nakano, Y., Guerrero, I., Hidalgo, A., Taylor, A., Whittle, J.R.S., and Ingham, P.W. (1989). A protein with several possible membrane-spanning domains encoded by the *Drosophila* segment polarity gene *patched. Nature,* **341,** 508–513.

Nauber, U., Pankratz, M.J., Kienlin, A., Seifert, E., Klenow, U., and Jäckle, H. (1988). Abdominal segmentation of the *Drosophila* embryo requires a hormone receptoı Le protein encoded by the gap gene *knirps. Nature,* **336,** 489–492.

Nelson, G.A., Lew, K.K., and Ward, S. (1978). Intersex, a temperature-sensitive mutant of the nematode *Caenorhabditis elegans. Dev. Biol.,* **66,** 386–409.

Nesbitt, M.N. (1971). X chromosome inactivation mosaicism in the mouse. *Dev. Biol.,* **26,** 252–263.

Nesbitt, M.N., and Gartler, S.M. (1971). The applications of genetic mosaicism to developmental problems. *Annu. Rev. Genet.,* **5,** 143–162.

Nigon, V. (1951). Polyploidie experimentale chez un nematode libre, *Rhabditis elegans* Maupas. *Bull. Biol. Fr. Belg.,* **85,** 187–225.

Nigon, V. (1965). Developpment et reproduction des nematodes. In P.-P. Grasse (Ed.), *Traité de Zoologie.* Paris: Masson et Cie, vol. 4, part 2, pp. 218–386.

Nijhout, H.F. (1990). Metaphors and the role of genes in development. *BioEssays,* **12,** 441–446.

Niki, Y., and Okada, M. (1981). Isolation and characterization of *grandchildless*-like mutants in *Drosophila melanogaster. Wilhelm Roux Arch.,* **190,** 1–10.

Nishida, Y., Hota, M., Ayaki, T., Ryo, H., Yamagata, M., Shimuzu, K., and Nishizuka, Y. (1988). Proliferation of both somatic and germ cells is affected in the *Drosophila* mutants of *raf* proto-oncogene. *EMBO J.,* **7,** 775–781.

Nissani, M. (1977). Cell lineage analysis of germ cells of *Drosophila melanogaster. Nature,* **265,** 729–731.

Niswander, L., Yee, D., Rinchik, E.M., Russell, L.B., and Magnuson, T. (1989). The albino-deletion complex in the mouse defines genes necessary for development of embryonic and extraembryonic ectoderm. *Development,* **105,** 175–182.

Nonet, M.L., and Meyer, B.J. (1991). Early aspects of *Caenorhabditis elegans* sex determination and dosage compensation are regulated by a zinc-finger protein. *Nature,* **351,** 65–68.

Nöthiger, R. (1972). The larval development of imaginal discs. In H. Ursprung and R. Nöthiger (Eds.), *The Biology of Imaginal Discs: Results and Problems in Cell Differentiation.* Berlin: Springer-Verlag, vol. 5, pp. 1–34.

Nöthiger, R. (1976). Clonal analysis in imaginal discs. In P.A. Lawrence (Ed.), *Insect Development.* Oxford: Blackwell Scientific Publications, pp. 109–118.

Nöthiger, R., Dubendorfer, A., and Epper, F. (1977). Gynandromorphs reveal two separate primordia for male and female genitalia in *Drosophila melanogaster. Wilhelm Roux Arch.,* **181,** 367–373.

Nöthiger, R., Leuthold, M., Andersen, N., Gerschwiler, P., Gruter, A., Keller, W., Leist, C., Roost, M., and Schmid, H. (1987). Genetic and developmental analysis of the sex-determining gene 'double sex' (*dsx*) of *Drosophila melanogaster. Genet. Res.,* **50,** 113–123.

Nöthiger, R., Jonglez, M., Leuthold, M., Meier-Gerschwiler, P., and Weber, T. (1989). Sex determination in the germ line of *Drosophila* depends on genetic signals and inductive somatic factors. *Development,* **107,** 505–518.

Nusbaum, C., and Meyer, B.J. (1989). The *Caenorhabditis elegans* gene sdc-2 controls sex determination and dosage compensation in XX animals. *Genetics,* **122,** 579–593.

Nüsslein-Volhard, C. (1977). Genetic analysis of pattern-formation in the embryo of *Drosophila melanogaster.* Characterization of the maternal-effect mutant *Bicaudal. Wilhelm Roux Arch.,* **183,** 249–268,

Nüsslein-Volhard, C. (1979a). Maternal effect mutations that alter the spatial coordinates of the embryo of *Drosophila melanogaster.* In S. Subtelny and I.R. Konigsberg (Eds.), *Determinants of Spatial Organization.* New York: Academic Press, pp. 185–211.

Nüsslein-Volhard, C. (1979b). Pattern mutants in *Drosophila* embryogenesis. In N. Le Douarin (Ed.), *Cell Lineage, Stem Cells, and Cell Determination.* Amsterdam: Elsevier/North-Holland, pp. 69–82.

Nüsslein-Volhard, C. (1991). Determination of the embryonic axes of Drosophila. *Development,* **1** (Suppl.), 1–10.

Nüsslein-Volhard, C., and Wieschaus, E. (1980). Mutations affecting segment number and polarity in *Drosophila. Nature,* **287,** 795–801.

Nüsslein-Volhard, C., Lohs-Schardin, M., Sander, K., and Cremer, C. (1980). A dorso-ventral shift of embryonic primordia in a new maternal effect mutant of *Drosophila. Nature,* **283,** 474–476.

Nüsslein-Volhard, C., Wieschaus, E., and Jurgens, G. (1982). Segmentation in *Drosophila,* a genetic analysis. *Verh. Dtsch. Zool. Ges.,* 91–104.

Nüsslein-Volhard, C., Wieschaus, E., and Kluding, H. (1984). Mutations affecting the pattern of the larval cuticle in *Drosophila melanogaster. Wilhelm Roux Arch.,* **193,** 267–282.

Nüsslein-Volhard, C., Frohnhöfer, H.G., and Lehmann, R. (1987). Determination of antero-posterior polarity in *Drosophila. Science,* **238,** 1675–1681.

O'Brien, S.J. (1973). On estimating functional gene number in eucaryotes. *Nature New Biol.,* **242,** 52–54.

O'Brochta, D.A., and Bryant, P.J. (1985). A zone of non-proliferating cells at a lineage restriction boundary in *Drosophila. Nature,* **313,** 138–141.

Odell, G.M., Oster, G., Albrech, P., and Burnside, B. (1981). The mechanical basis of morphogenesis. I. Epithelial folding and invagination. *Dev. Biol.,* **85,** 446–462.

Ohno, S. (1976). Major regulatory genes for mammalian sexual development. *Cell,* **7,** 315–321.

O'Kane, C.J., and Gehring, W.J. (1987). Detection *in situ* of genomic regulatory elements in *Drosophila. Proc. Natl. Acad. Sci., U.S.A.,* **84,** 9123–9127.

Olby, R.C. (1966). *Origins of Mendelism.* New York: Schocken Books.

Oliver, B., Perrimon, N., and Mahowald, A.P. (1988). Genetic evidence that the *sans fille* locus is involved in *Drosophila* sex determination. *Genetics,* **120,** 159–171.

Oppenheimer, J.M. (1967). *Essays in the History of Embryology and Biology.* Cambridge, MA: MIT Press.

Orenic, T.V., Slusarski, D.C., Kroll, K.L., and Holmgren, R.A. (1990). Cloning and characterization of the segment polarity gene, *cubitus interruptus Dominant* of *Drosophila. Genes Dev.,* **4,** 1053–1067.

Orr-Weaver, T.L. (1991). *Drosophila* chorion genes: Cracking the eggshell's secrets. *BioEssays,* **13,** 97–105.

O'Tousa, J.E., Baehr, W., Martin, R.L., Hirsh, J., Pak, W.L., and Applebury, M.L. (1985). The Drosophila *ninaE* gene encodes an opsin. *Cell,* **40,** 839–850.

Ozdzenski, W. (1967). Observations on the origin of the primordial germ cells in the mouse. *Zool. Pol.,* **17,** 367–379.

Padgett, R.W., St. Johnston, R.D., and Gelbart, W.M. (1987). A transcript from a *Drosophila* pattern gene predicts a protein homologous to the transforming factor-β family. *Nature,* **325,** 81–84.

Page, D.C., Mosher, R., Simpson, E.M., Fisher, E.M.C., Mordon, G., Pollack, J., McGillvray, B., De La Chapelle, A., and Brown, L.G. (1987). The sex determining region of the human Y chromosome encodes a finger protein. *Cell,* **51,** 1091–1104.

Page, D.C., Fisher, E., McGillivray, B., and Brown, L.G. (1990). Additional deletion in the sex-determining region of the human Y chromosome resolves paradox of X, t (Y;22) female. *Nature,* **346,** 279–281.

Palmer, M., Sinclair, A.H., Berta, P., Ellis, N., Goodfellow, P.N., Abbas, N., and Fellous, M. (1989). Genetic evidence that ZFY is not the testis-determining factor. *Nature,* **342,** 937–939.

Pampfer, S., Arceci, R.J., and Pollard, J.W. (1991). Role of colony stimulating factor-1 (CSF-1) and other lympho-hematopoietic growth factors in mouse preimplantation development. *BioEssays,* **13,** 535–540.

Pankratz, M.J., Hoch, M., Seifert, E., and Jäckle, H. (1989). *Krüppel* requirement for *knirps* enhancement reflects overlapping gene activities in the *Drosophila* embryo. *Nature,* **341,** 337–339.

Pankratz, M.J., Seifert, E., Gerwin, N., Billi, B., Nauber, U., and Jäckle, H. (1990). Gradients of *Krüppel* and *knirps* gene products direct pair-rule gene stripe patterning in the posterior region of the Drosophila embryo. *Cell,* **61,** 309–317.

Panthier, J.J., and Condamine, H. (1991). Mitotic recombination in mammals. *BioEssays,* **13,** 351–356.

Papaioannou, V.E. (1988). Investigation of the tissue specificity of the lethal *Yellow* (A)^y gene in mouse embryos. *Dev. Genet.,* **9,** 155–165.

Papaioannou, V., and Gardner, R.L. (1979). Investigation of the lethal yellow A^y/A^y embryo using mouse chimaeras. *J. Embryol. Exp. Morphol.,* **52,** 153–163.

Pardue, M.L., Lowenhaupt, K., Rich, A., and Nordheim, A. (1987). (dC-dA)n (dG-dT)n sequences have evolutionarily conserved chromosomal locations in *Drosophila* with implications for roles in chromosome structure and function. *EMBO J.,* **6,** 1781–1789.

Park, E.C., and Horvitz, H.R. (1986). Mutations with dominant effects on the behavior and morphology of the nematode *Caenorhabditis elegans. Genetics,* **113,** 821–852.

Parkhurst, S.M., Bopp, D., and Ish-Horowicz, D. (1990). X: A ratio, the primary sex-determining signal in Drosophila, is transduced by helix-loop-helix proteins. *Cell,* **63,** 1179–1191.

Parks, H.B. (1936). Cleavage patterns in *Drosophila* and mosaic formation. *Ann. Entomol. Soc. Am.,* **29,** 350–392.

Paro, R. (1990). Imprinting a determined state into the chromatin of *Drosophila. Trends Genet.* **6,** 416–421.

Paro, R., and Hogness, D.S. (1991). The Polycomb protein shares a homologous domain with a heterochromatin-associated protein of *Drosophila. Proc. Natl. Acad. Sci., U.S.A.,* **88,** 263–267.

Pauli, D., and Mahowald, A.P. (1990). Germ-line sex determination in *Drosophila melanogaster. Trends Genet.,* **6,** 259–264.

Pawson, T., and Bernstein, A. (1990). Receptor tyrosine kinases—Genetic evidence for their role in Drosophila and mouse development. *Trends Genet.,* **6,** 350–356.

Pedersen, R.A. (1986). Potency, lineage, and allocation in preimplantation mouse embryos. In J.R. Rossant and R.A. Pedersen (Eds.), *Experimental Approaches to Mammalian Embryonic Development.* New York: Cambridge University Press, pp. 3–33.

Pedersen, R.A., Wu, K., and Balakier, H. (1986). Origin of the inner cell mass in mouse embryos: Cell lineage analysis by microinjection. *Dev. Biol.,* **117,** 581–595.

Peifer, M., and Bender, W. (1986). The anterior and bithorax mutations of the bithorax complex. *EMBO J.,* **5,** 2293–2303.

Peifer, M., and Wieschaus, E. (1990). The segment polarity gene *armadillo* encodes a functionally modular protein that is the Drosophila homolog of human plakoglobin. *Cell,* **63,** 1167–1178.

Peifer, M., Karch, F., and Bender, W. (1987). The bithorax complex: Control of segmental identity. *Genes Dev.,* **1,** 891–898.

Perrimon, N., and Gans, M. (1983). Clonal analysis of the tissue specificity of recessive female sterile mutations of *Drosophila melanogaster* using a dominant female sterile mutation, *FS(1)K1237*. *Dev. Biol.,* **100,** 365–373.

Perrimon, N., and Mahowald, A.P. (1987). Multiple functions of segment polarity genes in *Drosophila. Dev. Biol.,* **119,** 587–600.

Perrimon, N., Engstrom, L., and Mahowald, A.P. (1984). The effects of zygotic lethal mutations on female germ-line functions in *Drosophila. Dev. Biol.,* **105,** 404–414.

Perrimon, N., Engstrom, L., and Mahowald, A.P. (1985a). Developmental genetics of the 2C-D region of the Drosophila X-chromosome. *Genetics,* **111,** 23–41.

Perrimon, N., Engstrom, L., and Mahowald, A.P. (1985b). A pupal lethal mutation with a paternally influenced maternal effect on embryonic development in *Drosophila melanogaster. Dev. Biol.,* **110,** 480–491.

Perrimon, N., Mohler, D., Engstrom, L., and Mahowald, A. (1986). X-linked female-sterile loci in *Drosophila melanogaster. Genetics,* **113,** 695–712.

Perrimon, N., Engstrom, L., and Mahowald, A.P. (1989). Zygotic lethals with specific maternal effect phenotypes in *Drosophila melanogaster*. I. Loci on the X chromosome. *Genetics,* **121,** 333–352.

Petersen, N.S., and Mitchell, H.K. (1982). Effects of heat shock on gene expression during development: Induction and prevention of the multihair phenocopy in *Drosophila*. In M.J. Schlesinger, M. Ashburner, and A. Tissières (Eds.), *Heat Shock*. Cold Spring Harbor, NY: Cold Spring Harbor Laboratory Press, pp. 345–352.

Phillips, R.G., Roberts, I.H., Ingham, P.W., and Whittle, J.R.S. (1990). The *Drosophila* segment polarity gene *patched* is involved in a position-signalling mechanism in imaginal discs. *Development,* **110,** 105–114.

Pignoni, F., Baldarelli, R.M., Steingrimsson, E., Diaz, R.J., Diaz, R.J., Patapoulian, A., Merriam, J.R., and Lengyel, J.A. (1990). The Drosophila gene *tailless* is expressed at the embryonic terminii and is a member of the steroid receptor superfamily. *Cell,* **62,** 151–163.

Pirotta, V., Steller, H., and Bozzetti, M.P. (1985). Multiple upstream regulatory elements control the expression of the *Drosophila white* gene. *EMBO J.,* **4,** 3501–3508.

Plenefisch, J.D., DeLong, L., and Meyer, B.J. (1989). Genes that implement the hermaphrodite mode of dosage compensation in *Caenorhabditis elegans. Genetics,* **121,** 57–76.

Politz, J.C., and Edgar, R.S. (1984). Overlapping stage-specific sets of numerous small collagenous polypeptides are translated *in vitro* from *Caenorhabditis elegans* RNA. *Cell,* **37,** 853–860.

Politz, S.M., Chin, K.J., and Herman, D.L. (1987). Genetic analysis of adult-specific surface antigenic differences between varieties of the nematode *Caenorhabditis elegans. Genetics,* **117,** 467–476.

Ponder, B.A.J., Wilkinson, M.M., and Wood, M. (1983). H-2 antigens as markers of cellular genotype in chimaeric mice. *J. Embryol. Exp. Morphol.,* **76,** 83–93.

Posakony, L.G., Raftery, C.A., and Gelbart, W.M. (1991). Wing formation in *Drosophila melanogaster* requires *decapentaplegic* gene function along the anterior-posterior compartment boundry. *Mech. Dev.,* **33,** 69–82.

Postlethwait, J.H. (1978). Clonal analysis of *Drosophila* cuticular patterns. In M. Ashburner and T.R.F. Wright (Eds.), *The Genetics and Biology of* Drosophila. London: Academic Press, pp. 359–441.

Postelthwait, J.H., and Schneiderman, H.A. (1969). A clonal analysis of determination in Antennapedia, a homeotic mutant of *Drosophila melanogaster Proc. Natl. Acad. Sci., U.S.A.,* **64,** 176–183.

Postlethwait, J.H., and Schneiderman, H.A. (1971). Pattern formation and determination in the antenna of the homoeotic mutant *Antennapedia* of *Drosophila melanogaster. Dev. Biol.,* **25,** 606–640.

Poulson, D.F. (1937). Chromosomal deficiencies and the embryonic development of *Drosophila melanogaster. Proc. Natl. Acad. Sci., U.S.A.,* **23,** 133–137.

Poulson, D.F. (1950). Histogenesis, organogenesis, and differentiation in the embryo of *Drosophila* Meigen. In M. Demerec (Ed.), *The Biology of* Drosophila. New York: John Wiley & Sons, pp. 168–274.

Pratt, H.P.M. (1989). Marking time and marking space: Chronology and typography in the early mouse embryo. *Int. Rev. Cytol.,* **117,** 99–130.

Pratt, H.P.M., Bolton, V.N., and Gudgeon, K.A. (1983). The legacy from the oocyte and its role in controlling early development of the mouse embryo. In R. Porter and J. Whelan (Eds.), *Molecular Biology of Egg Maturation*. London: Pitman, pp. 197–227.

Préat, T., Theraud, P., Lamouris, C., Limbourg, B., Tricoire, H., Erk, I., Mariol, M.C., and Busson, D. (1990). A putative serine-threonine protein kinase encoded by the segment-polarity *fused* gene of *Drosophila*. *Nature*, **347**, 87–89.

Price, J. (1987). Retroviruses and the study of cell lineage. *Development*, **101**, 409–419.

Price, J., Turner, D., and Cepko, C. (1987). Lineage analysis in the vertebrate nervous system by retrovirus-mediated gene transfer. *Proc. Natl. Acad. Sci., U.S.A.*, **84**, 156–160.

Price, J.V., Clifford, R.J., and Schüpbach, T. (1989). The maternal ventralizing locus *torpedo* is allelic to *faint little balls*, an embryonic lethal, and encodes the Drosophila EGF receptor homology. *Cell*, **56**, 1085–1092.

Priess, J., and Hirsh, D. (1986). *Caenorhabditis elegans* morphogenesis: The role of the cytoskeleton in the elongation of the embryo. *Dev. Biol.*, **117**, 156–173

Priess, J.R., and Thomson, J.N. (1987). Cellular interactions in ealry *C. elegans* embryos. *Cell*, **48**, 241–250.

Priess, J.R., Schnabel, H., and Schnabel, R. (1987). The *glp-1* locus and cellular interactions in early C. elegans embryos. *Cell*, **51**, 601–611.

Provine, W.B. (1971). *The Origins of Theoretical Population Genetics*. Chicago: University of Chicago Press.

Ptashne, M. (1986). *A Genetic Switch*. Cambridge: Cell Press.

Püschel, A., Balling, R., and Gruss, P. (1990). Position-specific activity of the *Hox-1.1* promoter in transgene mice. *Developement*, **108**, 435–442.

Rabinowitz, M. (1941). Studies on the cytology and early embryology of the egg of *Drosophila melanogaster*. *J. Morphol.*, **69**, 1–49.

Raff, R. (1992). Direct developing sea urchins and the evolutionary reorganization of early development. *BioEssays*, **14**,

Raftery, L.A., Sanicola, M., Blackman, R.K., and Gelbart, W.M. (1991). The relationship of *decapentaplegic* and *engrailed* expression in *Drosophila* imaginal discs: Do these genes mark the anterior-posterior compartment boundary? *Development*, **113**, 27–33.

Rassbash, P., Cooke, L.A., Herrman, B.G., and Beddington, R.G. (1991). A cell autonomous function of *Brachyury* in *T/T* embryonic stem cell divisions. *Nature*, **353**,

Ready, D.F. (1989). A multifaceted approach to neural development. *Trends Neurosci.*, **12**, 102–110.

Ready, D.F., Hanson, T.E., and Benzer, S. (1976). Development of the *Drosophila* retina, a neurocrystalline latice. *Dev. Biol.*, **53**, 217–240.

Reinke, R., and Zipursky, S.L. (1988). Cell-cell interaction in the *Drosophila* retina: The *bride of sevenless* gene is required in photoreceptor cell R8 for R7 cell development. *Cell*, **55**, 321–330.

Reith, A.D., Rottapel, R., Giddens, E., Brody, C., Forrester, L., and Bernstein, R. (1990). *W* mutant mice with mild or severe developmental defects contain distinct point mutations in the kinase domain of the c-*kit* receptor. *Genes Dev.*, **4**, 390–404.

Rice, R.B., and Garen, A. (1975). Localized defects of blastoderm formation in maternal effect mutants in *Drosophila*. *Dev. Biol.*, **43**, 277–286.

Richardson, H.E., Wittenberg, C., Cross, F., and Reed, S.I. (1989). An essential G1 function for cyclin-like proteins in yeast. *Cell*, **59**, 1127–1133.

Riddiford, L.M. (1985). Hormone action at the cellular level. In G.A. Kerkut and L.I. Gilbert (Eds.), *Comprehensive Insect Physiology, Biochemistry, and Pharmacology*. vol. 8. Oxford: Pergamon.

Richelle, J., and Ghysen, A. (1979). Determination of sensory bristles and pattern formation in *Drosophila*. I. A model. *Dev. Biol.*, **70**, 418–437.

Riddle, D.L. (1977). A genetic pathway for dauer larva formation in *Caenorhabditis elegans*. *Stadler Symp.*, **9**, 101–120.

Riddle, D.L. (1988). The dauer larva. In W.B. Wood (Ed.), *The Nematode* Caenorhabditis elegans. Cold Spring Harbor, NY: Cold Spring Harbor Laboratory Press, pp. 393–412.

Riddle, D.L., Swanson, M.M., and Albert, P. (1981). Interacting genes in nematode dauer larva formation. *Nature*, **290**, 668–671.

Riggleman, B., Wieschaus, E., and Schedl, P. (1989). Molecular analysis of the *armadillo* locus: Uniformly distributed transcripts and a protein with novel internal repeats are associated with a *Drosophila* segment polarity gene. *Genes Dev.,* **3,** 96–113.

Riggleman, B., Schedl, P., and Wieschaus, E. (1990). Spatial expression of the Drosophila segment polarity gene *armadillo* is posttranscriptionally regulated by *wingless. Cell,* **63,** 549–560.

Riggs, A.D. (1975). X inactivation, differentiation and DNA methylation. *Cytogenet. Cell Genet.,* **14,** 9–25.

Rinchik, E.M. (1991). Chemical mutagenesis and fine structure functional analysis of the mouse genome. *Trends Genet.,* **7,** 15–21.

Rinchik, E.M., Carpenter, D.A., and Selby, P.B. (1990). A strategy for fine-structure functional analysis of a 6- to 11-centimorgan region of mouse chromosome 7 by high efficiency mutagenesis. *Proc. Natl. Acad. Sci., U.S.A.,* **87,** 896–900.

Rijsewijk, F., Schuermann, M., Wagenaar, E., Parren, P., Weigel, D., and Nusse, R. (1987). The Drosophila homolog of the mouse mammary oncogene *int-1* is identical to the segment polarity gene *wingless. Cell,* **50,** 649–657.

Ripoll, P., and Garcia-Bellido, A. (1979). Viability of homozygous deficiencies in somatic cells of *Drosophila melanogaster. Genetics,* **91,** 443–453.

Roberts, P. (1961). Bristle formation controlled by the achaete locus in genetic mosaics of *Drosophila melanogaster. Genetics,* **46,** 1241–1243.

Rodgers, M.E., and Shearn, A. (1977). Patterns of protein synthesis in imaginal discs of *Drosophila melanogaster. Cell,* **12,** 915–921.

Rodriquez, I., Hernandez, R., Modolell, J., and Ruiz-Gomez, M. (1990). Competence to develop sensory organs is temporally and spatially regulated in *Drosophila* epidermal primordia. *EMBO J.,* **9,** 3583–3592.

Rohme, D., Fox, H., Herrmann, B., Frischauf, A.-M. Edstrom, J.-E., Mains, P., Silver, L.M., and Lehrach, H. (1984). Molecular clones of the mouse *t* complex derived from microdissected metaphase chromosomes. *Cell,* **36,** 783–788.

Romani, S., Campuzano, S., Macagno, E.R., and Modolell, J. (1989). Expression of *achaete* and *scute* genes in *Drosophila* imaginal discs and their function in sensory organ development. *Genes Dev.,* **3,** 997–1007.

Rosenburg, U.B., Schroder, C., Preiss, A., Kielin, A., Cote, S., Riede, I., and Jäckle, H. (1987). Finger protein of novel structure encoded by *hunchback,* a second member of the gap class of *Drosophila* segmentation genes. *Nature,* **319,** 336–339.

Rossant, J. (1975). Investigation of the determinative state of the mouse inner cell mass. *J. Embryol. Exp. Morphal.,* **33,** 979–990.

Rossant, J. (1986). Development of extraembryonic cell lineages in the mouse embryo. In J.R. Rossant and R.A. Pedersen (Eds.), *Experimental Approaches to Mammalian Embryonic Development.* New York: Cambridge University Press, pp. 97–120.

Rossant, J. (1987). Cell lineage analysis in mammalian embryogenesis. *Curr. Top. Dev. Biol.,* **23,** 115–146.

Rossant, J., and Joyner, A.L. (1989). Towards a molecular-genetic analysis of mammalian development. *Trends Genet.,* **5,** 277–283.

Roth, S., Stein, D., and Nüsslein-Volhard, C. (1989). A gradient of nuclear localization of the *dorsal* protein determines dorsoventral pattern in the Drosophila embryo. *Cell,* **59,** 1189–1202.

Roux, W. (1885). Beitrage zur entwicklungsmechanik des Embryo. Nr. 1 (Einleitung) *Z. Biol.,* **21,** 411–524.

Rubin, G.M., and Spradling, A.C. (1982). Genetic transformation of *Drosophila* with transposable element vectors. *Science,* **218,** 348–353.

Rugh, R. (1968). *The Mouse: Its Reproduction and Development.* Oxford: Oxford University Press.

Ruiz-Gomez, M., and Modolell, J. (1987). Deletion analysis of the *achaete-scute* locus of *Drosophila melanogaster. Genes Dev.,* **1,** 1238–1246.

Rushlow, C., Frasch, M., Doyle, H., and Levine, M. (1987). Maternal regulation of zerknullt: A homeobox gene controlling differentiation of dorsal tissues in *Drosophila. Nature,* **330,** 583–586.

Rushlow, C.A., Han, K., Manley, J.L., and Levine, M. (1989). The graded distribution of the *dorsal* morphogen is initiated by selective nuclear transport in Drosophila. *Cell,* **59,** 1165–1177.

Russell, E.S. (1979). Hereditary anemias of the mouse: A review for geneticists. *Adv. Genet.,* **20,** 357–455.

Russell, L.B., Montgomery, C.S., and Rayner, G.D. (1982). Analysis of the albino-locus region of the mouse: IV. Characterization of 34 deficiencies. *Genetics*, **100**, 427–453.

Russell, W.L., Kelly, E.M., Hunsicker, P.R., Bangham, J.W., Maddux, S.C., and Phipps, E.L. (1979). Specific-locus test shows ethylnitrosourea to be the most potent mutagen in the mouse. *Proc. Natl. Acad. Sci., U.S.A.*, **76**, 5818–5819.

Ruvkun, G., and Giusto, J. (1989). The *Caenorhabditis elegans* heterochronic gene *lin-14* encodes a nuclear protein that forms a temporal developmental switch. *Genetics*, **338**, 313–319.

Ruvkun, G., Ambros, V., Coulson, A., Waterston, R., Sulston, J., and Horvitz, H.R. (1989). Molecular genetics of the *Caenorhabditis elegans* heterochronic gene *lin-14*. *Genetics*, **121**, 501–516.

Ruvkun, G., Wightman, B., Burglin, T., and Aroan, P. (1991). Dominant gain-of-function mutations that lead to misregulation of the *C. elegans* heterochronic gene *lin-14* and the evolutionary implications of dominant mutations in pattern formation genes. *Development*, (Suppl.), 47–54.

Sadler, J.R., and Novick, A. (1965). Two properties of repressor and the kinetics of its action. *J. Mol. Biol.*, **12**, 305–327.

St. Johnston, D., and Gelbart, W.M. (1987). Decapentaplegic transcripts are localized along the dorsal-ventral axis of the *Drosophila* embryo. *EMBO J.*, **6**, 2785–2791.

St. Johnston, D., Beuchle, D., and Nüsslein-Volhard, C. (1991). *staufen*, a gene required to localize maternal RNAs in the Drosophila egg. *Cell*, **66**, 51–63.

Sakoyama, Y., and Okubo, S. (1981). Two dimensional gel patterns of protein species during development of *Drosophila* embryos. *Dev. Biol.*, **81**, 361–365.

Salser, S.J., and Kenyon, C. (1992). Activation of *C. elegans Antennapedia* homologue in migrating cells controls their direction of migration. *Nature*, **355**, 255–258.

Salz, H.K., Cline, T.W., and Schedl, P. (1987). Functional changes associated with structural alterations induced by mobilization of a P element inserted in the *Sex-lethal* gene of *Drosophila*. *Genetics*, **117**, 221–231.

Salz, H.K., Maine, E.M., Keyes, L.N., Samuels, E.M., Cline, T.W., and Schedl, P. (1989). The *Drosophila* female-specific sex-determination gene *Sex-lethal* has stage-, tissue-, and sex-specific RNAs, suggesting multiple modes of regulation. *Genes Dev.*, **3**, 708–719.

Sampedro, J., and Guerrero, I. (1991). Unrestricted expression of the *Drosophila* gene *patched* allows a normal segment polarity. *Nature*, **353**, 187–190.

Sanchéz, L., and Nöthiger, R. (1982). Clonal analysis of *Sex-lethal*, a gene needed for female sexual development in *Drosophila melanogaster*. *Wilhelm Roux Arch.* **191**, 211–214.

Sanchez-Herrero, E., and Crosby, M.A. (1988). The *Abdominal-B* gene of *Drosophila melanogaster*: Overlapping transcripts exhibit two different spatial distributions. *EMBO J.*, **7**, 2163–2173.

Sanchez-Herrero, E., Vernas, I., Marco, R., and Morata, G. (1985). Genetic organization of *Drosophila* bithorax complex. *Nature*, **313**, 108–113.

Sander, K. (1976). Specification of the basic body pattern in insect embryogenesis. *Adv. Insect Physiol.*, **12**, 125–238.

Sander, K. (1991). Wilhelm Roux and his programme in developmental biology. *Wilhelm Roux Arch. Dev. Biol.*, **200**, 1–3.

Sander, K., Lohs-Schardin, M., and Bowmann, M. (1980). Embryogenesis in a *Drosophila* mutant expressing half the normal segment number. *Nature*, **287**, 841–843.

Sandler, I., and Sandler, L. (1985). A conceptual ambiguity that contributed to the neglect of Mendel's paper. *Hist. Philos. Life Sci.*, **7**, 3–70.

Sanes, J.R., Rubenstein, J.L.R., and Nicolas, J.-F. (1986). Use of a recombinant retrovirus to study post-implantation cell lineage in mouse embryos. *EMBO J.*, **5**, 3133–3142.

Sanford, J.P., Clark, H.J., Chapman, V.M., and Rossant, J. (1987). Differences in DNA methylation during oogenesis and spermatogenesis and their persistence during early embryogenesis in the mouse. *Genes Dev.*, **1**, 1039–1046.

Santamaria, P., and Nüsslein-Volhard, C., (1983). partial rescue of *dorsal*, a maternal effect mutation affecting the dorso-ventral pattern of the *Drosophila* embryo, by the injection of wild-type cytoplasm. *EMBO J.*, **2**, 1695–1699.

Sato, T., and Denell, R.E. (1985). Homoeosis in *Drosophila*: Anterior and posterior transformations in Polycomb lethal embryos. *Dev. Biol.,* **110,** 53–64.

Sato, T., Russell, M.A., and Denell, R.E. (1983). Homoeosis in Drosophila: A new enhancer of *Polycomb* and related homeotic mutations. *Genetics,* **105,** 357–370.

Scalenghe, F., Turco, E., Edstrom, E., Pirrota, V., and Melli, M. (1981). Microdissection and cloning of DNA from a specific region of *Drosophila melanogaster* polytene chromosomes. *Chromosoma,* **82,** 205–216.

Schauer, I.E., and Wood, W.B. (1990). Early *C. elegans* embryos are transcriptionally active. *Development,* **110,** 1303–1317.

Scheidin, P., Hunter, C.P., and Wood, W.B. (1991). Autonomy and nonautonomy of sex determination in triploid intersex mosaics of *C. elegans. Development,* **112,** 863–879.

Schedl, T., and Kimble, J. (1989). *fog-2,* a germ-line specific sex determination gene required for hermaphrodite spermatogenesis in *Caenorhabditis elegans. Genetics,* **119,** 43–61.

Schejter, E.D., and Shilo, B-Z. (1989). The Drosophila EGF receptor homolog (DER) gene is allelic to *faint little ball,* a locus essential for embryonic development. *Cell,* **56,** 1093–1104.

Schierenberg, E. (1985). Cell determination during early embryogenesis of the nematode *Caenorhabditis elegans. Cold Spring Harbor Symp. Quant. Biol.,* **50, 5** 59–68.

Schierenberg, E. (1986). Developmental strategies during early embryogenesis of *Caenorhabditis elegans. J. Embryol. Exp. Morphol.,* **97,** (Suppl.), 31–44.

Schierenberg, E. (1987). Reversal of cellular polarity and early cell-cell interaction in the embryo of *Caenorhabditis elegans. Dev. Biol.,* **122,** 452–463.

Schierenberg, E. (1988). Localization and segregation of lineage-specific cleavage potential in embryos of *Caenorhabditis elegans. Wilhelm Roux Arch. Dev. Biol.,* **197,** 282–293.

Schierenberg, E., Miwa, J., and von Ehrenstein, G. (1980). Cell lineages and developmental defects of temperature-sensitive embryonic arrest mutants in *Caenorhabditis elegans. Dev. Biol.,* **76,** 141–159.

Schlicht, P., and Schierenberg, E. (1991). Altered establishment of cell lineages in the *Caenorhabditis elegans* embryo after suppression of the first cleavage supports a concentration-dependent decision mechanism. *Wilhelm Roux Arch. Dev. Biol.,* **199,** 437–448.

Schmid, W., Muller, G., Schutz, G., and Gluecksohn-Waelsch, S. (1985). Deletions near the albino locus on chromosome 7 of the mouse affect the level of tyrosine aminotransferase mRNA. *Proc. Natl. Acad. Sci., U.S.A.,* **82,** 2860–2869.

Schnabel, R., and Schnabel, H. (1990). Early determination in the *C. elegans* embryo: A gene, *cib-1,* required to specify a set of stem cell-like blastomeres. *Development,* **108,** 107–119.

Schneiderman, H., and Bryant, P.J. (1971). Genetic analysis of developmental mechanism in *Drosophila. Nature,* **234,** 187–194.

Schenuwly, S., Klemencz, R., and Gehring, W.J. (1986). Redesigning the body plan of *Drosophila* by ectopic expression of the homeotic *Antennapedia. Nature,* **325,** 816–818.

Schneuwly, S., Kuroiwa, A., and Gehring, W.J. (1987). Molecular analysis of the dominant homeotic *Antennapedia* phenotype. *EMBO J.,* **6,** 201–206.

Scholnick, S.B., Morgan, B.A., and Hirsh, J. (1983). The cloned dopa decarboxylase gene is developmentally regulated when reintegrated into the *Drosophila* genome. *Cell,* **34,** 37–45.

Schubiger, G. (1976). Adult differentiation from partial *Drosophila* embryos after egg ligation during stages of nuclear multiplication and cellular blastoderm. *Dev. Biol.,* **50,** 476–488.

Schubiger, G., and Schubiger, M. (1978). Distal transformation in *Drosophila* leg imaginal disc fragments. *Dev. Biol.,* **67,** 286–295.

Schubiger, G., and Newman, S.M. (1981). Determination in *Drosophila* embryos. *Am. Zool.,* **22,** 47–55.

Schubiger, G., Mosely, R.C., and Wood, W. (1977). Interaction of different egg parts in determination of various body regions in *Drosophila melanogaster. Proc. Natl. Acad. Sci., U.S.A.,* **74,** 2050–2053.

Schultz, R.M., LeTourneau, G.E., and Wassarman, P.M. (1979). Program of early development in the mammal: Changes in the patterns and absolute rates of tubulin and total protein synthesis during oocyte growth in the mouse. *Dev. Biol.,* **73,** 120–133.

Schultz, G.A. (1986). Utilization of genetic information in the preimplantation mouse embryo. In J. Rossant and R.A. Pedersen (Eds.), *Experimental Approaches to Mammalian Embryonic Development*. New York: Cambridge University Press, pp. 239–265.

Schultz, R.M. (1986). Molecular aspects of mammalian oocyte growth and maturation. In J. Rossant and R.A. Pedersen (Eds.), *Experimental Approaches to Mammalian Embryonic Development*. New York: Cambridge University Press, pp. 195–237.

Schüpbach, T. (1982). Autosomal mutations that interfere with sex determination in somatic cells of *Drosophila* have no direct effect on the germline. *Dev. Biol.*, **89**, 117–127.

Schüpbach, T. (1985). Normal female germ cell differentiation requires the female X chromosome to autosome ratio and expression of *Sex-lethal* in *Drosophila melanogaster*. *Dev. Biol.*, **109**, 529–548.

Schüpbach, T. (1987). Germ line and soma cooperate during oogenesis to establish the dorsoventral pattern of egg shell and embryo in *Drosophila melanogaster*. *Cell*, **49**, 699–707.

Schüpbach, T., and Wieschaus, E. (1986a). Germline autonomy of maternal-effect mutations altering the embryonic body pattern of *Drosophila*. *Dev. Biol.*, **113**, 443–448.

Schüpbach, T., and Wieschaus, E. (1986b). Maternal-effect mutations altering the anterior-posterior pattern of the *Drosophila* embryo. *Wilhelm Roux Arch. Dev. Biol.*, **195**, 2–317.

Scott, M.P., Wiener, A.J., Hazelrigg, T.I., Polisky, B.A., Pirrotta, V., Scalenghe, F., and Kaufman, T.C. (1983). The molecular organization of the *Antennapedia* locus of *Drosophila*. *Cell*, **35**, 763–776.

Searle, A.G. (1968). *Comparative Genetics of Coat Color in Mammals*. London: Logos Press.

Searle, A.G., and Beechey, C.V. (1978). Complementation studies with mouse translocations. *Cytogenet. Cell Genet.*, **20**, 282–303.

Sharrock, W.J. (1983). Yolk proteins of *Caenorhabditis elegans*. *Dev. Biol.*, **96**, 182–188.

Shearn, A. (1977). Mutational dissection of imaginal disc development in *Drosophila melanogaster*. *Am. Zool.*, **17**, 585–594.

Shearn, A., and Garen, A. (1974). Genetic control of imaginal disc development in *Drosophila*. *Proc. Natl. Acad. Sci., U.S.A.*, **71**, 1393–1397.

Shearn, A., Rice, R., Garen, A., and Gehring, W. (1971). Imaginal disc abnormalities in lethal mutants of *Drosophila*. *Proc. Natl. Acad. Sci., U.S.A.*, **68**, 2594–2598.

Shearn, A., Hersperger, G., and Hersperger, E. (1978a). Genetic analysis of two allelic temperature-sensitive mutants of *Drosophila melanogaster*, both of which are zygotic and maternal effect lethals. *Genetics*, **89**, 341–353.

Shearn, A., Hersperger, G., Hersperger, E., Pentz, E.S., and Denker, P. (1978b). Multiple allele approach to the study of genes in *Drosophila melanogaster* that are involved in imaginal disc development. *Genetics*, **89**, 355–370.

Shedlovsky, A., Guenet, J.L., Johnson, L.L. and Dove, W.F. (1988). Induction of recessive lethal mutations in the *t/T* H-2 region of the mouse genome by a point mutagen. *Genet. Res.*, **47**, 135–142.

Shedlovsky, A., King, T.R, and Dove, W.F. (1988). Saturation germ line mutagenesis of the mouse *t* region including a lethal allele of the *quaking* locus. *Proc. Natl. Acad. Sci., U.S.A.*, **85**, 180–184.

Shellenbarger, D.L., and Mohler, J.D. (1978). Temperature-sensitive periods and autonomy of pleiotropic effects of $1(1)N^{ts}$, a conditional *Notch* lethal in *Drosophila*. *Dev. Biol.*, **62**, 432–446.

Shibahara, S., Toguchi, H., Muller, R.M., Shibata, K., Cohen, T., Tomita, Y., and Togami, H. (1991). Structural organization of the pigment cell specific gene located at the borwn locus in mouse—Its promoter activity and alternatively spliced transcripts. *J. Biol. Chem.*, **266**, 5895–5901.

Silver, L.M. (1985). Mouse *t* haplotypes. *Annu. Rev. Genet.*, **19**, 179–208.

Silver, L.M. (1990). At the crossroads of developmental genetics: The cloning of the classical mouse T locus. *BioEssays*, **12**, 377–380.

Silvers, W.K. (1979). *The Coat Colors of Mice*. New York: Springer-Verlag, pp. 291.

Silvers, W.K., and Russell, E.S. (1955). An experimental approach to action of genes at the agouti locus in the mouse. *J. Exp. Zool.*, **130**, 199–220.

Silvers, W.K., Gasser, D.L., and Eicher, E.M. (1982). H-Y antigen, serologically detectable male antigen and sex determination. *Cell*, **28**, 439–440.

Simcox, A., and Sang, J.H. (1983). When does determination occur in *Drosophila* embryos? *Dev. Biol.*, **97**, 212–221.

Simcox, A.A., Roberts, I.J.H., Hersperger, E., Gribbin, M.C., Shearn, A., and Whittle, J.R.S. (1989). Imaginal discs can be recovered from cultured embryos mutant for the segment-polarity genes *engrailed, naked,* and *patched* but not *wingless. Development,* **107,** 715–722.

Simpson, P. (1983). Maernal-zygotic gene interations during formation of the dorso-ventral pattern in *Drosophila* embryos. *Genetics,* **105,** 615–632.

Simpson, P. (1990). Lateral inhibition and the development of the sensory bristles of the adult peripheral nervous system of *Drosophila. Development,* **109,** 509–519.

Simpson, P., and Grau, Y. (1987). The segment polarity gene *costal-2* in *Drosophila.* II. The origin of imaginal pattern duplications. *Dev. Biol.,* **122,** 201–209.

Sinclair, A.H., Berta, P., Palmer, M.S., Hawkins, J.R., Griffiths, B.L., Smith, M.J., Foster, J.W., Frischauf, A.-M., Lovell-Badge, R., and Goodfellow, P.N. (1990). A gene from the human sex-determining region encodes a protein with homology to a conserved DNA-binding motif. *Nature,* **346,** 240–244.

Siracusa, L.D., Russell, L.B., Eicher, E.M., Curran, D.J., Copeland, N.G., and Jenkins, N.A. (1983). Genetic organization of the *agouti* region of the mouse. *Genetics,* **117,** 93–100.

Skripsky, T., and Lucchesi, J.C. (1982). Intersexuality resulting from the interaction of sex-specific lethal mutations in *Drosophila melanogaster. Dev. Biol.* **94,** 153–162.

Skeath, J.B., and Carroll, S.B. (1991). Regulation of *achaete-scute* gene expression and sensory organ pattern formation in the *Drosophila* wing. *Genes Dev.,* **5,** 984–995.

Slack, J.M.W. (1991). *From Egg to Embryo,* 2nd ed. Cambridge: Cambridge University Press.

Smith, J.C., Price, B.M.J., Green, J.B.A., Weigel, D., and Herrmann, B.G. (1991). Expression of a Xenopus homolog of *Brachyury* (T) is an immediate-early response to mesoderm induction. *Cell,* **67,** 1–9.

Smith, P.D., and Lucchesi, J.C. (1969). The role of sexuality in dosage compensation in *Drosophila. Genetics,* **61,** 607–618.

Smith, R., and McLaren, A. (1977). Factors affecting the time of formation of the mouse blastocoele. *J. Embryol. Exp. Morphol.,* **41,** 79–92.

Snell, G.D., and Stevens, L.C. (1966). Early embryology. In E.L. Green (Ed.), *Biology of the Laboratory Mouse.* New York: McGraw-Hill, pp. 205–245.

Snow, M.H.L., Tam, P.P.L., and McLaren, A. (1981). On the control and regulation of size and morphogenesis in mammalian embryos. In A. McLaren and C.C. Wylie (eds.), *Current Problems in Germ Cell Differentiation.* Cambridge: Cambridge University Press.

Solter, D., Aronson, J., Gilbert, S.F., and McGrath, J. (1985). Nuclear transfer in mouse embryos—Activation of the embryonic genome. *Cold Spring Harbor Symp. Quant. Biol.,* **50,** 45–50.

Sonnenblick, B.P. (1950). The early embryology of *Drosophila melanogaster.* In M. Demerec (Ed.), *The Biology of Drosophila.* New York: John Wiley & Sons, pp. 62–167.

Sorrentino, V., Pepperko, R., Davis, R.L., Ansorge, W., and Philipson, L. (1990). Cell proliferation inhibited by MyoD1 independently of myogenic differentiation. *Nature,* **345,** 813–815.

Spencer, F.A., Hoffman, F.M., and Gelbart, W.M. (1982). Decapentaplegic: A gene complex affecting morphogenesis in *Drosophila melanogaster. Cell,* **28,** 451–461.

Spiegelman, S. (1948). Differentiation as the controlled production of unique enzymatic patterns. In *Growth in Relation to Differentiation and Morphogenesis.* Cambridge: Cambridge University Press, vol. II, pp. 286–325.

Spradling, A.C. (1986). P element-mediated transformation. In D.B. Roberts (Ed.), Drosophila, *a Practical Approach.* Oxford: IRL Press, pp. 185–197.

Spradling, A.C., and Mahowald, A.P. (1981). A chromosome inversion alters the pattern of specific DNA replication in *Drosophila* follicle cells. *Cell,* **27,** 203–209.

Spradling, A.C., and Rubin, G.M. (1981). *Drosophila* genome organization: Conserved and dynamic aspects. *Annu. Rev. Genet.,* **15,** 219–264.

Spradling, A.C., and Rubin, G.M. (1982). Transposition of cloned P elements into *Drosophila* germ line chromosomes. *Science,* **218,** 341–347.

Spradling, A.C., and Rubin, G.M. (1983). The effect of chromosomal position on the expression of the *Drosophila* xanthine dehydrogenase gene. *Cell,* **34,** 47–57.

Sprenger, F., Stevens, L.M., and Nüsslein-Volhard, C. (1989). The *Drosophila* gene *torso* encodes a putative receptor tyrosine kinase. *Nature*, **338**, 478–483.

Spurway, H. (1948). Genetics and cytology of *Drosophila pseudoobscura*. IV. An extreme example of delay in gene action, causing sterility. *J. Genet.*, **49**, 126–140.

Stanojevic, D., Hoey, T., and Levine, M. (1989). Sequence-specific DNA binding activities of the gap proteins encoded by *hunchback* and *Kruppel* in *Drosophila*. *Nature*, **341**, 331–335.

Stein, D., Roth, S., Vogelsang, E., and Nüsslein-Volhard, C. (1991). The polarity of the dorsoventral axis in the Drosophila embryo is defined by an extracellular matrix. *Cell*, **65**, 725–735.

Steiner, E. (1976). Establishment of compartments in the developing leg imaginal discs of *Drosophila melanogaster*. *Wilhelm Roux Arch.*, **180**, 9–30.

Steinmann-Zwicky, M. (1988). Sex determination in *Drosophila:* The X-chromosomal gene *liz* is required for *Sxl* activity. *EMBO J.*, **7**, 3889–3898.

Steinmann-Zwicky, M., and Nöthiger, R. (1985). A small region on the X chromosome of Drosophila regulates a key gene that controls sex determination and dosage compensation. *Cell*, **42**, 877–887.

Steinmann-Zwicky, M., Schmid, H., and Nöthiger, R. (1989). Cell autonomous and inductive signals can determine the sex of the germ line of Drosophila by regulating the gene *Sxl Cell*, **57**, 157–166.

Steinmann-Zwicky, M., Amrein, H., and Nöthiger, R. (1990). Genetic control of sex determination in *Drosophila*. *Adv. Genet.*, **27**, 189–237.

Steller, H., Fischbach, K.-F., and Rubin, G.M. (1987). *disconnected:* A locus required for neuronal pathway formation in the visual system of Drosophila. *Cell*, **50**, 1139–1153.

Stern, C. (1936). Somatic crossing over and segregation in *Drosophila melanogaster*. *Genetics*, **21**, 626–730.

Stern, C. (1954). Two or three bristles. *Am. Sci.*, **42**, 213–247.

Stern, C. (1955). Gene action. In B.H. Williams, P.A. Weiss, and V. Hamburger (Eds.), *Analysis of Development*. Philadelphia: W.B. Saunders, pp. 151–169.

Stern, C. (1956). The genetic control of developmental competence and morphogenetic tissue interactions in genetic mosaics. *Arch. Entwicklungs. Org.*, **149**, 1–25.

Sternberg, P.W. (1988). Lateral inhibition during vulval induction in *Caenorhabditis elegans*. *Nature*, **335**, 551–554.

Sternberg, P.W., and Horvitz, H.R. (1986). Pattern formation during vulval development in *C. elegans*. *Cell*, **44**, 761–772.

Stevens, L.M., Frohnhofer, H.G., Klinger, M., and Nüsslein-Volhard, C. (1990). Localized requirement for *torso-like* expression in follicle cells for development of terminal anlagen of the *Drosophila* embryo. *Nature*, **346**, 660–663.

Steward, R. (1987). *dorsal*, an embryonic polarity gene in *Drosophila*, is homologous to the vertebrate proto-oncogene, c-*rel Science*, **238**, 692–694.

Steward, R. (1989). Relocalization of the *dorsal* protein from the cytoplasm to the nucleus correlates with its function. *Cell*, **59**, 1179–1188.

Steward, R., Zusman, S.B., Huang, L.H., and Schedl, P. (1988). The *dorsal* protein is distributed in a gradient in early Drosophila embryos. *Cell*, **55**, 487–495.

Stewart, M., Murphy, C., and Fristrom, J.W. (1972). The recovery and preliminary characterization of X chromosome mutants affecting imaginal discs of *Drosophila melanogaster*. *Dev. Biol.*, **27**, 71–83.

Straney, S.B., and Crothers, D.M. (1987). Lac repressor is a transient gene-activating protein. *Cell*, **51**, 699–707.

Strecker, T.R., Halsell, S.R., Fisher, W.W., and Lipshitz, H.D. (1989). Reciprocal effects of hyper- and hypoactivity mutations in the *Drosophila* pattern gene *torso*. *Science*, **243**, 1062–1066.

Strome, S. (1986). Asymmetric movements of cytoplasmic components in *Caenorhabditis elegans* zygotes. *J. Embryol. Exp. Morphol.*, **97**, (Suppl.), 15–29.

Strome, S., and Wood, W.B. (1983). Generation of asymmetry and segregation of germ-line granules in early *C. elegans* embryos. *Cell*, **35**, 15–25.

Strome, S., and Hill, D.P. (1988). Early embryogenesis in *Caenorhabditis elegans*: The cytoskeleton and spatial organization of the zygote. *BioEssays*, **8**, 145–149.

Strub, S. (1977). Pattern regulation and transdetermination in *Drosophila* imaginal leg disc reaggregates. *Nature,* **269,** 688–691.

Struhl, G. (1977). Developmental compartments in the proboscis of *Drosophila. Nature,* **270,** 723–725.

Struhl, G. (1981a). Anterior and posterior compartments in the proboscis of *Drosophila. Dev. Biol.,* **84,** 372–385.

Struhl, G. (1981b). A gene product required for correct initiation of segmental determination in *Drosophila. Nature,* **293,** 36–41.

Struhl, G. (1981c). A homoeotic mutation transforming leg to antenna in *Drosophila. Nature,* **292,** 635–638.

Struhl, G. (1982a). Spineless-aristapedia: A homoeotic gene that does not control the development of specific compartments in *Drosophila. Genetics,* **102,** 737–749.

Struhl, G. (1982b). Genes controlling segmental specification in the *Drosophila* thorax. *Proc. Natl. Acad. Sci., U.S.A.,* **79,** 7380–7384.

Struhl, G. (1984). Splitting the bithorax complex of *Drosophila. Nature,* **308,** 454–457.

Struhl, G. (1989). Differing strategies for organizing anterior and posterior body patterns in *Drosophila* embryos. *Nature,* **338,** 741–744.

Struhl, G., and Akam, M. (1985). Altered distributions of *Ultrabithorax* transcripts in *extra sex combs* mutant embryos of *Drosophila. EMBO J.,* **4,** 3259–3264.

Struhl, G., and White, R.A.H. (1985). Regulation of the *Ultrabithorax* gene of *Drosophila* by other bithorax genes. *Cell,* **43,** 507–519.

Struhl, G., Struhl, K., and MacDonald, P.M. (1989). The gradient morphogen bicoid is a concentration-dependent transcriptional activator. *Cell,* **57,** 1259–1273.

Sturtevant, A.H. (1923). Inheritance of direction of shell coiling in *Limnaea. Science,* **58,** 269–270.

Sturtevant, A.H. (1929). The claret mutant type of *Drosophila simulans:* A study of chromosome elimination and of cell lineage. *Z. Wiss. Zool.,* **135,** 323–356.

Sturtevant, A.H. (1945). A gene in *Drosophila melanogaster* that transforms females into males. *Genetics,* **30,** 297–299.

Sturtevant, A.H. (1970). Studies on the bristle pattern of *Drosophila. Dev. Biol.,* **21,** 48–61.

Sullivan, W. (1987). Independence of *fushi tarazu* expression with respect to cellular density in *Drosophila* embryos. *Nature,* **327,** 164–167.

Sulston, J.E. (1976). Post-embryonic development in the ventral cord of *Caenorhabditis elegans. Proc. R. Soc. London. [Bol.],* **275,** 287–297.

Sulston, J.E. (1988). Cell lineage. In W.B. Wood (Ed.), *The Nematode* Caenorhabditis elegans. Cold Spring Harbor, NY: Cold Spring Harbor Laboratory Press, pp. 123–155.

Sulston, J.E., and Brenner, S. (1974). The DNA of *Caenorhabditis elegans. Genetics,* **77,** 95–104.

Sulston, J.E., and Horvitz, H.R. (1977). Post-embryonic cell lineages of the nematode, *Caenorhabditis elegans. Dev. Biol.,* **56,** 110–156.

Sulston, J.E., and White, J.G. (1980). Regulation and cell autonomy during postembryonic development of *Caenorhabditis elegans. Dev. Biol.,* **78,** 577–597.

Sulston, J.E., and Horvitz, H.R. (1981). Abnormal cell lineages in mutants of the nematode *Caenorhabditis elegans. Dev. Biol.,* **82,** 41–55.

Sulston, J.E., and Hodgkin, J. (1988). Methods. In W.B. Wood (Ed.), *The Nematode* Caenorhabditis elegans. Cold Spring Harbor, NY: Cold Spring Harbor Laboratory Press, pp. 587–606.

Sulston, J.E., Albertson, D.G., and Thomson, J.N. (1980). The *Caenorhabditis elegans* male: Postembryonic development of nongonadal structures. *Dev. Biol.,* **78,** 542–576.

Sulston, J.E., Schierenberg, E., White, J.G., Thomson, J.N., and von Ehrenstein, G. (1983). The embryonic cell lineage of the nematode *Caenorhabditis elegans. Dev. Biol.,* **100,** 64–119.

Sunkel, C.E., and Whittle, J.R.S. (1987). *Brista*—a gene involved in the specification and differentiation of distal cephalic and thoracic structures in *Drosophila melanogaster. Roux Archiv. Dev. Biol.,* **196,** 124–132.

Surani, M.A.H. (1986). Evidence and consequences of differences between maternal and paternal genomes during embryogenesis in the mouse. In J. Rossant and R.A. Pedersen (Eds.), *Experimental Approaches to Mammalian Embryonic Development*. New York: Cambridge University Press, pp. 401–435.

Surani, M.A.H., Barton, S.C., and Norris, M.L. (1984). Development of reconstituted mouse eggs suggests imprinting of the genome during gametogenesis. *Nature, 308,* 548–550.

Surani, M.A.H., Reik, W., Norris, M.L., and Barton, S.C. (1986a). Influence of germline modifications of homologous chromosomes on mouse development. *J. Embryol. Exp. Morphol., 97* (Suppl.), 123–136.

Surani, M.A.H., Barton, S.C., and Norris, M.L. (1986b). Nuclear transplantation in the mouse: Heritable differences between parental genomes after activation of the embryonic genome. *Cell, 45,* 127–136.

Surani, M.A.H., Barton, S.C., and Norris, M.L. (1987). Influence of parental chromosomes on spatial specificity in androgenetic ↔ parthenogenetic chimaeras in the mouse. *Nature, 326,* 395–397.

Surani, M.A., Reik, W., and Allen, N.D. (1988). Transgenes as molecular probes for genomic imprinting. *Trends Genet., 4,* 59–62.

Surani, M.A., Kothary, S., Allen, N.D., Singh, P.B., Fundele, A., Ferguson-Smith, A.C., and Barton, S.C. (1990). *Development,* (Suppl.), 89–98.

Suzuki, D.T. (1970). Temperature-sensitive mutations in *Drosophila melanogaster*. *Science, 170,* 695–706.

Suzuki, D.T., and Griffiths, A.J.F. (1981). *An Introduction to Genetic Analysis,* 2nd ed. San Francisco: W.H. Freeman.

Swanson, M.M., and Riddle, D.L. (1981). Critical periods in the development of the *Caenorhabditis elegans* dauer larva. *Dev. Biol., 84,* 27–40.

Sweeton, D., Parks, S., Costa, M., and Wieschaus, E. (1991). Gastrulation in *Drosophila:* The formation of the ventral furrow and posterior midgut invaginations. *Development, 112,* 365–370.

Szabad, J., Schupback, T., and Wieschaus, E. (1979). Cell lineage and development in the larval epidermis of *Drosophila melanogaster*. *Dev. Biol., 73,* 256–271.

Szabad, J., and Bryant, P.J. (1982). The mode of action of 'discless' mutations in *Drosophila melanogaster*. *Dev. Biol., 93,* 240–256.

Szabad, J., Schüpbach, T., and Wieschaus, E. (1979). Cell lineage and development in the larval epidermis of *Drosophila melanogaster*. *Dev. Biol., 73,* 256–271.

Takagi, N., and Sasaki, M. (1975). Preferential expression of the paternally derived X chromosome in the extra-embryonic membranes in the mouse. *Nature, 256,* 640–642.

Takeichi, M. (1988). The cadherins: Cell–cell adhesion molecules controlling animal morphogenesis. *Development, 102,* 639–655.

Tam, P.P.L. (1981). The control of somitogenesis in mouse embryos. *J. Embryol. Exp. Morphol., 65,* 103–128.

Tam, P.P.L., and Snow, M.H.L. (1981). Proliferation and migration of primordial germ cells during compensatory growth in mouse embryos. *J. Embryol. Exp. Morphol., 64,* 133–147.

Tanaka, S., Yamamoto, H., Takeuchi, S. and Takeuchi, T. (1990). Melanization in albino mice transformed by introducing cloned mouse tyrosinase gene. *Development, 108,* 223–227.

Tarkowski, A.K. (1961). Mouse chimaeras developed from fused eggs. *Nature, 190,* 857–860.

Tarkowski, A.K., and Wroblewska, J. (1967). Development of blastomeres of mouse eggs isolated at the 4- and 8-cell stage. *J. Embryol. Exp. Morphol., 18,* 155–180.

Tautz, D. (1988). Regulation of the *Drosophila* segmentation gene *hunchback* by two maternal morphogenetic centres. *Nature, 332,* 281–284.

Tautz, D. (1992). Redundancies, development and the flow of information. *BioEssays, 14,* 263–266.

Tautz, D., Lehmann, R., Schnurch, H., Schuh, R., Seifert, E., Kienlin, A., Jones, K., and Jäckle, H. (1987). Finger protein of novel structure encoded by *hunchback*, a second member of the gap class of *Drosophila* segmentation genes. *Nature, 327,* 383–389.

Tease, C., and Cattanach, B.M. (1986). Mammalian cytogenetics and genetics of nondisjunction. In F.J. de Serres (Ed.), *Chemical Mutagens*. New York: Plenum Press, pp. 215–283.

Technau, G.M. (1986). Lineage analysis of transplanted individual cells in embryos of *Drosophila melanogaster*. I. The method. *Wilhelm Roux Arch. Dev. Biol., 195,* 389–398.

Technau, G.M. (1987). A single cell approach to problems of cell lineage and commitment during embryogenesis of *Drosophila melanogaster*. *Development, 100,* 1–12.

Technau, G.M., and Campos-Ortega, J.A. (1986a). Lineage analysis of transplanted individual cells in embryos of *Drosophila melanogaster*. II. Commitment and proliferative capabilities of neural and epidermal cell progenitors. *Wilhelm Roux Arch. Dev. Biol.*, **195**, 445–454.

Technau, G.M., and Campos-Ortega, J.A. (1986b). Lineage analysis of transplanted individual cells in embryos. III. Commitment and proliferative capabilities of pole cells and midgut progenitors. *Wilhelm Roux Arch. Dev. Biol.*, **195**, 489–498.

Teugels, E., and Ghysen, A. (1983). Independence of the numbers of legs and leg ganglia in *Drosophila* bithorax mutants. *Nature*, **304**, 440–442.

Thierry-Mieg, D. (1976). Study of a temperature-sensitive mutant grandchildless-like in *Drosophila melanogaster*. *J. Microscop. B*, **25**, 1–6.

Thisse, R., Stoetzel, C., El Messal, M., and Perrin-Schmitt, F. (1987). Genes of the *Drosophila* maternal dorsal group control the specific expression of the zygotic gene *twist* in presumptive mesodermal cells. *Genes Dev.*, **1**, 709–715.

Thisse, B., Stoetzel, C., Gorostiza-Thisse, C., and Perrin-Schmitt, F. (1988). Sequence of the *twist* gene and nuclear localization of its protein in endomesodermal cells of early *Drosophila* embryos. *EMBO J.*, **7**, 2175–2183.

Thomas, K.R., Folger, K.R., and Capecchi, M.R. (1986). High frequncy targeting of genes to specific sites in the mammalian genome. *Cell*, **44**, 419–428.

Tickle, C. (1991). Retinoic acid and chick limb development. *Development*, (Suppl. 1), 113–121.

Tiong, S., Bone, L.M., and Whittle, J.R.S. (1985). Recessive lethal mutations within the biothorax complex in *Drosophila*. *Mol. Gen. Genet.*, **200**, 335–342.

Tobler, H. (1986). The differentiation of germ and somatic cell lines in nematodes. *Results Probl. Cell Differ.*, **13**, 1–69.

Tokunaga, C., and Gerhart, J.C. (1976). The effect of growth and joint formation on bristle pattern formation in *D. melanogaster*. *J. Exp. Zool.*, **198**, 79–86.

Tomlinson, A. (1985). The cellular dynamics of pattern formation in the eye of *Drosophila*. *J. Embryol. Exp. Morphol.*, **89**, 313–331.

Tomlinson, A. (1988). Cellular interactions in the developing *Drosophila* eye. *Development*, **104**, 183–193.

Tomlinson, A., and Ready, D.F. (1986). *sevenless*: A cell-specific homeotic mutation of the Drosophila eye. *Science*, **231**, 400–402.

Tomlinson, A., and Ready, D.F. (1987). Neuronal differentiation in the *Drosophila* ommatidium. *Dev. Biol.*, **120**, 366–376.

Tomlinson, A., Bowtell, D.D.L., Hafen, E., and Rubin, G.M. (1987). Localization of the *sevenless* protein, a putative receptor for positional information, in the eye imaginal disc of Drosophila. *Cell*, **51**, 143–150.

Tomlinson, A., Kimmel, B.E., and Rubin, G.M. (1988). *rough*, a Drosophila homeobox gene required in prhotoreceptors R2 and R5 for inductive interactions in the developing eye. *Cell*, **55**, 771–784.

Torres, M., and Sanchéz, L. (1989). The *scute* (T4) gene acts as a numerator element of the X: A signal that determines the state of activity of *Sex-lethal* in *Drosophila*. *EMBO J.*, **8**, 3079–3086.

Trent, C.N., Tsung, N., and Horvitz, H.R. (1983). Egg-laying defective mutants of the nematode *Caenorhabditis elegans*. *Genetics*, **104**, 619–647.

Trumbly, R.J., and Jarry, B. (1983). Stage-specific protein synthesis during early embryogenesis in *Drosophila melanogaster*. *EMBO J.*, **2**, 1281–1290.

Turing, A.M. (1952). The chemical basis of morphogenesis. *Phil. Trans. R. Soc.*, **273B**, 37–52.

Turner, D.L., and Cepko, C.L. (1987). A common progenitor for neurons and glia persists in rat retina late in development. *Nature*, **328**, 131–136.

Turner, F.R., and Mahowald, A.P. (1976). Scanning E.M. of *Drosophila* embryogenesis. I. The structure of the egg envelopes and the formation of the cellular blastoderm. *Dev. Biol.*, **50**, 95–108.

Underwood, E.M., Turner, F.R., and Mahowald, A.P. (1980a). Analysis of cell movements and fate mapping during early embryogenesis in *Drosophila melanogaster*. *Dev. Biol.*, **74**, 286–301.

Underwood, E.M., Caulton, J.H., Allis, C.D., and Mahowald, A.P. (1980b). Developmental fate of pole cells in *Drosophila melanogaster*. *Dev. Biol.*, **77**, 303–314.

van Blerkom, J. (1981). Structural relationships and post-translational modification of stage-specific proteins synthesized during early pre-implantation development in the mouse. *Proc. Natl. Acad. Sci., U.S.A.,* **78,** 7629–7633.

van Blerkom, J., Barton, S.C., and Johnson, M.H. (1976). Molecular differentiation in the pre-implantation embryo. *Nature,* **259,** 319–321.

van Blerkom, J., Janzen, R., and Runner, M.N. (1982). The patterns of protein synthesis during foetal and neonatal organ development in the mouse are remarkably similar. *J. Embryol. Exp. Morphol.,* **72,** 97–116.

van den Heuvel, M., Nüsse, R., Johnston, P., and Lawrence, P.A. (1989). Distribution of the *wingless* gene product in Drosophila embryos: A protein involved in cell-cell communication. *Cell,* **59,** 739–749.

Vanderslice, R., and Hirsh, D. (1976). Temperature-sensitive zygote defective mutants of *Caenorhabditis elegans. Dev. Biol.,* **49,** 236–249.

van Deusen, E.B. (1976). Sex determination in germ line chimaeras of *Drosophila melanogaster. J. Embryol. Exp. Morphol.,* **37,** 173–185.

Varmuza, S., Prideaux, V., Kothary, R., and Rossant, J. (1988). Polytene chromosomes in mouse trophoblast giant cells. *Development,* **102,** 127–134.

Vassin, H., Bremer, K.A., Knust, E., and Campos-Ortega, J.A. (1987). The neurogenic gene *Delta* of *Drosophila melanogaster* is expressed in neurogenic territories and encodes a putative transmembrane protein with EGF-like repeats. *EMBO J.,* **6,** 3431–3440.

Vestrecher, D., Gussler, A., Boller, K., and Kemler, R. (1987). Expression and distribution of cell adhesion molecule uvomorulin in mouse preimplantation embryos. *Dev. Biol.,* **124,** 451–456.

Villares, R., and Cabrera, C.V. (1987). The *achaete-scute* gene complex of D. melanogaster: Conserved domains in a subset of genes required for neurogenesis and their homology to *myc. Cell,* **50,** 415–424.

Villeneuve, A.M., and Meyer, B.J. (1987). *sdc-1:* A link between sex determination and dosage compensation in *C. elegans. Cell,* **48,** 25–37.

Villeneuve, A.M., and Meyer, B.J. (1990). The role of *sdc-1* in the sex determination and dosage compensation decisions in *Caenorhabditis elegans. Genetics,* **124,** 91–114.

von Ehrenstein, G., Schierenberg, E., and Miwa, J. (1979). Cell lineages of the wild-type and of temperature-sensitive embryonic arrest mutants of *Caenorhabditis elegans.* In N. Le Douarin (Ed.), *Cell Lineage, Stem Cells, and Cell Determination.* Amsterdam: Elsevier/North-Holland, pp. 49–58.

von Mende, N., Bird, D.M., Albert, P.S., and Riddle, D.L. (1988). *dpy-13:* A nematode collagen gene that affects body shape. *Cell,* **55,** 567–570.

Vowels, J., and Thomas, J.H. (1992). Genetic analysis of chemosensory control of dauer formation in *Caenorhabditis elegans. Genetics,* **130,** 105–123.

Waddington, C.H. (1940). The genetic control of wing development in *Drosophila. J. Genet.,* **41,** 75–139.

Waddington, C.H. (1941). Canalization of development and the inheritance of acquired chracteristics. *Nature,* **50,** 563–565.

Waddington, C.H. (1973). The morphogenesis of patterns in *Drosophila.* In S. Counce and C.H. Waddington (Eds.), *Developmental Systems: Insects.* London: Academic Press, pp. 499–535.

Wagner, E.F., Covarrubias, L., Stewart, T.A., and Mintz, B. (1983). Prenatal lethalities in mice homozygous for human growth hormone gene sequences integrated in the germ line. *Cell,* **35,** 647–655.

Wakimoto, B.T., and Kaufman, T.C. (1981). Analysis of larval segmentation in lethal genotypes associated with the Antennapedia gene complex in *Drosophila melanogaster. Dev. Biol.,* **81,** 51–64.

Wakimoto, B.T., Turner, F.R., and Kaufman, T.C. (1984). Defects in embryogenesis associated with the Antennapedia gene complex of *Drosophila melangaster. Dev. Biol.,* **102,** 147–172.

Wang, C., and Lehmann, R. (1991). Nanos is the localized posterior determinant in Drosophila. *Cell,* **66,** 637–647.

Ward, S. and Carrel, J.S. (1979). Fertilization and sperm competition in the nematode *Caenorhabditis elegans. Dev. Biol.,* **73,** 304–321.

Ward, S., Burke, D.J., Sulston, J.E., Coulson, A.R., Albertson, D.G., Ammons, D., Klass, M., and Hogan, E. (1988). The genomic organization of transcribed major sperm protein genes and pseudogenes in the nematode *C. elegans. J. Mol. Biol.,* **199,** 1–13.

Waring, D.A., and Kenyon, C. (1990). Selective silencing of cell communication influences anteroposterior pattern formation in *C. elegans. Cell,* **60,** 123–131.

Waring, G.L., and Mahowald, A.P. (1979). Identification and time of synthesis of chorion proteins in *Drosophila melanogaster. Cell,* **16,** 599–607.

Waring, G.L., and Pollack, J.C. (1987). Cloning and characterization of a dispersed multicopy X chromosome sequence in *Drosophila melanogaster. Proc. Natl. Acad. Sci., U.S.A.,* **84,** 2843–2847.

Wassarman, P.M. (1990). Profile of a mammalian sperm receptor. *Development,* **108,** 1–17.

Wassarman, P.M., and Mrozak, S.C. (1981). Program of early development in the mammal: synthesis and intracellular migration of histone H4 during oogenesis in the mouse. *Dev. Biol.,* **84,** 364–371.

Watanabe, T.K. (1975). A new sex-transforming gene on the second chromosome of *Drosophila melanogaster. Jpn. J. Genet.,* **50,** 269–271.

Weigel, D., Jurgens, G., Kuttner, F., Seifert, E., and Jäckle, H. (1989). The homeotic gene *fork head* encodes a nuclear protein and is expressed in the terminal regions of the Drosophila embryo. *Cell,* **57,** 645–658.

Weigel, D., Jurgens, G., Klingler, M., and Jäckle, H. (1990). Two gap genes mediate maternal terminal pattern information in *Drosophila. Science,* **245,** 495–498.

Weintraub, H. (1985). Assembly and propagation of repressed and derepressed chromosomal states. *Cell,* **42,** 705–711.

Welshons, W.J. (1971). Genetic basis for two types of recessive lethality at the *Notch* locus of *Drosophila. Genetics,* **68,** 259–270.

Welshons, W.J., and Russell, L.B. (1959). The Y-chromosome as the bearer of male determining factors in the mouse. *Proc. Natl. Acad. Sci., U.S.A.,* **45,** 560–566.

Wensink, P.C., Tabata, S., and Pachl, C. (1979). The clustered and scrambled arrangement of moderately repetitive elements in *Drosophila* DNA. *Cell,* **18,** 1231–1246.

West, J.D. (1978). Analysis of clonal growth using chimaeras and mosaics. In M.H. Johnson (Ed.), *Development in Mammals.* Amsterdam: Elsevier/North-Holland, vol. 3, pp. 413–416.

West, J.D., Frels, W.I., Chapman, V.M., and Papaioannou, V.E. (1977). Preferential expression of the maternally derived X chromosome in the mouse yolk sac. *Cell,* **12,** 873–882.

Wharton, K.A., Johansen, K.M., Xu, T., and Artavanis-Tsakonas, S. (1985). Nucleotide sequence from the neurogenic locus Notch implies a gene product that shares homology with proteins containing EGF-like repeats. *Cell,* **43,** 567–581.

Wharton, R.D., and Struhl, G. (1989). Structure of the Drosophila *BicaudalD* protein and its role in localizing the posterior determinant *nanos. Cell,* **59,** 881–892.

White, J., Albertson, D., and Anness, M. (1978). Connectivity changes in a class of motor neurones during the development of a nematode. *Nature,* **271,** 764–766.

White, R.A.H., and Wilcox, M. (1984). Protein products of the bithorax complex in Drosophila. *Cell,* **39,** 163–171.

White, R.A.H., and Lehmann, R. (1986). A gap gene, *hunchback,* regulates the spatial expression of *Ultrabithorax. Cell,* **47,** 311–321.

Wieschaus, E. (1978). The use of mosaics to study oogenesis in *Drosophila melanogaster.* In S. Subtelny and I. Sussex (Eds.), *The Clonal Basis of Development.* New York: Academic Press, pp. 23–43.

Wieschaus, E. (1979). *fs(1)K10,* a female-sterile mutation altering the pattern of both the egg coverings and the resultant embryos in *Drosophila.* In N. LeDouarin (Ed.), *Cell Lineage, Stem Cells, and Differentiation.* Amsterdam: Elsevier/North-Holland, pp. 291–302.

Wieschaus, E., and Gehring, W.J. (1976a). Gynandromorph analysis of the throacic disc primordia in *Drosophila melanogaster. Wilhelm Roux Arch.,* **180,** 31–46.

Wieschaus, E., and Gehring, W.J. (1976b). Clonal analysis of primordial disc cells in the early embryo of *Drosophila melanogaster. Dev. Biol.,* **50,** 249–263.

Wieschaus, E., and Nöthiger, R. (1982). The role of the transformer genes in the development of genitalia and analia of *Drosophila melanogaster. Dev. Biol.,* **90,** 320–324.

Wieschaus, E., and Nüsslein-Volhard, C. (1986). Looking at embryos. In D.B. Roberts (Ed.), Drosophila: *A Practical Approach.* Eynsham: IRL Press, pp. 199–227.

Wieschaus, E., Marsh, J.L., and Gehring, W.J. (1978). *fs(1)K10*, a germline-dependent female sterile mutation causing abnormal chorion morphology in *D. melanogaster. Wilhelm Roux Arch.*, **184,** 75–82.

Wieschaus, E., Nüsslein-Volhard, C., and Jurgens, G. (1984). Mutations affecting the pattern of the larval cuticle in Drosophila melanogaster. III. Zygotic loci on the X chromosome and fourth chromosome. *Wilhelm Roux Arch. Dev. Biol.,* **193,** 296–307.

Wilcox, M., and Smith, R.J. (1977). Regenerative interactions between *Drosophila* imaginal discs of different types. *Dev. Biol.,* **60,** 287–297.

Wilcox, M., and Smith, R.J. (1980). Compartments and distal outgrowth in the Drosophila imaginal wing disc. *Wilhelm Roux Arch. Dev. Biol.,* **188,** 157–161.

Wilkins, A.S. (1976). Replicative patterning and determination. *Differentiation, 5,* 15–19.

Wilkins, A.S. (1985). Expressed gene sets: How different in different tissues? *BioEssays, 2,* 80–82.

Wilkins, A.S. (1986). *Genetic Analysis of Animal Development.* 1st ed. New York: John Wiley & Sons.

Wilkins, A.S., and Gubb, D. (1991). Pattern formation in the embryo and imaginal discs of *Drosophila:* What are the links? *Dev. Biol.,* **145,** 1–12.

Wilkinson, D.G., Bhatt, S., Cook, M., Boncinelli, E., and Krumlauf, R. (1989). Segmental expression of Hox-2 homeobox-containing genes in the developing mouse hindbrain. *Nature,* **341,** 405–409.

Wilkinson, D.G., Bhatt, S., and Herrmann, B.G. (1990). Expression pattern of the mouse *T* gene and its role in mesoderm formation. *Nature, 343,* 657–659.

Williams, D.E., Eisenman, J., Baird, A., Rauch, C., Van Ness, K., March, C.J., Park, L.S., Martin, U., Machizuki, D.Y., Boswell, H.S., Burgess, G.S., Cosman, D., and Lyman, S.D. (1990). Identification of a ligand for the c-*kit* proto-oncogene. *Cell,* **63,** 167–174.

Williams, K.L., and Newell, P.C. (1976). A genetic study of aggregation in the cellular slime mould *Dictyostelium discoideum* using complementation analysis. *Genetics,* **82,** 287–307.

Wills, N., Gesteland, R.F., Karn, J., Barnett, L., Bolten, S., and Waterston, R.H. (1983). The genes *sup-7 X* and *sup-5 III* of *C. elegans* suppress amber nonsense mutations via altered transfer RNA. *Cell,* **33,** 575–583.

Wilson, E.B. (1898). Cell-lineage and ancestral reminiscence. In *Biological Lectures.* Woods Hole, MA: pp. 21–42.

Wilson, E.B. (1925). *The Cell in Development and Heredity.* New York: Macmillan.

Wilson, I.B., and Stern, M.S. (1975). Organization in the preimplantation embryo. In M. Balls and A.E. Wild (Eds.), *The Early Development of Mammals.* Cambridge: Cambridge University Press.

Winking, H., and Silver, L.M. (1984). Characterization of a recombinant mouse *t*-haplotype that expresses a dominant lethal maternal effect. *Genetics,* **108,** 1013–1020.

Wirz, J., Fassler, L.I., and Gehring, W.J. (1986). Localization of the *Antennapedia* protein in *Drosophila* embryos and imaginal discs. *EMBO J.,* **5,** 3327–3334.

Wolgemuth, D.J., Behringer, R.R., Mostoller, M.P., Brinster, R.L., and Palmiter, R.D. (1989). Transgenic mice overexpressing the mouse homeobox-containing gene *Hox-1.4* exhibit abnormal gut development. *Nature, 337,* 464–467.

Wolpert, L. (1969). Positional information and the spatial pattern of cellular differentiation: *J. Theor. Biol.,* **25,** 1–47.

Wolpert, L. (1971). Positional information and pattern formation. *Curr. Top. Dev. Biol.,* **7,** 183–224.

Wolpert, L. (1989). Positional information revisited. *Development* (Suppl.), 3–12.

Wood, W.B. (1988). *The Nematode* Caenorhabditis elegans. Cold Spring Harbor, NY: Cold Spring Harbor Laboratory Press.

Wood, W.B. (1991). Evidence from reversal of handedness in *C. elegans* embryos for early cell interactions determining cell fates. *Nature,* **349,** 536–538.

Wood, W.B., Hecht, R., Carr, S., Vanderslice, R., Wolf, N., and Hirsh, D. (1980). Parental effects and phenotypic characterization of mutations that affect early development in *Caenorhabditis elegans. Dev. Biol.,* **74,** 446–469.

Wood, W.B., Strome, S., and Laufer, J.S. (1983). Localization and determination in embryos of *Caenorhabditis elegans.* In W.R. Jeffery and R.A. Raff (Eds.), *Time, Space, and Pattern in Embryonic Development.* New York: Alan R. Liss, pp. 221–239.

Wright, S. (1917). Color inheritance in mammals. *J. Hered.,* **8,** 224–235.

Wright, T.R.F. (1970). The genetics of embryogenesis in *Drosophila. Adv. Genet.,* **15,** 262–395.

Xu, T., Rebay, I., Fleming, R.J., Scottgale, T.N., and Artavanis-Tsakonas, S. (1990). The *Notch* locus and the genetic circuitry involved in early *Drosophila* neurogenesis. *Genes Dev.,* **4,** 464–475.

Yochem, J., and Greenwald, I. (1989). *glp-1* and *li-12,* genes implicated in distinct cell–cell interactions in *C. elegans,* encode similar transmembrane proteins. *Cell,* **58,** 553–563.

Yochem, J., Weston, K., and Greenwald, I. (1988). The *Caenorhabditis elegans lin-12* gene encodes a transmembrane protein with overall similarity to *Drosophila Notch. Nature,* **335,** 547–550.

Young, B.D., Birnie, G.D., and Paul, J. (1976). Complexity and specifity of polysomal poly(A)+ RNA in mouse tissues. *Biochemistry,* **15,** 2823–2830.

Young, M.W., and Judd, B.H. (1978). Nonessential sequences, genes and the polytene chromosome bands of *D. melanogaster. Genetics,* **88,** 723–742.

Zalokar, M. (1976). Autoradiographic study of protein and RNA formation during early development of *Drosophila* eggs. *Dev. Biol.,* **49,** 425–437.

Zalokar, M., and Erk, I. (1976). Division and migration of nuclei during embryogenesis of *Drosophila melanogaster. J. Microsc. Biol. Cell,* **25,** 97–106.

Zalokar, M., Audit, C., and Erk, I. (1975). Developmental defects of female-sterile mutants of *Drosophila. Dev. Biol.,* **47,** 419–432.

Zalokar, M., Erk, I., and Santamaria, P. (1980). Distribution of ring-X chromosomes in the blastoderm of gynandromorphic *D. melanogaster. Cell,* **19,** 133–141.

Zimmerman, J.L., Fouts, D.L., and Manning, J.E. (1980). Evidence for a complex class of nonadenylated mRNA in *Drosophila. Genetics,* **95,** 673–691.

Zink, B., and Paro, R. (1989). *In vivo* binding pattern of a trans-regulator of homeotic genes in *Drosophila melanogaster. Nature,* **337,** 468–471.

Ziomek, C.A., Johnson, M.H., and Handyside, A.H. (1982). The roles of phenotype and position in guiding the fate of 16-cell mouse blastomeres. *Dev. Biol.,* **91,** 440–447.

Zuker, C.S., Cowan, A.F., and Rubin, G.M. (1985). Isolation and structure of a rhodopsin gene from D. melanogaster. *Cell,* **40,** 851–858.

INDEX